D1256657

Methods in Enzymology

Volume 336
MICROBIAL GROWTH IN BIOFILMS
Part A
Developmental and Molecular Biological Aspects

METHODS IN ENZYMOLOGY

EDITORS-IN-CHIEF

John N. Abelson   Melvin I. Simon

DIVISION OF BIOLOGY
CALIFORNIA INSTITUTE OF TECHNOLOGY
PASADENA, CALIFORNIA

FOUNDING EDITORS

Sidney P. Colowick and Nathan O. Kaplan

Methods in Enzymology

Volume 336

Microbial Growth in Biofilms

Part A
Developmental and Molecular Biological Aspects

EDITED BY

Ron J. Doyle

UNIVERSITY OF LOUISVILLE
LOUISVILLE, KENTUCKY

ACADEMIC PRESS

San Diego   London   Boston   New York   Sydney   Tokyo   Toronto

Academic Press
A Harcourt Science and Technology Company
525 B Street, Suite 1900, San Diego, California 92101-4495, USA
http://www.academicpress.com

Academic Press
Harcourt Place, 32 Jamestown Road, London NW1 7BY, UK
http://www.academicpress.com

International Standard Book Number: 0-12-182237-0

PRINTED IN THE UNITED STATES OF AMERICA
01 02 03 04 05 06 07 SB 9 8 7 6 5 4 3 2 1

Table of Contents

Section I. Developmental Processes in Biofilms

Section II. Signaling and Gene Expression in Biofilm

Section III. Growth of Bacteriophage in Bacteria in Biofilms

Section IV. Biofilms of Staphylococci

Section V. Metabolic Potential of Biofilms

Section VI. Extracellular Polymers

Section VII. Microbiological Aspects of Microbial Biofilm

Section VIII. Probiotics

Contributors to Volume 336

Article numbers are in parentheses following the names of contributors.
Affiliations listed are current.

WOLF-RAINER ABRAHAM (26), *Division of Microbiology, GBF, German Research Centre for Biotechnology, D-38124 Braunschweig, Germany*

MOHAMMAD ALAVI (3), *Center of Marine Biotechnology, University of Maryland Biotechnology Institute, Baltimore, Maryland 21202*

PATRIZIA ALBERTANO (28), *Department of Biology, University of Rome, I-00133 Rome, Italy*

MORITZ ALTEBAEUMER (8), *Department of Biological Sciences, University of Calgary, Calgary, Alberta, Canada T2N 1N4*

JENS BO ANDERSEN (12), *Department of Microbiology, Technical University of Denmark, Lyngby DK-2800, Denmark*

GARY M. ARON (16), *Department of Biology, Southwest Texas State University, San Marcos, Texas 78666*

RITA R. BACA-DELANCEY (11), *Departments of Medicine and of Molecular Biology and Microbiology, Case Western Reserve University, Cleveland, Ohio 44106*

GRANT J. BALZER (16), *Department of Civil Engineering, Northwestern University, Evanston, Illinois 60208*

KATRIN BARTSCHT (20), *Institut für Medizinische Mikrobiologie und Immunologie, Universitätsklinikum Hamburg-Eppendorf, 20246 Hamburg, Germany*

ACE M. BATY, III (24), *W. L. Gore & Associates, Inc., Flagstaff, Arizona 86002*

ROBERT BELAS (3), *Center of Marine Biotechnology, University of Maryland Biotechnology Institute, Baltimore, Maryland 21202*

ANTONIO BENNASAR (26), *Division of Microbiology, GBF, German Research Centre for Biotechnology, D-38124 Braunschweig, Germany*

RODRIGO BIBILONI (33), *Centro de Investigación y Desarrollo en Criotecnología de Alimentos, Facultad de Ciencias Exactas, 1900 La Plata, Argentina*

J. CLIFF BOUCHER (7), *Department of Microbiology, Harvard Medical School, Boston, Massachusetts 02115*

JAMES D. BRYERS (9), *The Center for Biomaterials, Department of Biostructure and Function, University of Connecticut Health Center, Farmington, Connecticut 06030*

JILLIAN CAPSTICK (17), *Department of Biotechnology and Environmental Biology, Royal Melbourne Institute of Technology, Bundoora 3083, Victoria, Australia*

HOWARD CERI (8), *Department of Biological Sciences, University of Calgary, Calgary, Alberta, Canada T2N 1N4*

TIMOTHY CHARLTON (12), *School of Microbiology and Immunology and The Centre for Marine Biofouling and Bio-Innovation, University of New South Wales, Sydney 2052, Australia*

KIMBERLY H. CHONG (1), *Departments of Medicine, and Microbiology and Immunology, Stanford University School of Medicine, Stanford, California 94305*

PATRICIA L. CONWAY (31), *CRC Food Industry Innovation, School of Microbiology and Immunology, University of New South Wales, NSW 2052, Sydney, Australia*

BRIAN D. CORBIN (16), *Department of Microbiology and Molecular Genetics, University of Texas–Houston Medical School, Houston, Texas 77030*

SARAH E. CRAMTON (21), *Mikrobielle Genetik, Universität Tübingen, D-72076 Tübingen, Germany*

PHILIPPE CRASSOUS (27), *Département d'Environnement Profond, DRO. IFRE-MER 29280, Plouzané, France*

PAUL N. DANESE (2), *Department of Microbiology and Molecular Genetics, Harvard Medical School, Boston, Massachusetts 02115*

GRACIELA L. DE ANTONI (33), *Centro de Investigación y Desarrollo en Criotecnología de Alimentos, Facultad de Ciencias Exactas, 1900 La Plata, Argentina*

CLEMENS DE GRAHL (20), *Institut für Medizinische Mikrobiologie und Immunologie, Universitätsklinikum Hamburg-Eppendorf, 20246 Hamburg, Germany*

MARGARET A. DEIGHTON (17), *Department of Biotechnology and Environmental Biology, Royal Melbourne Institute of Technology, Bundoora 3083, Victoria, Australia*

ROCKY DENYS (12), *The Centre for Marine Biofouling and Bio-Innovation, University of New South Wales, Sydney 2052, Australia*

VOJO DERETIC (7), *Department of Microbiology and Immunology, University of Michigan Medical School, Ann Arbor, Michigan 48109*

XUEDONG DING (11), *Departments of Medicine and of Molecular Biology and Microbiology, Case Western Reserve University, Cleveland, Ohio 44106*

E. ANIBAL DISALVO (33), *Cátedra de Química General e Inorgánica, Facultad de Farmacia y Bioquímica, Universidad, de Buenos Aires, 1113 Buenos Aires, Argentina*

ZHENJUN DIWU (24), *Molecular Devices Corp., Sunnyvale, California 94089*

SABINE DOBINSKY (20, 22), *Institut für Medizinische Mikrobiologie und Immunologie, Universitätsklinikum Hamburg-Eppendorf, 20246 Hamburg, Germany*

NADIA A. DOLGANOV (1), *Departments of Medicine, and Microbiology and Immunology, Stanford University School of Medicine, Stanford, California 94305*

EWA DOMALEWSKI (17), *Department of Biotechnology and Environmental Biology, Royal Melbourne Institute of Technology, Bundoora 3083, Victoria, Australia*

CORINNE DOREL (15), *Laboratoire de Microbiologie et Génétique (CNRS ERS 209), Institut National des Sciences Appliquées de Lyon, 69621 Villeurbanne, France*

GLEN DUNHAM (24), *Interfacial and Processing Sciences, Environmental Molecular Sciences Laboratory, Pacific Northwest National Laboratories, Richland, Washington 99352*

CALLIE C. EASTBURN (24), *Department of Microbiology and Center for Biofilm Engineering, Montana State University, Bozeman, Montana 59717-3980*

PAUL D. FEY (19), *Department of Internal Medicine, University of Nebraska Medical Center, Omaha, Nebraska 68198*

AARON M. FIROVED (7), *Department of Microbiology and Immunology, University of Michigan Medical School, Ann Arbor, Michigan 48109*

CLAUDIA FISCHER (20), *Institut für Medizinische Mikrobiologie und Immunologie, Universitätsklinikum Hamburg-Eppendorf, 20246 Hamburg, Germany*

HANS-CURT FLEMMING (25), *Department of Aquatic Microbiology, University of Duisburg, 47057 Duisburg, Germany*

BING S. GAN (32), *Lawson Health Research Institute, St. Joseph's Health Center, and Department of Surgery, University of Western Ontario, London, Ontario N6A 4V2, Canada*

GILLIAN GARDINER (32), *Lawson Health Research Institute, St. Joseph's Health Center, University of Western Ontario, London, Ontario N6A 4V2, Canada*

GRACIELA L. GARROTE (33), *Centro de Investigación y Desarrollo en Criotecnología de Alimentos, Facultad de Ciencias Exactas, 1900 La Plata, Argentina*

GILL G. GEESEY (24), *Department of Microbiology and Center for Biofilm Engineering, Montana State University, Bozeman, Montana 59717*

CHRISTIANE GERKE (21), *Mikrobielle Genetik, Universität Tübingen, D-72076 Tübingen, Germany*

MICHAEL GIVSKOV (12), *Department of Microbiology, Technical University of Denmark, Lyngby DK-2800, Denmark*

AMANDA E. GOODMAN (24), *Flinders University of South Australia, Adelaide 5001, South Australia*

FRIEDRICH GÖTZ (21), *Mikrobielle Genetik, Universität Tübingen, D-72076 Tübingen, Germany*

E. P. GREENBERG (4), *Department of Microbiology, College of Medicine, University of Iowa, Iowa City, Iowa 52242*

JEAN GUEZENNEC (27), *Laboratoire de Biochimie des Molecules Marines, DRV/VP, IFREMER 29280, Plouzané, France*

CLAUDIA GURTNER (29), *Institute of Microbiology and Genetics, University of Vienna, A-1030 Vienna, Austria*

JÖRG HACKER (18), *Institut für Molekulare Infektionsbiologie, D-97070 Würzburg, Germany*

MARTINA HAUSNER (13), *Institute of Water Quality Control and Waste Management, Technical University of Munich, D-85748 Garching, Germany*

CHRISTINE HEINEMANN (32), *Lawson Health Research Institute, St. Joseph's Health Center, University of Western Ontario, London, Ontario N6A 4V2, Canada*

LARISSA HENDRICKX (13), *Institute of Water Quality Control and Waste Management, Technical University of Munich, D-85748 Garching, Germany*

MORTEN HENTZER (12), *Department of Microbiology, Technical University of Denmark, Lyngby DK-2800, Denmark*

GUILLERMO HERNÁNDEZ-DUQUE (27), *Facultad de Ingenieria, Universidad del Mayab, Cordemex 97310, Mérida, Yucatán, Mexico*

MATTHIAS A. HORSTKOTTE (20), *Institut für Medizinische Mikrobiologie und Immunologie, Universitätsklinikum Hamburg-Eppendorf, 20246 Hamburg, Germany*

JEFFREY HOWARD (32), *Lawson Health Research Institute, St. Joseph's Health Center, University of Western Ontario, London, Ontario N6A 4V2, Canada*

JAN JORE (30), *Department of Applied Microbiology and Gene Technology, TNO Voeding Nutrition and Food Research Institute, 3700 AJ Zeist, The Netherlands*

SIBYLLE KALMBACH (23), *Studienstiftung des Deutschen Volkes, D-53173 Bonn, Germany*

DANIEL B. KEARNS (10), *Department of Molecular and Cellular Biology, Harvard University, Cambridge, Massachusetts 02138*

KATHRIN KIEL (20), *Institut für Medizinische Mikrobiologie und Immunologie, Universitätsklinikum Hamburg-Eppendorf, 20246 Hamburg, Germany*

STAFFAN KJELLEBERG (12), *The School of Microbiology and Immunology and The Centre for Marine Biofouling and Bio-Innovation, University of New South Wales, Sydney 2052, Australia*

JOHANNES K.-M. KNOBLOCH (20), *Institut für Medizinische Mikrobiologie und Immunologie, Universitätsklinikum Hamburg-Eppendorf, 20246 Hamburg, Germany*

ROBERTO KOLTER (2), *Department of Microbiology and Molecular Genetics, Harvard Medical School, Boston, Massachusetts 02115*

VANESSA KRIMMER (18), *Institut für Molekulare Infektionsbiologie, D-97070 Würzburg, Germany*

MARTIN KUEHN (13), *Institute of Water Quality Control and Waste Management, Technical University of Munich, D-85748 Garching, Germany*

RICHARD J. LAMONT (9), *Department of Oral Biology, University of Washington, Seattle, Washington 98195*

THANH-THUY LE THI (15), *Laboratoire de Microbiologie et Génétique (CNRS ERS 209), Institut National des Sciences Appliquées de Lyon, 69621 Villeurbanne, France*

ROB J. LEER (30), *Department of Applied Microbiology and Gene Technology, TNO Voeding Nutrition and Food Research Institute, 3700 AJ Zeist, The Netherlands*

ANDREW LEIS (25), *Department of Aquatic Microbiology, University of Duisburg, 47057 Duisburg, Germany*

PHILIPPE LEJEUNE (15), *Laboratoire de Microbiologie et Génétique (CNRS ERS 209), Institut National des Sciences Appliquées de Lyon, 69621 Villeurbanne, France*

ISABEL LOESSNER (18), *Institut für Molekulare Infektionsbiologie, D-97070 Würzburg, Germany*

ALEJANDRO LÓPEZ-CORTÉS (27), *Laboratorio de Ecologia y Biotecnologia Microbiana, CIBNOR, Mar Bermejo #195, La Paz, Baja California Sur, Mexico*

WERNER LUBITZ (29), *Institute of Microbiology and Genetics, University of Vienna, A-1030 Vienna, Austria*

HEINRICH LÜNSDORF (26), *Division of Microbiology, GBF, German Research Centre for Biotechnology, D-38124 Braunschweig, Germany*

DIETRICH MACK (20, 22), *Institut für Medizinische Mikrobiologie und Immunologie, Universitätsklinikum Hamburg-Eppendorf, 20246 Hamburg, Germany*

WERNER MANZ (23), *Technical University Berlin, Microbial Ecology, D-10587 Berlin, Germany*

BEATRIZ MARTINEZ (30), *Instituto de Productos Lacteos de Asturias, CSIC, 33300 Villaviciosa, Spain*

ROBERT J. C. MCLEAN (16), *Department of Biology, Southwest Texas State University, San Marcos, Texas 78666*

KARIN MEIBOM (1), *Departments of Medicine, and Microbiology and Immunology, Stanford University School of Medicine, Stanford, California 94305*

EDWARD R. B. MOORE (26), *Division of Microbiology, GBF, German Research Centre for Biotechnology, D-38124 Braunschweig, Germany*

PETER MULLANY (6), *Department of Microbiology, Eastman Dental Institute for Oral Health Care Sciences, University College London, London WC1X 8LD, United Kingdom*

BENJAMÍN OTTO ORTEGA-MORALES (27), *Laboratorio de Microbiologia Ambiental y Biotecnologia (CIET), Universidad Autonoma de Campeche, Lindavista CP 24030, Campeche, Camp., Mexico*

A. MARK OSBORN (26), *Division of Microbiology, GBF, German Research Centre for Biotechnology, D-38124 Braunschweig, Germany*

MICHAEL D. PARKINS (8), *Department of Biological Sciences, University of Calgary, Calgary, Alberta, Canada T2N 1N4*

MATTHEW R. PARSEK (4), *Department of Civil Engineering, Northwestern University, Evanston, Illinois 60208*

PABLO F. PÉREZ (33), *Centro de Investigación y Desarrollo en Criotecnología de Alimentos, Facultad de Ciencias Exactas, 1900 La Plata, Argentina*

GUADALUPE PIÑAR (29), *Institute of Microbiology and Genetics, University of Vienna, A-1030 Vienna, Austria*

JENS F. POSCHET (7), *Department of Microbiology and Immunology, University of Michigan Medical School, Ann Arbor, Michigan 48109*

PETER H. POUWELS (30), *Department of Applied Microbiology and Gene Technology, TNO Voeding Nutrition and Food Research Institute, 3700 AJ Zeist, The Netherlands, and Special Programme Infectious Diseases, TNO Prevention and Health, 2301 CE Leiden, The Netherlands, and Wageningen Centre for Food Sciences, 6700 AN Wageningen, The Netherlands*

LESLIE A. PRATT (2), *Department of Microbiology and Molecular Genetics, Harvard Medical School, Boston, Massachusetts 02115*

CLAIRE PRIGENT-COMBARET (15), *Laboratoire de Microbiologie et Génétique (CNRS ERS 209), Institut National des Sciences Appliquées de Lyon, 69621 Villeurbanne, France*

PHILIP N. RATHER (11), *Departments of Medicine and of Molecular Biology and Microbiology, Case Western Reserve University, and Research Service, Cleveland Veterans Affairs Medical Center, Cleveland, Ohio 44106*

GREGOR REID (32), *Lawson Health Research Institute, St. Joseph's Health Centre, and Departments of Microbiology and Immunology, and Surgery, University of Western Ontario, London, Ontario N6A 4V2, Canada*

SCOTT RICE (12), *The School of Microbiology and Immunology and The Centre for Marine Biofouling and Bio-Innovation, University of New South Wales, Sydney 2052, Australia*

ADAM P. ROBERTS (6), *Department of Microbiology, Eastman Dental Institute for Oral Health Care Sciences, University College London, London WC1X 8LD, United Kingdom*

ALEXANDER RODE (25), *Department of Aquatic Microbiology, University of Duisburg, 47057 Duisburg, Germany*

KARSTEN RODENACKER (13), *GSF National Research Centre for Environment and Health, Institute of Biomathematics and Biometry, D-85764 Neuherberg, Germany*

HOLGER ROHDE (20), *Institut für Medizinische Mikrobiologie und Immunologie, Universitätsklinikum Hamburg-Eppendorf, 20246 Hamburg, Germany*

MAURILIA ROJAS (31), *Department of General and Marine Microbiology, University of Göteborg, S-413 90 Göteborg, Sweden and Universidad Autonoma de B.C.S., La Paz, Baja California Sur, Mexico*

SABINE RÖLLEKE (29), *Institute of Microbiology and Genetics, University of Vienna, A-1030 Vienna, Austria*

UTE RÖMLING (5), *Research Group "Clonal Variability," Department of Cell Biology and Immunology, Gesellschaft für Biotechnologische Forschung, D-38124 Braunschweig, Germany*

MARK E. RUPP (19), *Department of Internal Medicine, University of Nebraska Medical Center, Omaha, Nebraska 68198*

AMY L. SCHAEFER (4), *Department of Microbiology, College of Medicine, University of Iowa, Iowa City, Iowa 52242*

DIRK SCHNAPPINGER (1), *Departments of Medicine, and Microbiology and Immunology, Stanford University School of Medicine, Stanford, California 94305*

GARY K. SCHOOLNIK (1), *Departments of Medicine, and Microbiology and Immunology, Stanford University School of Medicine, Stanford, California 94305*

JOS F. M. L. SEEGERS (30), *Special Programme Infectious Diseases, TNO Prevention and Health, 2301 CE Leiden, The Netherlands*

LAWRENCE J. SHIMKETS (10), *Department of Microbiology, University of Georgia, Athens, Georgia 30602*

SOOFIA SIDDIQUI (11), *Research Service, Cleveland Veterans Affairs Medical Center, Cleveland, Ohio 44106*

EGBERT SMIT (30), *Department of Applied Microbiology and Gene Technology, TNO Voeding Nutrition and Food Research Institute, 3700 AJ Zeist, The Netherlands*

DOUGLAS G. STOREY (8), *Department of Biological Sciences, University of Calgary, Calgary, Alberta, Canada T2N 1N4*

MARTIN STRATHMANN (25), *Department of Aquatic Microbiology, University of Duisburg, 47057 Duisburg, Germany*

CARSTEN STRÖMPL (26), *Division of Microbiology, GBF, German Research Centre for Biotechnology, D-38124 Braunschweig, Germany*

PETER A. SUCI (24), *Department of Microbiology and Center for Biofilm Engineering, Montana State University, Bozeman, Montana 59717*

SOMKIET TECHKARNJANARUK (24), *Biochemical Engineering and Pilot Plant Research and Development Unit, National Center for Genetic Engineering and Biotechnology, King Mongkut's University of Technology, Thonburi Thakham, Bangkhuntien, Bangkok, Thailand 10150*

FRANS J. TIELEN (30), *Special Programme Infectious Diseases, TNO Prevention and Health, 2301 CE Leiden, The Netherlands*

KENNETH N. TIMMIS (26), *Division of Microbiology, German Research Centre for Biotechnology, D-38124 Braunschweig, Germany*

CLARA URZÌ (28), *Department of Microbiological, Genetic, and Molecular Sciences, University of Messina, I-98166 Messina, Italy*

TRUNG VAN NGUYEN (17), *Department of Biotechnology and Environmental Biology, Royal Melbourne Institute of Technology, Bundoora 3083, Victoria, Australia*

MARTIN I. VOSKUIL (1), *Departments of Medicine, and Microbiology and Immunology, Stanford University School of Medicine, Stanford, California 94305*

ALDWIN VRIESEMA (30), *Department of Applied Microbiology and Gene Technology, TNO Voeding Nutrition and Food Research Institute, 3700 AJ Zeist, The Netherlands*

MICHAEL WAGNER (23), *Technical University Munich, Department of Microbiology, D-85350 Freising, Germany*

MICHAEL WILSON (6), *Department of Microbiology, Eastman Dental Institute for Oral Health Care Sciences, University College London, London WC1X 8LD, United Kingdom*

MICHAEL A. WILSON (1), *National Institute of Allergy and Infectious Diseases, National Institutes of Health, Bethesda, Maryland 20892*

JOST WINGENDER (25), *Department of Aquatic Microbiology, University of Duisburg, 47057 Duisburg, Germany*

DANIEL J. WOZNIAK (14), *Department of Microbiology and Immunology, Wake Forest University School of Medicine, Winston-Salem, North Carolina 27157*

STEFAN WUERTZ (13), *Institute of Water Quality Control and Waste Management, Technical University of Munich, D-85748 Garching, Germany*

TIMNA J. O. WYCKOFF (14), *Department of Microbiology and Immunology, Wake Forest University School of Medicine, Winston-Salem, North Carolina 27157*

FITNAT H. YILDIZ (1), *Departments of Medicine, and Microbiology and Immunology, Stanford University School of Medicine, Stanford, California 94305*

WILMA ZIEBUHR (18), *Institut für Molekulare Infektionsbiologie, D-97070 Würzburg, Germany*

Preface

Biofilms are usually characterized as a consortia of microorganisms surrounded by a protecting matrix of secreted polymers that are in most cases acidic polysaccharides, but may possess various functional groups other than carboxylate. In biofilms, the microorganisms possess regulatory molecules distinct from those produced by planktonic microorganisms. It is now possible not only to detect specific species and strains in a biofilm matrix, but also to identify which of their genes are up- or down-regulated in the planktonic cell to biofilm cell transition.

The advances in molecular biology methods have been paralleled by advances in instrumental and chemical probes used to define biofilm properties. Literature has burgeoned on all aspects of biofilms in the past few years. This growing literature, coupled with the recognition that biofilms are important in disease, industry, agriculture, and biotechnology, has prompted the development of a series of *Methods in Enzymology* volumes. The first, Volume 310, was concerned with the general approaches to biofilm molecular biology and the physical methods to probe biofilm structures. Volumes 336 and 337 focus on microbial growth in biofilms. In this volume emphasis is on the genetics and molecular biology of biofilm genesis. Its companion Volume 337 focuses on special environments and specific microorganisms contributing to biofilms. Collectively, the three volumes comprise methods from the leading researchers in the world. The following decades of research on biofilms will borrow heavily from these volumes.

I thank Shirley Light of Academic Press for her competent handling of numerous questions related to the development of the volumes on biofilms.

RON J. DOYLE

METHODS IN ENZYMOLOGY

Section I

Developmental Processes in Biofilms

[1] Whole Genome DNA Microarray Expression Analysis of Biofilm Development by *Vibrio cholerae* O1 El Tor

By Gary K. Schoolnik, Martin I. Voskuil, Dirk Schnappinger,
Fitnat H. Yildiz, Karin Meibom, Nadia A. Dolganov,
Michael A. Wilson, and Kimberly H. Chong

Introduction

The clinical and epidemiological features of Asiatic cholera are well known: it is caused by *Vibrio cholerae* O1, results in a purging diarrheal illness that can kill an adult within 24 hr, and has the capacity to spread from its epicenter in the Bengal region of Bangladesh and India to all continents of the habitable world. However, its most intriguing and least understood feature comes from the study of its annual epidemic profile in Bengal. There, nearly all cases each year occur in a synchronized, massive outbreak in the months of October and November, just as the monsoon rains decline.[1] During all other months of the year, cholera cases occur sporadically or not at all. This epidemic profile and its correlation with major transitions of climate point to the following: *V. cholerae* O1 resides in a stable environmental reservoir; then, seasonally determined changes in rainfall and sunlight trigger its periodic and transient emergence as a human pathogen.[2]

Between epidemics, *V. cholerae* O1 lives in natural aquatic habitats formed by the confluence of the Ganges and the Brahmaputra rivers. All the physicochemical and ecological features of this system are dramatically influenced by the monsoon climate. Within this ecosystem, four distinctive stages have been proposed to comprise the *V. cholerae* O1 life cycle: an independent, free-swimming form; a symbiont of phytoplankton[3]; a commensal of zooplankton[4]; and a surface-attached consortium or biofilm.[5,6]

Laboratory-based studies have demonstrated that *V. cholerae* O1, biotype El Tor, forms a typical three-dimensional biofilm on a variety of abiotic surfaces[7,8] and on biotic surfaces composed of chitin.[9] Investigation of this phenotype showed

[1] R. I. Glass, M. Claeson, P. A. Blake, R. J. Waldman, and N. F. Pierce, *Lancet* **338,** 791 (1991).

[2] R. R. Colwell, *Science* **274,** 2025 (1996).

[3] M. S. Islam, B. S. Drasar, and R. B. Sack, *J. Diarrhoeal Dis. Res.* **12,** 87 (1994).

[4] A. Huq, E. B. Small, P. A. West, M. I. Huq, R. Rahman, and R. R. Colwell, *Appl. Environ. Microbiol.* **45,** 275 (1983).

[5] S. N. Wai, Y. Mizunoe, A. Takade, S. I. Kawabata, and S. I. Yoshida, *Appl. Environ. Microbiol.* **64,** 3648 (1998).

[6] F. H. Yildiz and G. K. Schoolnik, *Proc. Natl. Acad. Sci. U.S.A.* **96,** 4028 (1999).

[7] P. I. Watnick, K. J. Fullner, and R. Kolter, *J. Bacteriol.* **181,** 3606 (1999).

[8] P. I. Watnick and R. Kolter, *Mol. Microbiol.* **34,** 586 (1999).

[9] D. R. Nalin, V. Daya, A. Reid, M. M. Levine, and L. Cisneros, *Infect. Immun.* **25,** 768 (1979).

that the rugose colonial variant forms a thicker and more differentiated biofilm than the smooth colonial variant.[5,6] This capacity was found to be associated with production of a glucose- and galactose-rich extracellular polysaccharide (EPS) by the rugose form.[6] Designated EPS^{ETr}, this compound was also shown to inactivate chlorine and protect the organism from the bactericidal action of hydrogen peroxide.[5,6] Transposon mutagenesis studies of the smooth colonial variant by Kolter and colleagues led to the identification of two other effectors of biofilm formation, a type IV pilus that confers mannose-sensitive hemagglutination and flagella.[7,8] From this information they proposed a three-step model of biofilm development: first, type IV pili and flagella facilitate the attachment of free-swimming bacteria to a surface; second, the motility function of flagella causes attached bacteria to spread across the surface; and third, EPS secretion occurs, providing the extracellular matrix of the three-dimensional structure of the mature biofilm.[7,8,10] With the completion of the entire *V. cholerae* O1 genome sequence, it has now become possible to conduct a whole-genome characterization of the biofilm phenotype through the use of microarray-based expression profiling. This method, applied to the biofilm phenotype of *V. cholerae* O1 El Tor, is described below.

Microarray Expression Analysis and Experimental Design

A microarray is a surface that contains representations of each open reading frame (ORF) of a sequenced and annotated genome.[11] The surface used, the method by which ORF-specific DNA is bound to the surface, and the overall arrangement of the array vary with the system employed. Of the several available formats, the most common, economical, and flexible—and the one used by the authors in their gene expression studies of *Mycobacterium tuberculosis*[12] and *Vibrio cholerae* O1—was developed by P. Brown and colleagues at Stanford University.[13–15] This array format consists of a microscope slide whose surface contains an *x* by *y* matrix of printed spots, each spot containing a polymerase chain reaction (PCR)-derived amplicon that corresponds to all or part of an ORF of the sequenced genome. Thus, each ORF of the genome is represented on the array as a separate spot, its location designated by its *xy* address. Additional spots are added and correspond to internal controls that monitor the printing and hybridization steps.

[10] P. I. Watnick and R. Kolter, *J. Bacteriol.* **182,** 2675 (2000).

[11] M. Schena, R. A. Heller, T. P. Theriault, K. Konrad, W. Lachenmeier, and R. W. Davis, *Trends Biotechnol.* **7,** 301 (1998).

[12] M. Wilson, J. DeRisi, H. H. Kristensen, P. Imboden, S. Rane, P. O. Brown, and G. K. Schoolnik, *Proc. Natl. Acad. Sci. U.S.A.* **96,** 12833 (1999).

[13] M. Schena, D. Shalon, R. W. Davis, and P. O. Brown, *Science* **270,** 467 (1995).

[14] D. Shalon, S. J. Smith, and P. O. Brown, *Genome Res.* **7,** 639 (1996).

[15] J. L. DeRisi, V. R. Iyer, and P. O. Brown, *Science* **278,** 680 (1998).

The principal innovation in gene expression profiling that was introduced by Brown and colleagues is "two-color" hybridization. This method employs two populations of cDNAs that have been differentially labeled with two different fluorochromes, the cDNAs having been derived from RNA prepared from the same organism cultivated under, or exposed to, two contrasting conditions. Equal masses of the two differentially labeled populations of cDNAs are combined, applied to the array surface, and allowed to hybridize to the corresponding ORF-specific targets. The array is then scanned and the intensity of each label for each ORF-specific spot is quantitated. These values are compared, yielding ratios that serve as a measure of the relative degree of expression or repression of each ORF for the two tested conditions.

The use of two or more contrasting conditions as the principal experimental paradigm is designed to learn more about the genome-wide transcriptional response of the organism to changes in the physicochemical features of its microenvironment. With respect to the study of the biofilm phenotype, the most basic dichotomy is the differential gene expression pattern between the sessile and planktonic populations of the same culture. The identification of genes selectively expressed during different stages of biofilm development and involution represents a second experimental goal. Finally, investigators will be interested in identifying genes differentially expressed during adaptation of a mature biofilm to changes in the fluid phase, including variations in shear, carbon and nitrogen availability, ionic strength, pH, and salinity and in the concentrations of phosphates, nitrates, and metals. Because it is proposed that biofilm formation enables *V. cholerae* O1 to persist in aquatic habitats, the physicochemical parameters that are chosen for the study of this organism are selected according to the range of conditions likely to be encountered by the organism in nature. On-the-ground and remote (satellite-derived) measurements of the Ganges–Brahmaputra delta system and of the adjacent Bay of Bengal are beginning to provide normative data of this kind.[16]

In general, of the thousands of ORFs monitored during experiments of this kind the vast majority show no differential expression for any particular tested condition. A much smaller set of genes exhibits selective expression or repression; of these, many, but not all, can be plausibly associated with an adaptive response of the organism that is specific to the condition under study. This appears to be particularly true early in the time course and under conditions that do not induce a generalized stress response. However, no rationale can be adduced, on the basis of current knowledge, for a fraction of the selectively induced or repressed genes. Some of these have no inferred function because, during the annotation process, they were not found to be homologous to genes of known function. For

---

[16] B. Lobitz, L. Beck, A. Huq, B. Wood, G. Fuchs, A. S. G. Faruque, and R. Colwell, *Proc. Natl. Acad. Sci. U.S.A.* **97**, 1438 (2000).

others, while their functions may be known or can be inferred, their regulation by the condition studied could not have been predicted by prior knowledge of the biology of the organism. Unexpected results of this kind are perhaps the most interesting and provide the basis for new hypotheses.

DNA Microarray Fabrication and Use

Detailed methods for microarray expression profiling are provided below. The protocol proceeds through the following steps: PCR amplification of ORF-specific amplicons; fabrication of the microarray, including slide preparation, robotic printing of the array, and postprinting processing of the array; RNA preparation and labeling; hybridization; scanning; image processing; and data mining. This is an evolving technology and technical refinements and innovations are frequent occurrences. Web-based resources have been commonly used to announce recent protocol changes in this field. Accordingly, these website addresses are referenced. A particularly useful general guide to the field, including protocols, software, and instructions for robotic printer construction, is MGuide 2.0, maintained by the Brown Laboratory at Stanford University (http://cmgm.stanford.edu/pbrown/mguide/index.html).

Protocol for Preparation of Open Reading Frame-Specific Amplicons

Identification of Forward and Reverse Primers for Each Identified Open Reading Frame in Annotated Genome Database. The entire genome sequence of *V. cholerae* O1, biotype EL Tor, strain N16961 has been completed by The Institute for Genomic Research (TIGR) and the unannotated nucleotide sequence has been made available prior to publication at http://www.tigr.org/tdb/mdb/mdb.html. Public presentations and abstracts describing this sequence indicate that it consists of 4,033,464 base pairs (bp) predicted to contain 4575 ORFs, divided between two circular chromosomes of 1,072,313 and 2,961,151 bp.[17] To prepare a *V. cholerae* O1 microarray prior to release of the annotated version of the sequence, the available nucleotide sequence was analyzed by two separate ORF-finding programs, one carried out by T. Gaasterland (Rockefeller University, New York, NY), using the Glimmer 2.0 program from TIGR (http://www.tigr.org/softlab/glimmer/glimmer.html), and the other carried out by S. Karlin and J. Mrazek, using a program developed by S. Karlin (Stanford University, Stanford, CA). Taken together, the results of these programs led to the identification of 4364 unique ORFs.

---

[17] J. F. Heidelberg, W. Nelson, J. D. Read, D. H. Haft, E. L. Hickey, M. L. Gwinn, R. J. Dodson, R. Clayton, C. M. Fraser, R. R. Cowell, K. J. Fullner, N. Judson, and J. J. Mekalanos, *in* "Proceedings of the 7th Conference on Small Genomes" (J. Zhou, ed.), paper 16. Oak Ridge National Laboratory, Oak Ridge, Tennessee, 1999.

To select optimal forward and reverse primers with equivalent predicted melting temperature, each of these ORFs was analyzed using Primer3 software developed by S. Rozen and H. J. Skaletsky (code available at http://www.genome. wi.mit.edu/genome_software/other/primer3.html). Batch processing and data formatting were performed by a script written by H. Salamon (Stanford University). The primers were synthesized by the Stanford University Protein and Nucleic Acid facility, using a 96-well, multiplex oligonucleotide synthesizer. These "master" plates were processed for PCR-amplification, using the 96-well format described below.

Polymerase Chain Reaction Amplification of Open Reading Frame-Specific Amplicons. Two master plates, one containing the forward primers (Master primer right) and the other the reverse primers (Master primer left), are delivered for each set of 96 ORFs. In preparation for PCR amplification to produce the ORF-specific amplicons, the microtiter wells of the master plates are diluted by the addition of water to yield a final concentration of 25 μM. The following procedure is used to determine the oligonucleotide concentration in each well of the master plates prior to this dilution step.

1. Using a 50- to 300-μl pre-PCR eight-channel micropipettor with plugged aerosol barrier tips, 100 μl of sterile (DNase- and RNase-free) distilled H_2O is added to each well of the Master primer left and Master primer right plates and mixed well without creating bubbles or aerosols.

2. Three UV-transparent microtiter plates are obtained and labeled UV Right, UV Left, and UV Blank.

3. Add 198 μl of $1\times$ TE buffer (10 mM Tris-HCl, 1.0 mM EDTA, pH 8.0) to each well of the plates labeled UV Right and UV Left.

4. Add 200 μl of the same $1\times$ TE buffer into each well of the plate labeled UV Blank. This will serve as a negative control for the spectroscopic determination of oligonucleotide concentration.

5. Using a 2- to 25-μl pre-PCR 12-channel pipettor with aerosol barrier tips, add to the plate labeled UV Right 2 μl of the primers from the plate labeled Master primer right. Repeat the procedure for the UV Left plate, using the plate labeled Master primer left.

6. To determine the oligonucleotide concentration in each well, obtain the $OD_{260\,nm}$ of the three plates (UV Right, UV Left, and UV Blank). From this information calculate the following values: stock concentration in each well of the master plate, dilution factors to obtain a 25 μM working primer concentration in the master plate, and the volume of water needed to dilute 5 μl of the original oligonucleotide stock solution to 25 μM. Add the prescribed volume of water to each well. Because the *V. cholerae* O1 genome sequence led to the prediction of 4364 ORFs, 46 master plates, each containing 96 forward PCR primers, and 46 master plates, each containing the corresponding 96 reverse PCR primers, were

prepared. Once the master plates were diluted as described above, the matched pairs of primer master plates were designated VcRight01 25 μM/VcLeft01 25 μM through VcRight46 25 μM/VcLeft46 25 μM.

7. Seven microliters of each primer from primer plate VcRight01 25 μM is added to another microtiter plate. Seven microliters of each primer from primer plate VcLeft01 25 μM is added to the same microtiter plate. Then, 21 μl of water is added to each of the wells. Each well thus contains a concentration of 5 μM of each forward and reverse primer in a volume of 35 μl; the resulting plate is designated "primer mix plate Vc01."

8. The amplification buffer was patterned after the PCR super mix by GIBCO-BRL (Gaithersburg, MD). A 1.1× concentration buffer mix was prepared, using the concentrations and the buffer described by the product sheet. To this buffer add 2 units of the *Taq* polymerase and 10 to 40 ng of genomic DNA per reaction. Dispense 60 μl of this mix into each well of the PCR plate, using the 50- to 300-μl pre-PCR multichannel pipettor. To each individual well, add 5 μl of the 5 μM primer mix just prior to performing the PCR.

9. Seal this plate with a mat that has been processed to be pre-PCR. Use the following thermocycler program:

Cycle	Temperature/time	No. of repetitions
1	94° for 60 sec	1
2	94° for 40 sec	
	52° for 40 sec	30
	72° for 75 sec	
3	72° for 300 sec	1
4	4°, hold	

Size Validation of Open Reading Frame-Specific Amplicons. Each of the 4364 amplicons is analyzed by agarose gel electrophoresis to assess the quality of the PCR product. Reactions that generated no PCR product, a product of the wrong size, or multiple products are flagged and repeated, and if the second reaction is still unsuccessful, the primers are either resynthesized or new primer pairs are selected for the ORF in question.

1. A 2.0% (w/v) agarose gel is cast in 1× TAE buffer (0.04 M Tris–acetate, 0.001 M EDTA, pH 8.0), using 26-well combs designed to load samples from 96-well plates.

2. Two microliters of sample from each PCR well is transferred to disposable 96-well plates containing 8 μl of distilled, deionized H_2O. Two microliters of 6× dye is added to each well.

3. The φX DNA molecular mass marker is added to the first and last wells of each of the four sections of the gel. Ten microliters of sample is added to every other well.

4. The gel is run at 100 V for 1 hr and stained by shaking for 15 min in ethidium bromide (EtBr) [50 μl of an EtBr stock solution (1.0 mg/ml) in 500 ml of water]. The gel is washed in water and photographed.

5. The image is stored as a TIF (tagged image format) file and transferred to a ZIP disk.

Protocol for Fabrication of Microarray, Using Open Reading Frame-Specific Amplicons

Preparation of Microscope Slides Prior to Array Printing. The ORF-specific amplicons are printed onto the surface of microscope slides that have been washed in an NaOH–ethanol solution and then coated with poly-L-lysine. The NaOH–ethanol solution is prepared by dissolving 70 g of NaOH in 280 ml of water to which 420 ml of 95% (v/v) ethanol is then added. This solution is caustic and should be used with protective clothing, eyewear, and gloves.

1. Place the microscope slides in metal slide racks. Place the racks in chambers.

2. Soak the slides in the NaOH–ethanol–distilled, deionized H_2O solution for 2 hr with gentle rotation.

3. Rinse with distilled H_2O for 5 min; transfer the slide racks to new chambers, filled with distilled H_2O; agitate on an orbital shaker for 10 min; repeat twice. It is critical to remove all traces of NaOH–ethanol.

4. Because it adheres to glass, prepare the poly-L-lysine solution [1% (w/v) poly-L-lysine in water; Sigma, St. Louis, MO] in a plastic container, using plastic graduated cylinders to measure volumes. Combine 30 ml of phosphate-buffered saline (PBS), 240 ml of distilled, deionized H_2O, and 30 ml of the poly-L-lysine solution.

5. Place one set of racks containing washed slides into the poly-L-lysine solution in the plastic container for 40 min with shaking. Then, gently rinse with distilled, deionized H_2O and place the remaining racks of washed slides in the poly-L-lysine solution. Store the first set of racks in water while the second set is incubated with poly-L-lysine.

6. Dry the slides by centrifugation, using a microtiter plate carrier (650 rpm for 2 min); place paper towels below the rack to absorb liquid.

7. Complete the drying process in a vacuum (45° for 10 min).

8. Store slides in a desiccator for 3 weeks prior to use; this aging process improves print quality, but the basis for this effect is not known. Slides that are older than 2 months will result in faint printing and higher background.

Transfer of Open Reading Frame-Specific Amplicons from 96-Well Microtiter Plates to 384-Well Microtiter Plates. The robotic printer described below prints from 384-well microtiter plates. Therefore, it is necessary to transfer the PCR products from the original 96-well amplicon plates to 384-well plates. This is accomplished with an automated 96-channel pipettor (Multmek 96; Beckman Instruments, Fullerton, CA) that aspirates 32 μl from the original PCR plate and dispenses 8 μl into four different 384-well plates. When the transfer is finished, four complete sets of 384-well plates are ready to be printed. These plates are stored in a −80° freezer.

1. On the day before printing, dry one set of plates in a SpeedVac (Savant, Holbrook, NY) for about 1 hr, using a medium setting.

2. Rehydrate the DNA in the 384-well plates using 5 μl of filter-sterilized 3× SSC (salt–sodium citrate) per well. Seal the plates with metal tape and store them overnight at room temperature.

3. Prepare plates about 15 min before printing. Take out two plates at a time and vortex them for 3 min at level 3.5, using a multitube vortexer (VWR Scientific, West Chester, PA). Centrifuge the plates for 1 min at 3000 rpm.

Printing of Open Reading Frame-Specific Amplicons onto Poly-L-Lysine-Coated Slide Surface. Microarrays are printed onto the surfaces of poly-L-lysine microscope slides by using a robotic spotting device fitted with stainless steel printing tips that enable production of replicate arrays for each printing session. Instructions for building the custom microarrayer used by the authors can be found at the MGuide website referred to above. Included with this guide are a parts list, software for controlling the motors, and the design specifications for custom-built parts. Several commercial arrayers are also available. The printing configuration is determined by the types of slides, coverslips, tips, and print head. A 16-tip print head equipped with ChipMaker 2 printing pins (TeleChem International, Sunnyvale, CA) can print a 20 × 20 mm array onto a standard microscope slide producing a configuration of 5000 spots with a spot density of 225 μm (center-to-center spacing). ChipMaker 3 uses smaller printing pins and can produce a 125-μm spot center-to-spot center spacing, yielding a high-density format so that each slide can accommodate ~15,000 ORF-specific spots and control spots. Each printing pin contains a slotted end that is filled by capillary action during contact with the liquid contents of the microtiter well. The manufacturer estimates an uptake volume per pin of 0.20 μl and an estimated delivery volume per spot printed of 0.5 nl.

1. Turn on the microarrayer, open up the ArrayMaker program from the PC window, click Connect, make sure that there is nothing blocking the pathway of the arrayer, and press OK. The numbers for the x, y, and z axes change as the robot arm finds its Home position; wait until these numbers stop changing before conducting the test print.

2. Test arrays are printed prior to the main print run to assess spot size quality and consistency and to detect possible carryover contamination. A test slide is

taped down in the first slot. The test 384-well plate containing printing solution is placed in the A1 position on the arrayer. Go to the Test Print window and set parameters for a test print. Push Start. Once the test print is completed, determine whether all the pins are printing. If some of the tips are not printing, wash by sonication according to the manufacturer recommendations and repeat the test print.

3. The main print run will yield 137 individual microarrays. Place the poly-L-lysine-coated glass slides on the microarrayer and tape them down on both sides to prevent motion of the slides during printing. Insert the 384-well plates in the correct orientation. Go to the Print window, set parameters for the run, and press Start. The principal parameters are Quad Width (the number of spots across one quadrant) and Spacing (the space between each individual spot). Periodically determine whether all the pins are printing and that they are not stuck in the print head bracket.

Postprocessing of Printed Array. After the array is printed, the slide is processed to effect a stable bond between the double-stranded DNA of the printed ORF-specific amplicon and the poly-L-lysine film on the glass surface. The postprocessing step also removes excess DNA and thus prevents "plumming" artifacts that can occur during slide hybridization.

1. Preparation for the postprocessing step entails the following steps: (a) bring the heating block to 90° and invert the insert; (b) place the slide-staining chambers in front of the heating block; (c) fill the slide staining chamber with warm water; (d) fill one of the slide dishes with 95% (v/v) ethanol; (e) in a fume hood place a 4-liter glass beaker containing 1 liter of steaming water, a stirring plate for preparation of the blocking solution, and a shaker; (f) obtain a clean 500-ml beaker to prepare the blocking solution; (g) assemble succinic anhydride (6.0 g), 1-methyl-2-pyrrolidinone (335 ml), and 15 ml of 1.0 M sodium borate, pH 8.0 (filtered).

2. Rehydrate the slides by inverting them, array side down, toward the steaming water (using a slide-staining chamber) until the spots become glistening and moist. Be careful not to allow the water to touch the array. Immediately flip the slides over, array side up, onto the heating block. When the array spots dry (approximately 5 sec), remove them from the heating block. Do only one slide at a time.

3. To UV cross-link the printed DNA to the poly-L-lysine film, place the slides, array side up, on a flat, dust-free board that fits into the UV cross-linker. Avoid the use of Saran Wrap because the slides stick to it. Irradiate with 600 μJ of UV light. If the spots are small and prior printings indicate that excess DNA is not present, delete the following three steps and proceed to step 4 below. Otherwise, perform the following wash steps: (a) wash the slides with $1 \times$ SSC–0.05% (w/v) sodium dodecyl sulfate (SDS) for 30 sec by plunging them in and out of the solution; (b) wash the slides with $0.06 \times$ SSC for 30 sec by plunging them in and out of the solution; (c) dry the slides in a centrifuge at 600 rpm for 1–2 min.

4. The free lysines are blocked through the use of succinic anhydride. Succinic anhydride (6 g; Aldrich, Milwaukee, WI) is added to 350 ml of 1-methyl-2-pyrrolidinone (Aldrich) with stirring. A laboratory coat should be worn while working with methyl pyrrolidinone. When the succinic anhydride has dissolved, quickly add 15 ml of 1.0 M sodium borate, pH 8.0 (prepare with boric acid, adjust to pH 8.0 with sodium hydroxide, and filter), and pour the mixed solution into a slide-washing tray to a level that will completely cover the slides when they are placed inside. Quickly place the slides into the succinic anhydride solution and vigorously plunge them up and down for 60 sec to remove any air bubbles. Rotate the slide rack at 60 rpm for 15 min. Remove the slide rack from the organic reaction mixture and plunge it immediately into a boiling water bath for 90 sec. Transfer the slide rack to a 95% (v/v) ethanol wash tray and carry it directly to a tabletop centrifuge. Spin dry the slides by centrifugation at 600 rpm for 2 min. Carefully transfer the slides to a dry slide box for storage in a desiccator.

Protocol for Isolation of Total RNA from Planktonic and Biofilm Bacteria

Whole genome expression profiling to identify genes differentially expressed by planktonic and sessile populations of the same culture is the most important single application of this method to biofilm research. In principle, any of the biofilm systems described in other contributions to this series could be used for this purpose, providing the biomass harvested from the device yields sufficient total RNA for microarray analysis. At present, the lower limit of total RNA for this purpose appears to be between 0.5 and 2.0 μg. On the basis of currently available evidence, there appears to be no practical advantage for the use of mRNA-enriched preparations and therefore no method for the removal of ribosomal RNA is given here. However, exact parallel processing of the two comparable conditions, for example, sessile versus planktonic, is of considerable importance. Slight differences in the early steps of the RNA preparation method, particularly before transcription has been terminated and the RNA pool stabilized, will result in method-induced gene expression patterns that may be confused with patterns directly related to the two conditions under study. This is especially of concern for the manipulation of sessile and planktonic populations because the former condition requires that bacteria be detached from a surface or that the RNA be preserved *in situ*. In contrast, planktonic cells are usually collected by centrifugation. To avoid procedure-related artifacts, the method described below emphasizes the need for rapid processing, freezing of cell pellets in liquid nitrogen, and the use of TRIzol solutions during the extraction process. Comparing expression profiles between different RNA preparation methods applied to the same population of bacteria is an important control experiment that reveals how much of an expression pattern is method dependent. For the isolation of planktonic bacteria, the following protocol for exponentially growing bacteria is recommended. However, the volume of sampled cells may have to be increased proportionately to adjust for biomass differences if the planktonic population from a biofilm reactor or chamber is less than the biomass available from exponentially growing cells cultivated in a rich medium.

1. Collect 10-ml volumes of an exponentially growing culture of *V. cholerae* O1 ($OD_{600\ nm}$ 0.6–1.0) in 15-ml sterile polypropylene Falcon tubes (Becton Dickinson Labware, Lincoln Park, NJ) and centrifuge [5 min, 5000 rpm, Sorvall (Newtown, CT) SA-600 rotor]. The temperature of the centrifuge should be warmed to the temperature of the biofilm experiment to avoid induction of cold-shock genes. Quickly remove the supernatant, freeze the cell pellet in liquid nitrogen, and store at $-80°$.

2. Add 10 ml of TRIzol reagent (GIBCO-BRL) to each bacterial pellet, lyse the cells in TRIzol by repetitive pipetting, and incubate the lysate for 10 min at room temperature.

3. Add 2 ml of chloroform to each sample, shake vigorously for 15 min, and incubate for 3 min at room temperature.

4. Centrifuge the samples (6000 rpm, Sorvall, SA-600 rotor, $4°$, 10 min), transfer the aqueous phase to sterile Oak Ridge tubes, and precipitate the RNA by adding 2.5 ml of 2-propanol and 2.5 ml of a salt solution (0.8 M sodium citrate, 1.2 M sodium chloride). Incubate the samples at room temperature for 10 min.

5. Centrifuge the samples (10,000 rpm, $4°$, 30 min), remove the 2-propanol, wash the RNA pellets with 75% (v/v) ethanol, and recover the washed pellet by centrifugation. Using 100% ethanol should be avoided in all steps of RNA preparation intended for array analysis because benzene contamination may fluoresce. Instead, use 95% (v/v) ethanol to prepare these solutions.

6. Remove the ethanol and air dry the RNA pellet.

7. Dissolve the RNA in 100 μl of RNase-free water and determine the RNA concentration.

8. Add 10× DNase I buffer to the RNA solution, followed by the addition of 4 units of RNase-free DNase (Ambion, Austin, TX) per 50 μg of total RNA. Incubate for 30 min at $37°$.

9. Clean up the RNA preparation, using a Qiagen (Valencia, CA) RNeasy kit. Add to the RNA sample 350 μl of RLT buffer (Qiagen) containing 10 μl of freshly added 2-mercaptoethanol per 1 ml of RLT buffer, vortex, add 265 μl of 95% (v/v) ethanol, vortex, and load onto an RNeasy spin column. Centrifuge (15 sec) and transfer the column to a new 2-ml collection tube. Add 500 μl of RPE buffer (Qiagen), centrifuge (15 sec), discard the flowthrough, add 500 μl of additional RPE buffer, and centrifuge (2 min). If the column is still wet on the sides, remove the wash solution from the tube and centrifuge (1 min) to dry. Transfer the column to a 1.5-ml collection tube and elute the RNA with 40 μl of RNase-free water and centrifuge (1 min). Determine the RNA concentration by A_{260}/A_{280} values. The quality of the RNA preparation should be determined

by agarose gel electrophoresis: load 1 μl of RNA onto a 2% (w/v) agarose TAE gel and run for 45 min at 100 V.

RNA is obtained from biofilm populations of bacteria by directly submerging the surface containing the attached bacteria in the TRIzol reagent. RNA purification then proceeds as described above.

Synthesis and Labeling of cDNA

Fluorescently labeled cDNA copies of the total RNA pool are prepared by direct incorporation of fluorescent nucleotide analogs during a first-strand reverse transcription (RT) reaction. The method described below employs Cy3 and Cy5 dyes because they have good spectral separation, high quantum efficiencies, minimal photobleaching, and they fluoresce when dry. In addition, most commercial scanners are equipped with lasers and barrier filters that are compatible with these dyes. Cy3 (excitation$_{max}$ of 550 nm; emission$_{max}$ of 570 nm) and Cy5 (excitation$_{max}$ of 649 nm; emission$_{max}$ of 670 nm) are available as dCTP or dUTP analogs (Amersham Pharmacia Biotech, Piscataway, NJ). One pool of RNA, derived from the sessile population, is labeled with Cy5, whereas the other pool, derived from the planktonic population, is labeled with Cy3.

1. Bring 0.5–5.0 μg of total RNA to 11 μl with distilled H$_2$O and add 2 μl (~2 mg/ml) of random hexamers. Heat for 10 min at 98°, and then snap cool on ice.

2. Add the following reaction components: 5 μl of 5× first-strand buffer, 2.5 μl of 100 mM of dithiothreitol (DTT), 2.3 μl of dNTP (5 mM each of dATP, dGTP, and dCTP and 0.2 mM dTTP) and 1.5 μl of Cy3 or Cy5; then add 2 μl of Superscript II RT.

3. Incubate for 10 min at 25° followed by 90 min at 42°, using a PCR thermocycler. The labeled cDNA can be frozen or left at 4°, overnight.

Microarray Hybridization: Application of Labeled cDNA to Array Surface

The "two-color" hybridization system employed here entails the use of two cDNAs derived from two contrasting conditions of growth, each cDNA having been labeled with a different dye (Cy3 or Cy5). These differentially labeled cDNAs are combined and applied to the array surface under conditions that favor hybridization. The resulting fluorescence intensity of each of the fluorescent labels for every ORF-specific spot is measured and compared. The resulting intensity ratio is derived and serves as a measure of the differential expression of each ORF at the transcriptional level. For these ratios to be quantitatively meaningful, the input total RNA for each of the two conditions should be equal. To achieve this, the following options are available: the number of organisms from each of the two conditions from which the RNA was prepared is equalized; or the mass of total RNA or of labeled cDNA prepared under each of the two conditions is adjusted to equivalence. The effect of these options on the ratios obtained has not been formally studied and this ambiguity is reflected in the method that follows.

1. Add both reverse transcriptase reactions, containing the two labeled cDNAs, to the same Microcon 10 (Amicon, Danvers, MA) tube together with 400 μl of TE buffer. Using a microcentrifuge, centrifuge at maximum velocity until ~25 μl remains (about 20 min at room temperature). Add 200 μl of TE to the same tube and centrifuge until almost dry (about 12 min at room temperature).

2. Recover the cDNA by inverting the Microcon 10 tube into a new 1.5-ml collection tube and centrifuge at one-half maximum velocity for 1 min at room temperature.

3. Bring the sample to 7 μl with TE buffer and transfer to a 0.5-ml tube. To this add 0.67 μl of tRNA (10 mg/ml), 1.9 μl of 20× SSC, and 1.35 μl of 2% (w/v) SDS.

4. Heat to 98° for 2 min. Centrifuge briefly (maximum velocity for 5 sec in a microfuge, room temperature). The supernatant is designated the hybridization solution.

5. Apply one small dot of rubber cement with a 25-gauge, blunt-ended needle and syringe to each corner of the array. Then, apply 10 μl of the hybridization solution to the microarray. Add a 22 × 22 mm coverslip, using bent precision forceps. Free any bubbles that may have collected under the coverslip by tapping the top of the coverslip with the forceps. Completely seal the edges of the coverslip with rubber cement and allow the cement to dry for a few minutes before closing the hybridization chamber. Conduct the hybridization at 63–65° for between 6 and 18 hr, using a dual hybridization chamber (HYB-03; Genemachines, San Carlos, CA), humidified by the addition of ~100 μl of water. Within this time interval, the duration of hybridization is not critical.

6. At the end of the hybridization period, remove the coverslip with the assistance of a razor blade while the array is submerged in 1× SSC containing 0.05% (w/v) SDS. Transfer the submerged slide to a microscope slide-staining rack that is submerged in the same solution.

7. Hand rinse the slide in 0.06× SSC, transfer it to another submerged staining rack, and wash for 2 min in fresh 0.06× SSC.

8. Centrifuge the slide for 1 min at 500 rpm to dry.

Scanning Microarray

The hybridized microarray is imaged by confocal laser scanning. For the studies described here, the authors have mainly used a ScanArray 3000 scanner manufactured by GSI Lumonics (Watertown, MA). This instrument provides 10-μm pixel resolution; for a 100-μm-diameter spot, this level of resolution yields 70 significant pixels. Newer models, including the ScanArray 4000 and 5000 instruments from the same manufacturer, provide 5-μm pixel resolution, allowing improved discrimination of signal and background values and the capacity to compensate for pixel heterogeneity within individual spots and for fabrication artifacts such as dust particles. Whereas the older generation of scanners is equipped with two lasers and thus is limited to two-color detection, the ScanArray 5000 instrument is equipped with four lasers and therefore can conduct four-color microarray analysis.

To maximize sensitivity, the laser intensity is increased to the point at which some spots on the array are saturated. Then, the photomultiplier gain settings are adjusted to normalize the two channels while maintaining a low overall background. Spots containing Cy3- and Cy5-labeled genomic DNA, printed at the same time as the ORF-specific amplicons, can be used both to detect carryover from incompletely washed printing pins and to normalize the two fluorescence channels at the time of scanning. The instrument settings for each scan should be selected quickly and the array not rescanned multiple times to avoid photobleaching and signal attenuation. The Cy5 dye is more susceptible to photobleaching than the Cy3 dye.

The process of scanner-derived data acquisition and storage entails the following steps.

1. Acquire the fluorescence signals from the array, following the instructions provided by the manufacturer of the scanner. Store the raw data files as 16-bit TIF (tagged image format) images.

2. Extract the numerical data from the spot intensities. Identify and mark any spots that are flawed technically and might yield misleading data. For this purpose, the authors have mainly used the ScanAlyze program written by M. Eisen (Stanford University) and available without cost to noncommercial users at http://rana.stanford.edu/software/. Manufacturers of scanners also provide image-processing software and the new ScanArray 5000 manufactured by GSI Lumonics is so equipped.

3. Store the extracted data in a database that links the spot positions to their identification and descriptive information. The large data sets generated by whole genome expression profiling, and the need to compare expression results within and between experiments, require a robust, scaleable, and flexible relational database. Several are in development, others are available commercially, and at least one can be downloaded without cost by academic institutions (http://genome.nhgri.nih.gov/arraydb/).

Data Analysis and Mining of Whole Genome Expression Results

Analysis of microarray expression profiling results involves three interconnected operations: image processing (discussed above), data management, and data mining. Preliminary data analysis of a microarray result can be undertaken as described below and will lead to the identification of differentially regulated

ORFs. However, further analysis will require additional data-mining programs. These are described in the last section of this chapter. Preliminary data analysis involves four steps.

1. Subtract the local background intensity from the intensity of each spot for both fluorescence channels. Estimate the sensitivity of the experiment by reference to internal standards and non-*V. cholerae* DNA spots on the array and assign a threshold value that defines the minimal detectable signal. Spots with values below the defined threshold may be omitted from further analysis. A publication from this laboratory using an *M. tuberculosis* microarray, arbitrarily, but sensibly, defined the "minimum specifically detectable signal" to be 1 standard deviation above the average signal of all spots containing non-*M. tuberculosis* DNA.[12] Alternatively, this value could be defined by the level that is 2 standard deviations above the background values for all spots on the array.

2. Normalize the Cy3 and Cy5 channels by identifying the set of labeled *V. cholerae* DNA spots that were printed as controls. The two channels are then normalized with respect to the median intensity values of these spots.

3. Calculate the Cy5/Cy3 ratios (channel 1/channel 2) and the $\log_2$ ratios for every spot. Assign a positive value to the $\log_2$ ratios that represent genes induced by the sessile compared with the planktonic population, and a negative value to the $\log_2$ ratios for genes repressed by the sessile compared with the planktonic population. The use of log-transformed values is appropriate for comparing ratios because the distance on the positive scale is equivalent to the distance of the inverse ratio on the negative scale. Accordingly, each integer represents a 2-fold increase in the ratio value: a 4-fold induction equals +2 using a $\log_2$ ratio, whereas a 4-fold repression equals −2 using a $\log_2$ ratio.

4. Apply statistical methods to identify ORFs with $\log_2$ ratios that are significantly different from the population of *V. cholerae* ORFs represented on the array. One appropriate statistical instrument for this purpose is the Student *t* test. Quantify the degree by which each ratio differs from the population mean for all spots on the array by calculating the number of standard deviations from the mean (z) for each spot:

$$z = (\log_2 \text{ ratio} - \text{mean})/\text{standard deviation}$$

Because z values are normalized to the standard deviation of a given experiment, they can be directly compared across different experiments. The z value can be used to deduce the probability (p) that a given spot is different from the population of other *V. cholerae* ORFs.

Advanced Data-Mining Methods

The preliminary level of analysis described above, while informative, will not necessarily lead to the identification of genes controlled by shared regulatory networks that underlie the capacity of the organism to respond in an adaptive manner to a particular condition or mode of growth. Although a detailed description of advanced data-mining programs is beyond the scope of this chapter, the following briefly describes and references two approaches to this challenge.

1. Hierarchical clustering has been employed in the analysis of microarray expression data in order to place genes into gene clusters on the basis of sharing similar patterns of expression.[18] If a variety of conditions is tested and the expression data are pooled and analyzed in this manner, the resulting clusters are likely to contain genes that mediate common cellular functions. This method yields a graphical display that resembles a kind of phylogenetic tree in which the relatedness of the expression behavior of each gene to every other gene is depicted by branch lengths. The programs Cluster and TreeView, both written by M. Eisen (Stanford University), are available without cost to academic institutions at http://rana.stanford.edu/software/.

2. Microarray expression data have also been analyzed by self-organizing maps (SOMs), a nonhierarchical method that involves selecting a geometry of nodes, in which the number of nodes defines that number of clusters.[19] Then, the number of genes analyzed and the number of experimental conditions that were used to provide the expression values of these genes are subjected to an iterative process (20,000–50,000 iterations) that maps the nodes and data points into multidimensional gene expression space. To facilitate this process, two preprocessing steps are used. First, variation filtering eliminates genes showing no significant differential expession for any of the tested conditions. Second, the expression level of each gene is normalized across experiments. As a result, the expression profile of the genome is highlighted in a manner that is relatively independent of the expression magnitude of each gene. Software for the GENECLUSTER SOM program for microarray expression analysis can be obtained without cost to academic institutions from the Whitehead/MIT Center for Genome Research at http://www.genome.wi.mit.edu/MPR/software.html.

Acknowledgments

Supported by the National Institutes of Health research grant RO1-AI43422.

[18] M. B. Eisen, P. T. Spellman, P. O. Brown, and D. Botstein, *Proc. Natl. Acad. Sci. U.S.A.* **95**, 14863 (1998).

[19] P. Tamayo, D. Slonim, J. Mesirov, Q. Zhu, S. Kitareewan, E. Dmitrovsky, E. S. Lander, and T. R. Golub, *Proc. Natl. Acad. Sci. U.S.A.* **96**, 2907 (1999).

[2] Biofilm Formation as a Developmental Process

By PAUL N. DANESE, LESLIE A. PRATT, and ROBERTO KOLTER

Introduction

The cellular constituents of biofilm communities are known to be physiologically distinct from their planktonic brethren.[1,2] One of the more medically relevant examples of such a physiological distinction is that biofilm cells often display hyperresistance to antimicrobial agents when compared with their planktonic counterparts.[2] In addition, biofilm cells will often exist in localized anoxic microenvironments and/or microenvironments that vary significantly in their pH and ionic strength when compared with cells in the bulk medium.[1] Clearly, these microenvironments must exert dramatic effects on the physiology of biofilm constituents.

Although these physiological characterizations of biofilms are well established, genetic and molecular biological approaches have only begun to reveal the molecular bases of the distinctions described above. For example, what molecular features of a biofilm confer the antimicrobial hyperresistance that is so often described? While this phenomenon is not completely understood, part of the answer can be attributed to increased production of exopolysaccharides (EPS) on interaction between colonizing bacteria and a suitable surface.[2] Indeed, several groups have shown that the expression of genes required for the biosynthesis of EPS is increased on surface attachment.[3,4] This increased production of EPS helps to establish the three-dimensional architecture of biofilm microcolonies[5] and presumably provides a physical barrier to external insults.[2] Taken together, these results imply that biofilm formation is not a haphazard process in which cells are strewn onto a surface for the simple purpose of remaining fixed in space. Instead, the biofilm is a highly structured community that forms for numerous rational reasons (protection from a host's immune system or predatory organisms, nutrient availability, protection from the harsh vicissitudes of the environment, etc.).

The findings described above have an additional implication: biofilm formation under various conditions is a developmentally regulated process. It is important to define what we mean by "developmental process," as the term may conjure up different meanings for different readers. In this case, we mean that a bacterial culture

[1] J. W. Costerton, Z. Lewandowski, D. E. Caldwell, D. R. Korber, and H. M. Lappin-Scott, *Annu. Rev. Microbiol.* **49,** 711 (1995).

[2] J. W. Costerton, P. S. Stewart, and E. P. Greenberg, *Science* **284,** 1318 (1999).

[3] D. G. Davies, A. M. Chakrabarty, and G. G. Geesey, *Appl. Environ. Microbiol.* **59,** 1181 (1993).

[4] C. Prigent-Combaret, O. Vidal, C. Dorel, and P. Lejeune, *J. Bacteriol.* **181,** 5993 (1999).

[5] P. N. Danese, L. A. Pratt, and R. Kolter, *J. Bacteriol.* **182,** 3593 (2000).

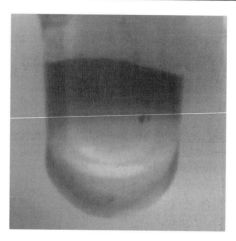

FIG. 1. Crystal violet-stained biofilm formed from *Escherichia coli* ZK2686 (W3110) after 24 hr of growth in LB at 30°.

monitors both its liquid environment and the surfaces available for colonization and it uses this information to determine whether it will form a biofilm and, if so, the type of biofilm that will be constructed. In addition, the developmental nature of biofilm formation also implies that different types of biofilms themselves are capable of monitoring changes in their environment and responding to these changes according to their best interests. In some cases, this may mean the dissolution of the biofilm and, in other cases, it may mean the conversion to a different type of biofilm. Below, we describe some of our laboratory's approaches to examining the developmental nature of biofilm formation.

Quantitative Macroscopic Assay for Biofilm Formation

Our laboratory has previously described a macroscopic visual assay for biofilm formation in polyvinylchloride (PVC) microtiter dishes.[6] This assay, which measures the degree of crystal violet (CV) staining of a biofilm, is rapid and simple, allowing examination of thousands of independently generated biofilms in one sitting. As shown in Fig. 1, this assay provides gross structural details of the formed biofilm. For example, in the case of Fig. 1, it can be seen that the biofilm forms most prominently near the air–water interface under the conditions tested. The amount of CV bound to the biofilm (and therefore the amount of biofilm material) can be quantified via solubilization of CV in dimethyl sulfoxide (DMSO)

---

[6] G. A. O'Toole, L. A. Pratt, P. I. Watnick, D. K. Newman, V. B. Weaver, and R. Kolter, *Methods Enzymol.* **310**, 91 (1999).

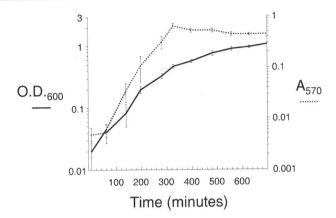

FIG. 2. Biofilm formation occurs during exponential phase growth in *E. coli* ZK2686. Cultures were grown at 30° in LB for the times indicated. Error bars represent the standard deviation from the mean of eight independent cultures.

and subsequently measuring the absorbance of each sample at 570 nm.[6] Note that both the gross visual characteristics and the quantitative measurement of CV-stained biofilms will often provide useful information about the nature of biofilm development (see below).

Examining Growth-Phase Control of Biofilm Formation

While the assay described above has numerous applications, one of the most important is the examination of biofilm formation as a function of growth phase. The growth phase(s) in which biofilms develop should provide information as to the overall purpose of the biofilm. For example, in our examination of biofilm development in *Escherichia coli,* we have found that saturated stationary-phase cultures of *E. coli* will not form biofilms when placed on an abiotic surface such as a PVC microtiter well. Rather, growth through exponential phase is required for this process (Fig. 2).

Measuring Biofilm Formation and Growth Phase as Function of Time

For measuring cell density and growth phase as a function of time, eight independent 5-μl samples of a saturated culture are inoculated into eight 125-μl samples of Luria–Bertani broth (LB) present in a polystyrene 96-well microtiter dish. This plate is used to determine cell density as a function of time by measuring the OD_{600}. Concomitant with the cell density cultures, 88 independent 5-μl samples of the same cultures (8 for each time point) are inoculated into 88 independent 125-μl samples of LB present in a PVC 96-well microtiter dish. At each point in

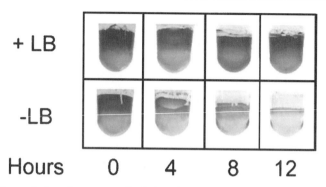

FIG. 3. Removal of nutrients causes dissolution of *E. coli* biofilms. Biofilms were formed with strain ZK2686 (W3110) for 24 hr in LB. The LB was removed and replaced with either fresh LB or distilled water, and the biofilms were then incubated for the given times. See text for details.

which an OD_{600} measurement is made, one set of eight PVC well cultures is also stained with CV and dried, and the amount of attached CV is then determined as described.[6] The amount of CV staining and the cell density values for one such experiment are plotted versus time in Fig. 2, clearly indicating that biofilm formation occurs during exponential phase growth under these conditions.

Examining the Effects of Nutrient Availability and Biofilm Formation

In addition to the growth-phase regulation of biofilm development, it is important to examine how preexisting biofilms adapt to changes in their environment. For example, one question concerns whether the sudden depletion of nutrients affects a biofilm in any manner. In our analyses of this issue we have allowed biofilms of *E. coli* ZK2686 (W3110) to develop over 24 hr of growth in LB and then replaced the nutrients with an equivalent volume of either sterile water or fresh LB. Figure 3 shows that over the course of approximately 12 hr the biofilm reverts almost completely to a planktonic state.

Specifically, biofilms of *E. coli* ZK2686 were allowed to develop in 125 µl of LB (in PVC microtiter dishes) at 30° for 24 hr. The liquid medium was removed and the biofilms were washed once with sterile water and then 125 µl of either sterile water or fresh LB was added to each biofilm. The state of biofilm development was then assessed over a 12-hr period via staining with CV.

Biofilm Development under Distinct Growth Conditions

As with the adaptation of existing biofilms to nutrient deprivation, we have found that *E. coli* will form visually and molecularly distinct biofilms under distinct

FIG. 4. Crystal violet-stained biofilm formed from *Escherichia coli* ZK2686 (W3110) after 24 hr of growth in M63 minimal glucose (0.8%, w/v) medium at 30°.

growth conditions. For example, in rich media, *E. coli* biofilms develop primarily near the air–liquid interface (Fig. 1). In contrast, the biofilms formed in minimal media are more uniformly distributed throughout the microtiter well, suggesting a molecular detailing that distinguishes these two types of biofilms (Fig. 4). Indeed, this is confirmed at the molecular level, as the outer membrane adhesin, antigen 43 (Ag43), is important for biofilm development in minimal media, but not in rich media.[7]

Moreover, the two types of biofilms can be interconverted by changing the growth medium (P. N. Danese, L. A. Pratt, and R. Kolter, unpublished observation, 1999), further implying that these biofilms are not terminally differentiated entities, and that they are capable of responding to environmental changes.

Conversion of Biofilms Formed in Rich Media (LB) to Minimal Media

Biofilms are initially formed by inoculating 125 μl of LB with a candidate culture and incubating for 24 hr in a PVC microtiter dish at 30°. The liquid medium is then removed and replaced with an equal volume of M63 minimal glucose (0.8%, w/v) supplemented with 1% (v/v) LB and biofilms are allowed to grow under these conditions for an additional 24 hr. The liquid medium is again removed and replaced with an equal volume of M63 minimal glucose (0.8%, w/v) and biofilms are allowed to grow for an additional 24 hr. CV staining of the biofilms is performed at each step in the transition from LB to minimal glucose to assess the alteration in biofilm structure.

[7] P. N. Danese, L. A. Pratt, S. Dove, and R. Kolter, *Mol. Microbiol.* **37,** 424 (2000).

Conversion of Biofilms Formed in Minimal Media to Rich Media

Biofilms are initially formed by inoculating 125 μl of minimal glucose (0.8%, w/v) with a candidate culture and growing for 24 hr in a PVC microtiter dish at 30°. The liquid medium is then removed and replaced with an equal volume of LB and biofilms are allowed to grow under these conditions for an additional 24 hr. CV staining of the biofilms is performed at each step to assess the alteration in biofilm structure.

Microscopic Analysis of Defective Steps in Biofilm Formation

On the basis of extensive genetic, physiological, and microscopic analyses, the entire process of biofilm formation can be described as a series of sequential steps.[8] For *E. coli,* biofilm development begins within initial attachment of planktonic cells to a surface, requiring flagellum-mediated motility and a surface adhesin (e.g., type I pili or antigen 43). After this initial attachment, the biofilm develops (through cell division and recruitment of additional planktonic cells) a complex three-dimensional structure consisting of exopolysaccharide-encased microcolonies punctuated by aqueous channels. This stage typically represents the mature biofilm structure. Finally, under certain circumstances (e.g., nutrient deprivation or nutritional changes) the biofilm can be dismantled or reorganized.

We have found that mutations that block the production of colanic acid (an exopolysaccharide produced by *E. coli*) also halt biofilm development at an initial stage during which cells are capable of attaching to a surface as a monolayer but fail to develop the characteristic three-dimensional structure of the mature biofilm.[5] This result implies that colanic acid production is critical for the development of the mature biofilm. However, because there may be other factors that are also important for this transition it may be useful to examine candidate mutations and various growth conditions to see if they also cause an arrest at an early step in biofilm development.

To examine the three-dimensional structure of a biofilm candidate strains are transformed with a green fluorescent protein expression plasmid in order to visualize the biofilm via fluorescence microscopy.[5] Five-milliliter cultures of candidate fluorescent strains are typically grown in the presence of a borosilicate glass coverslip (which serves as the attachment surface). We have found that the autofluorescence of PVC coverslips interferes with fluorescence microscopy. Cultures are typically grown for 72 hr, with the addition of fresh medium every 24 hr. The coverslips are then examined via fluorescence microscopy, and a series of z scans (step size, 0.2 μm) is collected for each sample. Sagittal (xz) images can be

---

[8] L. A. Pratt and R. Kolter, *Mol. Microbiol.* **30**, 285 (1998).

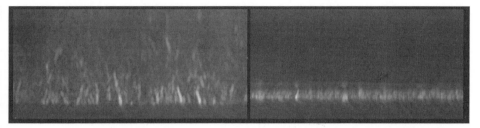

Wild Type   Colanic-acid deficient

FIG. 5. Sagittal (xz) images of wild-type and colanic acid-defective biofilms. Sagittal images were created from a collection of 130 consecutive z-series scans of the wild-type and colanic acid-defective biofilms. The step size between each z section was 0.2 μm. Sagittal images were generated with the Volume View program of the Deltavision software package (Applied Precision). Note that the ellipsoid shape of the cells along the z axis is normal for image reconstruction with an optical microscope. The conical extensions above and below an object represent the uncertainty involved with measuring light through a lens with less than a 90° cone angle. At present, the best available lenses have a cone angle of approximately 68°, thereby rendering all objects with an elongated appearance.

reconstructed from these z series scans, using the Volume View command in the Delta Vision software package (version 2.00; Applied Precision, Issaquah, WA). Figure 5 shows an example of a reconstructed sagittal image of biofilms formed by colanic acid-defective and wild-type strains of *E. coli*.

Conclusions

The methods and approaches outlined here were designed to assess the developmental nature of biofilm formation by *E. coli*. However, they are easily applied to other biofilm-forming organisms, and as such they should be widely applicable in the general analysis of prokaryotic biofilm formation. Specifically, we have focused on macroscopic and microscopic experiments that can be employed to answer three pertinent questions: (1) Do biofilms develop only under certain growth phases and growth conditions? In the case of *E. coli*, biofilm formation occurs during exponential phase growth, implying that this is a process that is favored during times of nutrient availability. (2) Can biofilm structure be altered by changing environmental conditions, and if so, what does this imply about biofilm formation for that particular organism? (3) Do any mutations that alter macroscopic biofilm structure also affect biofilm structure at the microscopic level? Again, in the case of *E. coli*, the developmental nature of the biofilm formation process is highlighted by mutants that fail to produce colanic acid. These mutants are arrested in their development of the typical and mature three-dimensional biofilm structure. Taken as a whole, the approaches outlined to address these questions should provide information on the developmental nature of biofilm formation for a variety of microbes.

Acknowledgments

This work was supported by NIH grant GM58213 to R.K. L.A.P. gratefully acknowledges support from the Jane Coffin Childs Memorial Fund for Cancer Research. P.N.D. gratefully acknowledges support from the Cancer Research Fund of the Damon Runyon-Walter Winchell Foundation.

Section II

Signaling and Gene Expression in Biofilm

[3] Surface Sensing, Swarmer Cell Differentiation, and Biofilm Development

By MOHAMMAD ALAVI and ROBERT BELAS

Introduction

Bacteria are versatile organisms that can quickly adapt to a changing environment that aids the cells in survival. A major adaptive strategy, and perhaps the predominant mode of growth for bacteria in nature, is biofilm formation. The formation of a biofilm often requires the interaction of groups of bacteria that exhibit a multicellular behavior reminiscent of the coordinated development associated with higher organisms. Predominant in these events are carefully orchestrated responses by the bacteria to cues derived from the surface and other bacteria. These signals are used to optimize bacterial survival and perpetuate life in the developing biofilm.

Bacteria in biofilms live a different existence than in the planktonic state, and this is reflected in their morphology and metabolic functions. These bacteria are often resistant to reagents that kill their planktonic counterparts and are thought to be the pathogenic forms of many disease-causing bacteria. This is demonstrated by the genus *Proteus,* in which swarmer cell differentiation and swarming motility have been correlated with expression of virulence genes[1,2] and urinary tract pathogenesis.[3] In *Proteus mirabilis,* biofilm formation is initiated after bacteria encounter a solid substratum. On this encounter *P. mirabilis* cells differentiate into elongated, hyperflagellated cells that can move collectively on the surface (Fig. 1). The phenomenon of cell elongation, hyperflagellation, and swarming has also been demonstrated in other bacteria such as *Vibrio*[4] and *Serratia*[5] and is referred to as swarmer cell differentiation and swarming behavior. Swarming is a multicellular behavior used by these bacteria not only to seek new energy sources but also to move away from a crowded center before it becomes metabolically deleterious.

Unlike chemotactic behavior, which is mediated through the sensing of chemical gradients in the surrounding environment, swarmer cell differentiation is mediated through a physical sensing of a surface by an individual bacterium. This surface sensing is mechanical and mediated through the inhibition of flagellar rotation as the cells near a submerged surface. Surface sensing is only the first component in swarming, which also requires active multicellular, coordinate movement of the

---

[1] C. Allison, N. Coleman, P. L. Jones, and C. Hughes, *Infect. Immun.* **60,** 4740 (1992).

[2] C. Allison, C. H. Lai, and C. Hughes, *Mol. Microbiol.* **6,** 1583 (1992).

[3] S. I. Bidnenko, E. P. Bernasovskaia, N. A. Iu, A. B. Iu, and E. V. Mel'nitskaia, *Mikrobiol. Zh.* **47,** 81 (1985).

[4] L. McCarter and M. Silverman, *Mol. Microbiol.* **4,** 1057 (1990).

[5] L. Alberti and R. M. Harshey, *J. Bacteriol.* **172,** 4322 (1990).

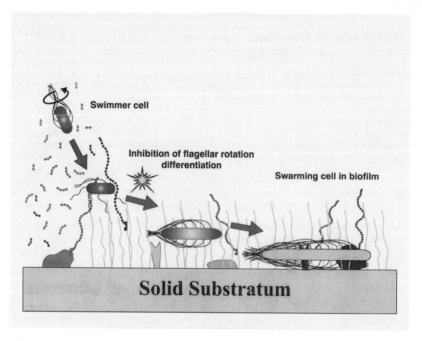

FIG. 1. Hypothetical model for surface sensing, differentiation, and swarmer cell formation. Swimmer cells are drawn to the substratum by chemotaxis toward food and energy sources. The boundary layer presents an environment that physically impairs flagellar rotation. Tethering of flagellar rotation will generate a signal that leads to differentiation of the swimmer cell into the swarmer cell. Swarmer cells then cooperate and move collectively on the surface within a biofilm.

cells across the surface to form a biofilm. Control of swarming migration requires additional signals to initiate morphological and physiological adaptation to a solid surface. For example, the amino acid glutamine is essential for *P. mirabilis* cell differentiation,[6] whereas in *Vibrio parahaemolyticus,* iron limitation appears to be required for induction of differentiation.[7] Swarming behavior is also regulated by cell density-sensing mechanisms that appear to be a common form of communication among bacteria in biofilms. In *Serratia liquefaciens,* the *swrI* gene encodes a putative *N*-acylhomoserine lactone (AHL) and is required for expression of swarmer cell differentiation and function.[8,9] In *P. mirabilis,* on the other hand, transposon inactivation of *rsbA*, a member of a two-component regulatory

[6] C. Allison, H. C. Lai, D. Gygi, and C. Hughes, *Mol. Microbiol.* **8,** 53 (1993).

[7] L. McCarter and M. Silverman, *J. Bacteriol.* **171,** 731 (1989).

[8] M. Givskov, J. Ostling, L. Eberl, P. W. Lindum, A. B. Christensen, G. Christiansen, S. Molin, and S. Kjelleberg, *J. Bacteriol.* **180,** 742 (1998).

[9] P. W. Lindum, U. Anthoni, C. Christophersen, L. Eberl, S. Molin, and M. Givskov, *J. Bacteriol.* **180,** 6384 (1998).

system, results in precocious swarming mutants that do not require surface contact or glutamine for expression of the swarmer cell phenotype,[10] suggesting that the Rsb A protein functions to control migration in this species.

Proteus mirabilis swarming behavior is punctuated by cycles of differentiation, migration, and dedifferentiation (consolidation) that give rise to a "bull's eye" colony on nutrient agar medium. In a liquid environment, the undifferentiated vegetative or "swimmer" cells are 1.5–2 μm long and possess 6 to 10 peritrichous flagella. When these cells are transferred to a solid substrate, that is, a nutrient agar surface, they sense the surface and begin differentiating into cells that are 60–80 μm in length and hyperflagellated, possessing between 10^3 and 10^4 flagella per cell.[11] After the cells have differentiated, a multicellular phase begins in which cooperation through cell-to-cell contact leads to migration of the colony, as a collective entity, on the surface of the substrate. Finally, after a period of migration on the surface, the cells stop and begin to dedifferentiate to swimmer cells and then multiply. This cycle repeats itself until the entire agar surface is covered by the colony.

The signals that control swarmer cell differentiation and behavior can be divided into those that induce the cellular differentiation process and those that control multicellular migration. Investigations into the molecular nature of these signals and the regulatory mechanisms underlying their control have led to exciting discoveries. It is now known that the signal each individual cell senses is mediated through the inhibition of its rotating flagella. The signal initiated by the tethered polar flagella is conveyed to the transcription machinery and leads to synthesis of new flagella, inhibition of cell wall septation, and other events that change the phenotype into the elongated hyperflagellar form. Thus, the first stimulus sensed by these cells as they form a biofilm is physical in nature, occurs at the level of the individual cell, and is transmitted into the cytoplasm via the inhibition of flagellar rotation.

In contrast, the signals controlling swarming migration occur at the level of the colony and function to coordinate multicellular behavior. Cell-to-cell contact and extracellular signaling by amino acids and other small molecules appear to play major roles in multicellular swarming behavior. These findings have come about through experiments that have employed a myriad of genetic and molecular techniques. In this chapter, we describe some of the methods developed to investigate the molecular mechanisms of surface sensing in gram-negative bacteria.

Methodology

Transposon Mutagenesis for Construction of Mutants Defective in Swarming and Transcriptional Fusions to Promoterless Reporter Genes

Transposon Mutagenesis in Vibrio parahaemolyticus. Transposon mutagenesis has been used to determine the role of different genetic loci and different

---

[10] R. Belas, R. Schneider, and M. Melch, *J. Bacteriol.* **180**, 6126 (1998).
[11] J. F. Hoeniger, *Can. J. Microbiol.* **12**, 113 (1966).

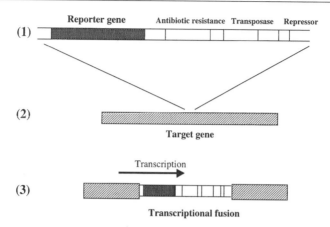

FIG. 2. Mutagenesis with mini-Mu transposon. Mini-Mu carrying an antibiotic marker and a reporter gene without a promoter (1) is delivered into the bacteria, using the general transducing phage P1. Mini-Mu inserts into the target gene on the chromosome by transposition (2). A transcriptional fusion between the target gene and the reporter gene is generated (3), whereby the target gene promoter controls expression of the reporter gene.

environmental signals in the development of swarmer cells and swarming behavior in *V. parahaemolyticus*. The transposable phage derivative mini-Mu is of practical use in studies of regulation of gene expression. The coding sequence for a reporter gene, but not its regulatory sequences, has been incorporated into mini-Mu at the right end and is flanked by only a few hundred base pairs. This arrangement allows transcription to proceed from genes outside of the transposon into the reporter gene. If the mini-Mu inserts in the correct orientation within a gene, it will generate a transcriptional fusion in which the promoter of the mutated gene controls the expression of reporter gene. Mini-Mu, therefore, can be used to investigate the expression of the target gene under different conditions (Fig. 2).

Transposon mini-Mu (dI1681 Kan^R *Bam*HI *cts*) is derived from the Mu d*Ilac* phage by deletion of the internal *Bam*HI fragment.[12] This transposon carries the structural genes for the *lacZ*, *Y*, and *A* genes at the right end and retains transposition replication genes *A* and *B* and the Mu repressor gene *c* on its left end. The *c* allele is heat sensitive and thus replication and transposition can be induced by temperature elevation. However, because mini-Mu lacks the structural genes for forming infective particles, a helper phage is required for packaging of mini-Mu DNA. *Escherichia coli* POI1681 (*araD araB*:: Mu *cts lac recA rpsL*) strain carries a Mu lysogen that can provide the helper functions. After introduction of mini-Mu into *E. coli* POI1681 and temperature induction, the help of resident Mu lysogen forms infective mini-Mu particles.

[12] B. A. Castilho, P. Olfson, and M. J. Casadaban, *J. Bacteriol.* **158**, 488 (1984).

Another derivative of the mini-Mu transposon harbors a promoterless bacterial luciferase gene cassette and is referred to as mini-Mu *lux*. Mini-Mu *lux* is derived from transposon mini-Mu *lac* (see above) by replacing the *lac* genes with the *lux* genes (*luxCDABE*) from *Vibrio fischeri*.[13] This transposon carries all the genes necessary to produce light in a variety of microorganisms. Mini-Mu *lux* mutagenizes by transposing into the target gene and because of its large size will generally cause a null mutation. Furthermore, insertion in the correct orientation will result in a transcriptional fusion in which the promoter of the target gene will control the expression of luminescence genes. The luminescence production can be easily detected in the dark in bacteria that do not naturally produce light. The mutants can be isolated by selection on tetracycline plates and the region flanking the transposon insertion site (target gene) can be cloned.

Mini-Mu *lux* is a powerful tool with which to study regulation of gene expression during swarmer cell differentiation and multicellular swarming behavior because bioluminescence resulting from the transcriptional fusion to *lux* genes can be measured in intact cells. This transposon has been successfully used to mutate and construct fusion strains in *V. parahaemolyticus*. Elegant studies to monitor reporter gene activity and swarmer cell differentiation under different conditions have indicated that lateral flagellar gene (*laf*) expression is critical in differentiation of swarmer cells. Furthermore, the signal for swarmer cell development is mediated by impairing the ability of the bacterium to move relative to its surroundings. Antibodies to the cell surface components of *V. parahaemolyticus* cause agglutination and subsequent tethering of polar flagella and result in the abnormal expression of swarmer cell differentiation.[14] Increasing the viscosity of nutrient broth to 40 cP by the addition of polyvinylpyrrolidone (PVP-360) to 10% (w/v) also results in swarmer cell differentiation and in the induction of luminescence in *laf::lux* fusion strains.[15] Similar studies have demonstrated that a group of genes encoding the structural proteins of the *V. parahaemolyticus* polar flagellum play a significant role in coupling the signal initiated from polar flagella to expression of the *laf* gene.[16] Taken together, these data, derived from work with mini-Mu *lux* transcriptional fusions, point out the pivotal role played by the polar flagellum in sensing the physical cue of viscosity and resulting inhibition of flagellar rotation.

In the following sections we provide a protocol for the use of mini-Mu *lac* to mutagenize *V. parahaemolyticus* and generate *lacZ* fusion strains. The same strategy may be used to construct transcriptional fusions with other derivatives such as mini-Mu *lux*.

[13] J. Engebrecht, M. Simon, and M. Silverman, *Science* **227,** 1345 (1985).
[14] L. McCarter, M. Hilmen, and M. Silverman, *Cell* **54,** 345 (1988).
[15] R. Belas, M. Simon, and M. Silverman, *J. Bacteriol.* **167,** 210 (1986).
[16] L. L. McCarter, *J. Bacteriol.* **177,** 1595 (1995).

Mutagenesis of Vibrio parahaemolyticus with mini-Mu (Tet^R). To mobilize mini-Mu from *E. coli* into *V. parahaemolyticus* coliphage P1 (P1 *clr*-100 CM) is used. Transducing phage P1 is capable of packaging mini-Mu *lux* DNA and infects but does not replicate in *V. parahaemolyticus.* The repressor of P1 and that of the mini-Mu *lux* transposon are both inactivated at 42°. Therefore, the lytic cycle of P1 and replication of mini-Mu *lux* can be coinduced in an *E. coli* strain that contains mini-Mu *lux* and is lysogenized by P1.

1. Transform a Mu *cts* lysogen of *E. coli* ED8654 (*supE supF met hsdR*) with pBA104,[17] which carries mini-Mu (Tet^R). Select the transformants on L agar containing 100 μg of ampicillin and 20 μg of tetracycline per milliliter.

2. Infect the cells carrying pBA104 with phage P1 (P1 *clr*-100 CM) and heat induce the infected cells at 42° to produce an infecting lysate of P1 that carries min-Mu (Tet^R).

3. Transduce the tetracycline-resistant gene to *E. coli* MC4100 [F⁻ *araD* Δ(*lacIOPZYA*) *U169 rpsL*].

4. Screen the transductants for ampicillin-sensitive colonies to obtain colonies that carry mini-Mu (Tet^R) in their chromosome with the concomitant loss of the plasmid pBA104.

5. Grow *E. coli* MC4100 harboring mini-Mu (Tet^R) overnight in L broth (10 g of tryptone, 5 g of yeast extract, and 10 g of NaCl per liter) containing 20 μg of tetracycline per milliliter. Infect with phage P1 (P1 *clr*-100 CM) to obtain a P1 lysogen of MC4100.

6. Grow wild-type *V. parahaemolyticus* BB22 overnight at 30° in Difco (Detroit, MI) marine broth 2216 at 75% (w/v) the recommended concentration (28 g of Difco 2216 in 1000 ml of distilled H$_2$O) or in heart infusion broth supplemented with 20 g of NaCl per liter supplemented with 10 μg of tetracycline per milliliter.

7. Temperature induce *E. coli* MC4100 [mini-Mu *lac* (Tet^R), P1 *clr*-100 CM] at 42° to prepare infecting lysate.

8. The infecting lysate are incubated with *V. parahaemolyticus* for 2 hr at 30° and transductants are plated on heart infusion agar (heart infusion broth containing 20 g of agar per liter) supplemented with 15 g of NaCl per liter and containing tetracycline at 10 μg/ml. Bacterial colonies do not swarm on this medium, thus allowing the selection of single clones of Tet^R transductants.

9. After overnight incubation at 30°, Tet^R colonies are picked onto a master array. To identify nonswarming colonies replica plate the transductants from the master array onto heart infusion agar containing 20 g of NaCl and 15 g of agar per liter. Measure swarming after 8 hr of incubation at 30° and save the nonswarmer colonies for further analysis.

[17] R. Belas, A. Mileham, M. Simon, and M. Silverman, *J. Bacteriol.* **158**, 890 (1984).

Measurement of β-Galactosidase Activity in laf::lac Fusion Strains. Mutant strains of *V. parahaemolyticus* that display β-galactosidase activity (Lac⁺) are detected by growing colonies on agar medium containing a 20-μg/ml concentration of chromogenic substrate 5-bromo-4-chloro-3-indolyl-β-D-galactopyranoside (X-Gal) and tetracycline at 10 μg/ml. To test the surface expression of *lacZ*, harvest cells from agar and liquid medium, adjust the optical densities to give equal readings at 600 nm, and perform the *o*-nitrophenyl-β-D-galactopyranoside (ONPG) assay as described by Miller.[18]

Measurement of Luminescence in Mutants with mini-Mu lux Transcriptional Fusion. The *laf::lux* fusion strains are sensitive tools for investigating the environmental signals that lead to *laf* gene expression and consequently to swarmer cell differentiation. Mutants that can produce light on agar but not in liquid can be detected in a dark room or by allowing the light-producing colonies to be exposed to an X-ray film in a dark room. The promoter of the target gene (*laf*) that controls expression of the *lux* genes is induced only on contact with the surface of agar but not in liquid. This activity can be quantified by measuring the level of luminescence. The following protocol describes the measurement of luminescence in *laf::lux* fusions of *V. parahaemolyticus* grown on a solid substrate and in liquid.

1. Grow *V. parahaemolyticus laf::lux* fusion strains at 30° overnight in 2216 marine broth at 75% of the recommended concentration supplemented with 10 μg of tetracycline per milliliter.

2. Dilute the culture 1 to 2000 in fresh 2216 medium (75% concentration) without tetracycline and incubate at 30° with shaking until an optical density of 0.05 at 600 nm is reached (about 2 hr). More dense sample does not produce light as well in response to surface contact in these fusion strains.

3. Inoculate 5 μl of the sample in 500 μl of liquid medium in a plastic 1.5-ml microcentrifuge tube. Inoculate another 5 μl of the sample on the surface of a cylindrical core of agar medium 1 cm in diameter and 1 cm in depth.

4. Place the fusion strain in liquid (microcentrifuge tube) and on solid substrate (agar core, or membrane filters) in different scintillation vials. Place the vials in a scintillation counter programmed to count each sample repetitively at 30-min intervals.

5. Measure luminescence as the output of the chemiluminescence channel of the scintillation counter in terms of light units.

6. To obtain values of light units per cell, remove the sample from the vial, dilute in 2216 broth, and spread on 2216 agar containing tetracycline at 10 μg/ml. After overnight incubation at 30°, count the number of colonies and calculate the average light units per cell by dividing the number of light units obtained for the

---

[18] J. H. Miller, "Experiments in Molecular Genetics." Cold Spring Harbor Laboratory Press, Cold Spring Harbor, New York, 1972.

sample by the number of colony-forming units on the agar multiplied by the dilution factor.

7. A microtiter dish luminometer may be used in place of a scintillation counter and the relative light units recorded accordingly.

Transposon Mutagenesis of Proteus mirabilis

Mini-Tn5 Chloramphenicol Mutagenesis in Proteus mirabilis. The genetic regulation and signal transduction events that control swarming cell differentiation and multicellular swarming behavior of *P. mirabilis* have been investigated in our laboratory by transposon mutagenesis. Tn5 has proved to be efficient in generating random mutagenesis in a variety of gram-negative bacteria and, unlike mini-Mu, does not require temperature induction or multiple phage systems for delivery and insertion. We have successfully utilized a suicide vector developed by de Lorenzo *et al.*[19] to generate random insertions in the *P. mirabilis* chromosome.[20] This vector is based on the pUT plasmid carrying a mini-Tn5 cassette and a chloramphenicol resistance gene (pUT/mini-Tn5/Cm). Plasmid pUT is a suicide vector derived from pGP704[21] and carries the π protein-dependent origin of replication from plasmid R6K.[22] Only bacteria that carry the *pir* gene encoding this protein can maintain an R6K plasmid. These strains include *E. coli* SM10 (λ*pir*) and S17-1 (λ*pir*), which are both capable of conjugal transfer of *mob*-containing vectors, such as pGP704. The successful insertion and maintenance of mini-Tn5 into the chromosomal DNA is achieved after conjugal transfer when the plasmid fails to replicate and is attacked by host restriction modification mechanisms. At that point, the Tn5 transposase gene carried on the pUT plasmid facilitates transposition of the mini-Tn5 into the target chromosome. The following protocol is used in our laboratory to mutagenize *P. mirabilis* BB2000 (rifampin resistant), using mini-Tn5.

1. Grow donor *E. coli* S17-1 (λ*pir*) cells harboring pUT/mini-Tn5/CM overnight at 37° in L broth containing 40 μg of chloramphenicol per milliliter.

2. Grow the recipient *P. mirabilis* BB2000 overnight in L broth containing 100 μg of rifampin per milliliter.

3. Place a sterile cellulose membrane filter (diameter, 45 mm; pore size, 0.2 μm) on the surface of LSW⁻ agar (10 g of tryptone, 5 g of yeast extract, 5 ml of ultrapure glycerol, 0.4 g of NaCl, and 20 g of agar per liter). Spot 100 μl of donor cells (~4.5 × 10⁸ bacteria) and 200 ml of recipient bacteria (~1.0 × 10⁹ cells) onto the membrane. LSW⁻ agar is used to prevent the phenotypic expression of swarming motility.

[19] V. de Lorenzo, M. Herrero, U. Jakubzik, and K. N. Timmis, *J. Bacteriol.* **172,** 6568 (1990).

[20] R. Belas, D. Erskine, and D. Flaherty, *J. Bacteriol.* **173,** 6289 (1991).

[21] V. L. Miller and J. J. Mekalanos, *J. Bacteriol.* **170,** 2575 (1988).

[22] R. Kolter, M. Inuzuka, and D. R. Helinski, *Cell* **15,** 1199 (1978).

4. Let the culture fluid adsorb and then incubate the membrane at 37° overnight.

5. Remove the filter from the agar surface and suspend the bacteria by vortexing in 1 ml of phosphate-buffered saline [PBS: 20 mM sodium phosphate (pH 7.5), 100 mM NaCl].

6. Spread 100 μl on LSW⁻ agar supplemented with rifampin and chloramphenicol and incubate at 37° for 18 to 36 hr.

7. Transfer the P. mirabilis Cm^R conjugants to a master plate for further analysis.

Motility and Differentiation Analysis

The swarming cell differentiation and multicellular behavior of *P. mirabilis* mutants generated by mini-Tn5 mutagenesis may be examined as follows. In all tests L agar is used after thorough drying at 42° to provide uniform conditions for swarming assays.

1. Replica plate *P. mirabilis* colonies from the master plate to L agar and incubate at 37° for 4 to 6 hr.

2. Score the potential nonswarming mutants and transfer them to a submaster plate.

3. Repeat the swarming analysis until no colonies exhibit swarming after 6 hr and designate them as Swr⁻.

4. Test the Swr⁻ mutants for swimming behavior on Mot semisolid agar (10 g of tryptone, 5 g of NaCl, and 3.5 g of agar per liter). Colonies that fail to move outward from the center of the Mot agar, after overnight incubation at 37°, are swimming mutants and are designated Swm⁻.

5. Grow the Swm⁻ mutants in Mot broth to a cell density of 10^7 to 10^8 cells/ml and examine them under a light microscope. Colonies that swim in the broth but fail to move outward on semisolid agar are chemotaxis mutants and are designated Che⁻.

6. The remaining colonies that fail to swim in broth or move on semisolid agar are defective in producing flagella (Fla⁻) or are defective in flagellar rotation (Mol⁻). The Fla⁻ strains can be separated from Mot⁻ strains by Western immunoblot analysis, using anti-flagellin antisera.

Examination of Swr⁻ Mutants

1. Inoculate an overnight culture of a *P. mirabilis* mutant on fresh L agar containing chloramphenicol and incubate at 37° for 6 hr.

2. Remove a loopful of cells from the edge of the colony, place in 1 ml of PBS, and vortex briefly.

3. Remove a sample immediately and examine by phase-contrast light microscopy. If more than 10% of the population of a strain display an average length

of at least 20 μm (the length of 10 vegetative swimmer cells) it is considered to be wild type in elongation phenotype (Elo⁺). Constitutive elongation mutants (Elo^c) can be detected if they display the Elo⁺ phenotype after growth under noninducing conditions (growth in L broth with shaking for 6 hr at 37°).

Construction of Fusion Strains for Study of Multicellular-Dependent Gene Expression

Experiments monitoring lateral flagellar gene expression have indicated that these genes are tightly coordinated with swarmer cell differentiation. For example, gene fusion experiments in *V. parahaemolyticus* indicates that *laf* gene expression is induced immediately after sensing of the surface and the cells require continuous contact with the surface to maintain *laf* transcription. Similarly, in *P. mirabilis* experiments with a *flaA::lacZ* fusion on a low copy number plasmid have indicated that *flaA* expression increased by 7.5-fold after surface contact.[23] These experiments have also indicated that induction or increased expression of flagellar genes is essential for differentiation and initiation of biofilm development.

In *P. mirabilis* transition from swimmer cells to swarmer cells also induces the expression of a host of different factors including the IgA-dependent metalloprotease, Zap A.[24] In a study by Walker *et al.*[25] a *zapA::lacZ* fusion strain was used to study the pattern of Zap A expression during swarmer cell differentiation and swarming motility. These data indicated an interesting pattern of gene regulation. Although *zapA* expression is tightly coordinated with swarmer cell differentiation, its expression is not required for differentiation and swarming motility. However, expression of *zapA* appears to be modulated by the cycles of multicellular swarming behavior. Careful analysis of β-galactosidase activity in swarming *zapA::lacZ* fusion strains indicates that maximum expression of *zapA* occurs at the boundary before the consolidation zone (zone of dedifferentiation) and is minimal during migration of swarmer cells. This suggests that specific genes involved in biofilm formation are expressed only at critical points during the growth of the biofilm.

In the following protocol, we describe a second approach to construct a transcriptional fusion between a chromosomal gene, in this case *zapA,* and a promoterless reporter gene, i.e., *lacZ*. This alternative method does not use random transposon mutagenesis, which may be a "plus" under certain circumstances, but does require that the target gene has been cloned and partially characterized.

[23] R. Belas and D. Flaherty, *Gene* **148,** 33 (1994).

[24] C. Wassif, D. Cheek, and R. Belas, *J. Bacteriol.* **177,** 5790 (1995).

[25] K. E. Walker, S. Moghaddame-Jafari, C. V. Lockatell, D. Johnson, and R. Belas, *Mol. Microbiol.* **32,** 825 (1999).

Construction of zapA'::lacZ::'zapA Transcriptional Fusion Strain

The *zapA* gene was cloned from an 8.6-kb insertion of *P. mirabilis* chromosomal DNA on the recombinant plasmid pCW101,[24] using the polymerase chain reaction (PCR). The oligonucleotide primers used in this amplification incorporated engineered *Kpn*I restriction sites at their 5' end. The resulting amplicon, containing the *zapA* ribosome-binding site and terminator, was cloned into the *Kpn*I site of pBluescript-II SK+ (Promega, Madison, WI) to produce pKW305. To construct a transcriptional fusion between *zapA* and *lacZ,* perform the following steps.

1. Digest pKW305 with *Bgl*II, which cuts *zapA* but not the plasmid vector.

2. Ligate the digested pKW305 to the 4.5-kb *Bam*HI *lacZ*-Kan^R transcriptional fusion cassette from plasmid pLZK83[26] to generate pKW314, which carries the *zapA'*::*lacZ*-Kan^R::*'zapA* cassette.

3. Digest pKW314 with *Kpn*I to remove the *zapA'*::*lacZ*-Kan^R::*'zapA* fragment and ligate the fragment to the *Kpn*I-digested suicide vector pGP704 (see above) to generate the mutator plasmid pKW350.

4. Transform *E. coli* SM10 λ*pir* with pKW350 and then mobilize the fusion cassette from *E. coli* to *P. mirabilis* BB2000 by filter conjugation as described above.

5. Select the mutants on LSW⁻ agar with rifampicin, tetracycline, and kanamycin. Select the representative strains that display β-galactosidase activity (Lac⁺) on LB agar containing X-Gal.

6. Check the mutant strains for loss of protease activity by streaking on skimmed milk agar (15 g of protease peptone, 10 g of skimmed milk powder, and 15 g of agar per liter) and incubating overnight at 37°. A positive reaction consists of a zone of clearing of the agar around the protease-positive colonies.

7. Screen the protease-negative colonies for the ability to degrade IgA by incubating the concentrated supernatant from *P. mirabilis* mutants with human IgA for 13 hr at 37°. The presence or absence of IgA-degrading protease activity can be detected by subjecting the incubation mixture to 12.5% (w/v) sodium dodecyl sulfate–polyacrylamide gel electrophoresis (SDS–PAGE).

Concluding Remarks

A great deal remains to be learned about the molecular machinery of surface sensing and multicellular swarming behavior leading to the development of a biofilm. We have described many of the methods that have been successfully

---

[26] G. J. Barcak, M. S. Chandler, R. J. Redfield, and J. F. Tomb, *Methods Enzymol.* **204,** 321 (1991).

employed in addressing these questions. Transposon mutagenesis is a powerful tool with which to generate mutants that are defective in surface sensing, swarming, and biofilm formation. Moreover, when coupled with transcriptional fusion to a promoterless reporter gene, transposon insertion into target genes can provide valuable information about the signals required for surface-induced gene expression. One possible limitation in using a transposon–reporter gene construct for gene expression studies is the large size of the reporter component. *lac* genes in mini-Mu *lac* and *lux* genes in mini-Mu *lux* add substantially to the size of the transposon. Integration of these transposons usually leads to insertional inactivation of the target gene and polar effects on downstream genes in an operon. If the functional loss of the target gene is deleterious to the organism, then mutations in this gene will not be detected. This is a limitation of any study in which random mutagenesis is used to identify important genetic elements.

A different reporter system is a variant of green fluorescent protein (GFPuv)[27] from *Aequorea victoria,* which has been used as a reporter gene in bacteria. Fluorescence production by GFP is species independent and does not require any substrate, cofactor, or other auxiliary genes. Furthermore, it has been possible to construct an N- or C-terminal fusion protein with the target gene while retaining the reporter activity of the GFP and function of the target gene.[28–30] Bacterial colonies expressing GFPuv can be detected on agar, using a UV transilluminator, and quantitation of promoter activity of the target gene can be measured with a fluorometer in a 96-well microtiter plate format. This system thus has the advantage of the transposon-generated transcriptional fusions, yet retains the activity of the target gene and lacks polar effects. This and other innovations will further our knowledge of the genes and signals involved in surface sensing and swarmer cell behavior that go into the formation of a swarming biofilm.

[27] A. Crameri, E. A. Whitehorn, E. Tate, and W. P. Stemmer, *Nature Biotechnol.* **14,** 315 (1996).

[28] J. Flach, M. Bossie, J. Vogel, A. Corbett, T. Jinks, D. A. Willins, and P. A. Silver, *Mol. Cell. Biol.* **14,** 8399 (1994).

[29] S. Wang and T. Hazelrigg, *Nature (London)* **369,** 400 (1994).

[30] T. Stearns, *Curr. Biol.* **5,** 262 (1995).

[4] Acylated Homoserine Lactone Detection in *Pseudomonas aeruginosa* Biofilms by Radiolabel Assay

By AMY L. SCHAEFER, E. P. GREENBERG, and MATTHEW R. PARSEK

Introduction

The ability of bacteria to sense one another and coordinate gene expression as a community has been identified in a wide range of species. Quorum sensing, or cell density-dependent gene expression, is one example of community-coordinated gene regulation. There are a variety of quorum sensing mechanisms. Gram-positive bacteria regulate gene expression in a cell density-dependent manner using small extracellular signaling peptides as an indicator of population numbers, whereas many gram-negative bacteria use freely diffusible signaling molecules, acylated homoserine lactones (acyl-HSLs), to monitor cell density (for reviews see Refs. 1–5). The acyl side chains of acyl-HSLs vary in length and degree of substitution depending on the particular organism and system. Cells possess an acyl-HSL synthase, also called an I-protein, which synthesizes acyl-HSLs from the common cellular metabolites *S*-adenosylmethionine (SAM) and acylated acyl carrier protein (acyl-ACP).[6–10] At low cell numbers, the acyl-HSL diffuses out of the cell, down the concentration gradient, and is lost into the environment. However, as the cell density increases, the local acyl-HSL concentration reaches a threshold level at which it interacts with a transcriptional regulatory protein (also called an R-protein).[11,12] This transcriptional regulatory protein/acyl-HSL

[1] B. L. Bassler, *Curr. Opin. Microbiol.* **2**, 582 (1999).

[2] G. M. Dunny and B. A. Leonard, *Annu. Rev. Microbiol.* **51**, 527 (1997).

[3] M. Kleerebezem, L. E. Quadri, O. P. Kuipers, and W. M. de Vos, *Mol. Microbiol.* **24**, 895 (1997).

[4] W. C. Fuqua, S. C. Winans, and E. P. Greenberg, *Annu. Rev. Microbiol.* **50**, 727 (1996).

[5] C. Fuqua and E. P. Greenberg, *Curr. Opin. Microbiol.* **1**, 183 (1998).

[6] T. T. Hoang, Y. Ma, R. J. Stern, M. R. McNeil, and H. P. Schweizer, *Gene* **237**, 361 (1999).

[7] Y. Jiang, M. Camara, S. R. Chhabra, K. R. Hardie, B. W. Bycroft, A. Lazdunski, G. P. Salmond, G. S. Stewart, and P. Williams, *Mol. Microbiol.* **28**, 193 (1998).

[8] M. R. Parsek, D. L. Val, B. L. Hanzelka, J. E. Cronan, Jr., and E. P. Greenberg, *Proc. Natl. Acad. Sci. U.S.A.* **96**, 4360 (1999).

[9] A. L. Schaefer, D. L. Val, B. L. Hanzelka, J. E. Cronan, Jr., and E. P. Greenberg, *Proc. Natl. Acad. Sci. U.S.A.* **93**, 9505 (1996).

[10] M. I. Moré, D. Finger, J. L. Stryker, C. Fuqua, A. Eberhard, and S. C. Winans, *Science* **272**, 1655 (1996).

[11] B. L. Hanzelka and E. P. Greenberg, *J. Bacteriol.* **177**, 815 (1995).

[12] A. M. Stevens and E. P. Greenberg, *J. Bacteriol.* **179**, 557 (1997).

complex then modulates expression of quorum sensing-regulated genes. More than 50 bacterial species produce acyl-HSLs to regulate a wide variety of physiological processes including bioluminescence, swarming motility, and conjugal plasmid transfer.

The current method of acyl-HSL detection uses biological reporter assays. Although these assays are critical for studying quorum sensing, they also have their limitations. They are time consuming, labor intensive, and biased toward detecting acyl-HSL signals of specific side chain lengths and/or substitutions. Most quorum sensing studies have focused on acyl-HSL production from planktonically grown bacteria. However, evidence demonstrating acyl-HSL involvement in development of *Pseudomonas aeruginosa* biofilms[13] highlights the need to monitor the quorum sensing mechanism (acyl-HSL production and reception) in nonplanktonic cultures. Unfortunately, many of the problems associated with bioassays limit their use for detecting acyl-HSLs produced by biofilm cultures. This chapter describes a new radiolabel assay for monitoring acyl-HSL production that is unbiased, faster, and more sensitive than traditional bioassays.[14] It also describes how this assay can easily be applied to both planktonic and biofilm cultures, using *P. aeruginosa* as a model system.

Acylated Homoserine Lactone Detection by Bioassays

Traditionally, acyl-HSL levels have been measured by a series of bioassays. Bioassays rely on reporter genes fused to promoters that are activated by specific R-proteins in the presence of specific acyl-HSLs. There are a number of bioassays described in the literature (for a review see Ref. 14). Bioassays can be used to quantitate the total amount of acyl-HSL present in a sample when a standard curve is generated with known amounts of purified acyl-HSL signal. Although bioassays are important for acyl-HSL detection, they are not without limitations. They are labor intensive and time consuming, and multiple reporters are necessary to detect the entire range of known acyl-HSL structures. Bioassays also require biological activity for detection, which may limit detection to those acyl-HSL molecules with structures similar to previously described signals. For example, if an organism makes an acyl-HSL with a novel structure that diverges significantly from other known acyl-HSL structures, it may not be detected with existing bioassays. These limitations led to the development of the radiolabel assay described below.

---

[13] D. G. Davies, M. R. Parsek, J. P. Pearson, B. H. Iglewski, J. W. Costerton, and E. P. Greenberg, *Science* **280,** 295 (1998).

[14] A. L. Schaefer, B. L. Hanzelka, M. R. Parsek, and E. P. Greenberg, *Methods Enzymol.* **305,** 288 (2000).

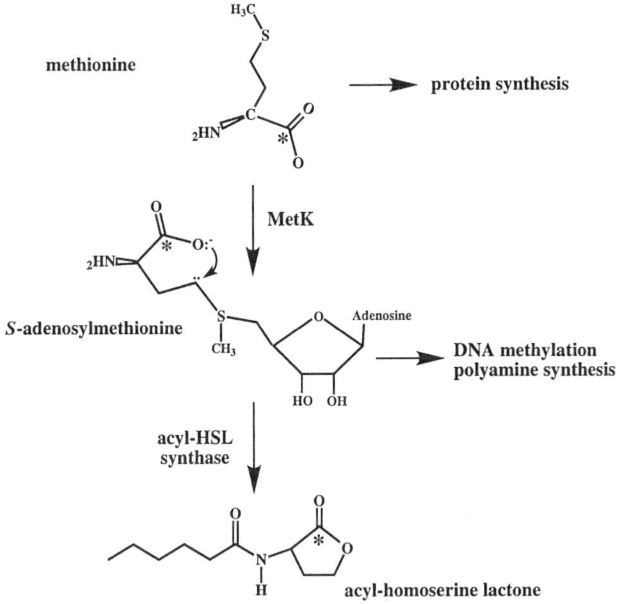

FIG. 1. Diagram showing flow of radiolabel into acyl-HSL from methionine through S-adenosylmethionine. The asterisk denotes the position of the radiolabel. The cells convert some of the methionine into SAM, using MetK, an SAM synthetase. Some of the SAM is then used by an acyl-HSL synthase as substrate for acyl-HSL synthesis. The acyl-HSL is labeled at the C-1 position of the homoserine lactone ring. The horizontal arrows indicate competing cellular reactions that use SAM or methionine as a substrate.

Detection of Acylated Homoserine Lactone Production by *Pseudomonas aeruginosa* Planktonic Cultures by Radiolabel Assay

The acyl-[^{14}C]HSL assay protocol is similar to that first described by Eberhard *et al.*[15] The assay is based on the uptake of radiolabeled methionine by living cells and incorporation of the radiolabel into acyl-HSL. Figure 1 illustrates this process. *Pseudomonas aeruginosa* is grown in 5 ml of methionine-free medium (in our case Jensen's medium plus glycerol[16]) at 37° with shaking to the appropriate culture density (usually early stationary phase). Cultures are then labeled with 5 μCi (50 μl) of [*carboxy*-^{14}C]methionine (American Radiochemical, St. Louis, MO) during a 10- to 30-min incubation (with shaking) at 37°. Cells are pelleted by

[15] A. Eberhard, T. Longin, C. A. Widrig, and S. J. Stanick, *Arch. Microbiol.* **155,** 294 (1991).
[16] S. E. Jensen, I. T. Fecycz, and J. N. Campbell, *J. Bacteriol.* **144,** 844 (1980).

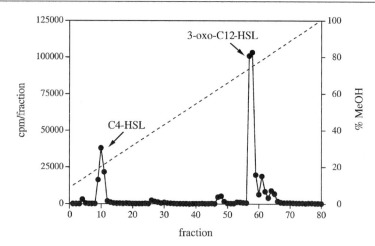

FIG. 2. An HPLC profile of acyl-[^{14}C]HSLs produced by planktonically grown *P. aeruginosa*. The *x* axis represents fraction numbers and the *y* axis denotes counts per minute of radiolabel in each fraction. The fractions were collected over a 10–100% methanol-in-water gradient. The two major peaks, C4-HSL and 3-oxo-C12-HSL, are the primary products of two *P. aeruginosa* acyl-HSL synthases, Rhl I and Las I, respectively. Radioactivity that eluted in the column void volume (fraction 4) is presumed to be unincorporated methionine.

centrifugation and the cell-free supernatant is extracted twice with equal volumes of acidified ethyl acetate (0.1 ml of glacial acetic acid per liter of solvent). The organic phase is collected and dried under a gentle stream of N_2 gas. The sample residue is dissolved in 200 µl of 50% methanol-in-water (adding 100 µl of methanol first, followed by an equal volume of water) and separated by high-performance liquid chromatography (HPLC) on a C_{18} reversed-phase column (200-µl loading loop), using a 10–100% (v/v) methanol-in-water gradient (0.5-ml/min flow, 140-min profile) as described previously.[14,17] Eighty 1-ml fractions are collected in scintillation vials, spiked with 4 ml of 3a70b scintillation cocktail (Research Products International, Mount Prospect, IL), and counted with a scintillation detector. Acyl-HSL assignment is determined by coelution with known acyl-HSL standards. Figure 2 illustrates the acyl-[^{14}C]HSL elution profile of radiolabeled acyl-HSLs produced by planktonically grown *P. aeruginosa*, using the radiolabel assay. *Pseudomonas aeruginosa* has two acyl-HSL synthases, Rhl I and Las I, whose primary acyl-HSL products are C4-HSL and 3-oxo-C12-HSL, respectively. Not surprisingly, these two acyl-HSLs correspond to the two major peaks produced in the radiolabel assay depicted in Fig. 2.

[17] J. P. Pearson, L. Passador, B. H. Iglewski, and E. P. Greenberg, *Proc. Natl. Acad. Sci. U.S.A.* **92,** 1490 (1995).

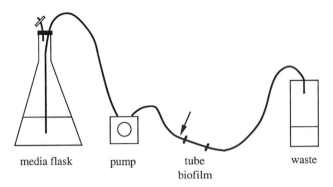

media flask   pump   tube   waste
                                biofilm

FIG. 3. Diagram of a tube biofilm reactor. The system consists of a 4-liter medium reservoir from which medium is pumped by a peristaltic pump. The medium flows from the reservoir through silicone tubing into the silicone biofilm reactor tubing on which the biofilms are grown. The medium then empties into an effluent waste flask. The site of inoculation is indicated by an arrow as described in text. The biofilm tubing is attached to the system tubing by polypropylene connectors (indicated by the intersecting dashes).

Detection of Acylated Homoserine Lactone Production by Biofilm Cultures

Considerations for Studying Acylated Homoserine Lactone Production by Biofilm Cultures

Bioassays are less useful for biofilm-grown samples because of the limited sample material and the number of bioassays required to detect all acyl-HSL signals present. The smaller volumes and shorter sampling times required for the radiolabel assay facilitate an analysis of acyl-HSL production by biofilm bacteria.

Biofilm Culture Apparatus

Biofilms are grown at 37° in silicone tubing (3.2-mm i.d.), using a once-through continuous-flow system.[13,18] This system (Fig. 3) consists of a growth medium reservoir (4-liter Erlenmeyer flask) from which medium is drawn through silicone tubing by a peristaltic pump (Watson-Marlow PumpPro MPL; Fisher Scientific, Pittsburgh, PA) at a rate of 0.2 ml/min. In addition to the outlet line, an air vent line is positioned at the top of the growth medium reservoir to maintain atmospheric pressure. A tubing section (1-ml volume) is joined to upstream tubing by polypropylene connectors, which are indicated by the dashes intersecting the line and flanking the biofilm tubing (Fig. 3) (Cole Parmer, Vernon Hills, IL). This section of tubing is inoculated with *P. aeruginosa* as described below. The growth

[18] M. R. Parsek and E. P. Greenberg, *Methods Enzymol.* **310**, 43 (1999).

medium proceeds through the silicone tubing and eventually passes through the tubing section where biofilm formation occurs. After passing through the biofilm tubing, the medium flows into an effluent reservoir. This system can be autoclaved and is completely closed to the environment.

Sample Preparation and Inoculation of System

Pseudomonas aeruginosa is grown in the appropriate methionine-free medium at 37° with shaking until midexponential phase. Cells are harvested by centrifugation (6000 g for 5 min) and resuspended in growth medium to an OD_{600} of 1.0. This ensures that the inoculum consists of a consistent number of cells at the same phase of growth.

Before inoculation, flow in the system is stopped and the tubing is clamped directly upstream of the biofilm tubing. The exterior tubing where injection occurs (Fig. 3) is first sterilized with ethanol. A 1-ml sample volume is then slowly injected into the upstream portion of the biofilm tubing, using a syringe and needle. After injection, the injection hole is immediately sealed with silicone. The bacteria are allowed to incubate for 30 min before flow is reinitiated, allowing bacterial attachment to the inside surface of the tubing. The biofilm culture develops on the inside of the tubing and can be assayed at selected time points after inoculation. The entire apparatus is incubated at 37°.

Use of Radiolabel Assay with Biofilm Cultures

The biofilm culture can be assayed for acyl-HSL production at selected time points (generally 6 to 72 hr) after inoculation, using the radiolabel assay. Medium flow is stopped and the biofilm tube section is separated from the connector tubing. Ten microcuries of [^{14}C]methionine (American Radiochemical) in 1 ml of fresh, prewarmed medium is then slowly injected into the 1-ml biofilm tubing section by a 1-ml syringe. Once medium is injected into the tube biofilm, the syringe remains attached to ensure no medium leakage. Biofilms are labeled for 30 min at 37°.

After the incubation period, collect the effluent by releasing the syringe. The tubing is washed once with an equal volume of phosphate-buffered saline (PBS, pH 7). Pool the collected effluent, extract twice with equal volumes of ethyl acetate, and analyze by HPLC as previously described. The remaining biofilm biomass can be scraped from the tubing and suspended in 2 ml of PBS. An OD_{600} reading of the suspended biomass can be taken and the biomass extracted twice with equal volumes of ethyl acetate. This allows for an estimate of cell number and of acyl-HSL retention within the biofilm biomass. After removal of the biomass, the silicone tubing can also be extracted with ethyl acetate to determine whether any acyl-HSLs adhered to the silicone, because certain acyl-HSLs have been reported to stick to silicone.[19]

[19] D. J. Stickler, N. S. Morris, R. J. McLean, and C. Fuqua, *Appl. Environ. Microbiol.* **64**, 3486 (1998).

Potential Applications of Radiolabel Assay

There are a number of potential applications for this assay. The absence of the bias for specific acyl-HSLs present in bioassays makes the radioactive assay ideal for screening new organisms for acyl-HSL production. This would require that the organism transport methionine and be cultivated on methionine-free medium. Another advantage would be the relative speed and ease of the assay, which allows for the comparison of acyl-HSL production by a large number of strains (e.g., clinical isolates of *P. aeruginosa*). This assay would also allow for the *in situ* analysis of acyl-HSL production in complex environments. For example, we have used this assay to detect acyl-HSL production by *P. aeruginosa* within the sputum of cystic fibrosis patients (P. Singh, personal communication, 1999).

Another benefit of this assay is the ability to generate significant amounts of radiolabeled acyl-HSL. Previous methods for chemically synthesizing radiolabeled acyl-HSL were cumbersome and involved many steps. Our laboratory has used this assay to generate large amounts of butyryl-HSL by labeling *Escherichia coli* overproducing the Rhl I acyl-HSL synthase.[20] These radiolabeled acyl-HSLs can be used in a variety of experiments, including studies examining acyl-HSL half-life, diffusion, and/or transport across membranes. Radiolabeled acyl-HSLs can also be used to monitor acyl-HSL catabolism when used as a carbon or nitrogen source by certain organisms.[20]

Summary

We describe the development of a new radioactive assay for acyl-HSL production by bacterial cultures. The assay is based on the uptake of radiolabeled methionine and conversion of the radiolabel into SAM. The radiolabeled SAM is then incorporated into acyl-HSL by an acyl-HSL synthase. This assay is faster than previously used bioassays and shows no bias for the detection of acyl-HSLs of a particular length or side chain substitution. Acyl-HSL production can be monitored over a wide range of growth conditions in liquid culture. This assay can also be used in conjunction with a tube biofilm reactor to monitor acyl-HSL production by biofilm cultures. Ultimately this assay will allow comparison of acyl-HSL production by cells subjected to a variety of physiological conditions.

[20] J. R. Leadbetter and E. P. Greenberg, submitted (2000).

[5] Genetic and Phenotypic Analysis of Multicellular Behavior in *Salmonella typhimurium*

By UTE RÖMLING

Introduction

According to Costerton *et al.*,[1] biofilms can be defined in a narrower sense as "a structured community of bacterial cells enclosed in a self-produced poly-meric matrix and adherent to an inert or living surface,"[1] or in a broader sense as "matrix-enclosed bacterial populations adherent to each other and/or to surfaces or interfaces."[2] To avoid confusion this chapter uses the term *multicellular behavior* for the latter definition, which describes the general phenomenon of bacteria behaving not as individual cells, but as a community. In *Salmonella typhimurium* a morphotype in which cells are acting as a community was identified. This morpho-type displayed various modes of multicellular behavior in stationary phase, such as cell clumping in aerated, liquid culture,[3] pellicle formation in standing liquid culture,[4] formation of a rigid cell network accompanied by surface translocation on solid medium,[3] adhesion to biological macromolecules,[5] and biofilm formation on various abiotic surfaces such as glass, polystyrene (PS), and stainless steel[3,6] (Fig. 1). On solid medium the colony bound the dye Congo red (CR) included in the agar plates, thereby leading to the pronounced display of a peculiar morphotype that is an irregularly shaped, flat, wrinkled colony with undulate margin (Fig. 1). Referring to the morphotype on plates, the whole multicellular behavior has been called rdar (*r*ed, *d*ry, *a*nd *r*ough) morphotype.[7]

Several components required for the expression of the rdar morphotype have already been identified. Genetic and electron microscopy analyses revealed that the extracellular matrix consists of at least two components; thin aggregative fimbriae (*agf*) and an unknown substance.[8] Thin aggregative fimbriae are 3- to 4-nm-wide filaments that are composed of polymerized Agf A, which can be ob-tained as a monomer only by treatment with 90% formic acid[9] (protocol 2, below).

[1] J. W. Costerton, P. S. Stewart, and E. P. Greenberg, *Science* **284**, 1318 (1999).

[2] J. W. Costerton, Z. Lewandowski, D. E. Caldwell, D. R. Korber, and H. M. Lappin-Scott, *Annu. Rev. Microbiol.* **49**, 711 (1995).

[3] U. Römling, W. D. Sierralta, K. Eriksson, and S. Normark, *Mol. Microbiol.* **28**, 249 (1998).

[4] U. Römling and M. Rohde, *FEMS Microbiol. Lett.* **180**, 91 (1999).

[5] A. Olsén, M. J. Wick, M. Mörgelin, and L. Björck, *Infect. Immun.* **66**, 944 (1998).

[6] J. W. Austin, G. Sanders, W. W. Kay, and S. K. Collinson, *FEMS Microbiol. Lett.* **162**, 295 (1998).

[7] U. Römling, Z. Bian, M. Hammar, W. D. Sierralta, and S. Normark, *J. Bacteriol.* **180**, 722 (1998).

[8] U. Römling, M. Rohde, A. Olsén, S. Normark, and J. Reinköster, *Mol. Microbiol.* **36**, 10 (2000).

[9] S. K. Collinson, L. Emödy, T. J. Trust, and W. W. Kay, *J. Mol. Biol.* **173**, 741 (1999).

Distinct features of cell–cell interaction can be assigned to the two extracellular components.[8] Thin aggregative fimbriae confer to the cells a community behavior that is characterized by easily breakable cell–cell interactions and an adhesive phenotype (adhesion to abiotic surfaces and biological macromolecules). In bacterial cells expressing the unknown extracellular substance elastic cell–cell interactions and loose adhesion to abiotic surfaces at the air–liquid interface are displayed. On agar plates the distinct binding of CR by thin aggregative fimbriae and the unknown substance leads to brown and pink colonies, respectively; the corresponding morphotypes are called bdar (*b*rown, *d*ry, *a*nd *r*ough; *rpoS* or *adrA* knockout[8]) and pdar (*p*ink, *d*ry, *a*nd *r*ough; *agfA* knockout). In addition, the unknown extracellular substance selectively binds the dye Calcofluor.[8] The dye-binding characteristics can be used for screening purposes (Fig. 1; protocol 1, below).

The rdar morphotype is regulated by *agfD*, a putative positive transcriptional regulator of the FixJ family of response regulators.[10] $AgfD^-$ cells possess the characteristic *S. typhimurium* colony morphology, which is a saw (*s*mooth *a*nd *w*hite) morphotype (Fig. 1). Point mutations in the *agfD* promoter decide about a highly regulated, temperature-dependent or constitutive expression of the rdar morphotype.[3] The upregulated mutants of the rdar morphotype are most suitable for the analysis of multicellular behavior, because almost all the cells are in the multicellular state in stationary phase.

Strategies for Analyzing Multicellular Behavior

Various aspects of the multicellular behavior of *S. typhimurium* can be investigated because so many characteristics change, in contrast to planktonic cells. Major aspects concern the production of the extracellular matrix and the arrangement of the cells in the matrix. Also, the cell physiology is suspected to change dramatically because of the tight cell–cell interactions and the expected microaerophilic or anaerobic conditions inside an aggregate of several hundreds or thousands of cells.[11] Using different strategies, features of multicellular behavior can be investigated by the traditional techniques of transposon mutagenesis.[12] On the other hand, transcriptional arrays and two-dimensional protein gels can be used to scan the whole genome/proteome repertoire of the cell. Both approaches have their advantages and disadvantages. Transposon insertion might lead to polar effects on downstream genes. When using genome/proteome scanning techniques,

[10] G. M. Pao and M. H. Saier, Jr., *J. Mol. Evol.* **40**, 136 (1995).

[11] J. W. Costerton, Z. Lewandowski, D. DeBeer, D. Caldwell, D. Korber, and G. James, *J. Bacteriol.* **176**, 2137 (1994).

[12] A. E. Altman, J. R. Roth, A. Hessel, and K. E. Sanderson, *in* "*Escherichia coli* and *Salmonella*: Cellular and Molecular Biology" (F. C. Neidhardt, ed.), p. 2613. ASM Press, Washington D.C., 1996.

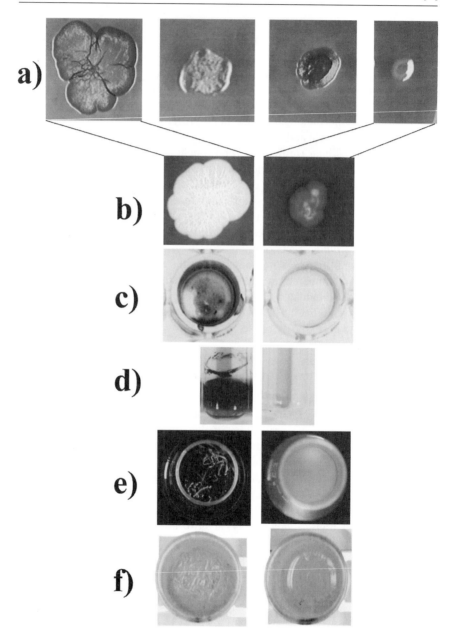

mutants must be successively constructed to investigate the phenotype conducted by the identified locus and the function of the respective gene in further detail.

Transposon mutagenesis can be used to look for the following.

1. Colony morphology mutants of the rdar morphotype, by screening for an altered phenotype on CR/Calcofluor plates: This strategy covers pathways involved in the production of the extracellular matrix.

2. *agfD*-regulated genes: *agfD* is known to be responsible for all aspects of multicellular behavior investigated so far[8] and is therefore considered to be the major regulatory switch point. This strategy reveals the subset of genes that are regulated by *agfD*.

3. Adhesion (biofilm) mutants: e.g., components of the extracellular matrix involved in the adhesion process can be determined by this screening process. The methodology to analyze biofilm mutants is described in this series by O'Toole *et al.*[13] for bacteria such as *Pseudomonas aeruginosa*, but can be directly transferred to *Salmonella* species by using the growth media indicated in protocol 1. Alternatively, other modes of multicellular behavior can be used for screening purposes (e.g., cell clumping, pellicle formation).

4. Differentially regulated genes in planktonic versus biofilm cells: This approach will lead to the detection of genes involved in changes in cell physiology that take place because of altered environmental conditions. This methodology is described in this series by Prigent-Combaret and Lejeune, who applied it to *Escherichia coli* biofilms.[14]

Genome/proteome arrays are most useful for looking at differences between wild-type (rdar morphotype) and mutants deleted for a regulatory gene. Most

---

[13]G. A. O'Toole, L. A. Pratt, P. I. Watnick, D. K. Newman, V. B. Weaver, and R. Kolter, *Methods Enzymol.* **310**, 91 (1999).

[14]C. Prigent-Combaret and P. Lejeune, *Methods Enzymol.* **310**, 56 (1999).

FIG. 1. The multicellular morphotype of *S. typhimurium*. MAE52, a strain with a constitutive rdar morphotype, and its isogenic mutants are shown.[3] (a) The known multicellular morphotypes of *S. typhimurium* on CR plates incubated at 37° for 24 hr. Left to right: rdar (*agfD+*), pdar (*agfBA*), bdar (*adrA*), and saw (*agfD−*). (b) The *agfD+/agfD−* morphotypes on LB without salt agar containing Calcofluor, incubated at 37° for 24 hr. (c) The adhesion phenotype of *agfD+/agfD−* morphotypes on PS grown in M9 minimal medium at 37° for 24 hr. Cells adhere at the bottom and to the walls of the well. (d) The adhesion phenotype of *agfD+/agfD−* morphotypes on glass, grown in LB without salt medium at 37° for 24 hr. Most *agfD+* cells adhere at the air–liquid interface and slightly below the interface. (e) Cell clumping of the *agfD+/agfD−* morphotypes in LB without salt medium, incubated at 37° for 24 hr with shaking. (f) Pellicle formation of the *agfD+/agfD−* morphotypes in standing culture medium (LB without salt), incubated at 37° for 24 hr. AgfD+ cells form a tight pellicle (view from the top).

useful is a deletion mutant in *agfD*, which is the lowest regulatory level known to be responsible for the expression of the whole multicellular behavior. Using this type of analysis with cells in solution, all aspects that discriminate planktonic (*agfD*⁻) from aggregated cells (*agfD*⁺) will be covered. The same techniques applied to wild-type/mutant cells on plates will be restricted to detecting matrix-specific changes.[15]

In this chapter the focus is on the use of transposon mutagenesis for screening of mutants in multicellular behavior.

The MudJ (originally called Mud I1734[16]) element is used for transposon mutagenesis. This phage Mu derivative carries a kanamycin (Km) resistance gene and the promoterless ′*trpB trpA*′-*lacZ* element, which can be used to determine the transcriptional activity of the affected gene by the measurement of ß-galactosidase activity.[17]

The transduction of the MudJ element into the recipient strain by phage P22 is conveniently carried out with an *S. typhimurium* strain in which the localization of the transposition-defective MudJ element at a certain distance from the temperature-dependent transposition-proficient MudI enables the transitory *cis* complementation of the transposition function.[18] A detailed protocol, describing the use of the MudJ element, is found in the laboratory manuals *Experimental Techniques in Bacterial Genetics*[19] and *Genetic Analysis of Pathogenic Bacteria*.[17] The use of phage P22 is described in Volume 204 of this series.[20] Strains and phages can be obtained from the *Salmonella* Genetic Stock Center (SGSC) run by K. E. Sanderson (http://www.ucalgary.ca/~kesander).

Strategy to Screen for Colony Morphology Mutants of the rdar Morphotype

1. Perform random mutagenesis with MudJ, using as recipient a strain positive for the rdar morphotype at 37°.

2. Screen for colony morphology mutants of the rdar morphotype on CR plates at 37° (for screening medium see protocol 1).

3. Check the mutants for prototrophy (optional).

4. Back transduce the MudJ vector integrated into genes affecting the rdar morphotype into the wild-type strain.

5. Determine the sequence of the knocked-out gene by inverse polymerase chain reaction (PCR) or cloning.

---

[15] U. Römling, unpublished results (1999).

[16] B. A. Castilho, P. Olfson, and M. J. Casadaban, *J. Bacteriol.* **158**, 488 (1984).

[17] S. R. Maloy, V. J. Stewart, and R. K. Taylor, "Genetic Analysis of Pathogenic Bacteria." Cold Spring Harbor Laboratory Press, Cold Spring Harbor, New York, 1996.

[18] K. T. Hughes and J. R. Roth, *Genetics* **119**, 9 (1988).

[19] S. R. Maloy, "Experimental Techniques in Bacterial Genetics." Jones and Bartlett, Boston, 1990.

[20] N. L. Sternberg and R. Maurer, *Methods Enzymol.* **204**, 18 (1991).

6. Create an in-frame deletion in the gene (optional).

7. Check the features of the mutant; e.g., the production of polymerized thin aggregative fimbriae on the bacterial surface (protocol 2), adhesion behavior of the mutant (biofilm formation), detection of environmental conditions that influence the transcriptional activity of the gene of interest.[17]

Comment on Step 1. The virulent *S. typhimurium* strain ATCC 14028, which is used in a number of laboratories, shows a temperature-regulated multicellular behavior that is "on" at 28° and "off" at 37°. However, constitutive mutants in the multicellular behavior were isolated by screening the original culture purchased by the American Type Culture Collection (ATCC, Manassas, VA) on CR plates (see protocol 1) at 37°. Upregulated mutants were found at a frequency of 10^{-3}. The constitutive mutant used for further analysis was subsequently created by transferring the point mutation in the *agfD* promoter responsible for the upregulated multicellular behavior into the ATCC 14028 wild-type strain.[3]

Comment on Step 2. Only the mutants with a blue phenotype on Luria–Bertani (LB) agar without salt plates containing 5-bromo-4-chloro-3-indolyl-β-D-galactopyranoside (X-Gal) are taken for further analysis to preselect for insertions of the MudJ element that enable the use of *'trpB trpA'-lacZ* for the measurement of transcriptional activity.

Comment on Step 3. The *agfD* mutant is a prototroph. Therefore, this screen will eliminate precursor pathways that are connected with cell physiology.

Comment on Step 4. This step is essential, because we observed spontaneous secondary mutations in the rdar morphotype in more than 50% of our transposon mutants.[21]

Comment on Step 5. Several easily accessible strategies can be used to determine the sequence of the genes affected by the MudJ element.

In the cloning procedure, advantage is taken of the fact that the Km cassette in the MudJ element is flanked by a *Sal*I site. Therefore the adjacent genomic DNA can be cloned after *Sal*I digestion by selecting for Km resistance and the antibiotic resistance marker of the cloning vector (Fig. 2a).

For the technique of inverse PCR[22] the genomic DNA is cleaved with the 4-bp cutter *Taq*I or *Alu*I to create fragments of an average size of ∼250 bp. After circularization of the fragments by intramolecular ligation primers in the left end of phage Mu (MudOut/MudTaq or MudOut/MudAlu; Table I) are used to amplify the adjacent unknown gene sequence[23] (Fig. 2). As was reported previously,[22] we observed that the DNA concentration for the intramolecular ligation reaction had

[21] X. Zogay, M. Nimtz, M. Rohde, and U. Römling, *Mol. Microbiol.,* in press (2001).

[22] H. Ochman, F. J. Ayala, and D. L. Hartl, *Methods Enzymol.* **218,** 309 (1993).

[23] B. M. M. Ahmer, J. van ReeuwijK, C. D. Timmers, P. J. Valentine, and F. Heffron, *J. Bacteriol.* **180,** 1185 (1998).

TABLE I

PRIMERS USED TO AMPLIFY UNKNOWN SEQUENCES ADJACENT TO LEFT END OF PHAGE Mu IN MudJ[a]

Primer	Sequence	Location
MudOut	CCGAATAATCCAATGTCCTCCCGGT	Nucleotides 54–30 of Mu left end (M64097)
MudTaq	AGTGCGCAATAACTTGCTCTCGTTC	Nucleotides 701–725 of Mu left end (M64097)
MudAlu	CGAAAAACAAAAACACTGCAAATCATTTCAATAAC	Nucleotides 167–201 of Mu left end (M64097)

[a] Taken from B. M. M. Ahmer, J. van Reeuwijk, C. D. Timmers, P. J. Valentine, and F. Heffron, *J. Bacteriol.* **180,** 1185 (1998).

to be optimized to ensure subsequent amplification. In our hands 1–3.5 ng of DNA per microliter of ligation reaction worked best.[21]

Comment on Step 6. An in-frame deletion is done by the allelic exchange of a gene that contains a deletion and is located on a plasmid, with the respective gene on the chromosome. For this purpose the pMAK plasmid, which carries a temperature-sensitive pSC101 replicon, is used[24] (protocol 3). The application of the pMAK plasmid for allelic exchange has proved to be successful in several strains of *Salmonella, E. coli,* and other bacteria.[3,8,24–26] An advantage of the pMAK system is that the strain used for allelic exchange does not need to fulfill any prerequisites besides being chloramphenicol sensitive (Cm).

Comment on Steps 6 and 7. Step 6 is optional. Checking the features of multicellular behavior can also be done with the transposon mutant, particularly when a polar effect of the transposon on downstream genes can be excluded.

Comment on Step 7. The methodology to analyze the adhesion phenotype is described in this series by O'Toole *et al.*[13] Besides the quantification of the amount of adhesive cells, the kinetics of biofilm formation and the analysis of the distribution of cells on the surface by light and electron microscopy are first steps in characterizing the adhesion phenotype.

Strategy to Screen for agfD-Regulated Genes

This strategy is generally applicable to screen for genes regulated by a global regulator. For this purpose an in-frame deletion in the regulator of interest must be carried out.

[24] C. M. Hamilton, M. Aldea, B. K. Washburn, P. Babitzke, and S. R. Kushner, *J. Bacteriol.* **171,** 4617 (1989).

[25] A. U. Kresse, K. Schulze, C. Deibel, F. Ebel, M. Rohde, T. Chakraborty, and C. A. Guzman, *J. Bacteriol.* **180,** 4370 (1998).

[26] J.-F. Viret, S. J. Cryz, Jr., and D. Favre, *Mol. Microbiol.* **19,** 949 (1996).

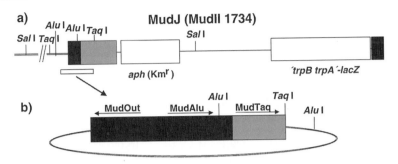

FIG. 2. (a) The MudJ element (not drawn to scale). Only elements important for the cloning and amplification of the genomic DNA are shown. The black boxes indicate the left and right ends of bacteriophage Mu. The double line indicates the genomic DNA where at some distance from the Mu left end restriction sites for *Alu*I, *Taq*I, and *Sal*I are located. The open box below the MudJ element indicates the DNA fragment received after cutting a strain with a MudJ insertion with *Taq*I. (b) After intramolecular ligation of this DNA fragment inverse PCR can be carried out with the primers MudOut and MudTaq. If the DNA was cut with *Alu*I the DNA fragment will be religated at the *Alu*I sites. [Adapted from B. A. Castilho, P. Olfson, and M. J. Casadaban, *J. Bacteriol.* **158**, 488 (1984).]

1. Create an in-frame deletion in *agfD* (Δ*agfD* strain).

2. Clone *agfD* in pBAD30[27] and introduce the plasmid into the Δ*agfD* strain.

3. Perform random mutagenesis with the MudJ element, using the Δ*agfD* pBAD (*agfD*) strain.

4. Screen for differentially regulated genes on +/− induction medium in the presence of X-Gal (40 μg/ml).

5. Back transduce the MudJ vector integrated into differentially regulated genes into *agfD*[+]/Δ*agfD* strains. Verify the differential expression by the measurement of ß-galactosidase activity.[17]

6. Determine the sequence of the knocked-out gene by cloning or inverse PCR (see Comment on Step 5, above).

7. Create an in-frame deletion in the gene regulated by *agfD* (protocol 3).

8. Phenotypical analysis of the in-frame deletion mutant [e.g., for the production of polymerized thin aggregative fimbriae (protocol 2); and for the production of the extracellular substance by Calcofluor binding (protocol 1)].

Comment on Step 2. A low copy number replicon in combination with a tight promoter, such as in the vector pBAD30, where the P15A replicon from pACYC and the *araBAD* promoter are used, is frequently needed to ensure regulated expression and successful complementation of genes involved in the multicellular behavior of *S. typhimurium.*

---

[27] L. M. Guzman, D. Belin, M. J. Carson, and J. Beckwith, *J. Bacteriol.* **177**, 4121 (1995).

Comment on Step 4. In the case of the *araBAD* promoter in pBAD30 up to 0.4% (w/v) L-arabinose can be used for induction.[27] To complement an *agfD* deletion with the gene on the pBAD plasmid 0.1% (w/v) arabinose at 28° is sufficient to achieve the full rdar morphotype.

Protocol 1: Screening Media for rdar Morphotype

The rdar morphotype has been found to be optimally expressed on LB agar plates without salt (per liter of distilled H_2O: 10 g of tryptone, 5 g of yeast extract, 15 g of agar). An agar concentration of at least 1.5% (w/v) should be used; a higher concentration of the agar (2%, w/v) does not counteract the development of the rdar morphotype. Addition of a 2% (w/v) CR solution [CR at 2 mg/ml, Brilliant Blue G at 1 mg/ml, in 70% (v/v) ethanol] before pouring of plates enhances the discrimination of morphotypes; in the case of *agfBA* mutants colony spreading occurs only when CR has been added to the medium.[28] Rather thick plates (~1 cm) incubated in a humid atmosphere over several days give an optimal expression of the morphotype of single colonies (Fig. 1). After inoculation of a colony with the flat side of a toothpick on the surface of the plate, an rdar morphotype colony has an average diameter of 7 cm after 3 days, in contrast to the saw morphotype (1.5 cm).

LB agar without salt plus Calcofluor (fluorescent brightener 28) at 50 μg/ml can be used as a screening medium for the unknown extracellular substance. It must be noted that Calcofluor in the medium slightly inhibits cell growth. If the cells bind the dye, fluorescence can be detected by observing the colonies under a long-wavelength UV light source (366 nm).

Biofilm formation on PS, glass, and other surfaces can be monitored by growing the cells in LB without salt medium or M9 minimal medium,[29] using 0.2% (w/v) glucose as carbon source. Various biofilm-forming phenotypes are found depending on the respective medium (see Fig. 1), but the effect of various genes on the ability to form biofilms was comparable when using the two media.

Protocol 2: Detection of Thin Aggregative Fimbriae by Western Blotting

Intact thin aggregative fimbriae have been found to (partially) depolymerize only after harsh acidic treatment[30] into monomers of an apparent size of 15.3 kDa, which can be detected by Western blot analysis with a specific antibody. Therefore,

[28] U. Römling, unpublished observation (1998).

[29] F. M. Ausubel, R. Brent, R. E. Kingston, D. D. Moore, J. G. Seidman, J. A. Smith, and K. Struhl, "Current Protocols in Molecular Biology." John Wiley & Sons, New York, 1994.

[30] S. K. Collinson, L. Emödy, K.-H. Müller, T. J. Trust, and W. W. Kay, *J. Bacteriol.* **173,** 4773 (1991).

5 mg of cells collected from a plate or from liquid medium is suspended in 100 μl of 99% (v/v) formic acid and incubated on ice for 10 min. After removal of the liquid by evaporation for 2–4 hr in a SpeedVac (Savant, Holbrook, NY) the pellet is suspended in 200 μl of sodium dodecyl sulfate (SDS) sample buffer, Tris crystals are added when the buffer turns yellow, and 6 μl is loaded on an SDS-polyacrylamide gel. As a control untreated cells are suspended directly in SDS sample buffer. It must be noted that boiling of the SDS sample buffer as well as its freezing favor repolymerization of the fimbriae, leading to nonreproducible results. Completely polymerized fimbriae stuck in the well are hardly detected by the AgfA-specific antibody. Therefore, freshly grown cells (or frozen cell pellets) are treated just before application onto the gel.

Protocol 3: Construction of Chromosomal Deletion without Antibiotic Resistance Marker *in Salmonella typhimurium*, Using pMAK Suicide Vector System

Principle. The plasmids pMAK700 and pMAK705 [chloramphenicol resistance marker (Cm^r)] contain a temperature-sensitive replicon (pSC101 origin) that is active at 28° and inactive at 42 to 44°.[24]

1. Construct the appropriate plasmid in *E. coli* by standard molecular biology techniques[29,31] (growth at 28°, Cm at 20 μg/ml).

2. Transform or electroporate the respective plasmid into *S. typhimurium* LB5010. Isolate the plasmid and characterize by restriction digest.

3. Electroporate the respective plasmid into the target *Salmonella* strain. Purify the strain by streaking it one or two times over a plate.

4. Pick a single colony of the target strain. Inoculate in 10 ml of LB containing Cm (20 μg/ml) and incubate at 28° overnight.

5. Plate 150 μl of a 10^{-5} dilution on LB–Cm plates (prewarmed) at 44° and 28° to check insertion into the chromosome. Plate 150 μl of a 10^{-3} dilution on two plates and incubate at 44° overnight.

6. Take approximately 10 colonies and purify them for 1 day at 44° on LB–Cm plates (prewarmed).

7. Check the insertion of the plasmid into the chromosome by PCR. The strategy for choosing the primers is outlined in Fig. 3. Take a positive colony (plasmid integrated into the chromosome) and inoculate overnight in 100 ml of LB–Cm (batch 1) at 28°.

8. In the morning make a dilution in 100 ml of LB–Cm (take 100 μl from batch 1). Incubate at 28° for 24 hr (batch 2).

[31] J. Sambrook, E. F. Fritsch, and T. Maniatis, "Molecular Cloning: A Laboratory Manual." Cold Spring Harbor Laboratory Press, Cold Spring Harbor, New York, 1989.

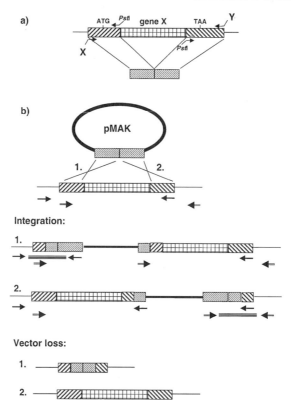

FIG. 3. Creation of an in-frame deletion without an antibiotic resistance marker in a gene of interest. (a) Strategy for the amplification of DNA fragments, creating an in-frame deletion in gene X. For the description of the strategy see text. (b) Homologous recombination between the sequences of the plasmid and the chromosome can occur at the 5′ end (1) or at the 3′ end (2) of the gene at an incubation temperature of 42°. Depending on the site of recombination the integration of the plasmid into the chromosome can be detected by one of the two primer sets indicated. Each primer set contains a primer adjacent to the region cloned in the plasmid and a second primer at the opposite side within the region cloned in the plasmid. If the plasmid is integrated into the chromosome the length of the amplified product will be shortened by the length of the deletion relative to the full-length gene. The second possible product is too long to be amplified under the standard PCR conditions. After a second recombination event at 28°, the permissive temperature for plasmid excision, the deletion will stay in the chromosome or be cut out again with the plasmid, maintaining the full-length gene on the chromosome.

9. Make another dilution in 50 ml of LB–Cm (100 μl from batch 2) and incubate for 1 day or overnight (batch 3). The frequency of resolution of the plasmid is usually greater than 50% after three rounds of incubation.

10. Plate batch 2 and batch 3 on LB plates at 44° without any antibiotics. Use 100 μl from the 10^{-6} and 10^{-7} dilutions; use two plates for each dilution.

11. Replica plate at 44° on a Cm plate and a plate without antibiotics (in this order).

12. From the plate without antibiotics streak Cm^s colonies on a CR plate and incubate at the appropriate temperature overnight.

13. From the CR plate: restreak 4–10 colonies with the desired phenotype; as a control some colonies with the wild-type phenotype can be taken.

14. Test the colonies for the deletion by PCR.

Comment on Step 1. Several strategies can be used to create the gene with a deletion.[32] In the scheme outlined in Fig. 3 two PCR fragments that overlap the beginning and the end of the gene of interest are amplified. For the ease of cloning two different restriction sites, X and Y, respectively, are used for cloning and are introduced by the outward primers. The primers flanking the deletion carry another, but common restriction site, e.g., a *Pst*I site. In that way, an additional 6 bp is introduced at the side of the deletion. However, the same plasmid construct can be used to introduce a resistance marker into the gene of interest by cloning, e.g., a kanamycin cassette[33] into the *Pst*I site flanking the deletion. A resistance marker is useful when infection experiments are carried out in order to recover the mutant strain.

Comment on Step 2. The LB5010 strain is deficient for all three restriction systems in *Salmonella,* but proficient for the respective methylation systems.[34]

Comment on Step 5. The plate incubated at 28° is used as a control. Because the frequency of plasmid insertion is less than 1 : 1000 a 10^{-5} dilution of an overnight culture gives a blank plate or only a few colonies at 44°. At the 10^{-3} dilution there are enough colonies.

Acknowledgments

I acknowledge a fellowship from the program Infektionsbiologie from the Bundesministerium für Forschung und Technologie (BMFT). I thank G. Maaß and J. Wehland for generous support and interest. The research described here has been aided in part by grant RO 2023/3-1 from the Deutsche Forschungsgemeinschaft and by grants from the Medicinska Forskningsrådet and Bristol-Myers Squibb to Staffan Normark.

[32] S. N. Ho, H. D. Hunt, R. M. Horton, J. K. Pullen, and L. R. Pease, *Gene* **77,** 51 (1989).

[33] Amersham Pharmacia Biotech, "BioDirectory 2001," p. 327 (2001).

[34] L. R. Bullas and J.-I. Ryu, *J. Bacteriol.* **156,** 471 (1983).

[6] Gene Transfer in Bacterial Biofilms

By Adam P. Roberts, Peter Mullany, and Michael Wilson

Introduction

For many years it has been assumed gene transfer will take place within biofilms and that the frequency of such transfer will be greater than that which occurs between planktonic organisms. The latter assertion is largely intuitive and based on the observations that the bacteria in a biofilm are tightly packed and that they display a variety of phenotypes (with an increased likelihood of some cells displaying suitable "gene transfer" phenotypes) due to physicochemical and nutritional gradients within the biofilm. However, few articles have been published that actually demonstrate gene transfer in biofilms in their natural environment or in laboratory models, which mimic such conditions. The study of gene transfer in bacterial biofilms is of great importance because of the increasing production and use of genetically modified microorganisms in the environment and because of concerns over the spread of antibiotic resistance in clinically important organisms. Also, as the molecular genetics of the biofilm phenotype are being determined and differential expression of genes in bacterial biofilms is being shown, the possibilities of transfer of a particular gene or set of genes within a biofilm remain to be fully explored. Knowing how gene transfer occurs in biofilms produced *in vitro* will help us to understand the spread of genes in the natural environment. This chapter concentrates on *in vitro* methods, which utilize laboratory-based methods to analyze gene transfer in biofilms, and not the more complex *in vivo* models such as animals, riverbeds, and soils.

It is likely that transformation, transduction, and conjugation all occur in bacterial biofilms. For example, many oral streptococcal species are naturally competent[1] and these may utilize free DNA that occurs in the mixed oral biofilm environment of humans and other animals. Although the level of free DNA in the oral environment remains to be determined, the presence of large quantities of extracellular DNA in soil environments, for example, has been demonstrated previously.[2] Indeed, it has been shown that plasmid DNA can be taken up, and successfully expressed, by bacteria in epilithic environments.[3] Bacteriophages also have a part to play in the biofilm environment; phage-mediated transfer of tetracycline resistance in both liquid[4,5] and biofilm[5] environments has been shown for

---

[1] R. D. Lunsford, *Plasmid* **39**, 10 (1998).

[2] M. G. Lorenz and W. Wackernagel, *Microbiol. Rev.* **58**, 563 (1994).

[3] H. G. Williams, M. J. Day, J. C. Fry, and G. J. Stewart, *Appl. Environ. Microbiol.* **62**, 2994 (1996).

[4] E. E. Udo and W. B. Grubb, *Curr. Microbiol.* **32**, 286 (1996).

[5] M. S. Pereira, V. P. Barreto, and J. P. Siqueira-Junior, *Microbios* **92**, 147 (1997).

Staphylococcus aureus. Of the three mechanisms of gene transfer, conjugation is likely to be the most promiscuous. Indeed, transfer by this mechanism has been demonstrated between bacteria of different genera and even interkingdom transfer has been demonstrated.[6] Conjugative plasmids,[7,8,9] mobilizable plasmids,[10,11] and conjugative transposons[12] have all been shown to transfer intergenerically in biofilms. Conjugative transposons are probably the most promiscuous of all mobile elements. For example, members of the Tn*916* family have been found naturally in, or have been transferred into, both gram-positive and gram-negative organisms.[13] These elements integrate into the host cell genome and, unlike plasmids, do not have to provide their own replication machinery or rely on interactions with the host cell for stable replication and maintenance.

The methods described in this chapter have been designed for detecting conjugative gene transfer, although gene transfer by other mechanisms could be detected and this must be borne in mind when designing experiments and interpreting results.

Single-Recipient Biofilm Experiments

Bacteria will form biofilms on membrane filters (0.45-μm pore size) and it is possible simply to mix a donor and recipient organism on a filter and subsequently resuspend and plate onto selective agar. This technique, termed *filter mating,* is the simplest way of assaying for gene transfer in a bacterial biofilm. Filter mating is flexible in that oxygen concentration and temperature and nutrient content can all be easily controlled. The reason for using the 0.45-μm pore filter, as opposed to plating directly on agar, is that the filter is believed to bring the bacteria into close proximity and hold them in intimate contact.[14] It has been shown that a filter with a pore size of 0.45 μm results in more transconjugants than a filter mating carried out on a 0.22-μm pore size filter, supplied by the same company, when *Enterococcus faecalis* and *Lactococcus plantarum* were mated together.[14] The side of the filter on which the mating is carried out is important, as the three-dimensional structure of

[6] G. T. Haymen and P. L. Bolen, *Plasmid* **30,** 251 (1993).

[7] C. Dahlberg, M. Bergström, and M. Hermansson, *Appl. Environ. Microbiol.* **64,** 2670 (1998).

[8] B. B. Christensen, C. Sternberg, J. B. Andersen, L. Eberl, S. Moller, M. Givskov, and S. Molin, *Appl. Environ. Microbiol.* **64,** 2247 (1998).

[9] T. R. Licht, B. B. Christensen, K. A. Krogfelt, and S. Molin, *Microbiology* **145,** 2615 (1999).

[10] D. L. Beaudoin, J. D. Bryers, A. B. Cunningham, and S. W. Peretti, *Biotechnol. Bioeng.* **57,** 280 (1998).

[11] P. Lebaron, P. Bauda, M. C. Lett, Y. Duval-Iflah, P. Simonet, E. Jacq, N. Frank, B. Roux, B. Baleux, G. Faurie, J. C. Hubert, P. Normand, D. Prieur, S. Schmitt, and J. C. Block, *Can. J. Microbiol.* **43,** 534 (1997).

[12] A. P. Roberts, J. Pratten, M. Wilson, and P. Mullany, *FEMS Microbiol. Lett.* **177,** 63 (1999).

[13] L. B. Rice, *Antimicrob. Agents Chemother.* **42,** 1871 (1998).

[14] Y. Sasaki, N. Taketomo, and T. Sasaki, *J. Bacteriol.* **170,** 5939 (1988).

each side is different. It is recommended that filter mating be carried out on the front side of a filter, which exhibits a more spongelike appearance under the scanning electron microscope. Bacteria are held in close contact within the spongelike matrix more effectively on the front side than on the back side of the filter. We routinely use 0.45-μm pore size filters supplied by Fisher Scientific UK (Loughborough, UK). The front side of this filter is marked with a grid, which is placed facing upward on the agar plate. It has also been reported that embedding filters in agar increases the conjugation efficiency when mating strains of *Streptococcus pneumoniae*.[15] The filter-mating technique is a relatively simple and quick way of looking for gene transfer between bacteria in a simple model of a biofilm. More sophisticated models that more closely mimic the *in vivo* situation are discussed below.

Filter Mating

Single well-isolated colonies of donor and recipient organisms are taken from agar plates containing relevant antibiotics and/or growth supplements. These are then used to inoculate 1 ml of appropriate broth. The broth culture is incubated for 18 hr at 37°. This culture is used to inoculate 10 ml of prewarmed (37°) liquid medium. The cultures are allowed to grow to late exponential phase and are mixed together and centrifuged at 1000 *g* for 5 min at room temperature. The supernatant is discarded and the cells are suspended in 1 ml of prewarmed nonselective liquid medium. Aliquots (100 μl) are dispensed onto, and spread over, sterile 0.45-μm pore filters that have previously been placed on antibiotic-free agar plates, prewarmed, and incubated overnight at 37°. The filters are then aseptically removed from the plates and placed in a sterile universal container containing 1 ml of prewarmed antibiotic-free liquid medium. The filters are vortexed vigorously for 20 sec and the resultant bacterial suspension is aliquoted onto, and spread over, agar plates containing the appropriate antibiotics and/or components to select for transconjugants. Background growth of donor or recipient can sometimes occur because of the high numbers of cells that are present. For this reason, it is necessary to plate out small volumes of suspended bacteria. We routinely spread a 100-μl aliquot onto a plate. Plates are incubated for 24–48 hr depending on the species of bacteria. Both donors and recipients should be plated on the selective medium to measure the frequency of spontaneous mutation. The frequency of gene transfer is usually expressed as the number of transconjugants per donor cell present in the original mating mix, although it can also be expressed as the number of transconjugants per recipient cell.[16] When examining transconjugants from the same filter, be aware of the possibility that siblings may be present. If a

[15] M. D. Smith and W. R. Guild, *J. Bacteriol.* **144,** 457 (1980).
[16] T. Netherwood, R. Bowden, P. Harrison, A. G. O'Donnel, D. S. Parker, and H. J. Gilbert, *Appl. Environ. Microbiol.* **65,** 5139 (1999).

transfer event occurred early in the mating process, that transconjugant will have a chance to undergo several cycles of cell division and will be represented more than once in the transconjugant pool. This needs to be borne in mind when calculating the transfer frequency; therefore, look at the data from several independent filters before coming to conclusions regarding the frequency of gene transfer.

Selection and Analysis of Recipients

Recipients can usually be selected simply by plating on antibiotic-containing medium to which the mobile element, from the donor, confers resistance plus a counterselective agent to kill donors. The latter include aerobic/anaerobic environments, metabolic limitations, or other intrinsic or acquired antibiotic resistance present in the recipient cells. An example is the selection of *Clostridium difficile* in a mixed culture using *C. difficile* selective supplement (Oxoid, Basingstoke, UK), which contains D-cycloserine and cefoxitin. *Clostridium difficile* is intrinsically resistant to both of these antimicrobial agents.

Transconjugants should be speciated and, in some cases, typed to the strain level to rule out contamination. This can be done by standard biochemical tests and/or 16S rRNA sequencing. Once the identity of the transconjugant has been confirmed, analysis can begin. A physical analysis of transconjugants can be undertaken by plasmid isolation, Southern blotting (if a probe for the genes that are being transferred is available), polymerase chain reaction (PCR; if the DNA sequence of part of the transferred material is known), and subsequent sequencing of the PCR products to confirm that the expected DNA sequence has been amplified. Furthermore, particularly if the transferred element integrates into the chromosome or is a large plasmid, the location can be determined by pulsed-field electrophoresis.

Multispecies (Recipient) Experiments

Many of the same initial considerations apply that have already been detailed for the single recipient biofilm experiments. However, when studying a multispecies biofilm careful consideration must also be given to the choice of inoculum and apparatus. Before the experiment, it is necessary to know the growth requirements of donor and recipient cells and to have a selective agar that will kill, or prevent the growth of, both donor and recipient, allowing only transconjugants to grow. It has also been observed that transfer of both conjugative transposons and conjugative plasmids appears to take place soon (between 0 and 6 hr) after the addition of the donor strain to the experimental system.[9,11] We have studied gene transfer in multispecies oral biofilms (specifically, microcosm dental plaques) *in vitro* and so will use this to illustrate an approach that could be used to investigate gene transfer in other types of biofilms.

Growth of Biofilm and Inoculation of Donor Strain

Setting up a microcosm dental plaque in a constant depth film fermenter (CDFF) is detailed extensively in Volume 310 of this series.[17] In our studies, saliva is used as an inoculum to provide a multispecies biofilm (a microcosm dental plaque) consisting of organisms found in the oral cavity. The saliva is collected from 10 healthy individuals who have had no antibiotics within the last 12 months. Equal volumes from each person are pooled and 1 ml of this saliva is added to 500 ml of an artificial saliva,[17] mixed, and pumped into the CDFF for 8 hr. The inoculum flask is then disconnected and the CDFF fed from a medium reservoir of sterile artificial saliva as described previously.[18] Steady state biofilms form within approximately 1 week and the system can be maintained in this state for long periods of time (several weeks or months).

The donor strain(s) are grown overnight on appropriate selective agar. One colony from this plate is used to inoculate 10 ml of medium, which is grown for 18 hr. The culture is diluted 1 in 10 with fresh prewarmed medium and incubated for 1 hr at 37°. The resulting suspension is then used as an inoculum by pumping it into the CDFF at the desired rate.

Assessing Antibiotic Resistance Profile of Biofilm

To assess the antibiotic resistance profile of the biofilm, one pan containing five biofilms is removed and the disks on which the biofilms have formed are placed into individual aliquots of 100 μl of sterile phosphate-buffered saline (PBS). These are vortexed for 1 min and a 10^{-1} dilution is prepared. One hundred microliters of this dilution is spread onto blood agar containing an appropriate amount of antibiotic and onto blood agar plates containing no antibiotics. These are incubated under both aerobic and anaerobic conditions for 48 hr, checking for growth at 24 and 48 hr. This is repeated with five replicate biofilms. The antibiotic resistance profile should be established for the initial inoculum and for the biofilm before and after inoculation of the donor strain(s).

Identification and Analysis of Recipients

Morphologically distinct organisms that grow on the selective plates are identified and assayed for the presence of the mobile element(s) under study, using the methods described above. In our laboratory, we routinely use PCR for the amplification and sequencing of the 16S rRNA gene for identification purposes.[19]

[17] M. Wilson, *Methods Enzymol.* **310,** 264 (1999).

[18] J. Pratten, A. W. Smith, and M. Wilson, *J. Antimicrob. Chemother.* **42,** 453 (1998).

[19] D. J. Lane, *in* "Nucleic Acid Techniques in Bacterial Systematics" (E. Stackebrandt and M. Goodfellow, eds.), p. 115. John Wiley & Sons, Chichester, 1996.

The sequences obtained can be analyzed at the Ribosome Project Database II site,[20] which facilitates sequence analysis and the production of rRNA-derived phylogenetic trees, and provides aligned and annotated rRNA sequences.

Conclusions

With increasing recognition that the usual mode of growth for most bacteria in their natural environments is as a biofilm, it is important that more studies of the potential for gene transfer in such environments, or in appropriate laboratory models, be undertaken. The information gained from such studies is important for understanding, and ultimately controlling, the spread of antibiotic resistance genes and those genes incorporated into genetically modified organisms released into the environment.

[20] This site can be found at http://cme.msu.edu/rdp/html/index.html.

[7] Conversion to Mucoidy in *Pseudomonas aeruginosa* Infecting Cystic Fibrosis Patients

By J. F. POSCHET, J. C. BOUCHER, A. M. FIROVED, and V. DERETIC

Introduction

Pseudomonas aeruginosa is a potent human opportunistic pathogen[1] causing chronic respiratory infections in cystic fibrosis (CF).[2] The persistent infection of the CF lung is facilitated by a biofilm mode of growth[3] of mucoid, exopolysaccharide alginate-overproducing mutants of *P. aeruginosa* selected during the process of colonization of the CF respiratory tract.[4] The molecular mechanism of conversion to mucoidy involves two pathways.[5] The more common pathway depends on the activation of the alternative sigma factor Alg U via mutations in the gene termed *mucA*.[4,6] In addition to bacterial genetic factors, the host environment contributes to the selection of *P. aeruginosa* mutants and formation of biofilms because

[1] V. Deretic, *in* "*Pseudomonas aeruginosa* Infections" (J. P. Nataro, M. J. Blaser, and S. Cunningham-Runddles, eds.), p. 305. ASM Press, Washington, D.C., 2000.

[2] J. R. Govan and V. Deretic, *Microbiol. Rev.* **60**, 539 (1996).

[3] J. Lam, R. Chan, K. Lam, and J. W. Costerton, *Infect. Immun.* **28**, 546 (1980).

[4] J. C. Boucher, H. Yu, M. H. Mudd, and V. Deretic, *Infect. Immun.* **65**, 3838 (1997).

[5] J. C. Boucher, M. J. Schurr, and V. Deretic, *Mol. Microbiol.* **36**, 341 (2000).

[6] D. W. Martin, M. J. Schurr, M. H. Mudd, J. R. W. Govan, B. W. Holloway, and V. Deretic, *Proc. Natl. Acad. Sci. U.S.A.* **90**, 8377 (1993).

of mutations in the human gene encoding the cystic fibrosis transmembrane conductance regulator (CFTR). The alginate biofilm may play a role in resistance to phagocytosis,[7] scavenging of reactive oxygen intermediates,[8,9] and a number of other biofilm-related phenomena.[2,3] The alginate biofilm has been shown to be of importance in reducing *P. aeruginosa* clearance from the lungs in animal models.[4,10] Alginate biofilms counteract processes of *P. aeruginosa* elimination via uptake and killing by host macrophages and neutrophils[7] and potentially by respiratory epithelial cells.[11] Here we describe genetic methods, molecular approaches, cell culture techniques, and animal models used to examine *P. aeruginosa* conversion to mucoid phenotype and biofilm mode of growth and their role in the pathogenesis of CF.

Bacterial Strains

Pseudomonas aeruginosa PAO1 is the standard genetic strain for which the complete genomic sequence is available (www.pseudomonas.com). PAO1 and its leucine auxotrophic derivative PAO381 are nonmucoid when grown on all media including Lennox L broth (LB) or *Pseudomonas* isolation agar (PIA). Strain PAO578I, a mucoid derivative of PAO381, carries the *mucA22* mutation[6] and constitutively expresses alginate on all growth media. Strain PAO578II, a derivative of PAO578I, carries an additional *sup-2* mutation attenuating the mucoid phenotype and shows nonmucoid growth on LB media but is mucoid on PIA. Strain PAO578III, a nonmucoid derivative of PAO578II, carries an additional mutation in *algU*, which eliminates the activation of the key biosynthetic gene *algD* and is nonmucoid on all media. Strains PAO568 (*mucA2*), PAO581 (*muc-25*), and PAO579 (*muc-23*) are all mucoid derivatives of PAO381 and are mucoid on all media. As yet, the genes corresponding to the *muc-25* and *muc-23* mutations have not been identified.[5] The CF clinical strains CF1–CF55 are *P. aeruginosa* isolates purified from CF patients sputa.[4]

Detection of Alginate

Alginate Assay

The quantification of alginate production by mucoid *P. aeruginosa* strains is carried out by measuring uronic acids, using a modification of the carbazole reaction.[12]

---

[7] S. Schwarzmann and J. R. Boring, *Infect. Immun.* **3**, 762 (1971).

[8] D. B. Learn, E. P. Brestel, and S. Seetharama, *Infect. Immun.* **55**, 1813 (1987).

[9] J. A. Simpson, S. E. Smith, and R. T. Dean, *Free Radic. Biol. Med.* **6**, 347 (1989).

[10] H. Yu, M. Hanes, C. E. Chrisp, J. C. Boucher, and V. Deretic, *Infect. Immun.* **66**, 280 (1998).

[11] G. B. Pier, M. Grout, T. S. Zaidi, J. C. Olsen, L. G. Johnson, J. R. Yankaskas, and J. B. Goldberg, *Science* **271**, 64 (1996).

[12] C. A. Knutson and A. Jeanes, *Anal. Biochem.* **24**, 470 (1976).

Procedures

1. Grow *P. aeruginosa* strain for 24–48 hr on appropriate medium.

2. Suspend the bacteria in 5 to 20 ml of sterile 0.85% (w/v) sodium chloride (saline) by scraping the plates. The volume of saline used depends on the level of mucoidy. It is important to make sure that all the alginate is in suspension.

3. Spin the bacteria at 7000*g* for x min at xx° and transfer the supernatant to a new tube. Determine the wet cell weight of the bacterial pellet. Sometimes it is necessary to dilute alginate–bacterial suspensions to pellet all bacteria, as the viscosity of alginate solutions can trap a significant portion of bacterial mass.

4. Layer 350 μl of the alginate-containing supernatant solution onto 3 ml of uronic acid detection reagent (2.93 ml of sulfuric acid and 70 μl of borate solution containing 45 m*M* potassium hydroxide and 1 *M* boric acid) on ice and mix by vortexing.

5. Add 100 μl of 0.1% (w/v) carbazole (made fresh in ethanol), vortex, and incubate at 55° for 30 min.

The amount of alginate produced is determined by spectrophotometric measurement (UV-1601; Shimadzu, Columbia, MD) at 530 nm against a standard curve generated by using 5 to 1000 μg of alginic acid (Sigma, St. Louis, MO) and is expressed as micrograms of alginate (uronic acid) per milligram of wet cell weight.[1]

Quantification of *algD* Transcriptional Activity

The expression of *algD* is presently the best defined molecular correlate of alginate production and alginate biofilm formation. The expression levels are routinely determined by transcriptional fusions using *xylE* as a reporter gene or direct detection of RNA levels by S1 nuclease protection or reverse transcription (primer extension) assays.

xylE Assay

Catechol 2,3-deoxygenase (CDO), encoded by *xylE*, catalyzes the formation of yellow-greenish chromogenic product 2-hydroxymuconic semialdehyde from catechol. A *xylE* gene cassette is fused to the promoter of the critical alginate biosynthetic gene *algD*,[13] allowing the direct detection of the transcriptional activity of *algD* under various growth conditions or mutation backgrounds, which may affect the mucoid phenotype of *P. aeruginosa*.

---

[13] W. M. Konyecsni and V. Deretic, *Gene* **74**, 375 (1988).

Procedure

1. Grow *P. aeruginosa* strain on medium allowing expression of the mucoid phenotype. For solid media/substrates incubate for 24–48 hr.
2. Suspend bacteria by scraping (e.g., plates) in 5 ml of sterile saline and pellet by centrifugation at 7000g in an SM24 rotor (Sorvall, Wilmington, DE) for 10 min at room temperature.
3. Suspend the bacterial pellet in 5 ml of 50 mM potassium phosphate buffer (pH 7.5) containing 10% (v/v) acetone.
4. Sonicate on ice for 1 min, using a VirSonic sonicator (VirTis, Gardiner, NY) at power level 2.5.
5. Centrifuge the extracts at 7000g in an SM24 rotor for 10 min at room temperature.
6. Determine protein concentrations in supernatants by Bradford protein assay (Bio-Rad, Hercules, CA), using bovine serum albumin (BSA) for the standard curve.
7. Assay CDO, using equal amounts of protein added to 3 ml of 0.3 mM catechol in 50 mM potassium phosphate buffer (pH 7.5) at 24°.
8. Record the CDO-catalyzed reaction rate by spectrophotometer (Shimadzu UV-1601), monitoring the absorbance change at 375 nm for 1 min.

One unit of CDO is defined as the amount of enzyme that oxidizes 1 mmol of catechol per minute at 24°. The molar extinction coefficient for the product 2-hydroxymuconic semialdehyde is 4.4×10^4.

Transcriptional Analysis by S1 Nuclease Protection and Primer Extension Assays

Total bacterial RNA is isolated from *P. aeruginosa* for promoter (mRNA 5′ end) mapping and for determining expression levels.

Procedures

RNA ISOLATION

1. Inoculate 100 ml of LB (GIBCO-BRL/Life Technologies, Gaithersburg, MD) with 1 ml of a 2-ml overnight culture and shake for 4 hr at 37°.
2. Rapidly chill the bacterial suspension by swirling in a dry ice–ethanol bath for 2–5 min (do not allow the suspension to freeze), followed by centrifugation at 13,700g for 5 min at 4°.
3. Suspend the bacterial pellet in 10 ml of ice-cold lysis buffer consisting of 50 mM Tris-HCl at pH 7.5 in diethyl pyrocarbonate (DEPC)-treated water.
4. Spin the suspension at 7000g in an SM24 rotor for 5 min at 4°.
5. Resuspend the pellet in 5 ml of lysis buffer at room temperature and add 1 ml of 20% (w/v) sodium dodecyl sulfate (SDS).

6. Mix gently and place at 67° for 1–5 min, until the solution becomes clear and viscous, indicating that complete lysis has occurred.

7. Add 4 g of cesium chloride and mix gently on a rotating platform for 30 min.

8. When dissolved, add a further 5 ml of the above-described lysis buffer.

9. Mix gently and spin at 7000g in an SM24 rotor at room temperature for 10 min.

10. Prepare 13 × 51 mm swinging bucket ultracentrifuge tubes (Beckman, Palo Alto, CA) by forming a cushion of 2 ml of 5.7 *M* CsC1 at the bottom of each tube.

11. Carefully layer supernatant from step 9 on top of the 2-ml CsCl cushion and centrifuge overnight at 150,000g in a swinging bucket rotor (AH-650; Sorvall) at 17°.

12. Discard the supernatant, invert the centrifuge tubes, and dry for 30 min at room temperature.

13. Suspend the translucent RNA pellet in 300 μl of DEPC-treated water and extract with 200 μl of chloroform. If multiple tubes are used to contain sample, suspend the first pellet in 300 ml and then transfer the contents to the next tube. Centrifuge at maximal speed for 4 min in a microcentrifuge at 4°.

14. Remove the top layer, containing the aqueous phase, avoiding the debris at the interphase. Add 30 μl of 3 *M* sodium acetate and 750 μl of ethanol, and place at −20° for at least 2 hr. (*Note:* At this step RNA kept in ethanol is stable for several months when stored at −20°. Once RNA is dissolved in aqueous buffers it is stable only for a limited period.)

15. Pellet precipitated RNA by centrifugation in a microcentrifuge at maximal speed for 15 min at 4° and discard the resulting supernatant.

16. Dry the RNA pellet for 4 min and dissolve the RNA in 20 μl of DEPC-treated water.

17. Use 1 μl to determine RNA concentration by spectrophotometry at 260 nm (Shimadzu UV-1601). One microgram of RNA corresponds to 0.027 optical units (A_{260}).

S1 NUCLEASE PROTECTION ASSAY. For S1 nuclease protection assays RNA is hybridized to radiolabeled single-stranded DNA probes followed by digestion with S1 nuclease. The nuclease-resistant products are then analyzed by electrophoresis on sequencing gels.

1. To generate a single-stranded uniformly labeled DNA probe, anneal 14 μl of template containing putative mRNA start site with 4 μl of oligonucleotide (see Ref. 14 for templates and oligonucleotides).

2. Add 1 μl of 100 m*M* dithiothreitol (DTT), 2 μl of 3000 mix (5 m*M* dGTP, 5 m*M* dATP, and 5 m*M* dTTP in 1× Rxn mix [50 m*M* Tris-HCl (pH 8.0), 10 m*M* MgCl$_2$, and 50 m*M* NaCl]), 20 μCi of [α-^{32}P]dCTP (NEN-Du Pont, Boston, MA),

[14] M. J. Schurr, H. Yu, J. C. Boucher, N. S. Hibler, and V. Deretic, *J. Bacteriol.* **177,** 5670 (1995).

and 1 U of Klenow enzyme (GIBCO-BRL/Life Technologies) to 5 μl of annealed template.

3. Incubate for 20 min at 42° followed by the addition of 2 μl of chase (5 mM dGTP, 5 mM dATP, 5 mM dTTP, 5 mM dCTP, and 1 U of Klenow enzyme) and a further incubation of 10 min at 42°.

4. Heat inactivate the Klenow enzyme by incubating the mix at 65° for 5 min.

5. Add 1 μl of a restriction endonuclease to generate a 50- to 250-bp DNA fragment and incubate at 37° for 1 hr.

6. Add 30 μl of FAM [95% (v/v) formamide, 20 mM EDTA, 0.05% (w/v) bromphenol blue, 0.05% (w/v) xylene cyanol], boil for 3 min to denature DNA, and load onto a 5% (w/v) nondenaturing polyacrylamide gel.

7. Run the gel until the dye is approximately three-quarters of the way into the gel, separate the plates, and detect single-stranded probe by autoradiography. Align the film with the gel and cut out the appropriate band for electroelution.

8. Electroelute the probe from the gel into dialysis tubing containing 300 μl of 1× Tris–acetate–EDTA (TAE) for 1 hr at 150 V. Reverse the polarity and run for an additional 40 sec.

9. Remove the dialyzed probe and add 30 μl of 3 M sodium acetate, 5 μl of tRNA (1 μg/ml), and 750 μl of ethanol. Precipitate the probe overnight at −20°.

10. Centrifuge for 10 min and dissolve the pellet in 0.1% (w/v) SDS, 10 mM piperazine-N,N'-bis(2-ethanesulfonic acid) (PIPES) buffer at pH 6.5. The volume depends on the number of RNA samples to be analyzed (e.g., to process five RNA samples the volume should be 50 μl). *Note:* The exposure time for band detection in step 16 will be affected by the dilution of the probe.

11. Add 10 μl of probe to 50 μg of RNA and bring the volume to 20 μl with DEPC-treated water.

12. Add 3 μl of annealing buffer (25 μl of 5 M NaCl and 5 μl of 500 mM PIPES buffer at pH 6.5), boil for 3 min, and incubate for 1 hr at 67°.

13. Add 300 μl of ice-cold S1 nuclease buffer [280 mM NaCl, 50 mM sodium acetate, 4.5 mM zinc sulfate, and single-stranded DNA (20 μg/μl)], and 1000 U of S1 nuclease (GIBCO-BRL/Life Technologies) and incubate for 30 min at 37°.

14. Stop the reaction by adding 150 μl of phenol saturated with 50 mM Tris at pH 8.0. Centrifuge for 4 min at 4° in a microcentrifuge at maximal speed, remove the supernatant, and add 750 μl of ice-cold ethanol, followed by overnight precipitation at −20°.

15. Centrifuge for 20 min at 4° in a microcentrifuge at maximal speed, dry, and suspend in 10 μl of FAM.

16. Boil the probe for 3 min and load onto a 7.6% (w/v) sequencing polyacrylamide gel alongside a sequencing ladder generated with the same primer and template as in step 1. The band of protection is visualized by autoradiography.

PRIMER EXTENSION ASSAY. The primer extension analysis provides an alternative method to S1 nuclease protection assay and is applied where confirmation is necessary or when multiple mRNA 5' ends are detected.

Unless stated otherwise, all steps are carried out at 4°.

1. End label the primer by adding 1 μl of 0.1 M DTT, 1 μl of 5 μM primer, 20 μCi of [^{32}P]ATP (NEN-Du Pont) in 1× T4 polynucleotide kinase buffer (Boehringer Mannheim, Mannheim, Germany), and 20–30 U of polynucleotide kinase (Boehringer Mannheim).

2. Incubate for 1 hr at 37° and stop the reaction by adding 2 μl of 0.5 M EDTA.

3. Add 50 μl of Tris–EDTA (TE) buffer and incubate for 5 min at 65°.

4. Remove unincorporated nucleotides by using a G-25 Sephadex spin column (Boehringer Mannheim).

5. Anneal 3 μl of end-labeled primer with 10–20 μg of RNA by adding 2 μl of 5× hybridization buffer (0.5 M KCl, 0.25 M Tris-HCl at pH 8.3) followed by the addition of DEPC-treated water to bring the volume up to 10 μl. Mix on ice and incubate for 1 min at 95°.

6. Immediately transfer the tubes to 55° for 2 min, followed by placing the tubes on ice for a further 15 min.

7. Extend primer by adding 5.0 μl of annealing mix containing 1 μl of 0.1 M DTT, 2 μl of 2.5 mM dNTPs, 1 μl of 5× reverse transcriptase buffer, and 1 μl of Superscript II reverse transcriptase (all from GIBCO-BRL/Life Technologies) and incubate for 45 min at 44°.

8. Add 5 μl of sequencing buffer [0.05% (w/v) bromphenol blue, 0.05% (w/v) xylene cyanol, 20 mM EDTA in deionized formamide], boil, and load onto a 7.6% (w/v) sequencing polyacrylamide gel alongside a sequencing ladder generated with the same oligonucleotide as used for the primer extension and a template, which contains the promoter region generated either by polymerase chain reaction (PCR) or from a plasmid construct containing the promoter region. Visualize the extension product by autofluorography.

Mucoid *Pseudomonas aeruginosa* and Host–Pathogen Interactions

Mucoid *P. aeruginosa* have been detected in the lungs of CF patients.[3] The interactions of mucoid *P. aeruginosa* with the host are modeled with a murine inhalation exposure infection system[4,10,15] and *in vitro* cellular models of interactions with phagocytic and epithelial cells.

Animal Model of Respiratory Infection

Using the inhalation exposure system AF212 (Glas-Col, Terre Haute, IN), mice are exposed to aerosols containing droplet nuclei 1 to 2 μm in size, resulting

[15] H. Yu, S. Z. Nasr, and V. Deretic, *Infect. Immun.* **68,** 2142 (2000).

in deep lung infections. Mucoid or nonmucoid *P. aeruginosa* suspensions are aerosolized via a glass nebulizer, allowing the control of bacterial deposition into the lungs of mice by varying the exposure time, air pressure, and cloud decay time. Mice are placed in mesh chambers allowing five different experimental groups of up to 20 mice each to be exposed simultaneously. The strains of mice that have been used for this type of study[4, 10, 15] are BALB/c, DBA/2NHsd, C57BL/6/J, interferon γ knockout mice (JR2287), inducible nitric oxide synthase 2 knockout mice (JR2609), and the CFTR transgenic mice *CFTR^mlUnc−/−* and FABP-hCFTR *CFTR^mlUnc−/−*. All mice are from Jackson Laboratory (Bar Harbor, ME), except for DBA/2NHsd mice, which are from Harlan-Sprague (Indianapolis, IN).

Procedure

1. Inoculate *P. aeruginosa* into 200 ml of medium and grow for 12 hr at 37°.
2. Centrifuge the culture at 13,700*g* for 15 min at 4°.
3. Wash the pellet with 5 ml of cold 1% (w/v) proteose peptone (Difco, Detroit, MI) in phosphate-buffered saline (PBS, pH 7.5). The wet mass of the pellet should be 1.2 g and is used to ensure that culture conditions are consistent. *Note*: Proteose peptone improves *P. aeruginosa* viability.
4. Suspend the pellet thoroughly in 5 ml of 1% (w/v) proteose peptone–PBS. Determine the colony-forming units (CFU) by plating (for viability record).
5. Place the mice (up to 20 mice per group) into labeled mesh compartments.
6. Aerosolize *P. aeruginosa* for 30 min with the air pressure set at 20 ft^3/hr and the vacuum pressure set at 60 ft^3/hr followed by 15 min of cloud decay and subsequent decontamination by germicidal UV lamp irradiation for 15 min prior to opening of the chamber.
7. On completion of the exposure, kill mice immediately for initial bacterial deposition determination. Other mice are killed at various time points postinfection by exposure to CO_2 gas for 5 min. Remove *en bloc* the trachea and the right and left lobes of the lung and clamp the trachea bifurcation, using a hemostat.
8. Remove the left lung, record the weight, place it in 0.5 ml of 1% (w/v) proteose peptone–PBS and use a homogenizer to break up the lung tissue. Determine colony-forming units by plating and report colony-forming units per gram of lung weight.
9. Insufflate the right lung with alcoholic 2% (w/v) paraformaldehyde solution, using a needle placed inside the trachea. Remove the trachea and place the lung in 10% (w/v) paraformaldehyde in PBS overnight and process the tissue for lung histopathology the following day.

Macrophage/Neutrophil Killing Assay

The murine macrophage cell line J774 (ATCC TIB-67) or murine bone marrow-derived macrophages prepared as described elsewhere[16] are used to determine the rate of macrophage-mediated killing of mucoid *P. aeruginosa* strains.

Procedure

1. Grow and maintain J774 cells in Dulbecco's modified Eagle's low-glucose medium (DMEM; BioWhittaker, Walkersville, MD) supplemented with 5 mM L-glutamine and 5% (v/v) fetal bovine serum (FBS; HyClone, Logan, UT) at 37° under 5% CO_2.

2. Collect midexponential phase *P. aeruginosa* by centrifugation at 13,700g for 10 min and suspend in PBS at pH 7.4.

3. Incubate J774 macrophages at a multiplicity of infection (MOI) of 1 with *P. aeruginosa* in DMEM supplemented with 5 mM L-glutamine and 5% (v/v) fetal bovine serum for 30 min at 37° and 5% CO_2.

4. Remove extracellular *P. aeruginosa* with three washes in Hanks' buffered saline solution (HBSS; BioWhittaker).

5. Incubate J774 cells in DMEM supplemented with 5 mM L-glutamine, 5% (v/v) fetal bovine serum. Include gentamicin (400 μg/ml) to kill the extracellular bacteria at 37°, 5% CO_2 as needed. The incubation time with gentamicin should be determined empirically to ensure that only the extracellular bacteria are killed and that most of them are eliminated.

6. Wash once with medium to remove gentamicin.

7. Harvest macrophages by scraping, suspend in equal volumes of cold sterile water for each sample, and lyse by vortexing three times for 1 min at maximal output.

8. Determine bacterial viability by plating and counting colony-forming units.

Bactericidal Assay with Human Neutrophils

Procedure

1. For each sample, mix heparinized blood by gentle inversion after addition of an equal volume of 3% (w/v) dextran (Pharmacia Fine Chemicals, Uppsala, Sweden) in saline solution at room temperature.

2. Sediment erythrocytes by gravity for 18 min at room temperature.

3. Collect leukocytes by centrifugation at 500g for 10 min at 4° and suspend in 40 ml of ice-cold saline.

4. Make a 9.97% (w/v) Hypaque (Winthrop, New York, NY), 6.35% (w/v) Ficoll (Pharmacia Fine Chemicals) gradient with a density of 1.08. Layer 10 ml of the Ficoll–Hypaque mix underneath the suspended leukocytes.

5. Centrifuge the cell suspension on the Ficoll–Hypaque mix at 250g for 40 min at 20°.

---

[16] L. E. Via, R. A. Fratti, M. McFalone, E. Pagan-Ramos, D. Deretic, and V. Deretic, *J. Cell Sci.* **111**, 897 (1998).

6. Suspend the neutrophil pellets in 10 ml of 0.2% (w/v) NaCl and incubate for 20 sec to lyse any remaining erythrocytes.

7. After lysis, immediately add 10 ml of 1.5% (w/v) NaCl and pellet neutrophils by centrifugation at 500g for 6 min at 4°.

8. Suspend the neutrophil pellet in 10 ml of HBSS and incubate at an MOI of 10 with midexponential phase *P. aeruginosa* in the presence of 10% (v/v) heat-inactivated autologous human serum in a total volume of 10 ml.

9. For each time point, remove a 1-ml sample and centrifuge at 900g, followed by determination of the bacterial viability by plating and counting colony-forming units.

Interaction of *Pseudomonas aeruginosa* with Cystic Fibrosis Epithelial Cells and Initial Stages in Biofilm Formation

Sialylation Levels of Glycoconjugates on Cystic Fibrosis Cells

Host cells contribute to *P. aeruginosa* biofilm formation in the CF lung. The state of sialylation of host epithelial cell plasma membrane proteins contributes to the adhesion of *P. aeruginosa*.[2] Changes in sialylation of these membrane proteins can be detected by the use of lectins conjugated to fluorescent fluorophores. Correlations between CFTR expression, sialylation of glycoconjugates, and *P. aeruginosa* adhesion/biofilm formation are examined using the following methods.

Procedure

1. The cell lines used are IB3-1, a human bronchial cell line derived from a CF patient with a ΔF508/W1282X *CFTR* mutant genotype,[17] and C-38 and S-9, which are stably transfected derivatives of IB3-1 cells corrected for CFTR function by introduction of a functional CFTR.[18]

2. Maintain IB3-1, C-38, and S-9 cells in LHC-8 medium (Biosource International, Rockville, MD), 10% (v/v) FBS, and penicillin (50 U/ml) and streptomycin (50 μg/ml) (GIBCO-BRL/Life Technologies) in a humidified CO_2 incubator at 37°, 5% CO_2.

3. Split the cells and seed at 10^5 cells per well on 18-mm sterile No. 1 coverslips placed in 12-well plates (Corning, Corning, NY). Grow the cells for 48 hr to 80–90% confluence in a CO_2 incubator.

4. Wash the cells carefully three times with PBS at pH 7.4.

[17] P. L. Zeitlin, L. Lu, J. Rhim, G. Cutting, G. Stetten, K. A. Kieffer, R. Craig, and W. B. Guggino, *Am. J. Respir. Cell Mol. Biol.* **4,** 313 (1991).

[18] M. Egan, T. Flotte, S. Afione, R. Solow, P. L. Zeitlin, B. J. Carter, and W. B. Guggino, *Nature (London)* **358,** 581 (1992).

5. Fix the cells for 2 hr at room temperature with freshly made 4% (w/v) paraformaldehyde in PBS containing 5% (w/v) sucrose.

6. Wash the cells three times with PBS followed by incubation for 2 hr in a CO_2 incubator with either PBS or neuraminidase type X (0.4 U/ml; Sigma) in 0.1 M acetate buffer, pH 5.0.

7. Wash the cells three times with PBS and incubate for 30 min in a CO_2 incubator with a 0.1-mg/ml concentration of tetramethyl rhodamine isothiocyanate (TRITC)-conjugated peanut agglutinin (PNA; Sigma) in PBS.

8. Wash the cells carefully five times with PBS to remove all unbound conjugated lectin.

9. Remove the coverslips, dry, and mount the cells facing down with PermaFluor (Shandon, Pittsburgh, PA) onto microscopy slides. Leave the mounted cells overnight in the dark before visualization.

10. IB3-1, C-38, and S-9 cells treated with PNA are visualized with an Olympus (Melville, NY) IX-60 microscope at ×10, an Olympix KAF1400, CCD camera (LSR; Olympus), and a 545/25 excitation filter and a dichroic mirror/emitter cube set 5100 > sbx 784-3 filter (Chroma Technology, Brattleboro, VT) or, alternatively, a more common 570/20 excitation filter and a dichroic mirror/emitter cube set 8300 (Chroma Technology). The setup is controlled with a Dell (Dallas, TX) Optiplex GX1p computer running Esprit software version 1.20 (LSR; Olympus).

Note. This method can be adapted to screen for the binding of other lectins to either sialylated or unsialylated glycoproteins or glycolipids, using different CF or mutant CFTR-transfected cell lines. However, it is of importance to make sure that the lectin employed binds specifically to the targeted oligosaccharide moiety. It is also important to ensure that a correct concentration of conjugated lectin is used to avoid nonspecific binding.

Association of Pseudomonas aeruginosa with Host Cells

For adhesion and uptake studies of *P. aeruginosa* by bronchial epithelial cells, use the same protocol for maintenance, cell splitting, and seeding as described above. Sixteen hours before the experiment, remove antibiotics by incubating all cells in LHC-8 medium (Biosource international), 10% (v/v) FBS.

1. Grow *P. aeruginosa* in an appropriate medium promoting mucoid or nonmucoid phenotype to a late exponential phase. Centrifuge the bacterial culture for 15 min at 13,700g at 4°.

2. Suspend the bacteria in 1% (w/v) proteose peptone in PBS to a final concentration of 10^8 bacteria/ml.

3. Wash IB3-1, C-38, and S-9 cells carefully and incubate monolayers with 10^6 to 10^7 bacteria/ml for 1 hr at 37°, 5% CO_2.

4. Wash the cells three times carefully with PBS to remove nonadherent bacteria, followed by lysis of cells with 1 ml of cold 0.1% (v/v) Triton X-100 in sterile water.

5. Make serial dilutions of the cell lysates and plate 100 μl onto appropriate media. Incubate overnight at 37° and count colony-forming units the following day.

[8] Subtractive Hybridization-Based Identification of Genes Uniquely Expressed or Hyperexpressed during Biofilm Growth

By Michael D. Parkins, Moritz Altebaeumer, Howard Ceri, and Douglas G. Storey

Introduction

It has been demonstrated that biofilms display several phenotypes fundamentally different from those of planktonically grown bacteria, including decreased susceptibility to antimicrobial agents and host immune response. It is our belief that the biofilm represents a unique mode of growth distinct from planktonic bacteria. Furthermore, it is the difference in the type and level of gene expression during biofilm growth that imparts the unique phenotypic characteristics associated with biofilms. We have developed a subtractive hybridization procedure to identify genes uniquely expressed or hyperexpressed during the biofilm mode of growth.

Subtractive hybridization library construction is one of many molecular approaches that can identify genes specifically induced under defined growth conditions. In its simplest form subtractive hybridization involves hybridizing nucleic acid from an induced population of cells (termed the tester population) with excess nucleic acid from an uninduced population (termed the driver population), with the goal of eliminating transcripts common to both populations and thereby increasing the concentration of sequences unique to the tester population.

Subtractive hybridization has a number of inherent features that distinguish it from traditional genetic approaches used to identify genes induced under specific growth conditions. Traditional genetic approaches such as transposon mutagenesis require that a vast number of transposon mutants be screened and are limited by the frequency of transposon insertion. Furthermore, transposon mutagenesis would be effective only for identification of factors absolutely required for biofilm formation. Subtractive hybridization, however, has the potential for identifying all factors that are uniquely expressed or hyperexpressed by cells during the biofilm mode of growth in a single assay.

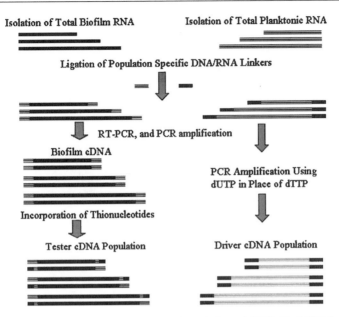

FIG. 1. Generation of biochemically differentiated populations of cDNA. Total RNA is extracted from biofilm and planktonic populations grown in the CBD, and ligated to population-specific DNA or DNA/RNA oligonucleotides, using T4 RNA ligase. Each population is then reverse transcribed, using population-specific primers, to generate full-length cDNA to all transcripts expressed. Planktonic cDNA (driver population) is PCR amplified, using dUTP in place of dTTP, resulting in susceptibility to UNG digestion. Biofilm cDNA (tester population) is PCR amplified, and then treated with Klenow enzyme to incorporate thionucleotides into tester cDNA, resulting in exonuclease digestion resistance.

Traditional subtractive hybridization procedures often require a large amount of starting material, and are inefficient for the identification of low-abundance transcripts because of prolonged hybridization and inefficient separation steps. As such, we have incorporated several previously identified steps shown to increase the efficacy of subtractive hybridization, as well as developed an initial T4 RNA ligation step of our own design. Our subtractive hybridization protocol is dependent on differential polymerase chain reaction (PCR) amplification of biofilm (tester) and planktonic (driver) populations of cDNA, resulting in population-specific biochemical properties (Fig. 1). This enables selection against transcripts commonly expressed in both populations of cells, and allows for selection of transcripts unique to the tester population (Fig. 2).

Biofilm and Planktonic Growth

The Calgary biofilm device (CBD) (MBEC Biofilm Technologies, Calgary, AB, Canada) allows for the simultaneous formation of biofilm and planktonic

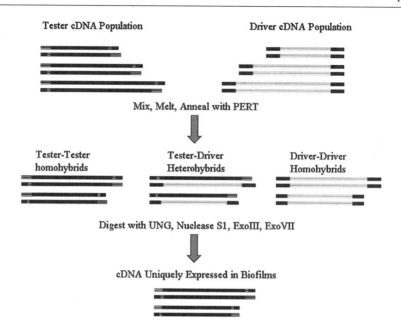

FIG. 2. Subtractive hybridization and differential amplification. Tester and driver total cDNA are mixed at a 1 : 10 ratio, melted to separate duplex cDNA, and then reassociated using PERT. Tester–driver heterohybrid and driver–driver homohybrid populations of cDNA are eliminated by enzymatic digestion with UNG, nuclease S1, and exonucleases III and VII. Three rounds of subtraction are performed. Remaining tester–tester homohybrid duplex cDNA molecules are cDNA to genes uniquely expressed or hyperexpressed in biofilms. These are PCR amplified and ligated into cloning vectors for further characterization.

populations from a common initial inoculum and for continued growth under the same conditions. [1] Sampling of biofilm populations can be achieved by dislodging a peg from the 96-peg lid, whereas planktonic populations can be sampled by removing a sample from the trough. The number of pegs or volume of planktonic culture required for RNA extraction is adjusted to a final cell equivalent of 10^8–10^9. Biofilm samples taken at early time points will contain cells expressing genes involved in initial adhesion and biofilm development, whereas late time point samples will contain cells expressing genes involved in mature biofilm maintenance.

RNA Extraction and Purification

Total cellular RNA is extracted from biofilm and planktonic cells by conventional techniques and separately stored at −80° in diethylpyrocarbonate (DEPC)-treated water.[2,3]

[1] H. Ceri, M. E. Olson, C. Stremick, R. R. Read, D. Morck, and A. Buret, *J. Clin. Microbiol.* **37,** 1771 (1999).

Ligation of Population-Specific Linkers to RNA

Previous subtractive hybridization procedures have utilized reverse transcription (RT)-PCR performed with random primers to create short segments of cDNA and inevitably result in the loss of significant amounts of nucleic acid sequences. To overcome this, we have devised a scheme in which we utilize T4 RNA ligase to ligate population-specific single-stranded DNA and DNA/RNA oligonucleotides to the 3' and 5' ends, respectively, of total cellular RNA.[4–6] As such, subsequent RT-PCR will generate cDNA to full-length transcripts. The ligation reaction is performed as follows: add 10 μl of 10× T4 RNA ligase buffer [50 mM Tris-HCl, 10 mM dithiothreitol (DTT), bovine serum albumin (BSA, 30 μg/ml), 10 mM MgCl$_2$, 1 mM ATP], 1 μl of 4% (w/v) bentonite, 1 μl of hexamine CoCl$_3$ (1 mM), 10 μl of dimethyl sulfoxide (DMSO), 1 μl of RNasin® (40 units/μl; Promega, Madison, WI), and 62.5 μl of 40% (w/v) polyethylene glycol (PEG) to 5 μg of total cellular RNA (tester or driver). Add 3 μg of 5' linker to the reaction (tester = GTG TTG TGT GTT TGG TGG GTG TTG GCT CTC TTA AGG TAG CGC GUC; driver = GGT GTG TGG TTT GTG GTT GGT TTG GCUG) (underlined nucleotides are ribonucleotides) and 3 μg of 3' linker (tester = GCC TCT CTT AAG GTA GCC CAA CAC CCA CCA AAC ACA CAA CAC; driver = CAG CCA AAC CAA CCA CAA ACC ACA CAC C). The total reaction volume should be brought to 100 μl with DEPC-treated distilled deionized water. The reaction is catalyzed with 30 units of T4 RNA ligase (Boehringer Mannheim, Indianapolis, IN) and incubated for 10–14 hr at 17°. Excess unligated linkers are removed from the DNA-ligated RNA molecules, using Sephacryl 400 microspin columns (Pharmacia Biotech, Piscataway, NJ) according to the manufacturer instructions. Precipitate RNA with ethanol overnight at −20°. Pellet the RNA, and wash once with 70% (v/v) ethanol. Suspend the pellet in 10 μl of DEPC-treated distilled deionized water.

Generation of cDNA to Total Cellular RNA

Add 5 μl of linker-ligated RNA to 5 μl of PCR primer (10 pmol/μl) (tester = GTG TTG TGT GTT TGG TGG GTG TTG G; driver = GGT GTG TGG TTT GTG GTT GGT TTG GCU G) and 1 μl of DMSO. Incubate at 70° for 10 min to allow for relaxation of RNA molecules. Cool to room temperature to allow for primer annealing and add 1 μl of RNasin (Promega), 1 μl of dNTPs (10 mM), 4 μl of first-strand synthesis buffer (GIBCO-BRL, Gaithersburg, MD), and 2 μl of 0.1 M DTT. Incubate at 48° for 2 min to allow for temperature equilibration. Add 1 μl of Superscript II reverse transcriptase (GIBCO-BRL) and incubate for

---

[2] D. W. Frank, D. G. Storey, M. S. Hindahl, and B. H. Iglewski, *J. Bacteriol.* **171,** 5304 (1989).

[3] D. G. Storey, E. E. Ujack, I. Mitchell, and H. R. Rabin, *Infect. Immun.* **65,** 4061 (1997).

[4] M. I. McCoy and R. I. Gumport, *Biochemistry* **19,** 635 (1980).

[5] D. C. Tessier, R. Brousseau, and T. Vernet, *Anal. Biochem.* **158,** 171 (1986).

[6] L. G. Tyulkina and S. Mankin, *Anal. Biochem.* **138,** 285 (1984).

1 hr to allow for elongation. Denature the reverse transcriptase by heat treatment at 70° for 15 min.

Polymerase Chain Reaction Amplification

Each population can be independently PCR amplified using population-specific primers to increase the abundance of low-copy transcripts, thereby reducing the possibility of false negatives.

Tester cDNA Polymerase Chain Reaction Amplification

Tester cDNA is amplified under standard PCR conditions. Briefly, 5 μl of template from the reverse transcription reaction is added to 10 μl of 5× WB buffer [335 mM Tris-HCl (pH 8.8), 80 mM ammonium sulfate, 20 mM MgCl$_2$, 50 mM 2-mercaptoethanol, BSA (5 mg/ml)], 2 μl of tester population-specific primer (10 pmol/μl) (GTG TTG TGT GTT TGG TGG GTG TTG G), 1 μl of dNTPs (2.5 mM dATP, 2.5 mM dCTP, 2.5 mM dGTP, 2.5 mM dTTP), 2.5 μl of DMSO, 29 μl of DEPC-treated distilled deionized water, and 0.5 μl of Taq polymerase (GIBCO-BRL). PCR amplification of tester cDNA is performed as follows: a 5-min denaturation step at 94° followed by 30 cycles of 94° for 30 sec, 55.5° for 45 sec, and 72° for 4 min. Finally, a 15-min elongation step at 72° is performed. Precipitate the cDNA overnight at −20° with ethanol. Pellet the cDNA and suspend in distilled deionized water. Measure tester cDNA concentration by UV absorbance at 260 nm.

Driver cDNA Polymerase Chain Reaction Amplification

Driver cDNA is PCR amplified under conditions that allow for differentiation from tester cDNA. Uridine nucleotides are incorporated in place of thymidine, thereby creating driver cDNA that is susceptible to uracil-DNA glycosylase digestion.[7] Briefly, 5 μl of template from the reverse transcription reaction is added to 10 μl of 5× WB buffer, 2 μl of driver population-specific primer (GGT GTG TGG TTT GTG GTT GGT TTG GCU G, 10 pmol/μl), 1 μl of dNTPs (2.5 mM dATP, 2.5 mM dCTP, 2.5 mM dGTP, 2.5 mM dUTP), 2.5 μl of DMSO, 29 μl of DEPC-treated distilled deionized water, and 0.5 μl of Taq polymerase. PCR amplification is performed as follows: a 5-min initial denaturation step at 94°, followed by 30 cycles of 94° for 30 sec, 55.5° for 45 sec for primer annealing and 72° for 4 min to allow for primer extension. Finally, a 15-min elongation step at 72° is performed. Precipitate the cDNA with ethanol overnight at −20°. Pellet the cDNA and suspend in prehybridization buffer [40 mM Tris-HCl (pH 8.8), 4 mM EDTA]. Measure driver cDNA concentration by UV absorbance.

---

[7] M. Sugai, S. Kondo, A. Shimizu, and T. Honjo, *Nucleic Acids Res.* **26,** 911 (1998).

Incorporation of Thionucleotides into Tester cDNA

Thiodiester bonds formed by the incorporation of thionucleotides into the 3′ termini of tester cDNA will impart exonuclease digestion resistance. Resistance to exonuclease digestion can then be used to further distinguish between tester and driver cDNA.[7,8] Thionucleotides are incorporated into tester cDNA as follows: treat 5 μg of tester cDNA with 50 units of Klenow (Boehringer Mannheim) in 10 μl of 10× Klenow buffer in a total volume of 100 μl at 37° for 45 min. This reaction depends on the 3′ → 5′ exonuclease activity of Klenow to create recessed 3′ ends. Next, add 10 μl of dNTP-α-S (1.6 mM; Amersham, Arlington Heights, IL) and incubate for 30 min at 37°. The Klenow fragment will incorporate thionucleotides into the 3′ recessed termini of tester cDNA, using 5′ → 3′ polymerase activity. To terminate the reaction and purify resultant thionucleotide-incorporated cDNA, extract once with an equal volume of phenol–chloroform–isoamyl alcohol (25 : 24 : 1, v/v/v), and once with an equal volume of chloroform–isoamyl alcohol (24 : 1, v/v). Ethanol precipitate thionucleotide-incorporated tester cDNA overnight at −20°. Pellet tester cDNA and suspend the resultant pellet in prehybridization buffer. Measure tester cDNA concentration by UV absorbance.

Strand Separation and Reassociation

Duplex cDNA from both populations is separated and then allowed to re-anneal, using a phenol emulsion reassociation technique (PERT) that optimizes strand reassociation.[9] This step ensures that single-stranded cDNA found in both populations can reanneal, creating tester–driver heterohybrids, driver–driver homohybrids, and tester–tester homohybrids. Mix 0.5 μg of thionucleotide-incorporated tester cDNA with a 10-fold excess of driver cDNA in 126 μl of prehybridization buffer. Denature duplex cDNA by heating to 95° for 5 min. To the mixture add 334 μl of 3 M NaSCN and 40 μl of phenol to a final concentration of 2 M NaSCN and 8% (v/v) phenol in a final volume of 500 μl. Continuous agitation is required for optimal strand reassociation. To this end the use of a Fisher (Pittsburgh, PA) Vortex Genie 2 Turbo Mixer operating at half-maximum speed is recommended. Allow 24 hr for reassociation of single-stranded cDNA. Extract once with an equal volume of chloroform–isoamyl alcohol (24 : 1, v/v). Ethanol precipitate the duplex cDNA overnight at −20°. Pellet cDNA and suspend the resultant pellet in distilled deionized water.

Three separate populations of cDNA should be present: tester–driver heterohybrids, driver–driver homohybrids, and tester–tester homohybrids. Tester–driver heterohybrids are duplex cDNA in which one strand is derived from the driver

[8] J. Zeng, R. A. Gorski, and D. Hamer, *Nucleic Acids Res.* **22**, 4381 (1994).
[9] E. Kohne, S. A. Levison, and M. Byers, *Biochemistry* **16**, 5329 (1977).

population and one strand from the tester population. Tester–driver heterohybrids are cDNA to genes commonly expressed in both biofilm and planktonic cells. Tester and driver homohybrids are duplex cDNA molecules in which both strands have been derived from the same population. Driver–driver homohybrids should be the most abundant of the three populations owing to the excess of driver cDNA used in the hybridization. After several rounds of subtraction the tester–tester homohybrid population should contain cDNA corresponding to transcripts unique to or hyperexpressed within a biofilm.

Uracil-DNA Glycosylase and Nuclease S1 Digestion

Treatment with uracil-DNA glycosylase (UNG) will result in the removal of uracil residues from driver cDNA. Dissolve the 5.5-μg mix of tester–driver heterohybrid, driver–driver homohybrid, and tester–tester homohybrid cDNA in 20 μl of uracil-DNA glycosylase buffer [60 mM Tris-HCl (pH 8.8), 1 mM EDTA, 1 mM DTT, and BSA (0.1 mg/ml)]. Digest with 8 units of uracil-DNA glycosylase (Boehringer Mannheim) for 30 min at 37° to remove uracil residues from driver cDNA. A subsequent digestion with nuclease S1 will remove all resultant single strands of tester cDNA formed after UNG digestion. Add 20 μl of 10× nuclease S1 reaction buffer [330 mM sodium acetate (pH 4.5), 500 mM NaCl, 0.3 mM ZnCl$_2$], and bring the reaction volume to 200 μl with distilled deionized water. Digest with 60 units of nuclease S1 for 30 min at 37°. To terminate the reaction extract once with an equal volume of phenol–chloroform–isoamyl alcohol (25 : 24 : 1, v/v/v), and once with an equal volume of chloroform–isoamyl alcohol (24 : 1, v/v). Ethanol precipitate the samples overnight at −20°. Pellet the cDNA and dissolve the uracil-DNA glycosylase/nuclease S1-digested cDNA in 10 μl of prehybridization buffer. After this procedure the only intact DNA molecules remaining will be tester–tester homohybrid duplex cDNA.

Subsequent Rounds of Subtraction

After only one round of subtracting tester cDNA from driver cDNA (using a 10-fold excess of driver cDNA) it is still possible that complementary tester cDNA strands will anneal to each other despite the same strands being present in the driver population. This could result in a high proportion of false positives. As such, two additional rounds of subtractive hybridization are performed by adding new driver cDNA to previously subtracted tester cDNA. With each round of subtraction the likelihood of identifying false positives decreases. Add 5 μg of the original uracil nucleotide-incorporated driver cDNA to the already subtracted tester cDNA in a total volume of 24.4 μl of prehybridization buffer. As before, heat to 95° to dissociate duplex cDNA. Add 68 μl of 3 M NaSCN and 8 μl of phenol and perform the phenol emulsion reassociation technique as previously described to promote new duplex formation. As previously, digest the resultant tester–tester homohybrid, driver–driver homohybrid, and tester–driver heterohybrid cDNA populations with uracil-DNA glycosylase and nuclease S1. Perform a total of three rounds of subtraction, each time adding 5 μg of the original uracil-incorporated driver cDNA to the subtracted tester cDNA. With each round of subtraction the proportion of tester–driver heterohybrid cDNA molecules will decrease while the driver–driver homohybrid population will increase. This occurs because tester cDNA to genes expressed also during planktonic growth is progressively eliminated. After three rounds of subtraction have been completed, ethanol precipitate the remaining cDNA overnight at −20°. Pellet the remaining cDNA and dissolve the three round-subtracted tester cDNA pellet in 10 μl of distilled deionized water.

Exonuclease Digestion

To eliminate contaminating driver cDNA molecules, two exonuclease digestions are performed. Tester cDNA 3′ termini are protected from exonuclease activity by thiodiester bonds, whereas driver cDNA molecules remain susceptible. In a total volume of 200 μl of exonuclease III reaction buffer [66 mM Tris-HCl (pH 7.6), 0.66 mM MgCl$_2$, 1 mM 2-mercaptoethanol], digest the subtracted tester cDNA with 300 units of exonuclease III (Boehringer Mannheim) at 37° for 10 min. Exonuclease III rapidly degrades all unprotected cDNA molecules by cleaving phosphodiester bonds through 3′ → 5′ exonuclease activity. After 10 min, add 3 μl of 0.5 mM EDTA and 20 μl of 0.5 M potassium phosphate buffer, pH 7.6, to inhibit the activity of exonuclease III. Add 20 units of exonuclease VII (Amersham-Life Science) and digest for 30 min at 37°. Exonuclease VII specifically degrades residual single-stranded cDNA molecules released after exonuclease III digestion. To terminate the reaction and isolate resultant tester–tester homohybrid cDNA, extract once with an equal volume of phenol–chloroform–isoamyl alcohol (25 : 24 : 1, v/v/v) and once with an equal volume of chloroform–isoamyl alcohol (24 : 1, v/v). Ethanol precipitate the sample overnight at −20°. Pellet tester–tester homohybrid cDNA and dissolve the pellet in 5–20 μl of distilled deionized water.

Cloning Subtracted Tester cDNA

After several rounds of subtractive hybridization the number of duplex tester cDNA molecules will be greatly diminished. These remaining tester–tester cDNA homohybrid molecules represent genes uniquely expressed or hyperexpressed during biofilm growth. To increase the copy number and therefore the likelihood of cloning these fragments, one final round of PCR amplification must be performed. Five microliters of subtracted cDNA is added to 10 μl of 5× WB buffer, 2 μl of tester population-specific primer (10 pmol/μl) (GTG TTG TGT GTT TGG TGG GTG TTG G), 1 μl of dNTPs (2.5 mM dATP, 2.5 mM dCTP, 2.5 mM dGTP, 2.5 mM dTTP), 2.5 μl of DMSO, 29 μl of DEPC-treated distilled deionized water and 0.5 μl of *Taq* polymerase (GIBCO-BRL). PCR amplification is performed as follows: a 5-min denaturation step at 94°, followed by 30 cycles of 94° for 30 sec, 55.5° for 45 sec, and 72° for 4 min. Finally, a 15-min elongation step at 72° is required to ensure complete elongation.

Run 5–10 μl of the PCR product on an agarose gel to verify that amplification has occurred. After successful amplification, cloning of the products is necessary for further characterization. Ligating the subtracted tester cDNA into a general cloning vector will allow for sequencing as the first step in characterization. As *Taq* polymerase adds a 3' A tail to all PCR products, cloning PCR products is most efficient when using T-tailed cloning vector kits.

Preliminary results indicate that this technique can be successfully used to identify genes uniquely expressed or hyperexpressed during the biofilm mode of growth. This technique has identified several genes of known function whose expression is upregulated during biofilm growth. Furthermore, this technique has implicated a role for a number of genes of unknown function whose expression is limited to biofilm growth. Because of its ability to identify genes hyperexpressed during biofilm growth, subtractive hybridization may represent a powerful tool with which to dissect the genetics of biofilm formation.

[9] Biofilm-Induced Gene Expression and Gene Transfer

By RICHARD J. LAMONT and JAMES D. BRYERS

Introduction

The transition of bacterial cells from a planktonic state to a biofilm state, involving adhesion to a substrate and to other cells along with production of extracellular polysaccharides, represents a major physiological change for bacteria and is the process that drives bacterial colonization of inert and biological surfaces. Evidence is accumulating that bacteria in biofilms are not simply planktonic cells adherent to a surface, but rather exhibit considerable biofilm-induced phenotypic variation dependent on highly orchestrated changes in gene expression.[1] Furthermore, the phenotypic properties of biofilm-mode bacteria can depend on plasmid stability and transfer with the biofilm. The close proximity of cells within a biofilm, combined with the presence of matrix material that can concentrate compounds (such as plasmid and chromosomal DNA, pheromones, and bacteriophage), is conducive to the exchange of mobile genetic elements by processes such as conjugation,

---

[1] J. W. Costerton, P. S. Stewart, and E. P. Greenberg, *Science* **284,** 1318 (1999).

transposition, and transduction. Molecular definition of biofilm dependent changes in gene expression and plasmid transfer is thus necessary to understand the unique properties of biofilms.

Reporter Gene Systems

Fusions of promoter-containing regions to reporter genes can be used to monitor expression of a specific gene, or as a promoter trap to screen for promoters active under biofilm conditions. A number of reporter systems are available and those with fluorescent or chemiluminescent substrates (such as *lacZ*, *lux*, or *luc*), or that express a fluorescent protein directly (such as *gfp*), are well described in a previous volume of this series.[2,3] Another useful system, particularly for the oral streptococci, is the *cat* reporter, which is based on acetylation and inactivation of chloramphenicol. The following methodology has been used successfully in our laboratory and is based on protocols developed by H. Jenkinson (University of Bristol, Bristol, UK) and R. McNab (Eastman Dental Institute, London, UK).

To generate a promoter–*cat* fusion for a specific gene, the upstream region containing the consensus RNA polymerase recognition sites, but excluding the ribosome-binding site, is amplified by polymerase chain reaction (PCR). The extent of the amplification product can be increased to encompass potential *cis*-acting elements as necessary. To facilitate subsequent cloning, the upstream primer is designed with a *Sma*I restriction enzyme site and the downstream primer with a *Sac*I site. The PCR product is digested with a mixture of *Sma*I and *Sac*I and ligated into the pMH109 plasmid, similarly digested. pMH109 contains a promoterless gram-positive *cat* gene along with its ribosome-binding site and a transcriptional terminator upstream of the cloning site to reduce transcriptional readthrough.[4] The ligation mixture is transformed into *Escherichia coli* HB101 with selection for tetracycline (10 μg/ml) and chloramphenicol (5 μg/ml). If no colonies are obtained, the promoter may be weak and lower concentrations of chloramphenicol should be tried. As pMH109 can replicate in some streptococci, it is necessary to remove the construct and clone into a nonreplicating plasmid such as pSF143.[5] Recombinant pMH109 is cut with *Bam*HI, and the required fragment is isolated and ligated into *Bam*HI-cut pSF143 plasmid. Transform first into HB101 with selection for tetracycline and chloramphenicol and then into the relevant streptococcus with selection for tetracycline (10 μg/ml). The construct will integrate into the chromosome with a single (Campbell) cross-over, which can be confirmed by Southern hybridization and PCR sequencing.

[2] D. R. Korber, G. M. Wolfaardt, V. Brozel, R. MacDonald, and T. Niepel, *Methods Enzymol.* **310**, 3 (1999).

[3] C. Prigent-Combaret and P. Lejeune, *Methods Enzymol.* **310**, 56 (1999).

[4] M. C. Hudson and G. C. Stewart, *Gene* **48**, 93 (1986).

[5] L. Tao, D. J. LeBlanc, and J. J. Ferretti, *Gene* **120**, 105 (1992).

After construction of the recombinant strain, promoter activity can be determined by a chloramphenicol acetyltransferase (CAT) assay.[6] Streptococcal cells grown in standard medium (10 ml) are harvested by centrifugation and washed once in TPE buffer [100 mM Tris-HCl (pH 7.8), 1 mM phenylmethylsulfonyl fluoride (PMSF), 1 mM EDTA]. Cells are suspended in 0.1 ml of spheroplast buffer [20 mM Tris-HCl (pH 6.8), 10 mM MgCl$_2$, 1 mM PMSF, 26% (w/v) raffinose, mutanolysin (500 U/ml)] and the suspension is incubated at 37° for 30 min. To disrupt cells, 0.1 ml of glass beads (0.10–0.11 mm in diameter) and 0.4 ml of ice-cold TPE buffer are added and the cells vortex mixed twice for 30 sec. Suspensions are then centrifuged (12,000g, 20 min, 4°) to pellet beads and cell debris, and supernatants are removed for protein assay and CAT enzyme assay. CAT enzyme activity is determined spectrophotometrically, utilizing a recording spectrophotometer with temperature-controlled cuvette chamber [such as a Beckman (Fullerton, CA) DU-70]. A 0.05-ml volume of streptococcal cell extract (containing 40–60 μg of protein) is added to a 0.5-ml reaction mixture [100 mM Tris-HCl (pH 7.8), 0.1 mM acetyl-coenzyme A, 1 mM 5,5′-dithiobis-2-nitrobenzoic acid] in 1-ml glass cuvettes. The reaction mixture is warmed at 37° for 2 min, and the background change in absorbance at 412 nm is recorded for a further 2 min. Chloramphenicol (16 μl, 0.1 mM) is added and the change in absorbance is recorded for 2 min. The reaction rate is determined from the linear portion of the graph, corrected for background change in A_{412}, and divided by 0.0136 to yield CAT activity expressed as nanomoles of chloramphenicol acetylated per minute at 37°.

CAT activity thus reflects promoter activity and the level of gene expression. It should be remembered, however, that gene transcriptional activity is only one factor in determining the level and activity of a particular protein. Wherever possible it is prudent to confirm results at the mRNA level and protein level.[7]

Analysis of mRNA: Transcriptomics

It is becoming increasingly apparent that the characterization of cellular events, such as occur in biofilm-mode bacteria, requires analysis not only of individual gene expression but also of the interrelationships among expressed genes. For example, a particular protein that is expressed under certain environmental conditions may function differently when in the presence or absence of a repressor protein. Thus, it is the pattern of expression of a series of genes that is more informative. A variety of powerful tools are now available, often in kit form, to analyze steady state mRNA levels. Conventional DNA arrays are useful where the genome sequence is known, so in future an increasing number of bacteria can be studied in this manner. In the absence of a complete genome sequence, there are a number of

---

[6] W. V. Shaw, *Methods Enzymol.* **43**, 737 (1975).

[7] C.-T. Huang, S. W. Peretti, and J. D. Bryers, *Biotechnol. Lett.* **16**, 903 (1994).

high-throughput systems for transcriptional analysis with theoretical total genome coverage, one of which is arbitrary primed differential display reverse transcription DD-RTPCR.[8]

The technique of differential display of mRNA is based on reverse transcription of RNA isolated from biofilm or planktonic bacterial cells, and overcomes the requirement (in alternative techniques such as subtractive hybridization) for high levels of transcript. A number of commercial kits are available for isolation of RNA, including TRIzol reagent (GIBCO-BRL, Gaithersburg, MD), and ToTALLY (Ambion, Austin, TX), which is then quantitated by spectrophotometric measurement at OD_{260} and used immediately when possible. Arbitrary primers are random hexamers or oligonucleotides up to about 20-mer. Homology to high-abundance structural RNA should be avoided to reduce the number of false positives. If the genome sequence is known, primer sequences should be chosen from the most frequent oligonucleotides of the coding genome. RNA (5 μg in 10 μl) is DNase I treated at 25° for 15 min. DNase is inactivated with 1 μl of 25 mM EDTA at 65° for 10 min. First-strand cDNA is synthesized by annealing 50 ng of primers in the presence of 10 mM dithiothreitol (DTT, 1 μl), 10 mM dNTP mix (1 μl), 40 U of RNase inhibitor, and the recommended amount of reverse transcriptase [such as SuperScript or Moloney murine leukemia virus (Mo-MuLV)]. The synthesized cDNA (2 μl) is amplified by PCR with a random primer (50 pmol), 200 μM dNTP mix, and *Taq* DNA polymerase. The annealing and cycling conditions along with the $MgCl_2$ concentration (2.5–4 mM) should be determined empirically. PCR products can be analyzed on agarose gels, with bands of interest excised and eluted and cloned into a TA cloning vector (InVitrogen, San Diego, CA).

The advantages of differential display are that it allows for mass screening of bacterial transcripts, it can detect transcripts present in low copy number, and it is not limited to genes that are exclusively expressed or repressed under the test conditions but will detect genes that are expressed at different levels. The disadvantages are that false positives are common, because of the relatively large proportion of structural RNA species in the total RNA (although these are not usually differentially expressed) and because bacterial RNA often has structural instability and a short half-life.

Analysis of Proteins: Proteomics

It is the expressed proteins that are ultimately responsible for most of the processes that are important for bacterial function. Unfortunately, the relationship between mRNA and protein levels is not simple. Degradation rates of mRNAs and

[8] R. Fislage, *Electrophoresis* **19**, 613 (1998).

proteins differ and, even in prokaryotes, many proteins are modified posttranslationally. Direct measurement of proteins will thus provide the most accurate representation of biofilm phenotype.

High-resolution two-dimensional gel electrophoresis with large-sized gels is now considered the method of choice for the separation and analysis of complex protein mixtures, such as those present in bacterial cells. Separation of proteins occurs on the basis of both the isoelectric points and molecular sizes of the proteins. The following technique is based on established protocols as developed by W. Chen in the laboratory of M. Hackett (University of Washington, Seattle, WA).

Bacterial cells are harvested by centrifugation and lysed with a 100° solution of 0.3% (w/v) sodium dodecyl sulfate (SDS), 1% (v/v) 2-mercaptoethanol, in 50 mM Tris-HCl, pH 8.0. The cells are placed quickly in boiling water for 1–1.5 min. The sample is frozen in a liquid nitrogen bath, and freeze-dried for a minimum of 2 hr to dryness. After drying, the sample is redissolved to the original volume with a 1D buffer [7.95 M urea, 2 M thiourea, 4.0% (w/v) 3-[(3-cholamidopropyl)-dimethyl-ammonio]-1-propanesulfonate (CHAPS), 2% (w/v) pH 6–8 ampholytes, 100 mM DTT]. The sample can then be stored at −80°. The first dimension of two-dimensional electrophoresis is performed with a Pharmacia Biotech (Piscataway, NJ) Multiphor II electrophoresis system. Nonlinear pH 3–10 Immobiline DryStrips (18 cm) are rehydrated for 4–6 hr at room temperature in a rehydration buffer [8 M urea, 10 mM DTT, 2% (w/v) CHAPS, 0.8% (w/v) carrier ampholyte pH 3–10, and bromophenol to color]. Bacterial cell lysate is applied during the rehydration step. The first dimension is run for 65.5 kV · hr (electrophoresis power supply EPS 3501 XL; Pharmacia Biotech), using the following program at 22°: 0–300 V, 2.5 V · hr; 300 V, 600 V · hr; 300–3500 V, 5.7 kV · hr; 3500 V, 59.2 kV · hr. The gels are soaked for 10 min in each of the following buffer solutions before transfer to the second dimension [solution A: 2% (w/v) DTT, 2% (w/v) SDS, 6 M urea, 30% (v/v) glycerol, 0.05 M Tris-HCl (pH 6.8); solution B: 2.5% (w/v) iodoacetamide, 2% (w/v) SDS, 6 M urea, 30% (v/v) glycerol, 0.05 M Tris-HCl (pH 6.8)]. The second dimension is run on a Bio-Rad (Hercules, CA) Protean II xi cell, with a 10.5% (w/v) polyacrylamide slab gel. The Immobiline DryStrips are laid directly on the top of the slab gel without a stacking gel. The gels are placed in the electrophoresis cell chamber with 10× running buffer [0.25 M Tris base, 1.92 M glycine, and 1% (w/v) SDS] in both chambers. Electrophoresis is performed at constant current and the cooling core is cooled to 10° with circulating water. Power is supplied in two steps, low current at the beginning (10 mA/gel for 20 min) and then 25 mA/gel. The run is stopped when the solvent front reaches the bottom edge of the gel.

After electrophoresis, gels are stained with silver. First, gels are fixed with 50% (v/v) ethanol–10% (v/v) acetic acid for 30 min or overnight and then incubated with 10% (v/v) ethanol for 15 min. After three changes of water, the gels are sensitized by incubating in sodium thiosulfate (0.2 mg ml^{-1}) followed by several washes with water. The gels are incubated in a silver nitrate (2 mg ml^{-1}) solution for 25 min followed by two washes with a large volume of water. The gels are finally developed in 0.02% (v/v) formaldehyde, sodium carbonate (60 mg ml^{-1}) to the desired level of staining. The reaction is terminated with an EDTA (14 mg ml^{-1}). solution.

One of the attractive features of two-dimensional electrophoresis is that profiled proteins, can subsequently be identified. Spots are excised from the gel and digested with trypsin. The proteolytic fragments are then analyzed by mass spectroscopy in, for example, a matrix-assisted laser desorption ionization time-of-flight (MALDI-TOF) mass spectrometer that ionizes peptides with a laser. Mass values are then compared with predicted values of proteins in the databases.

Plasmid Segregation Loss in Biofilms

The amount of biofilm present at any one instant is the net result of several simultaneous dynamic processes, some contributing to the biofilm and some removing biofilm from the substratum. Consequently, it is not possible simply to "grind up" a sample from biofilm periodically to measure the distribution of plasmid-bearing and plasmid-free phenotypes. Any study investigating plasmid dynamics (segregation, structural instability, plasmid transfer) must mathematically account for the processes of bacterial cell adsorption, desorption, bacterial growth, and biofilm detachment in order to quantify plasmid-specific processes. Here, we provide an example for plasmid segregational loss in a biofilm.

Two mathematical models describing plasmid loss in both freely suspended and biofilm cultures have been derived to calculate from experimental data the probabilities of plasmid segregational loss in biofilm cultures. It was presumed that biofilm formation was a higher imperative on an energetic and genetic basis for the culture in comparison with expression of a heterologous gene.

Batch Suspended Cell Plasmid Loss Model

Assume a sterile complete nutrient solution, with no selection pressure, is initiated with a small inoculum of 100% plasmid-bearing cells at time $t = 0$. As the batch suspended culture is allowed to proceed, plasmid-bearing cells can grow and replicate on essential nutrients according to simple Monod kinetics. Without a selection pressure, plasmid-bearing cells have a certain probability of losing the plasmid, by segregation instability. The loss of plasmid by segregation is assumed here to be proportional to the growth rate of the plasmid-bearing population. Thus, material balances for the change in plasmid-bearing cells, X^+ (cell no./L^3), and plasmid-free cells, X^-, in a batch suspended culture as a function of time can be written, respectively, as follows:

$$\text{Plasmid-bearing cells: } dX^+/dt = \mu^+ X^+ - \mathcal{P}\mu^+ X^+ \qquad (1)$$

$$\text{Plasmid-free cells: } \quad dX^-/dt = \mu^- X^- + \mathcal{P}\mu^+ X^+ \qquad (2)$$

By measuring experimentally the X^+ and X^- populations and having independent measures of their growth rates (μ^+, μ^-), Eqs. (1) and (2) can be solved[9,10] to provide estimates of the probability of segregational loss, $\mathcal{P}$, in suspended culture.

Biofilm Plasmid Loss Model

Consider that a substratum, within a well-mixed continuously fed reactor, is inoculated with a small population of 100% plasmid-bearing cells. The reactor is subsequently operated at a residence time that is less than the generation time of the bacterial culture, which eliminates suspended cell growth and replication. The system is fed a sterile nutrient solution promoting growth only of adherent bacteria. As the biofilm develops, plasmid-bearing cells lose their plasmid by segregation, thus giving rise to adherent populations of both plasmid-bearing and plasmid-free cells. Because suspended growth is experimentally minimized, the only source of the suspended cells living the reactor is through the detachment of bacteria from the developing biofilm. In this model, Huang *et al.*[9–11] assume that the ratio of plasmid-bearing to plasmid-free cells detaching from the biofilm is equal to the same ratio of cells in the entire biofilm (i.e., no population gradients in the biofilm; this may not be true for thick biofilms). Thus, material balances for the change in plasmid-bearing (X^+) and plasmid-free cells (X^-; cell no./L^3) in liquid phase, and plasmid-bearing cells (B^+; cell no./L^2) and plasmid-free cells (B^-) in the biofilm, are as follows:

$$\text{Biofilm plasmid bearing: } \quad dB^+/dt = \mu^+ B^+ - \mathcal{P}\mu^+ B^+ - k_{\text{det}} B^+ \qquad (3)$$

$$\text{Biofilm plasmid free: } \quad dB^-/dt = \mu^- B^- - \mathcal{P}\mu^+ B^+ - k_{\text{det}} B^- \qquad (4)$$

$$\text{Suspended plasmid bearing: } dX^+/dt = \mu^+ X^+ - \mathcal{P}\mu^+ X^+ + k_{\text{det}} B^+ A/V \qquad (5)$$

$$\text{Suspended plasmid free: } \quad dX^-/dt = \mu^- X^- + \mathcal{P}\mu^+ X^+ + k_{\text{det}} B^- A/V \qquad (6)$$

where k_{det} is the rate of detachment of each population from the biofilm directly into the fluid phase (t^{-1}), A is reactor surface area (L^2), and V is reactor volume

[9] C.-T. Huang, S. W. Peretti, and J. D. Bryers, *Biotechnol. Bioeng.* **41**, 211 (1993).
[10] C.-T. Huang, S. W. Peretti, and J. D. Bryers, *Biotechnol. Bioeng.* **44**, 329 (1994).
[11] C.-T. Huang, Ph.D. dissertation. Duke University, Durham, North Carolina, 1993.

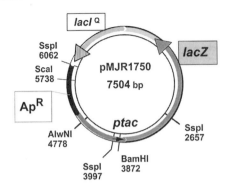

FIG. 1. Base map of pMJR1750 used for segregational loss studies.[9,10]

(L^3). Huang *et al.*[9,10] provide details of the solution technique to directly estimate $\mathcal{P}$ and k_{det} from experiments. Without correcting for the loss of both plasmid-bearing and plasmid-free cells from the biofilm by detachment process, an erroneous value for the plasmid segregational loss term will be derived.

Huang *et al.*[9,10] show how these models and the experimental system described below may be used to ascertain whether suspended cultures lose plasmid at a higher probability than biofilm culture. Suspended cultures are cultivated batchwise. Biofilm cultures are cultivated, under controlled hydrodynamic conditions, in a parallel-plate flow cell reactor. Two reporter genes are introduced into *Escherichia coli* DH5α by recombinant plasmid pMJR1750 (Fig. 1), for (1) the constitutive secretion of β-lactamase and (2) the inducible expression of β-galactosidase; both proteins foreign to *E. coli* DH5α. Thus, pMJR1750 can provide information about the kinetics of plasmid segregation loss, in suspension and biofilm *E. coli* DH5α cultures.

Plasmid Expression for the Study of Biofilm Bacterial Physiology

Unfortunately, any unstable plasmid expression system is not suitable for a quantitative study of gene expression in any culture type (suspended or biofilm). Parameters related to cloned gene protein synthesis, using the pMJR1750 plasmid, cannot be specified on a per cell basis because the destructive sampling of cells (biofilm or suspended) and subsequent determination of cellular metabolites related to the induced expression of β-galactosidase cannot distinguish between plasmid-bearing and plasmid-free cells.

To circumvent this problem, a genetically stabilized version of the plasmid pMJR1750 was fabricated (pTKW106; Fig. 2) containing the *hok/sok* loci (formerly the *parB* locus), which encodes the two small genes, *hok* and *sok*. The *hok* gene product is a potent cell-killing protein, the expression of which is regulated

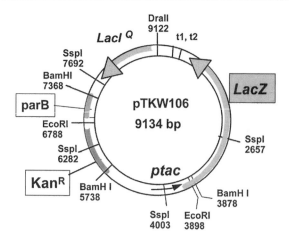

FIG. 2. Base map of pTKW106 used for protein expression studies.[9,10]

by the *sok* product, an antisense RNA complementary to the *hok* mRNA.[12] *hok* mRNA is extraordinarily stable, while the *sok* RNA is rapidly degraded. Postsegregational killing occurs because of the differences in decay rates of the *hok* and *sok* RNAs. As long as a cell maintains a plasmid with the *hok/sok* loci, the effects of the Hok protein are suppressed. However, in plasmid-free progeny the prolonged persistence of the *hok* mRNA leads to the synthesis of the Hok protein, thus initiating cell membrane potential breakdown and rapid cell lysis (Fig. 3). While the *hok/sok* locus does not eliminate segregation, it does prevent the establishment of a finite plasmid-free population. Thus, experiments with the segregationally stabilized plasmid pTKW106 can be carried out to compare protein expression levels between suspended and biofilm cells. For example, Huang *et al.*[10] found expression levels (now on a per plasmid-bearing cell number basis) of β-galactosidase (β-Gal) on induction with isopropyl-β-D-thiogalactopyranoside (IPTG) were dramatically different in biofilm-bound bacteria when compared with freely suspended cultures. The amount of β-galactosidase produced by biofilm cells at the peak of induction, using the maximum amount of IPTG, was 0.052 versus 0.50 pg/cell for suspended cells. Furthermore, the metabolic response time to initiate induction was different in the two cases. For biofilm cells, β-Gal production began ∼10 hr after induction versus 1 hr for suspended cells to begin β-Gal production. In addition, synthesis rates of total RNA, rRNA, and β-Gal-specific mRNA in *E. coli* DH5α (pTKW106) all increased after induction with IPTG, but levels of each parameter seen in biofilm cultures were 10–20 times lower than those values observed in suspended cell cultures and the time of onset of a response was slower in the biofilm cultures. Note

[12] K. Gerdes, *Bio/Technology* **6**, 1402 (1998).

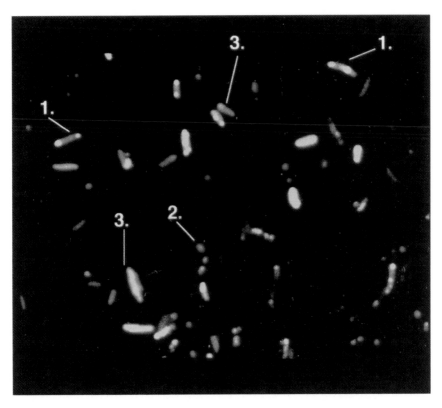

FIG. 4. Confocal laser scanning micrograph of a 72 μm *E. Coli* DH5α (pTKW106) expressing β-galactosidase under IPTG induction. Bar is 1.5 μm.

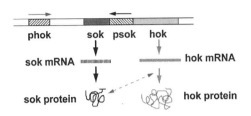

FIG. 3. Schematic of the interaction of *hok/sok* suicide genes.

that the β-Gal amounts per cell pass through a maximum after induction, which is attributed to plasmid instability. Once induced, the cell spends little energy on anything other than β-Gal synthesis. Thus, β-Gal synthesis per cell initially increases because of induction but does begin to decrease as the number of living cells decreases. These results reiterate that, for these two commercially available unstable plasmids, the expression of the chromosomal gene for synthesis of the polysaccharide matrix of the biofilm commands a higher priority in cellular activity than the expression of foreign genes.

Readers should be cautioned that most estimations of plasmid-bearing versus plasmid-free cells (like those described above) are made using selective plate counting techniques. The recombinant cells (suspended or biofilm) described above that express β-galactosidase form blue colonies on plate medium containing 5-bromo-4-chloro-3-indolyl-β-D-galactopyranoside (X-Gal) and IPTG, whereas plasmid-free cells from white colonies. The selective plate counting method to estimate plasmid retention, expression, or transfer in biofilm systems should be used with caution, because the plates are themselves artificial biofilms. As the plate method requires cells to proliferate as a biofilm (a colony-forming unit) for significant time periods, we have found that selective plate counts allow for further plasmid transfer to occur, even in the presence of antibiotic counterselection against donors and recipients.[13]

Also, any procedure to assess plasmid presence, expression, or stability that requires destructive sampling of an entire section of biofilm is, by its very nature, inherently unable to analyze "spatially pertinent" events. Destructive sampling and sample preparations destroy any spatial gradients that may exist within a biofilm. The reader should realize that most of the destructive sampling procedures described above give only results that are "homogenized" over the entire biofilm sample volume.

A more direct, noninvasive method of detecting reporter genes would be preferable in order to provide a more accurate estimate of plasmid-bearing populations. This would also find utility in real-time plasmid gene transfer studies. Such a

[13] J. Bertram, M. Stratz, and P. Durre, *J. Bacteriol.* **170**, 443 (1991).

technique is shown in Fig. 4 (see color insert), where we have used confocal scanning laser microscopy (CSLM) to quantify the number of plasmid-bearing cells in a biofilm directly, by fluorescently staining with fluorescein di-β-D-galactopyranoside (FDG-60) the marker protein, β-galactosidase, on induction with IPTG. The *E. coli* DH5α pMJR1750 biofilm was also stained with 5-cyano-2, 3-ditolyl tetrazolium chloride (CTC), which gives an indication of those cells that were carrying out respiration at the time of plasmid expression. The image is of an x–y horizontal plane within the biofilm at an altitude of 35 μm above the glass substratum surface in the z-height direction; overall average depth of the biofilm was 64 μm. The active respiration stain CTC fluoresces red–orange, indicating respiratory activity, and FDG-60 fluoresces green on binding to β-galactosidase, the gene product expressed by the plasmid pMJR1750. Referring to Fig. 4, **1** represents cells that are viable (respiring) and expressing β-galactosidase, **2** represents cells that are viable but not expressing β-galactosidase, and **3** represents cells having expressed β-galactosidase but that are not viable (not respiring). Similar combinations of species-specific fluorescent 16S rRNA oligonucleotide probes with plasmid gene markers could allow an on-line noninvasive method to quantify plasmid transfer kinetics. These methodologies will allow direct interrogation of biofilms for gene expression and transfer and will begin to define the molecular basis of the biofilm phenotype on an individual cellular level.

[10] Directed Movement and Surface-Borne Motility of *Myxococcus* and *Pseudomonas*

By DANIEL B. KEARNS and LAWRENCE J. SHIMKETS

Introduction

Biofilms have elaborate internal architectures composed of pillars of biomass surrounded by regular vertical and horizontal channels. These channels carry currents that deliver essential nutrients and oxygen for bacterial growth.[1] The complex morphology seems vital for the survival of the component bacteria, but how biofilms are constructed and become organized is not known. Are these intricate structures shaped by purely physical forces or does directed movement of bacteria play a role in biofilm assembly?

Myxococcus xanthus is a soil bacterium that associates in biofilms and glides over solid surfaces by an unknown mechanism. When *M. xanthus* is starved for

---

[1] J. W. Costerton, Z. Lewandowski, D. E. Caldwell, D. R. Korber, and H. M. Lappin-Scott, *Annu. Rev. Microbiol.* **49,** 711 (1995).

nutrients, complex motility patterns are initiated to convert a flat two-dimensional swarm to a spherical fruiting body. At least six intercellular signals and gliding motility are essential to complete morphogenesis.[2] While the *M. xanthus* fruiting body is a dramatic and highly specialized type of biofilm, similar regulatory systems appear to control biofilms in other organisms. For instance, *Pseudomonas aeruginosa* forms a biofilm during infections of the human lung[3] and on artificial catheters.[4] Just as with *M. xanthus,* cell-to-cell chemical communication[5] and a form of surface motility called twitching are essential for *Pseudomonas* biofilm development.[6] Twitching and gliding both require type IV pili.[7-9] How is motility regulated during biofilm formation and are the same processes at work in both twitching and gliding?

Chemotaxis, directed movement up chemical gradients, may play a role in both *P. aeruginosa* and *M. xanthus* surface translocation. Type IV pili have been implicated as the *P. aeruginosa* twitching motor[7] and some pilus-deficient strains have mutations in genes similar to those of *Escherichia coli che* genes, which encode proteins that transmit information about chemoattractants.[10] The phenotypes of these mutants suggest that type IV pilius production may be regulated by chemical signals. In *M. xanthus,* mutations in *che* homologs called *frz* and *dif* perturb gliding behavior to the point of abolishing fruiting body formation.[11-13] While the *frz* and *dif* phenotypes strongly suggested that chemotaxis plays a role in fruiting body formation, only recently has a discrete chemoattractant been discovered for *M. xanthus.* The attractant, phosphatidylethanolamine (PE), was isolated from the *M. xanthus* cell membrane, and cells directed motility up PE gradients.[14]

Hydrophobic surfactants provide an ideal venue for regulating surface motility in submerged biofilms, and it seems plausible that chemotaxis could be an important factor in the organization of biofilm tertiary structure. Gliding and twitching motility are similar in that both require a solid surface and type IV pili. In this

[2] L. J. Shimkets, *Microbiol. Rev.* **54,** 473 (1990).

[3] J. Lam, R. Chan, K. Lam, and J. W. Costerton, *Infect. Immun.* **28,** 546 (1980).

[4] D. J. Stickler, N. S. Morris, R. J. C. McLean, and C. Fuqua, *Appl. Environ. Microbiol.* **64,** 3486 (1998).

[5] G. A. O'Toole and R. Kolter, *Mol. Microbiol.* **30,** 295 (1998).

[6] D. G. Davies, M. R. Parsek, J. P. Pearson, B. H. Iglewski, J. W. Costerton, and E. P. Greenberg, *Science* **280,** 295 (1998).

[7] D. E. Bradley, *Can. J. Microbiol.* **26,** 146 (1980).

[8] A. Rosenbluh and M. Eisenbach, *J. Bacteriol.* **174,** 5406 (1992).

[9] S. S. Wu and D. Kaiser, *Mol. Microbiol.* **18,** 547 (1995).

[10] A. Darzins, *Mol. Microbiol.* **11,** 137 (1994).

[11] B. D. Blackhart and D. R. Zusman, *Proc. Natl. Acad. Sci. U.S.A.* **82,** 8767 (1985).

[12] M. J. McBride, R. A. Weinberg, and D. R. Zusman, *Proc. Natl. Acad. Sci. U.S.A.* **86,** 424 (1989).

[13] Z. Yang, Y. Geng, D. Xu, H. B. Kaplan, and W. Shi, *Mol. Microbiol.* **30,** 1123 (1998).

[14] D. B. Kearns and L. J. Shimkets, *Proc. Natl. Acad. Sci. U.S.A.* **95,** 11957 (1998).

review, we describe techniques used to isolate the *M. xanthus* gliding chemoattractant and apply them to identify a *P. aeruginosa* twitching chemoeffector.

Lipid Extraction

During fruiting body formation, *M. xanthus* cells are in continuous contact and
chemoattractants can be sequestered in the cell envelope or extracellular matrix.[14]
These attractants are extracted by a procedure modified from Bligh and Dyer[15]
that enriches for lipids.

Hydrophobic Extraction

A 3.75-ml volume of methanol–chloroform (2 : 1, v/v) is added to 0.4 g of cells
(wet weight) and vortexed for 1 hr. The suspension is centrifuged at 10,000g for
5 min at room temperature and the supernatant is saved. The pellet is reextracted
with 4.75 ml of methanol–chloroform–water (2 : 1 : 0.8, v/v/v), centrifuged at
10,000g for 5 min at room temperature, and the supernatants are combined. Chloroform (2.5 ml) and 2.5 ml of water are added to the combined supernatants, vortexed,
and centrifuged as described above to separate the chloroform and aqueous layers.
The chloroform layer is collected and dried under nitrogen in a preweighed tube.
The mass of the residue is determined and suspended in an appropriate solvent.

Chemotaxis Assay

Chemotaxis has been most extensively studied in the swimming bacterium
E. coli. The earliest studies demonstrated that when a capillary tube is filled with
a solution containing a chemoattractant and placed in a culture, diffusion will
create a chemical gradient, and the *E. coli* cells would preferentially swim into
the capillary tube.[16] This assay provided a quantitative evaluation of chemotaxis
because the degree to which cells accumulated within the tube was correlated
with the attractant concentration. Unfortunately, the dependence of this assay on
swimming motility makes it intractable to the study of surface motile organisms.

The *M. xanthus* chemotaxis assay (modified from a protocol by Dworkin and
Eide[17]) measures preferential colony migration up lipid gradients. A lipid sample
is applied to the center of a petri plate, time is allowed for diffusion to establish a
gradient, cells are placed within the gradient, and their behavior is recorded several
hours later. A chemoattractant will bias motility farther up the gradient than down
the gradient.

[15] E. G. Bligh and W. J. Dyer, *Can. J. Biochem. Physiol.* **37**, 911 (1959).
[16] W. Pfeffer, *Untersuch. Bot. Inst. Tübingen* **1**, 363 (1884).
[17] M. Dworkin and D. Eide, *J. Bacteriol.* **154**, 437 (1983).

Applying Lipid

Lipid is applied to the center of a petri plate containing TPM agar [10 mM Tris (hydroxymethyl)aminomethane HCl (pH 7.6), 8 mM MgSO$_4$, 1 mM KHPO$_4$– KH$_2$PO$_4$, and 1.5% (w/v) Bacto-agar] in one of two ways. First, a 6-mm-diameter Whatman (Clifton, NJ) No. 1 filter paper disk is created, using a standard paper hole punch, and mounted on pins. The disk is washed by the application of 10 μl of chloroform and allowed to dry. The lipid of interest is dissolved in an appropriate solvent to the desired concentration and 10 μl is applied to the disk. After the solvent has dried, forceps are used to place the disk on the petri plate. This method of application is recommended for samples of 25 μg or greater. Alternatively, 10 μl of the lipid solution may be spotted directly on the petri plate and allowed to dry. The outline of the lipid origin is marked on the bottom of the plate. This method of application is recommended for samples of 25 μg or less. Mock gradients generated with solvent alone are used as negative controls.

Generating Gradient

The gradient is generated by an 18 hr incubation at 32°. Test gradients generated with 1 μg of fluorescent dansyl-labeled phosphatidylethanolamine (Molecular Probes, Eugene, OR) spotted directly in 10 μl of chloroform are short and steep. After 18 hr of incubation, agar slices are collected every 3 mm along a radial transect beginning at the origin. The PE is extracted from the agar with chloroform and quantified, using a luminescence spectrometer. With this protocol, the PE concentration drops 10-fold within 3 mm of the origin (Fig. 1).

Introducing Cells

Exponential phase cells are suspended in India ink–morpholinopropane-sulfonic acid (MOPS) buffer [10 mM MOPS (pH 7.6), 8 mM MgSO$_4$, 10% (v/v) India ink] to a final density of 5×10^9 cells/ml. Samples (1 μl) of the cell suspension are spotted in a circle between 3 and 10 mm away from the lipid origin. As the suspension dries, the India ink absorbs to the agar surface and delineates the boundary of the cell spot. The plates are incubated at a temperature appropriate for the organism of interest. While the cells move across the surface of the agar, the India ink remains at the origin and can be used as a reference for recording motility within the gradient.

Measuring Chemotaxis

The plate is incubated until the cells have migrated approximately 1 mm from the India ink origin in negative controls. Surface motile organisms travel at a wide range of speeds and therefore the incubation time for cells within the gradient is

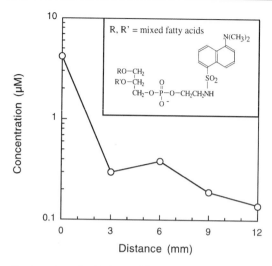

FIG. 1. Lipid gradient constructed with 1 μg of fluorescent dansyl-labeled PE. *Inset:* Dansyl–PE chemical structure.

determined empirically for each species. For example, *M. xanthus* cells produce a measurable zone of gliding motility after 6 hr, whereas *P. aeruginosa* twitching motility requires 24 hr of incubation. The plate is placed under a light microscope and observed at ×100 magnification. The distance the cells move toward and away from the lipid origin is quantified with an ocular micrometer. The preferential migration value is calculated:

$$\text{Preferential migration} = \text{distance traveled up gradient/distance traveled down gradient}$$

When measuring preferential migration, observation at ×100 magnification is strongly recommended. While the motility bias created by a chemoattractant is visible to the naked eye, in the case of *M. xanthus* many single cells may actually travel farther than the swarm front. Therefore, the preferential migration can be underestimated without microscopic observation.

Sample Results

Lipids are extracted from stationary phase *P. aeruginosa* PAO1 cells. Bulk extract (100 μg) is dissolved in chloroform and applied to a Whatman No. 1 paper disk. A gradient is generated as described above. *Pseudomonas aeruginosa* PAO1 is then mixed with India ink and spotted within the gradient. In a mock gradient generated by solvent alone, the cells expand from their origin evenly in all directions (Fig. 2a). Therefore, the control preferential migration is 1. However, the

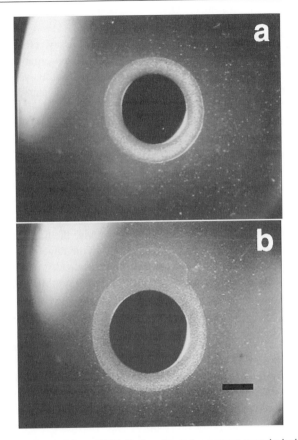

FIG. 2. *Pseudomonas aeruginosa* PAO1 displays directed movement toward a hydrophobic PAO1 cellular extract. Cells were spotted in a mock gradient generated by chloroform (a) or a gradient generated by a 100-μg (disk-delivered) PAO1 hydrophobic extract (b). Test compounds originate at the top of each panel. The dark area is the India ink; the bright halo is the twitching zone of cells. Scale bar: 1 mm.

cells migrate farther up the extract gradient than down it and in this example the preferential migration is nearly 2.5 (Fig. 2b). These results suggest that *P. aeruginosa* contains an endogenous lipid chemoeffector. Adsorption chromatography[18] and thin-layer chromatography[19] were conducted to further characterize the active chemical as phosphatidylethanolamine.[20]

[18] J. G. Hamilton and K. Comai, *Lipids* **23,** 1146 (1988).
[19] W. G. Jennings, *in* "Thin Layer Chromatography: Techniques and Applications" (B. Fried and J. Sherma, eds.), p. 245. Marcel Dekker, New York, 1994.
[20] D. B. Kearns, J. Robinson, and L. J. Shimkets, *J. Bacteriol.* **153,** 763 (2001).

Comments

While any agar composition could be used in this assay, TPM offers two distinct advantages for *M. xanthus*. First, it is a nonnutrient agar and reduces the possibility of growth as a potential bias. Second, and fortuitously, we have discovered that chemotaxis toward dilauroyl PE in *M. xanthus* is starvation dependent. When nutrient-rich CTT medium [Casitone (10 g/liter), 10 mM Tris (hydroxymethyl)aminomethane HCl (pH 7.6), 8 mM MgSO$_4$, 1 mM KHPO$_4$–KH$_2$PO$_4$, 1.5% (w/v) Bacto-agar] is used, preferential motility up a PE gradient is completely inhibited.[21]

Reversal Period Assay

Escherichia coli swims through liquid media, utilizing peritrichous flagella that turn in two directions. When rotating counterclockwise the flagella form a bundle and propel the cell in a straight line. However, when the flagella rotate clockwise, the bundle falls apart and causes the cell to tumble erratically.[22] Runs and tumbles alternate approximately every second and the period of tumbling allows *E. coli* to reorient itself in a new direction prior to initiating a new period of movement.[23]

Chemotaxis in *E. coli* has been attributed to a biased random walk.[24] By biasing the periods of runs and tumbles, *E. coli* is able to produce net movement in the direction of a chemoattractant. If the cell happens to be traveling toward an attractant, the tumble behavior is suppressed and the cell will run for a longer period of time up the gradient. Furthermore, if *E. coli* is rapidly presented with a uniform concentration of attractant, tumbles will be suppressed and this increase in the period between direction changes is considered the hallmark of a chemotactic response.[23]

While *M. xanthus* moves on solid surfaces and does not utilize flagella, the *E. coli* chemotaxis paradigm is still applicable. *Myxococcus xanthus* cells glide in a linear path and reverse direction on average every 6.8 min.[11] However, in the presence of the chemoattractant PE, the reversal periods can increase to more than 20 min. Just as in *E. coli, M. xanthus* direction reversals are suppressed in the presence of an attractant.[14]

The reversal period assay measures the amount of time an average *M. xanthus* cell travels before changing direction. To test the effect of a putative chemoattractant, a lipid sample is spotted on a petri plate and allowed to dry. Cells are then applied to the top of the compound and exposed to a uniform, high concentration of the chemical, and cell movement is observed microscopically and recorded.

[21] D. B. Kearns, B. D. Campbell, and L. J. Shimkets, *Proc. Natl. Acad. Sci. (USA)* **97,** 11505 (2000).
[22] S. H. Larsen, R. W. Reader, E. N. Kort, W. W. Tso, and J. Adler, *Nature (London)* **249,** 74 (1974).
[23] H. C. Berg and D. A. Brown, *Nature (London)* **239,** 500 (1972).
[24] R. M. Macnab and D. E. Koshland, Jr., *Proc. Natl. Acad. Sci. U.S.A.* **69,** 2509 (1972).

A chemoattractant will increase the reversal period relative to the controls. The following protocol is modified from Blackhart and Zusman.[11]

Applying Lipid

Because the compounds to be tested are hydrophobic, a TPM agar petri plate is incubated at $37°$ for 10 min to evaporate water from the surface prior to lipid application. The lipid is dissolved in an appropriate solvent to the desired concentration and 4 μl of the solution is spotted on the TPM agar. The plate is then incubated at $37°$ for 20 min to evaporate the solvent and allow the lipid to spread within the spot. The lipid should be dried under conditions of low humidity.

Introducing Cells

Myxococcus xanthus cells are grown to exponential phase, diluted to 5×10^7 cells/ml in MOPS buffer [10 mM MOPS (pH 7.6), 8 mM MgSO$_4$], and 5 μl of the cell suspension is then spotted on top of the lipid. The plate is incubated at $32°$ for 15 min to allow the cells to adjust to the surface.

Observing Motility

The petri plate is observed at $\times 400$ with a phase-contrast microscope and a time-lapse video recorder for 45 min. We employ a Sony Power HAD 3CCD digital video camera, and a Power Macintosh 9500 with Adobe Premier moviemaking software set at a frame capture rate of 12 frames/min. The film file is played back to manually enumerate cell reversals. While monitoring cell motility, the microscope environment is maintained at $28°$. While thin agar slide cultures can be utilized as a substrate for *M. xanthus* motility, over the course of 45 min the agar dries dramatically. Even using petri plates as a substrate, the microscope must be continually focused to maintain the focal plane as the agar shrinks during dehydration.

Calculating Reversal Period

The reversal period is the number of minutes an average cell travels prior to reversing direction. Because the value is an average, microscopic fields must be chosen such that 20–40 individual cells remain within the field of view for the duration of the experiment. Twenty cells are chosen at random and the film is reviewed. The number of times a cell reverses direction is manually enumerated and the average reversal period is calculated.

$$\text{Reversal period} = \text{duration of experiment}/(\text{total number of}$$
$$\text{reversals}/\text{number of cells})$$

The 20 individual cells should be chosen randomly and cells are excluded if (1) the cell leaves the field of view, (2) the cell does not move continuously for the duration of the experiment, or (3) the cell comes into contact with other cells.

Reversal period and reversal frequencies are inverse values; period is expressed as minutes per reversal and frequency is expressed as reversals per minute. While *M. xanthus* behavior has been described in the literature using both formats, the two units can be easily interconverted.

This assay clearly depends on organisms that move as single cells on solid surfaces. While *P. aeruginosa* moves on solid surfaces, it appears to move only in groups and isolated cells may be nonmotile.[25] Thus far, attempts to adapt the reversal period assay for *P. aeruginosa* behavioral studies have been unsuccessful.[20] Techniques that allow the study of cells within groups, such as tetrazolium staining,[26] may need to be developed to study the behavior of individual *P. aeruginosa* cells in response to chemoeffectors.

Acknowledgments

This material is based on work supported by the National Science Foundation under Grant MCB0090946 and by a Grant-in-Aid of Research from the National Academy of Sciences, through Sigma Xi, the Scientific Research Society.

[25] A. B. T. Semmler, C. B. Whitchurch, and J. S. Mattick, *Microbiology* **145**, 2863 (1999).
[26] W. Shi, F. K. Ngok, and D. R. Zusman, *Proc. Natl. Acad. Sci. U.S.A.* **93**, 4142 (1996).

[11] A General Method to Identify Bacterial Genes Regulated by Cell-to-Cell Signaling

By XUEDONG DING, RITA R. BACA-DELANCEY, SOOFIA SIDDIQUI, and PHILIP N. RATHER

Introduction

The importance of cell-to-cell signaling in the formation of *Pseudomonas aeruginosa* biofilms has led to an interest in addressing the general role of intercellular signaling in bacterial biofilm formation.[1] Several studies have demonstrated

[1] D. G. Davies, M. R. Parsek, J. P. Pearson, B. H. Iglewski, J. W. Costerton, and E. P. Greenberg, *Science* **280**, 295 (1998).

that quorum sensing signals are produced in aquatic biofilms and on catheter-associated biofilms from patients.[2,3] Furthermore, quorum sensing signals have been shown to promote recovery from starvation in *Nitrosomonas europaea*.[4] These data raise the possibility that cell-to-cell signaling is an important requirement for biofilm development in many types of bacteria. The close association of cells in a biofilm community creates an ideal environment for effective communication between cells. To study the relationship between cell-to-cell signaling and biofilm development, a fundamental understanding of the signals and the response pathways is required. The first step to this understanding is the identification of genes regulated by cell-to-cell signaling.

The analysis of bacterial transcriptional and translational control signals has been greatly simplified by the use of reporter gene fusions.[5] In bacteria, the *lacZ* gene encoding β-galactosidase has been widely used for this purpose. The advantages of this reporter include the ease of enzyme analysis and the wide variety of plasmid- and transposon-based systems for creating *lacZ* fusions.[6–9] An important expansion of this technology was the use of dual reporter genes to identify promoters active at different phases of growth. The initial description of such a system was by Youngman *et al.*, who utilized a Tn*917* derivative containing promoterless *lacZ* and chloramphenicol acetyltransferase (*cat*) genes in tandem for the identification of genes induced during sporulation in *Bacillus subtilis*.[7] Using this transposon, colonies with random insertions were allowed to grow for an extended period of time. Blue colonies represented insertions within a transcriptional unit. At this point, the utility of the second reporter for chloramphenicol acetyltransferase becomes important for the identification of fusions activated at high density. Colonies with a blue phenotype are screened for chloramphenicol resistance and those that exhibit chloramphenicol resistance contain insertions within a constitutively expressed gene. However, blue colonies with a chloramphenicol-sensitive phenotype represent insertions that are expressed at a point beyond early exponential phase.

This powerful screening technique was then adapted for use in gram-negative bacteria by deLorenzo *et al.*, using the transposon mini-Tn*5lacZ-tet/1*.[8] This transposon was used to identify *Pseudomonas putida* genes expressed in postexponential phase.[8] A useful property of the mini-Tn*5* transposons is that they have

[2] D. J. Stickler, N. S. Morris, R. J. McLean, and C. Fuqua, *Appl. Environ. Microbiol.* **64**, 3486 (1998).

[3] R. J. McLean, M. Whiteley, D. J. Stickler, and W. C. Fuqua, *FEMS Microbiol. Lett.* **154**, 259 (1997).

[4] S. E. Batchelor, M. Cooper, S. R. Chhabra, L. A. Glover, G. S. Stewart, P. Williams, and J. I. Prosser, *Appl. Environ. Microbiol.* **63**, 2281 (1997).

[5] M. J. Casadaban and S. N. Cohen, *J. Mol. Biol.* **138**, 179 (1980).

[6] J. H. Miller, "Experiments in Molecular Genetics." Cold Spring Harbor Laboratory Press, Cold Spring Harbor, New York, 1972.

[7] P. Youngman, P. Zuber, J. B. Perkins, K. Sandman, M. Igo, and R. Losick, *Science* **228**, 285 (1985).

[8] V. deLorenzo, I. M. Cates, M. Herrero, and K. N. Timmis, *J. Bacteriol.* **175**, 6902 (1993).

[9] V. deLorenzo, M. Herrero, U. Jakubzik, and K. N. Timmis, *J. Bacteriol.* **172**, 6568 (1990).

been shown to transpose randomly in a variety of gram-negative bacteria.[9] These transposons are carried on a plasmid R6K derivative missing the replication protein π, encoded by the *pir* gene. These plasmids are maintained in *Escherichia coli* strains, such as SM10 or S17.1, that contain the *pir* gene on a λ prophage and also provide mobilization functions.[9,10] The transposase gene is also on the pUT plasmids, but it is physically separate from the mini-Tn5 element.[9] Therefore, insertions of mini-Tn5 are stable.

Methods

Overview of General Strategy to Identify Genes Activated by Cell-to-Cell Signaling

The basic strategy outlined in Fig. 1 has been used in our laboratory to identify genes activated by extracellular signals in *E. coli*, *Providencia stuartii*, and *Proteus mirabilis*.[11–13] This method uses bicistronic reporter transposons and is designed for use in organisms in which there are no mutants available in the quorum sensing signal. Although alternative screening strategies are possible, such as screening random *lacZ* fusions on agar plates with or without extracellular signal, we have found that this is of limited use. The self-production of signals by cells in the growing colony activates the fusion and limits the effectiveness of the preexisting signal that was incorporated into the agar.

The strategy in Fig. 1 makes the assumption that genes activated by cell-to-cell signaling will be preferentially expressed at midexponential phase or later because of the requirement for a threshold concentration of signal to be present. In the first step, colonies with representative mini-Tn5*lacZ-tet/1* insertions are allowed to grow to stationary phase (36–48 hr) on plates containing 5-bromo-4-chloro-3-indolyl-β-D-galactopyranoside (X-Gal). This allows for the selection of fusions expressed at all stages of growth. Blue colonies are then tested for expression of an antibiotic resistance phenotype by the ability to grow as single colonies in the presence of a drug. Blue and antibiotic-resistant cells contain a fusion to a constitutively expressed promoter and are not studied further. However, blue and antibiotic-sensitive colonies represent fusions to a promoter that is activated beyond early exponential phase or requires physical contact between cells for activation. These blue/antibiotic-sensitive colonies are then tested individually for β-galactosidase activation in the presence of extracellular signals in crude conditioned medium.

[10] V. H. Miller and J. J. Mekalanos, *J. Bacteriol.* **170**, 2575 (1998).

[11] R. R. Baca-DeLancey, M. M. T. South, X. Ding, and P. N. Rather, *Proc. Natl. Acad. Sci. U.S.A.* **96**, 4610 (1999).

[12] P. N. Rather, X. Ding, R. R. Baca-DeLancey, and S. Siddiqui, *J. Bacteriol.* **181**, 7185 (1999).

[13] P. N. Rather, unpublished (1999).

Step 1. Construct an insertional library with a bicistronic reporter transposon in a given bacterial strain. Allow colonies to grow for 36 hours on X-gal plates to identify fusions expressed at all growth phases.

Step 2. Test blue colonies for antibiotic resistance.

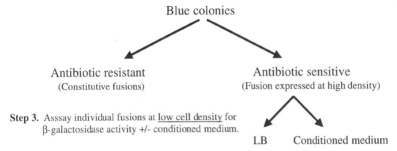

Blue colonies

Antibiotic resistant
(Constitutive fusions)

Antibiotic sensitive
(Fusion expressed at high density)

Step 3. Asssay individual fusions at <u>low cell density</u> for
β-galactosidase activity +/- conditioned medium.

LB   Conditioned medium

FIG. 1. Basic strategy to identify genes activated by extracellular signals in *Escherichia coli, Providencia stuartii,* and *Proteus mirabilis.*

The following sections detail methods to isolate genes activated by cell-to-cell signaling, using bicistronic reporter transposons.

Creating Library of Transposon Insertions Using Mini-Tn5 Derivatives. The following procedures have been used in our laboratory to create reporter *lacZ* fusions using mini-Tn5 derivatives in various *Enterobacteriaceae.*[11,12] In general, these methods can be applied to any lactose-negative gram-negative bacterium in which Tn5 will transpose. The pUT plasmids containing various mini-Tn5 transposon derivatives are maintained in *E. coli* S17.1 λ *pir* or SM10 λ *pir.* A consideration for using these strains to create insertions in *E. coli* is that donor strains can spontaneously release λ *pir* that can lysogenize the recipient strain. This will allow for pUT replication in the recipient and give rise to colonies without transposon insertions. Therefore, an *E. coli* recipient that is λ immune, such as MC1061.LR, should be used.[11] In general, the introduction of pUT::mini-Tn5 derivatives into the recipient strains is done by plate mating. The appropriate antibiotics for counterselection of the *E. coli* donor will have to be individually determined, depending on the resistances of the recipient strain. Typically, 3-ml cultures of donor and recipient strains are grown overnight to saturation with ampicillin added at 100 μg/ml to donor *E. coli* strains to maintain the pUT plasmids. Cells of the donor *E. coli* culture are washed three times with an equal volume of sterile Luria–Bertani medium (LB) to remove any remaining ampicillin and equal volumes of both donor and recipient cells (100 μl) are mixed in an Eppendorf tube and spotted onto the surface of a dried LB agar plate. Plate matings are typically conducted for 8–12 hr at 37°. Cells are then scraped into 4 ml of LB containing 20% (v/v) glycerol, suspended, and aliquots of 500 μl are prepared and stored at $-80°$. A test plating is done with 50, 100, and 200 μl to determine the optimal amount to obtain 100–200 colonies per plate.

Screening Colonies for Fusions Activated by Cell-to-Cell Signaling. Colonies from the above matings are allowed to grow for 36 hr at 37° to allow for the identification of *lacZ* fusions expressed at all phases of growth. Blue colonies are then picked with a toothpick and transferred to a master plate with X-Gal and to plates containing tetracycline (mini-Tn*5lacZ-tet/1* fusions) or chloramphenicol (mini-Tn*5cat/lacZ* fusions). Because of the wide range in promoter strength driving expression of the dual reporters, we test antibiotic sensitivity at several different concentrations of drug. The actual concentrations of antibiotic must be individually determined for each organism before screening is initiated. We initially begin screening at a concentration of antibiotic that is 2-, 4-, and 8-fold above the determined minimum inhibitory concentration (MIC). A single colony is picked up with a flat toothpick and transferred to the appropriate plates as a small patch. Fresh toothpicks are then used to streak out each patch for single colonies. Using this technique, a density of 30 colonies can be obtained per plate.

The interpretation of antibiotic sensivity can vary from a complete lack of growth to poor growth of single colonies. In general, any observable defect in growth is compared with the master plate containing X-Gal. This master plate facilitates the interpretation of colonies with borderline sensitivity. For example, consider colony "A" that appears to exhibit borderline antibiotic sensitivity, yet is dark blue on the master plate with X-Gal. If other colonies that are lighter blue on the master plate exhibit greater resistance on the antibiotic plates, then colony A is a good candidate for a fusion activated in postexponential phase. Colonies with a blue (lac^+) and antibiotic-sensitive phenotype are then rescreened to confirm the sensitivity phenotype. Although false positives are obtained, this screening procedure greatly enriches for fusions that are activated in postexponential phase.

Testing Fusions for Activation by Conditioned Medium. The testing of potential *lacZ* fusions for activation by extracellular signals is performed with spent culture supernatants from high-density cultures (conditioned medium). Although fusions activated by extracellular signals will vary in their responsiveness to conditioned medium from different cell densities, we use conditioned medium from early stationary phase cultures for screening. Typically a dilute culture (1 : 10,000) is allowed to grow overnight in LB broth at 37° and cells are harvested at early stationary phase by pelleting at 4300*g* for 10 min. Depleted nutrients are replenished by the addition of concentrated tryptone/yeast extract solution (20×) that is added to a final concentration of 0.5×. The pH is adjusted to pH 7.5 and the medium is filter sterilized. Preparations of conditioned medium from *E. coli*, *P. stuartii*, and *P. mirabilis* are stored at $-80°$ and are stable to repeated freeze–thaw cycles.

Screening individual *lacZ* fusions for activation by extracellular signals can be done in microtiter plates or in small test tubes. The inoculation into conditioned medium and the control media should be done at low density, typically a 1 : 1000 dilution of an early exponential phase starter culture. Cells are allowed to grow to early exponential phase (OD_{600} of 0.3) and assayed for β-galactosidase by the method of Miller.[6]

Fusions that display activation by conditioned medium should be retested with freeze-dried conditioned medium that is added back in concentrated form to fresh media. This will rule out activation by oxygen depletion. In addition, a number of parameters can influence the activation of a fusion by extracellular signals. These parameters include pH, nutrient concentration, and the growth phase of cells. Therefore, any fusion that displays a 3-fold or greater activation by conditioned medium should be retested with supplemented conditioned medium at pH 7, 7.5, and 8. In addition, each fusion may display optimal activation by extracellular signals at different stages of growth.

Limitations of This Approach and Potential Problems

An important limitation of this approach is that several classes of genes may be missed. First, this procedure will not allow for the identification of genes that are repressed by cell-to-cell signaling. Second, essential genes may not be identified because of the creation of null alleles by insertion of the reporter transposon. To overcome this problem, we have created an integrational plasmid based on pKNG101[14] that contains the *lacZ/tet* reporters. Random chromosomal fragments are cloned in front of the reporter genes and the library is integrated into the chromosome of the desired strain by homologous recombination. If the cloned region contains the 5′ or 3′ end of the essential gene, then the integration will restore a functional copy and also create a *lacZ/tet* fusion to this gene. It is important to note that multicopy plasmids containing the *lacZ/tet* dual reporters do not appear to be useful as conditioned medium alters the copy number of plasmids.[13] The mechanism for the changes in copy number is unknown.

An additional limitation is that genes with high constitutive levels of expression further activated by cell-to-cell signaling may be eliminated during the first step of the screen. Fusions to genes with this expression profile will display an antibiotic-resistant phenotype even though they may be substantially activated by cell-to-cell signaling in later stages of growth. The use of several levels of antibiotic during step 2 of the screening process (Fig. 1) may help to identify these fusions. For example, these fusions may display a prominent blue color on X-Gal plates. However, on the antibiotic selection plates, they may display a defect in the ability to grow at higher levels of drug. Comparisons of blue colony phenotype of this type of fusion with other fusions on the master plate will help identify constitutive fusions that are further activated by cell-to-cell signaling, such as the *cysK* and *tnaB* fusions in *E. coli*.[11]

[14] K. Kaniga, I. Delor, and G. Cornelis, *Gene* **109**, 137 (1991).

A final consideration is that this method will identify genes activated by all types of extracellular signals. These will include authentic quorum sensing signals such as N-acylhomoserine lactones and peptides and simple metabolites such as fermentation products. Therefore, a subset of genes identified by this approach may not have functions related to quorum sensing. However, this unbiased approach has the advantage of allowing for the identification of new classes of signaling molecules that may coordinate multicellular activities such as biofilm formation.

Clearly, future studies of biofilm development in bacteria will involve organisms in which there is little or no information on the mechanisms of cell-to-cell signaling. The procedures outlined here will allow for the identification of these genes and, subsequently, permit further investigations to address the relationship between cell-to-cell signaling and biofilm development.

Acknowledgments

Our studies have been supported by Grants MCB9405882 and MCB9904766 from the National Science Foundation. We are grateful to P. deHaseth and R. Hogg for comments on this manuscript.

[12] Genetic and Chemical Tools for Investigating Signaling Processes in Biofilms

By Timothy Charlton, Michael Givskov, Rocky deNys, Jens Bo Andersen, Morten Hentzer, Scott Rice, and Staffan Kjelleberg

Introduction

The vast majority of microbial interactions in the environment center around surfaces, where it has been estimated that approximately 99% of all bacteria reside.[1] Of those surface-associated organisms, many are found in large aggregates or communities that are recognized as biofilms. The formation of biofilms appears to offer cells several advantages over planktonic cells in that the biofilm protects the cells from many adverse conditions including, but not limited to, changes in pH, low nutrients, predation, and antimicrobial compounds. It is in part the organization of biofilm cells as a community that enhances the ability of bacteria growing in a biofilm to proliferate under stressful conditions with which isolated

---

[1] J. W. Costerton, K. J. Cheng, G. G. Geesy, T. I. Ladd, N. C. Nickel, M. Dasgupta, and T. J. Marine, *Annu. Rev. Microbiol.* **41,** 325 (1987).

planktonic cells are unable to cope.[2] Thus, the development of cells into a structured community, such as a biofilm, represents one adaptive strategy for bacteria when they encounter stressful conditions. Given high population densities, up to 10^9 cells/mm^3,[3] and evidence of structured communities within biofilms, it is not surprising that biofilm formation has been demonstrated to be regulated in some bacteria through the use of extracellular signaling molecules, the N-acylhomoserine lactones.[4]

Communication systems based on N-acylhomoserine lactones (AHLs) have been described in a range of gram-negative bacteria and are proposed to be involved in causing such diverse problems as food spoilage and infection of the lungs. This bacterial communication system functions via small diffusible AHL signal molecules. The signal molecules are synthesized from precursors by a synthase protein, "I," and they interact with a transcriptional activating "R" protein to induce expression of different target genes. Such regulatory systems allow bacteria to sense and express target genes in relation to their population density. AHL systems are referred to as quorum-sensing (QS) systems, that is, they express target genes in relation to the quorum size (or density) of the population.[5] This renders target gene expression a group phenomenon. In most known cases, QS systems control expression of virulence factors and hydrolytic enzymes (for reviews see Refs. 6 and 7). In addition, there is growing evidence that the ability to form surface-associated, structured, and cooperative consortia (referred to as biofilms) may also be QS controlled (for reviews see Refs. 6 and 8). It has been suggested that QS phenotypes play significant roles in bacterial pathogenesis. In favor of this suggestion is the finding that communication-deficient strains are less pathogenic. This is true for *Erwinia carotovora,* which causes soft rot,[9] as well as for the opportunistic human pathogen *Pseudomonas aeruginosa.* AHL-deficient mutants of *P. aeruginosa,* which are impaired in their ability to form biofilms, have been demonstrated to have decreased ability to bind host epithelial cells, and are significantly impaired in virulence when tested in a mouse infection model.[4,9,10] In addition, QS mutants of *Serratia liquefaciens* have been demonstrated to be

[2] A. T. Nielsen, S. Tolker-Nielsen, K. B. T. Barken, and S. Molin, *Environ. Microbiol.* **2,** 59 (2000).

[3] M. Kolari, K. Mattila, and M. S. Salkinoja-Salonen, *J. Indust. Microbiol. Biotech.* **21,** 261 (1998).

[4] D. G. Davies, M. R. Parsek, J. P. Pearson, B. H. Iglewski, J. W. Costerton, and E. P. Greenberg, *Science* **280,** 295 (1998).

[5] W. C. Fuqua, S. C. Winans, and E. P. Greenberg, *J. Bacteriol.* **176,** 269 (1994).

[6] L. Eberl, *System. Appl. Microbiol* **22,** 493 (1999).

[7] S. Swift, P. Williams, and G. S. A. B. Stewart, *in* "Cell–Cell Signaling in Bacteria" (G. M. Dunny and S. C. Winans, eds.), p. 297. American Society for Microbiology, Washington, D.C., 1999.

[8] J. W. Costerton, P. S. Stewart, and E. P. Greenberg, *Science* **284,** 1318 (1999).

[9] M. Pirhonen, D. Flego, R. Heikinheimo, and E. T. Palva, *EMBO J.* **12,** 2467 (1993).

[10] H. B. Tang, D. DiMango, R. Bryan, M. Gambello, B. H. Iglewski, J. B. Goldberg, and A. Prince, *Infect. Immun.* **64,** 37 (1996).

defective in swarming motility, which normally would assist the bacterial population in the colonization of surfaces (for reviews see Refs. 11 and 12).

On the basis of the involvement of QS systems in colonization phenotypes (e.g., biofilm formation and surface translocation) and virulence factor production, it seems logical that interference with the quorum-sensing process would be one strategy for a host to prevent the detrimental association of microorganisms with its surface. Not surprisingly, some organisms have evolved a range of defenses designed to cope with bacterial colonization of their surfaces. While most defenses are relatively nonspecific, we have identified one example in which a marine organism, *Delisea pulchra,* produces chemical compounds that have been demonstrated to act as AHL antagonists and are proposed to be important for protection of the plant from colonization by bacteria.[13–16] These chemicals, halogenated furanones, are produced within the plant and are secreted onto the surface of the plant. Furanones inhibit AHL-mediated quorum sensing, can prevent colonization, and inhibit the expression of virulence factors such as exoenzymes that might damage the alga host.[15–18] On the basis of the use of signaling molecules by some bacteria to regulate biofilm formation and the ability of some organisms to generate AHL antagonists, the surface chemistry of organisms and the signal chemistry of bacteria growing at surfaces are particularly relevant to signal-regulated phenotypes, including bacterial colonization and the expression of virulence factors on surfaces and in biofilms.

Our laboratories have studied several of these phenomena, with particular emphasis on how interactions between bacteria and higher organisms involve signal antagonists that modify the expression of traits controlled by bacterial communication systems. To do this, we have developed genetic tools to monitor AHL signals and to measure AHL and antagonist concentrations under environmentally relevant conditions in a highly sensitive and quantitative fashion. We present three methods for the detection and/or quantification of bacterial signals and eukaryote-generated signal antagonists. First, we have developed a modified green fluorescent

[11] M. Givskov, J. Ostling, L. Eberl, P. W. Lindum, A. B. Christensen, G. Christiansen, S. Molin, and S. Kjelleberg, *J. Bacteriol.* **180,** 742 (1998).

[12] L. Eberl, M. K. Winson, C. Sternberg, G. S. A. B. Stewart, G. Christiansen, S. R. Chhabra, B. Bycroft, P. Williams, S. Molin, and M. Givskov, *Mol. Microbiol.* **20,** 127 (1996).

[13] R. deNys, A. D. Wright, G. M. Konig, and O. Sticher, *Tetrahedron* **49,** 11213 (1993).

[14] R. deNys, P. D. Steinberg, P. Willemsen, S. A. Dworjanyn, C. L. Gableish, and R. J. King, *Biofouling* **8,** 259 (1995).

[15] M. Givskov, R. deNys, M. Manefield, L. Gram, R. Maximilien, L. Eberl, S. Molin, P. D. Steinberg, and S. Kjelleberg, *J. Bacteriol.* **178,** 6618 (1996).

[16] S. Kjelleberg, P. Steinberg, M. Givskov, L. Gram, M. Manefield, and R. deNys, *Aquat. Microb. Ecol.* **13,** 85 (1997).

[17] S. A. Rice, M. Givskov, P. Steinberg, and S. Kjelleberg, *J. Mol. Microbiol. Biotechnol.* **1,** 23 (1999).

[18] L. Gram, R. deNys, R. Maximillien, M. Gisgov, P. D. Steinberg, and S. Kjelleberg, *Appl. Environ. Microbiol.* **62,** 4284 (1996).

protein–AHL sensor as a state-of-the-art tool for studies of communication be-tween the individuals present in mixed bacterial communities. This sensor can also be used to visualize bacterial signaling processes "on-line" in biofilm-grown bacteria. Second, we have developed highly sensitive and quantitative methods for the identification of AHLs from biofilm-grown bacteria. Third, we present methods for the extraction and identification of signal antagonists presented on the surfaces of eukaryotes.

Live Bacterial Monitors

AHL monitor constructs have greatly facilitated the identification of AHL-producing organisms, have been used to estimate the number and structure of AHL molecules produced by a given organism, and have proved highly useful for the identification and cloning of I and R genes. Many such AHL monitors have been described, for example, the purple-pigment producing *Chromobacterium violaceum* CVO26[19] and the plasmid-based biosensors carrying *luxRI'::luxAB* (*Vibrio harveyi*),[20] *luxRI'::lux CDABE* (*Photorhabdus luminescens*),[12,21,22] *rhlRI'::luxCDABE*,[22] *ahyRI''::luxCDABE*,[21] *lasRI'::luxCDABE*,[22] *traR traG::lacZ*,[23] *rhlA::lacZ*,[24] and *lasB::lacZ*.[25] AHL-detecting monitors are established either in AHL signal generation mutants (AHL⁻ background) or in *Escherichia coli*. The *luxRI'::luxCDABE* (*P. luminescens*) construct carried on the broad host range plasmid pSB403 has been used successfully to identify a number of AHL-producing organisms,[22] to identify the *swrRI* genes of *Serratia liquefaciens,* and to describe the role of quorum sensing in prokaryotic multicellular behavior.[6,12,26]

While the above-described AHL reporter constructions have been particularly useful for the identification and detection of AHLs, and AHL-producing organ-isms, especially in conjunction with thin-layer chromatography,[27] monitors based on bioluminescence or LacZ expression have their individual drawbacks. The

[19] K. H. McClean, M. K. Winson, L. Fish, A. Taylor, S. R. Chhabra, M. Camara, M. Daykin, J. H. Lamb, S. Swift, B. W. Bycroft, G. S. A. B. Stewart, and P. Williams, *Microbiology* **143**, 3703 (1997).

[20] S. Swift, M. K. Winson, P. F. Chan, N. J. Bainton, M. Birdsall, P. J. Reeves, C. E. D. Rees, S. R. Chhabra, P. J. Hill, J. P. Throup, W. Byeroft, G. P. C. Salmond, P. Williams, and G. S. A. B. Stewart, *Mol. Microbiol.* **10,** 511 (1993).

[21] S. Swift, A. V. Karlyshev, L. Fish, E. L. Durant, M. K. Winson, S. R. Chhabra, P. Williams, S. Macintyre, and G. S. A. B. Stewart, *J. Bacteriol.* **179,** 5271 (1997).

[22] M. K. Winson, S. Swift, L. Fish, J. P. Throup, F. Jorgensen, S. R. Chhabra, B. W. Bycroft, P. Williams, and G. S. A. B. Stewart, *FEMS Microbiol. Lett.* **163**, 185 (1998).

[23] K. Piper, S. Beck von Bodnan, and S. Farrand, *Nature (London)* **362,** 448 (1993).

[24] E. C. Pesci, J. P. Pearson, P. C. Seed, and B. H. Iglewski, *J. Bacteriol.* **179**, 3127 (1997).

[25] K. M. Gray, L. Passador, B. H. Iglewski, and E. P. Greenberg, *J. Bacteriol.* **176**, 3076 (1994).

[26] M. Givskov, L. Eberl, and S. Molin, *FEMS Microbiol. Lett.* **148**, 115 (1997).

[27] P. D. Shaw, G. Ping, S. L. Daly, C. Cha, J. E. Cronan, K. L. Rinehart, and S. K. Farrand, *Proc. Natl. Acad. Sci. U.S.A.* **94**, 6036 (1997).

expression of bioluminescence is a complex biochemical phenotype and it is sensitive to many factors, such as the availability of the aldehyde substrate, and requires high concentrations of ATP as well as oxygen. Bioluminescence works extremely well in population analysis under aerobic conditions but does not allow detection at the single-cell level in our biofilm setup (as described below). Furthermore, addition of the long-chain aldehyde, n-decanal, is required as a substrate for bioluminescence in constructions that have only $luxAB$. The utility of $lacZ$ as a reporter gene is limited to the assessment of single time points because of the stability of LacZ and the product of the reaction between LacZ and 5-bromo-4-chloro-3-indolyl-β-D-galactopyranoside (X-Gal), which forms an insoluble precipitate that is not easily removable from induced cells.

Modification of Green Fluorescent Protein for Use as Reporter Gene

An ideal, live bacterial AHL sensor, would be one that is not dependent on the addition of substrates and is robust under a variety of physiological conditions, and with the reporter product being removed from the cells, so that multiple events in gene regulation can be observed over time. The reporter should also be sensitive enough to be detected in a small number of cells, or ideally, at the single-cell level. The green fluorescent protein (GFP) fulfills these requirements. GFP, obtained from the jellyfish *Aequorea victoria,* requires only trace amounts of oxygen to mature, that is, no external compounds need to be added to organisms expressing GFP in order to detect green fluorescence.[28] GFP works well as a reporter for monitoring gene expression at the single-cell level in biofilms.[29,30] New variants of GFP (*mut3**) have been constructed that exhibit excitation maxima at about 470 nm, with little excitation in the UV band, a significant improvement in quantum efficiency, and a shorter maturation time, with typically a $t_{1/2}$ of 10–30 min compared with the 2 hr for wild-type *Aequorea* GFP. The problem with the extreme stability of GFP has been overcome by the construction of versions that possess a species-independent instability.[31,32] A collection of *gfp* genes, encoding proteins with various half-lives ranging from minutes to days,[32] were constructed by manipulating the *gfp* gene so the resultant protein carried C-terminal peptide tags such as AANDENYALAA. This peptide tag targets the GFP for degradation

[28] M. Chalfie, Y. Tu, G. Euskirchen, W. W. Ward, and D. C. Prasher, *Science* **263**, 802 (1994).

[29] S. Moller, C. Sternberg, J. B. Andersen, B. B. Christensen, J. L. Ramos, M. Givskov, and S. Molin, *Appl. Environ. Microbiol.* **64**, 721 (1998).

[30] C. Sternberg, B. B. Christensen, T. Johansen, A. T. Nielsen, J. B. Anderson, M. Givskov, and S. Molin, *Appl. Environ. Microbiol.* **65**, 4108 (1999).

[31] R. Tombolini, A. Unge, M. E. Davey, F. J. de Bruijn, and J. K. Jansson, *FEMS Microbiol. Ecol.* **22**, 17 (1997).

[32] J. B. Andersen, C. Sternberg, L. K. Poulsen, S. P. Bjorn, M. Givskov, and S. Molin, *Appl. Environ. Microbiol.* **64**, 2240 (1998).

by the tail-specific CLP proteases.[32-35] Modification of the tag determines the efficiency of targeting for degradation, and hence plays a role in determination of the half-life of the targeted protein. The basic medium for growth of GFP-expressing bacteria was either modified Luria–Bertani (LB) medium[36] containing 4 g of NaCl per liter instead of the normal 10 g of NaCl per liter or ABT minimal medium (AB minimal medium[37] containing 2.5 mg of thiamine per liter). For identification of GFP-expressing colonies we used a charge-coupled device camera mounted on an epifluorescence microscope (Zeiss Axioplan; Zeiss, Oberkochen, Germany) equipped with a ×2.5 lens.

Construction of Green Fluorescent Protein-Based N-Acylhomoserine Lactone Monitors

The $luxR-P_{luxI}$ quorum sensor derived from *Vibrio fischeri* has been used as the molecular component governing control over expression of our first collection of GFP reporters.[38] LuxR monitors the concentration of AHL and accordingly drives the expression of the *luxI* promoter (P_{luxI}), which in turn controls production of the *gfp* reporter. LuxR functions in a number of different bacterial strains[9,12,20,25] and is responsive to a variety of AHL molecules,[22,39] which makes it suitable as a broad-range monitor for the presence of AHLs. A fragment encoding $luxR-P_{luxI}$ and the adjacent region preceding the start codon of *luxI* is fused to *gfp* and introduced into high and low copy number broad host range plasmids. There are several issues that need to be addressed when designing *gfp* gene fusions. A major obstacle in the design of GFP (unstable)-based sensor systems is to match the transcriptional and/or translational output of the quorum sensor-regulated promoter (in this case P_{luxI}) with accumulation of GFP in the cells. Therefore, if AHL-induced gene expression must be detected at the single-cell level it puts a constraint on the GFP versions that can be used: expression of the construct must match the sensitivity of the state-of-the-art epifluorescence microscopes and laser confocal microscopes. If the stable GFP is employed, the cells express background fluorescence, so even in the absence of AHL molecules the bacteria will eventually fluoresce green. If, on the other hand, a version with a short half-life is used, almost no fluorescent signal can be detected from single cells. To obtain the lowest possible background signal, we have constructed a transcriptional fusion between $luxR-P_{luxI}$ and *gfp*(ASV), which encodes an unstable version of GFP exhibiting a half-life of approximately 40 min in

[33] S. Gottesman, E. Roche, Y. N. Zhou, and R. T. Sauer, *Genes Dev.* **12**, 1338 (1998).
[34] K. C. Keiler, P. R. H. Waller, and R. T. Sauer, *Science* **271**, 990 (1996).
[35] K. C. Keiler and R. T. Sauer, *J. Biol. Chem.* **271**, 2589 (1996).
[36] G. Bertani, *J. Bacteriol.* **62**, 293 (1951).
[37] D. J. Clark and O. Maaløe, *J. Mol. Biol.* **23**, 99 (1967).
[38] J. H. Devine, C. Countryman, and T. O. Baldwin, *Biochemistry* **27**, 837 (1988).
[39] A. L. Schaefer, B. L. Hanzelka, A. Eberhard, and E. P. Greenberg, *J. Bacteriol.* **178**, 2897 (1996).

exponentially growing *E. coli* cells.[39a] For efficient initiation of translation, we have equipped the fusion with an optimized *E. coli* ribosome-binding sequence (RBSII). *E. coli* harboring this *luxR*–P_{luxI}–RBSII–*gfp*(ASV) cassette on a high copy number plasmid appears completely dark under the epifluorescence microscope after 20 hr of incubation at 30° on LB plates but produces bright green fluorescent cells when cultivated under identical conditions in the presence of 5 n*M* *N*-(3-oxohexanoyl)-L-homoserine lactone (OHHL). This is referred to as the LuxR-based AHL monitor hereafter.

Detection and Quantification of Green Fluorescent Protein Expression

The green fluorescence phenotypes of colonies on semisolid surfaces, of single cells in colonies on semisolid surfaces, or of single cells from liquid cultures are recorded with a charge-coupled device camera mounted on an epifluorescence microscope (Zeiss Axioplan) equipped with a ×2.5 lens, a ×50 lens, or a ×63 lens, respectively. Detection of GFP is accomplished with Zeiss filter set 10 (excitation 480/410 nm, emission 540/525 nm, dichromic 510 nm), and a Zeiss HBO-100 mercury lamp. A Photometrics CH250 charge-coupled device (CCD) camera, equipped with an Eastman Kodak (Rochester, NY) KAF1400 liquid-cooled chip, is used for image acquisition (Roper Scientific, Bogart, GA). For detection of GFP signals from cultures grown in microtiter trays, samples are illuminated with a halogen lamp (Intralux 5000-1; Volpi, Schlieren, Switzerland) equipped with a 480/440 excitation filter (F44-001; AF Analysentechnik, Tübingen, Germany). Samples are placed in a light-sealed dark box (UnitOne, Birkeroed, Denmark) and images are captured with a Hamamatsu C2400-47 double-intensified CCD camera (Hamamatsu, Herrsching, Germany), using a 532/510 emission filter (Melles Griot 03 FIV 111; Melles Griot, Irvine, CA). A PC computer in part controls the camera and the images are saved in 16-bit format, using ARGUS-50 software (Hamamatsu).

A calibration curve relating the concentration of AHL to expression of fluorescence is made from liquid cultures of bacteria harboring the LuxR-based AHL monitor. An overnight culture is diluted 4-fold in fresh medium and incubated for 1 hr. The culture is split into subcultures (in microtiter trays), each receiving different AHL concentrations (1 to 1000 n*M*) and subsequently incubated for 2 hr. Fluorescence is determined on a fluorometer (model RF-1501; Shimadzu, Tokyo, Japan) set at an excitation wavelength of 475 nm and emission detection at 515 nm and related to concentration. Samples are placed in the light-sealed dark box and images are captured with the Hamamatsu C2400-47 camera as described above. A correlation between the 16-bit format color code of the ARGUS-50 software (Hamamatsu) and concentration of AHL can then be obtained. This system

[39a] J. B. Andersen, A. Heydorn, M. Hentzer, L. Eberl, O. Geisenberger, B. B. Christensen, S. Molin, and M. Givskov, *Appl. Environ. Microbiol.* **67**, 575 (2001).

can detect the presence of 3 nM OHHL, 15 nM N-hexanoyl-L-homoserine lactone (HHL), and 15 nM N-(3-oxododecanoyl)-L-homoserine lactone (OdDHL) in 1 hr.[39a] This calibration forms the basis for a high-throughput analysis of AHL induction and inhibitor action, where a number of combinations of furanone and AHLs can be measured in the ARGUS setup in relatively short time.

Application of Green Fluorescent Protein-Based N-Acylhomoserine Lactone Monitors to Biofilm-Attached Bacteria

For the study of laboratory cultures of biofilms we use a flow cell system. A single flow cell is composed of three separated flow chambers, which are small parallel channels cut into a Plexiglas block covered with a microscope coverslip glass that serves as substratum. The channel dimensions are 1 × 4 × 40 mm and the biofilms are supplied with a flow of 50× diluted LB medium. The flow systems are assembled and prepared as described previously.[29] The medium flow is typically 3 ml/hr or 0.2 mm/sec, corresponding to a laminar flow with a Reynolds number of 0.02. To inoculate the chamber, bacteria are injected with a syringe and needle through the connecting silicone tubing. The biofilm growing on the substratum in the flow chambers can be readily observed microscopically without obstruction. An epifluorescence microscope requires a "flat" sample, as the instrument is able to record only a single focal plane. Because biofilm samples can reach 100 μm in thickness, the confocal laser scanning microscope (CLSM) is required to discriminate and capture sharp images from within such structures. All microscopic observations of biofilms and subsequent image acquisitions are performed on a Leica TCS4D confocal microscope (Leica Lasertechnik, Heidelberg, Germany). For detection of GFP, we use a single laser line from an ArKr laser (Omnichrome T643; Omnichrome, Chino, CA) at 488 nm for emission and the Leica TCS4D fluorescein isothiocyanate (FITC) emission standard filter set for detection. Using confocal microscopy, only a single focal plane will be recorded by the detectors at one time, providing a means of virtual slicing of the object. Slices from other focal planes are acquired by moving the sample in the vertical direction. Using advanced computer software (e.g., Imaris; Bitplane, Zurich, Switzerland) on an SGI Indigo2 workstation (Silicon Graphics, Sunnyvale, CA) it is possible to reconstruct a three-dimensional representation of the sample by assembling the virtual slices obtained from the confocal microscope. This setup enables the formation and examination of biofilm structures in which AHL signaling and reception of signals take place. In a wild-type background this system offers a unique opportunity to watch the process of quorum sensing taking place in relation to biofilm build-up and organization. Wild-type as well as signal generation mutants can be investigated for their signal responses in relation to colony morphology (Fig. 1, see color insert). Cells of *Pseudomonas aeruginosa* (*lasI, rhlI*[40]) harboring the LuxR-based

[40] J. P. Pearson, E. C. Pesci, and B. Iglewski, *J. Bacteriol.* **179,** 5756 (1997).

monitor system organized in a microcolony-based biofilm respond within 15 min to the addition of 250 nM OHHL (Fig. 2, see color insert). We also find this system particularly useful for the study of quorum-sensing antagonists, as it allows for analysis of the efficiency of antagonist compounds against cells organized into biofilm structures. Furthermore, the use of the GFP-based single-cell technology in combination with confocal microscopy enables us to determine the penetration and half-life of both AHL and antagonist compounds such as furanones in the biofilm model systems. Figure 2 demonstrates the utility of this system to evaluate the role of signals and signal antagonists in regulating QS phenotypes in a live biofilm system. This in turn gives us the opportunity to redesign the chemistry and optimize the functionality of the compounds to comply with the requirements for biological activity under complex *in vivo* conditions. We have expanded this research into the lungs of mice infected with *P. aeruginosa* and our Lux R-based AHL monitor bacteria. Green fluorescent monitor cells are easily identified, indicating that cell-to-cell signaling occurs in the lungs of infected animals, and highlighting the usefulness of GFP-based, single-cell technology in a variety of complex biological systems.[40a]

Chemical Identification of *N*-Acylhomoserine Lactone Signals in Biofilms

Bioassay-based methods have proved to be useful tools for detecting signaling events in biofilms from ecologically relevant settings, including the rhizosphere,[41] a freshwater stream,[42] biofilm-coated urinary catheters,[43] packaged fish,[44] and the lungs of mice infected with *P. aeruginosa* (see above). However, none of these studies identified AHLs in the biofilm sample directly. Where AHLs have been identified, the usual approach has been to grow planktonic cultures of environmental isolates to generate a sample for analysis by thin-layer chromatography[27,42] or high-performance liquid chromatography (HPLC[45]) in combination with a bioassay. AHL identification is therefore based on comparison of retention factor or retention time, and the bioassay response obtained from AHL standards and the bacterial extract. Chromatographic resolution by thin-layer chromatography is, however, relatively poor, in comparison with other commonly used

[40a] H. Wu, Z. Song, M. Hentzer, J. Andersen, A. Heydorn, K. Mathee, C. Moser, L. Eberl, S. Molin, N. Hoiby, and M. Givskov, *Microbiology* **146**, 2481 (2000).

[41] L. I. Pierson and E. Pierson, *FEMS Microbiol. Lett.* **136**, 101 (1996).

[42] R. J. C. McClean, M. Whiteley, D. J. Stickler, and W. C. Fuqua, *FEMS Microbiol. Lett.* **154**, 259 (1997).

[43] D. J. Stickler, N. S. Morris, R. J. C. McLean, and C. Fuqua, *Appl. Environ. Microbiol.* **64**, 3486 (1998).

[44] L. Gram, A. B. Christensen, L. Ravn, S. Molin, and M. Givskov, *Appl. Environ. Microbiol.* **65**, 3458 (1999).

[45] J. P. Pearson, K. M. Gray, L. Passador, K. D. Tucker, A. Eberhard, B. H. Iglewski, and E. P. Greenberg, *Proc. Natl. Acad. Sci. U.S.A.* **91**, 197 (1994).

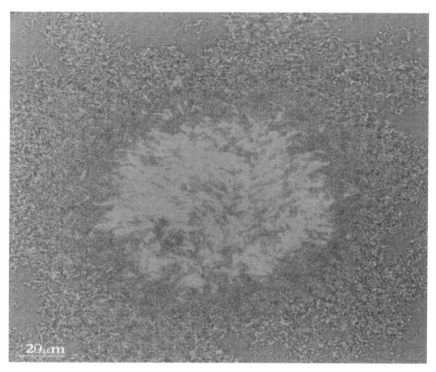

Fig. 1. Diffusion of AHL into the matrix of a biofilm. The image shows a microcolony of *Pseudomonas aeruginosa* (*lasI, rhlI*) carrying the Lux R AHL monitor. The flow cell has been perfused with 70 n*M* OdDHL for 6 hr and demonstrates that the autoinducer molecules are capable of diffusing into the center of a microcolony.

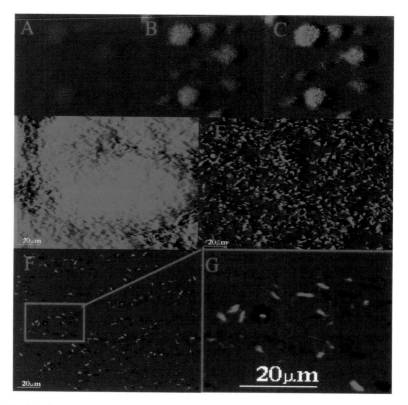

Fɪɢ. 2. Induction and repression of fluorescence in biofilm-grown *Pseudomonas aeruginosa* (*lasI, rhlI*) harboring the Lux R-based AHL monitor system. The flow chamber was perfused with 250 n*M* OHHL at time 0 and microcolonies were inspected at (A) *t* = 0 min, (B) *t* = 12 min, and (C) *t* = 30 min. (D) Closer examination of a fully induced microcolony at *t* = 60 min. The AHL antagonist, the halogenated furanone, was fed into the flow chambers and 3 hr later the effect on the quorum sensor was inspected. (E) Partial inhibition with furanone (5 μg/ml). (F) Almost complete inhibition with fu-ranone at 15 μg/ml. (G) A more highly magnified view of (F), showing few green cells in the pres-ence of furanone at 15 μg/ml.

A.

B.

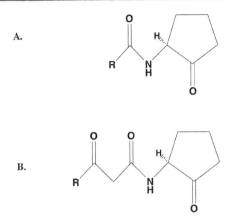

FIG. 3. The structures of the two major classes of *N*-acylhomoserine lactones: (A) those without the ketone group, such as *N*-butyrolactone homoserine lactone (BHL), and (B) those with the ketone group (3-oxo AHLs), such as *N*-(3-oxohexanoyl)-L-homoserine lactone (OHHL).

chromatographic techniques.[46] Bioassays can be compromised by the presence of other signaling compounds[17,47] and toxic compounds can compromise the integrity of the bioassay.[48] We therefore sought an alternative to the bioassay-based methodologies that was sensitive enough to detect AHLs directly in biofilm samples, provide a high degree of confidence in the identity of the AHL, and could be used as a quantitative assay. One technique, which has been used for analysis of trace levels of organic compounds in environmental samples, is gas chromatography–mass spectrometry (GC–MS). Gas chromatography has a higher chromatographic separation efficiency than thin-layer chromatography[46] and mass spectrometry has the potential for a high degree of specificity in identification of compounds. GC–MS has been applied to a wide range of complex environmental samples including pesticides,[49] steroids,[50] and microbial chemical markers such as carbohydrates[51] and fatty acids.[52]

AHLs can be split into two chemical classes based on the presence or absence of the ketone group on carbon-3 of the alkyl chain (Fig. 3). We developed a quantitative GC–MS method for the AHLs containing the ketone group, referred

[46] K. Robards, P. Haddad, and P. Jackson, "Principles and Practice of Modern Chromatographic Methods." Academic Press, London, 1994.

[47] M. T. G. Holden, S. R. Chhabra, R. deNys, P. Stead, N. J. Bainton, P. J. Hill, M. Manefield, N. Kumar, M. Labbate, D. England, S. A. Rice, M. Givskov, G. Salmond, G. S. A. B. Stewart, B. W. Bycroft, S. Kjelleberg, and P. Williams, *Mol. Microbiol.* **33**, 1254 (1999).

[48] J. Boutin, P. Lambert, S. Bertin, J. Volland, and J. Fauchere, *J. Chromatogr.* **725**, 17 (1999).

[49] J. A. Laramée, P. C. H. Eichinger, P. Mazurkiewicz, and M. L. Deinzer, *Anal. Chem.* **67**, 3476 (1995).

[50] B. G. Wolthers and G. P. B. Kraan, *J. Chromatogr.* **843**, 247 (1999).

[51] A. Fox, *J. Chromatogr.* **843**, 287 (1999).

[52] L. Larsson, *APMIS* **102**, 161 (1994).

to below as 3-oxoacylhomoserine lactones (3-oxo AHLs). The 3-oxo group was used to convert the AHL to an electron-capturing derivative. In addition to allowing gas chromatography of 3-oxo AHLs these derivatives can be used to exploit one of the most sensitive and specific mass spectrometric techniques available, i.e., electron capture–negative ionization (EC–NI).[53]

Measurement of 3-Oxo N-Acylhomoserine Lactones by Gas Chromatography–Mass Spectrometry: Electron Capture–Negative Ionization

Pentafluorobenzyl derivatives have been applied to a range of compounds for sensitive detection by EC–NI.[54] A method developed for steroids by Koshy *et al.* (1975)[55] is applied to the 3-oxo AHLs: *N*-(3-oxohexanoyl)-L-homoserine lactone (OHHL), *N*-(3-oxooctanoyl)-L-homoserine lactone (OOHL), *N*-(3-oxodecanoyl)-L-homoserine lactone (ODHL), *N*-(3-oxododecanoyl)-L-homoserine lactone (OdDHL), and *N*-(3-oxotetradecanoyl)-L-homoserine lactone (OtDHL).[52a]

The accuracy of quantitative analysis can be improved with an internal standard that is a stable isotope of the analyte. As organic synthesis of a 3-oxo AHL labeled with either deuterium [^{2}H] or carbon-13 [^{13}C] is difficult, a method has been developed to generate [^{13}C$_{16}$]OdDHL in a planktonic bacterial culture for use as the internal standard. An overnight planktonic culture of *P. aeruginosa* (6294)[56] is grown in 100 ml of the defined medium M9[57] containing [^{13}C$_6$]glucose (0.2%, w/v) as the only carbon source. The supernatant obtained after centrifugation (26,000 g for 10 min) (Avanti J-251; Beckman Instruments, Fullerton, CA) is extracted (three times) with dichloromethane (50 ml). The organic extracts are combined and the volume reduced *in vacuo* (Rotovapor; Büchi, Flawil, Switzerland), filtered (Teflon, 0.2-μm membrane; SGE Pty Ltd, Ringwood Victoria, Australia) and then taken to dryness *in vacuo* (SpeedVac-SVC200; Savant, Holbrook, NY). The sample is reconstituted in ethyl acetate (150 ml) and stored at −20°. Although this method does not provide a pure sample of [^{13}C$_{16}$]OdDHL it fulfills the specifications required of an internal standard, i.e., that a precise amount can be added to the known and unknown standards and there is no significant contribution to the signal for the 3-oxo AHLs.

[52a] T. S. Charlton, R. deNys, A. Netting, N. Kumar, M. Hentzer, M. Givskov, and S. Kjelleberg, *Environ. Microbiol.* **2**, 530 (2000).

[53] J. T. Watson, "Introduction to Mass Spectrometry." Lippincott-Raven, Philadelphia, 1997.

[54] D. A. Cancilla and S. S. Que Hee, *J. Chromatogr.* **627**, 1 (1992).

[55] K. T. Koshy, D. G. Kaiser, and A. L. VanDerSlik, *J. Chromatogr.* **13**, 97 (1975).

[56] I. Kudoh, J. P. Wiener-Kronish, S. Hashimoto, J. F. Pittet, and D. Frank, *Am. J. Physiol.* **267**, L551 (1994).

[57] J. Sambrook, E. Fritsch, and T. Maniatis, "Molecular Cloning: a Laboratory Manual." Cold Spring Harbor Laboratory Press, Cold Spring Harbor, New York, 1989.

Workup of the standards is a simple one-step derivatization reaction.[52a] A stock solution of pentafluorobenzylhydroxylamine hydrochloride (50 mg/ml pyridine; stored at $-20°$) is diluted to 1 mg/ml with pyridine. Aliquots (10 μl) of the diluted reagent are added to glass microreaction vials (100 μl; Hewlett-Packard, Palo Alto, CA), which contain previously dried samples or standards. The vials are capped with polytetrafluoroethylene (PTFE)-lined crimp-top caps. The solution is vortexed briefly and heated (90° for 0.5 hr) in a dry block heater. The samples are taken to dryness *in vacuo* (SpeedVac-SVC200; Savant), reconstituted in dichloromethane (100 μl), and then washed with HCl (1 M; 50 μl) and deionized water (50 μl). The washing procedure consists of recapping the vials after addition of the dichloromethane and 1 M HCl, vortexing briefly, and separating the organic and aqueous layers by centrifugation (1 min at 6000 g; Biofuge pico; Heraeus Instruments, Hanau, Germany). The aqueous layer is withdrawn from the vial with a syringe (100-μl barrel Teflon-tipped plunger; SGE Pty Ltd) and discarded. Water is then added with the syringe and the washing procedure repeated up to, and including, centrifugation. The organic layer is dried over sodium sulfate and transferred to a glass insert vial (100 μl; Alltech Pty Ltd, Baulkham Hills, N SW, Australia). The samples are taken to dryness *in vacuo* (SpeedVac) and reconstituted in ethyl acetate (50 μl), as chlorinated compounds can result in significant background for EC–NI mass spectrometry. The inserts are placed in vials (2 ml) and capped for GC–MS (EC–NI) analysis. Samples are usually assayed directly after derivatization or the following day (store at $-20°$).

The pentafluoro benzyloxime (PFBO) derivatives of the 3-oxo AHLs are separated on a gas chromatograph (5890 Series II; Hewlett-Packard) and detected by a mass spectrometer (MS Engine 5989B; Hewlett-Packard) configured for EC–NI. An aliquot (2 μl) of the sample is injected in the splitless mode (3 min), using helium ("ultra high purity") and a pressure program (head pressure, 0–2 min at 15 psi; 2–25 min at 5 psi). The injection port (300°) contains a gooseneck injection port liner with a tapered lower section (4-mm i.d., deactivated; SGE Pty Ltd). The liner is packed with stationary phase (1-1781, 1% SP 2100, 100/120 Supelcoport; Supelco, Bellefonte, PA) with a small amount of quartz wool at the top and bottom. The capillary column (cross-linked 5% PHME siloxane, 0.25-μm film depth, 0.25-mm i.d., 15 m, HP5-MS; Hewlett-Packard) is connected directly from injection port to mass spectrometer via the interface region (200°). The oven temperature is programmed for 50° (3 min) and then increased at 20°/min up to 310° (5 min). Mass spectrometer ion source conditions are as follows: 200° and 1.3 atm (methane). For quantification the mass spectrometer is configured for selected ion monitoring. The ions selected and the respective 3-oxo AHLs are as follows: m/z 227 (OHHL), m/z 255 (OOHL), m/z 283 (ODHL), m/z 311 (OdDHL), m/z 327 ([$^{13}C_{16}$]OdDHL), and m/z 339 (OtDHL).[52a]

The ion chromatograms for a standard spiked with [$^{13}C_{16}$]OdDHL show a pair of peaks for each derivatized 3-oxo AHL (Fig. 4). The two peaks are characteristic

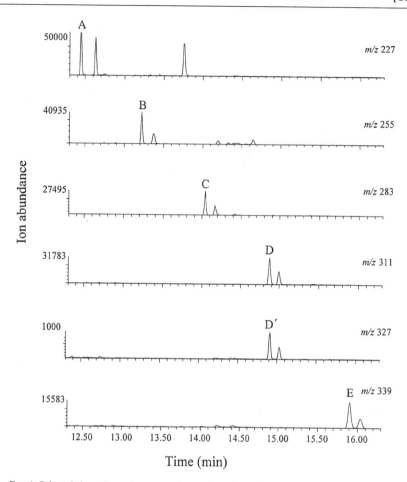

FIG. 4. Selected ion chromatograms of pentafluorobenzyloxime derivatives of 3-oxo AHL standards. The first peak of each derivative is (A) N-(3-oxohexanoyl)-L-homoserine lactone (OHHL); (B) N-(3-oxooctanoyl)-L-homoserine lactone (OOHL); (C) N-(3-oxo-decanoyl)-L-homoserine lactone (ODHL); (D) N-(3-oxododecanoyl)-L-homoserine lactone (OdDHL); (D′) [$^{13}C_{16}$]-N-(3-oxododecanoyl)-L-homoserine lactone ([$^{13}C_{16}$]OdDHL); and (E) N-(3-oxotetradecanoyl)-L-homoserine lactone (OtDHL).

of PFBO derivatives and are due to *syn/anti* isomerization.[54] While this reduces the signal-to-noise ratio by splitting the sample, the presence of two peaks in an unknown sample increases confidence in identification of the compound. The other features of the ion chromatograms are the good separation between the 3-oxo AHL peaks, a characteristic mass for each 3-oxo AHL, and the relative lack of other chromatographic peaks.

The sensitivity and quantitative validity of the assay are tested by assaying cell and supernatant fractions from a planktonic culture of the *lasI* mutant *P. aeruginosa* (PAO-JP1)[40] that have been spiked with a known amount of 3-oxo AHLs.[52a] The assay is sensitive, with detection of all of the derivatized 3-oxo AHLs at 1 ng/sample. Although the method is less accurate for the shorter chain AHLs, i.e., OHHL and OOHL, levels are still acceptable. Accuracy and precision of the assay for these shorter chain AHLs could be improved with the use of an isotopically labeled analog of OHHL or OOHL, as has been done for *P. aeruginosa*.

3-Oxo N-Acylhomoserine Lactone Concentration in Biofilm Culture

The 3-oxo AHL assay was then applied to a biofilm of *P. aeruginosa*. Two primary considerations were to obtain sufficient biofilm to detect the 3-oxo AHLs and to measure the biofilm volume. The strain *P. aeruginosa* (6294)[56] carrying a constitutive GFP-producing plasmid, pMF230 (M. Franklin, unpublished, 2000), was chosen because a number of 3-oxo AHLs, including OdDHL, were detected in this strain in planktonic culture, using the above-described assay (data not shown). Furthermore, OdDHL has been shown to play a role in biofilm structure for *P. aeruginosa*.[4] Constitutive expression of GFP provided a method to visualize the cells for *in situ* measurement of biofilm volume prior to destructive sampling for the 3-oxo AHL assay.

There has been little consideration given in the literature to the problem of measuring biofilm volume on a large-enough scale to provide enough sample for measurement of signals. For our study glass flow cells fed by sterile $2M$ medium[58] are used to grow the biofilm for 3-oxo AHL analysis.[52a] Biofilm volume is determined *in situ*, which allows a comparison of 3-oxo AHL concentration for a known amount of biofilm and effluent. An alternative method could utilize the constant depth film fermenter (CDFF), most recently reviewed by Dibdin and Wimpenny (1999).[59] This provides a volume of fixed depth in which biofilms can grow and under ideal conditions completely fill the recessed space.

Prior to destructive sampling for 3-oxo AHL analysis, biofilm volume on the baseplate and coverslip (Fig. 5) is determined by measuring the surface area of the biofilm at regular intervals of the biofilm. A confocal scanning laser microscope (CSLM) (Fluoview FV 300; Olympus, Tokyo, Japan) is used to image the biofilm with a ×10 objective to give the widest possible field of view and to have the focal length necessary to image the biofilm on the baseplate. Fields of view of the biofilm are randomly selected and the surface area of the biofilm is measured at 5-μm steps in the z plane from the top of the biofilm to the substratum. While this method allows calculation of volume for a biofilm with an uneven structure, only one flow

[58] C. Paludan-Muller, D. Weichart, D. McDougald, and S. Kjelleberg, *Microbiology* **142**, 1675 (1996).
[59] G. Dibdin and J. Wimpenny, *Methods Enzymol.* **310**, 296 (1999).

FIG. 5. A cutaway cross-sectional view of a flow cell for the growth of biofilms. The biofilm grows on the baseplate (B) and coverslip (C) of the flow cell. The depth of the flow cell (1 mm) was set by the thickness of the microscope slides (S).

cell can be analyzed each day because of the large number (ca. 60) of fields of view taken for each flow cell. The number of sample points may be reduced, however, depending on the degree of biofilm heterogeneity throughout the flow cell.

Application of the 3-oxo AHL assay to flow cell cultures detects a range of these signaling compounds in both the biofilm and effluent. Ion chromatograms of a biofilm sample show peaks for ODHL, OdDHL, $[^{13}C_{16}]$OdDHL, and OtDHL (Fig. 6). Three of the longer chain AHLs, OtDHL, OdDHL, and ODHL, are detected in the biofilm and effluent at micromolar and nanomolar concentrations, respectively. OdDHL is the most abundant of the 3-oxo AHLs. The next most abundant AHL is OtDHL, which has not previously been reported as an endogenous signal for *P. aeruginosa* and has been shown by us to have the same activity as OdDHL for activation of *lasR* in an *E. coli* reporter strain.[52a]

Surface-Associated Eukaryotic Signal Antagonists

Colonization of biological surfaces by microorganisms is often detrimental to the eukaryotic host surface. This is evident in the marine setting, where the colonization of algal surfaces leads to a reduction of available area for photosynthesis and increased drag, which can lead to mechanical shearing of the plant under high flow conditions (e.g., tidal changes or wave action). Further, the colonizing microorganisms may secrete products such as exoenzymes that are destructive to the plant surface. Therefore, the ability to regulate the development of bacterial biofilms, and the expression of bacterial phenotypes by plants and animals, confers a strong advantage to these organisms. A mechanism by which higher organisms may be able to select for bacteria and regulate the expression of phenotypes in some bacteria is through interference with the genetic regulatory systems used by these bacteria. For example, organisms may produce antagonists of the AHL signaling systems that some bacteria use to regulate phenotypes of colonization and exoenzyme production.

The sole example of specific interference with the AHL regulatory system by eukaryote-produced compounds comes from the red alga *Delisea pulchra* (Greville) Montagne. *Delisea pulchra* produces a range of halogenated furanones (Fig. 7),[13,60] which have a broad range of biological activities. Laboratory studies

[60] R. Kazlauskas, P. Murphy, R. Quinn, and R. Wells, *Tetrahedr. Lett.* **1**, 37 (1977).

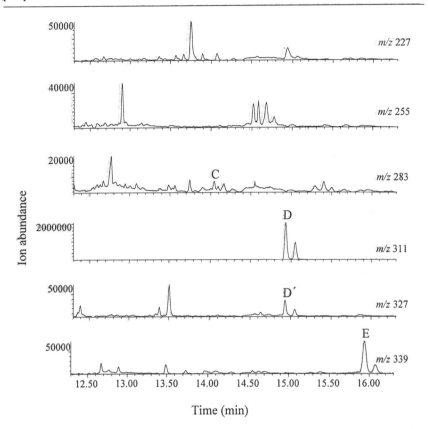

FIG. 6. Selected ion chromatograms of a biofilm sample taken from a flow cell and treated as per the pentafluorobenzyloxime derivatization method. The 3-oxo AHLs detected are C, ODHL; D, OdDHL; D', [^{13}C$_{16}$]OdDHL (internal standard), and E, OtDHL. Abbreviations for the 3-oxo AHLs are listed in full in the legend to Fig. 4.

have demonstrated that furanones are capable of inhibiting a range of AHL-regulated phenotypes[15] and that the furanones act as antagonists by either displacing or preventing the AHL signal molecule from interacting with the signal receptor protein, the R protein.[61] To determine whether furanone-mediated inhibition of quorum sensing is ecologically relevant, it is important to be able to determine how they are presented and their concentration on the surface of the plant.

[61] M. Manefield, R. deNys, N. Kumar, M. Givskov, P. Steinberg, and S. Kjelleberg, *Microbiology* **145**, 283 (1999).

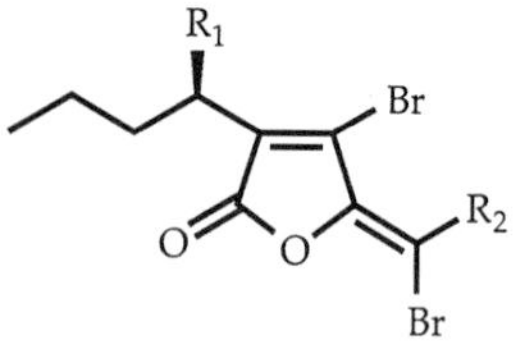

FIG. 7. Structures of furanones from *Delisea pulchra*. Four of the furanones produced by *D. pulchra* are presented. Each compound has the same basic ring structure, and varies only in substitutions at the positions designated as R_1 and R_2. *Inset:* The various constituents for the R groups for the four compounds.

Localization of Chemical Signals

Understanding how and where metabolites are localized has been essential in determining the ecological roles of plant-associated secondary metabolites.[62] Terrestrial and marine plants store and transport secondary metabolites in structures such as vacuoles and glands to protect against autotoxicity and present compounds at plant surfaces.[63,64] Studies of the localization of metabolites combined with quantitative analyses of these compounds have been essential in determining the ecological role of compounds.[62,65] We have developed sensitive methods to quantify furanones on the surface of *D. pulchra,* which has allowed an assessment of their ecological relevance as AHL antagonists.[16,66,67] The methodologies used in these studies are equally applicable to metabolites produced by eukaryotes, which act as signal agonists or antagonists in bacterial regulatory systems.

Light and Epifluorescence Microscopy

The ultrastructure of *D. pulchra* and the localization of furanones within the plant are determined by light and epifluorescence microscopy.[63] Light micrographs are made from unfixed and unstained hand-cut sections, using a Leitz Orthoplan microscope with bright-field optics and T-64 color reversal film. They show the presence of large numbers of gland cells among cortical cells and on the surface of the plant. The use of epifluorescence microscopy to localize compounds within these gland cells relies on the presence of conjugated double bonds in the compounds being investigated. All furanones have conjugated double bonds and

[62] R. F. Chapman and E. A. Bernays, *Experientia* **45**, 215 (1989).

[63] S. A. Dworjanyn, R. deNys, and P. D. Steinberg, *Mar. Biol.* **133**, 727 (1999).

[64] M. Wink, *Adv. Bot. Res.* **25**, 141 (1997).

[65] A. E. Stapleton, *Plant Cell* **4**, 1353 (1992).

[66] R. Maximilien, R. deNys, C. Holmstrom, L. Gram, M. Givskov, K. Crass, S. Kjelleberg, and P. Steinberg, *Aquat. Microb. Ecol.* **15**, 233 (1998).

[67] P. D. Steinberg, R. Schneider, and S. Kjelleberg, *Biodegradation* **8**, 211 (1997).

therefore fluoresce when irradiated by near-ultraviolet light. The fluorescence excitation and emission spectra of furanones are measured in ethanol at 100 μg/ml, using a Perkin-Elmer (Norwalk, CT) Luminescence LS 50B spectrophotometer with an excitation peak of 387 nm and an emission peak in the white–blue region at 435 nm. This characteristic fluorescence is used to localize compounds in hand-cut sections viewed under a Zeiss Axiophot microscope with a 360- to 395-nm long-pass filter, dichromatic mirror at 395 nm, and band-pass filter of 420 nm. These techniques demonstrate that furanones are localized in specialized gland cells within the thallus of *D. pulchra*, and at the surface of the plant. They also allow the distribution of gland cells to be quantified within the plant to determine the distribution of compounds internally. The subsequent step in the process of determining the surface concentration of furanones and other chemical signals is surface extraction and quantitative analysis.

Quantitative Analysis of Chemical Signals

To be able to quantify compounds on a surface, they first need to be removed from the surface without damaging the surface tissue. We have developed a technique to optimize the extraction and quantification of furanones from the surface of *D. pulchra*, while minimizing damage to cells.[68] We have also applied this technique to other algal species[68] and to culture studies of *D. pulchra*.[63] The technique is adapted from methodologies developed for the quantitative analysis of secondary metabolites on the surface of terrestrial plants.[69,70] Therefore, the techniques described below are equally applicable to future signal agonists and antagonists that may be identified from other eukaryotes.

Extraction Procedure for Surface Metabolites

Furanones are extracted from the surface of *D. pulchra* by selected solvent extraction. The choice of solvent and the time of extraction are critical factors ensuring that the solvent does not lyse cells, thereby removing compounds present within cells. We have assessed five organic solvents for their effects on the lysis of epithelial cells, using epifluorescence microscopy. The solvents [analytical reagent (AR) grade], in order of increasing polarity, are hexane (least polar), dichloromethane, diethyl ether, ethyl acetate, and methanol (most polar). Pieces of *D. pulchra* (1.5 g) dried in a salad spinner are held in each of the solvents for times of 10, 20, 30, and 60 sec, vortexed, and quantitatively assessed for the presence of lysed cells, using epifluorescence microscopy. Pigments in chloroplasts of algae autofluoresce when excited under near-UV to blue light, emitting a yellow to red fluorescence.

[68] R. deNys, S. A. Dworjanyn, and P. D. Steinberg, *Mar. Ecol. Prog. Ser.* **162**, 79 (1998).
[69] A. M. Zobel and S. A. Brown, *J. Nat. Prod.* **51**, 941 (1988).
[70] A. M. Zobel and S. A. Brown, *J. Chem. Ecol.* **19**, 939 (1998).

Healthy living cells have discrete chloroplasts that are identified by their fluorescence. When the plasmalemma of a cell is disrupted the chloroplasts of the cell lyse, filling the whole cell with pigment. The surface cells of whole axes are observed for fluorescence on a Leitz epiflourescence microscope with filter block D (BP 355-425, dichromic mirror 455, LP 460 nm). The effect of extraction time is directly assessed by counting the number of lysed cells per unit field (0.1 mm^2) for tissue from each extraction time. Of the five solvents tested for extraction of surface metabolites, all except hexane cause lysis of surface cells of *D. pulchra* at all the times tested.[68] Cells of *D. pulchra* dipped in hexane for 10, 20, and 30 sec remain intact but exposure for 60 sec causes lysis of cells.

The broader application of this technique is assessed with five species of brown algae, three other species of red algae, and one green alga. Whole plants are collected from the field and samples are taken from the top portions of the thallus for surface extraction in hexane for 30 and 60 sec, and in methanol for 30 sec. Extraction in methanol causes widespread lysis of cells for all nine algal species tested, whereas the cells of all species remain intact after extraction in hexane for 30 sec.[68] While hexane is a suitable extraction solvent for furanones, as they are hexane soluble, more polar compounds may require a more polar solvent for extraction. For surface-associated signal agonists or antagonists, a validation of extraction methods is necessary for each organism being studied prior to devising a quantitative analysis.

Quantitative Analysis of Surface Secondary Metabolites

The measurement of low levels of secondary metabolites requires a sensitive quantitative methodology. To quantify compounds it is necessary to have previously identified the metabolites being measured and to have purified standards available to construct accurate calibration curves. The methodology we have used for these sensitive studies is selected ion monitoring gas chromatography–mass spectrometry. This technique has been used to quantitatively verify surface extraction methodologies[68] and to measure the distribution of furanones both on and within *D. pulchra*.[63,68] In both studies, extracts for gas chromatography–mass spectrometry (GC–MS) are dissolved in ethyl acetate containing naphthalene as an internal standard at a concentration of 10 µg/ml. Surface extracts for GC–MS are made up in a volume of 10 to 400 µl of ethyl acetate (plus internal standard), dependent on the plant mass being extracted. Standards are isolated from dichloromethane extracts of *D. pulchra* by established protocols.[13] All standards are identified by comparison of ^{1}H and ^{13}C nuclear magnetic resonance (NMR) data with published data. Gas chromatography is performed with a Hewlett-Packard (HP) 5980 series II gas chromatograph and a polyimide-coated fused-silica capillary column (BP1 or BPX5, 12-m length, 0.15-mm i.d., 0.22-µm modified siloxane stationary phase; SGE Pty Ltd). All injections are performed in the split-less mode with an inlet pressure of 100 kPa. The injection port is held at 280° and the interface at 300°. The GC is held at 50° for 1.5 min and ramped at 20°/min to 300° (and held there for 10 min). Helium is used as the carrier gas. The mass spectrometry is performed on an HP 5971 mass selective detector (MSD). Ions characteristic of the internal standard and furanones (Fig. 7) are monitored in the selected ion monitoring (SIM) mode and are quantitatively analyzed using purified standards. Quantification is performed by measuring peak areas for each compound and the internal standard. The ratio of peak areas (compound to internal standard) is calculated for each metabolite and converted to concentration by reference to standard curves. To determine the surface concentration of secondary metabolites, the surface areas of pieces of D. pulchra are calculated, using a wet weight (mg) to surface area (cm^2) conversion factor. To calculate the conversion factor, fresh pieces of D. pulchra, between 5 and 500 mg wet weight, are dried and weighed before being placed on white paper. The algae are then immediately scanned and analyzed for surface area, using the National Institutes of Health (NIH) Imaging Program. Surface area is calculated to account for the area on both sides of the alga. The regression between wet weight and surface area has an associated R^2 value greater than 0.97 and probability less than 0.001.[68]

Surface concentrations of total furanones on D. pulchra are equal to or greater than 100 ng/cm^2 and levels of the most abundant furanone range from 50 to 125 ng/cm^2.[63,68] These ecologically relevant levels have significant effects on bacterial colonization phenotypes[16,66,71] and deter the settlement and growth of epiphytes.[14] We are now investigating how surface-associated metabolites affect the species composition and abundance of bacteria on D. pulchra. Through manipulation of plant chemistry, using culture techniques,[63] and analysis of the bacterial community by denaturing gradient gel electrophoresis (DGGE)[72] we are aiming to study the specific ecological effects of furanones on bacterial colonization.

Further Applications for Mass Spectrometric Analysis of Biofilms for Signaling Compounds

The sensitivity and structural information that can be obtained from mass spectrometry offer considerable analytical power for understanding the role of signaling compounds in biofilms. The 3-oxo AHL assay described above could be applied to any signaling compound with a ketone group, e.g., the γ-butyrolactone A-factor produced by *Streptomyces griseus*.[73] Pentafluorobenzyl oxime derivatives have

[71] P. D. Steinberg, R. deNys, and S. Kjelleberg, *Biofouling* **12**, 227 (1998).

[72] I. Dahllöf, H. J. Baillie, and S. Kjelleberg, *Appl. Environ. Microbiol.* **66**, 3376 (2000).

[73] S. Horinouchi, *in* "Cell–Cell Signalling in Bacteria" (G. M. Dunny and S. C. Winans, eds.), p. 291. ASM Press, Washington, D.C., 1999.

also been used to determine the structure of carbonyl compounds in environmental samples.[74]

A number of other analytical methods employing sensitive mass spectrometric detection of environmental samples are currently being explored for both AHLs and furanones. The AHLs that do not contain the 3-oxo group (e.g., BHL, HHL, and OHL) are also amenable to GC–MS analysis. Without an electron-capturing group, however, EC–NI mass spectrometry is not an appropriate detection method. These AHLs can, however, be detected by positive electron impact (EI^+) mass spectrometry and for standard solutions we have achieved limits of detection close to those obtained for the 3-oxo AHL assay and detected BHL in the supernatant of planktonic culture (data not shown). Similarly, we have detected cyclic dipeptides, another class of signaling compound,[47] by GC–MS (EI^+). EI^+ mass spectrometry is a form of universal detector and consequently it is likely that sensitivity would drop in complex samples such as biofilms. An alternative approach is HPLC directly coupled to an electrospray ionization mass spectrometer (HPLC–ESI–MS). Advantages of this method relevant to detection of AHLs in biofilm are the relatively high sensitivity (femtomoles to low picomoles) of ESI–MS[75] and the ability to chromotograph 3-oxo AHLs by HPLC without derivatization.

The sensitivity of EC–NI can also be exploited for detection of furanones because these AHL antagonists contain bromine, which is an electron-capturing group. We have used EC–NI mass spectrometry to detect furanones in *D. pulchra* in the nanogram per milliliter range (our unpublished data, 2000). This level of sensitivity is two orders of magnitude below that obtained by the EI^+ method used to measure furanones from the surface of *D. pulchra*.

Acknowledgments

We gratefully acknowledge the help of those members of our research teams who contributed to the development and refinement of these techniques. We thank our colleagues, whose work contributed to this research: Alex Toftgaard Nielsen, Arne Heydorn, Allan B. Christensen, Thomas B. Rasmussen, Claus Sternberg, Søren Molin, Peter Steinberg, Mike Manefield, Lyndal Thompson, and Simon Dworjanyn. T. Charlton was supported in part by an Australian Postgraduate Award. R. deNys was supported by an Australian Research Council Research Fellowship.

[74] R. M. Le Lacheur, L. B. Sonnenberg, P. C. Singer, R. F. Christman, and M. J. Charles, *Environ. Sci. Technol.* **27**, 2745 (1993).

[75] G. Siuzdak, "Mass Spectrometry for Biotechnology." Academic Press, San Diego, California, 1996.

[13] *In Situ* Quantification of Gene Transfer in Biofilms

By STEFAN WUERTZ, LARISSA HENDRICKX, MARTIN KUEHN,
KARSTEN RODENACKER, and MARTINA HAUSNER

Gene Transfer in Biofilms

In most natural and engineered systems microorganisms grow at interfaces as spatially organized multispecies communities of microorganisms embedded in a matrix of extracellular polymeric substances (EPS), forming biofilms or aggregates (e.g., sludge flocs, river or marine snow). Biofilms are defined as a collection of microorganisms and their EPS associated with an interface.[1] EPS are organic polymers of biological origin, composed of polysaccharides, proteins, nucleic acids, and other amphiphilic polymeric compounds,[2,3] which in biofilm systems are responsible for the interaction with interfaces[4] and for binding together cells and other particulate materials.[5]

The establishment of novel genetic traits in diverse natural environments such as biofilms is to a large extent the result of horizontal gene transfer. Transfer of plasmids by conjugation, a gene transfer mechanism that requires cell-to-cell contact in order to transfer plasmids or transposons from a donor to a recipient cell, is likely to be prevalent in biofilms because of the relative spatial stability of bacterial cells within the EPS matrix. Similarly, natural genetic transformation, in which competent recipient organisms take up free DNA liberated by excretion or by lysis of dying cells, may readily occur in biofilms, because the accessibility of free DNA may be increased as a result of its sorption to and accumulation in the EPS matrix, where it may be bound and protected. For example, it is known that DNA adsorbs to mineral surfaces such as mineral clays,[6] which protect it against degradation by DNases.

The spread of genetic information in biofilms is certainly a beneficial process when the accruement of new metabolic capabilities within a microbial community is desired. Such a situation may arise in the case of waste water treatment or

[1] K. C. Marshall, *Life Sci. Res. Rep.* **31** (1984).

[2] T. R. Neu, *Microbiol. Rev.* **60**, 151 (1996).

[3] T. R. Neu and J. R. Lawrence, *in* "Microbial Extracellular Polymeric Substances" (J. Wingender, T. R. Neu, and H.-C. Flemming, eds.), p. 21. Springer, Berlin, 1999.

[4] K. E. Cooksey, *in* "Biofilms—Science and Technology" (L. F. Melo, T. R. Bott, M. Fletcher, and B. Capdeville, eds.), Vol. 223, p. 137. NATO ASI Series. Kluwer Academic Publishers, Dordrecht, The Netherlands, 1992.

[5] W. G. Characklis and P. A. Wilderer, "Structure and Function of Biofilms." John Wiley & Sons, Chichester, 1989.

[6] G. Romanowski, M. G. Lorenz, and W. Wackernagel, *Appl. Environ. Microbiol.* **57**, 1057 (1991).

the bioremediation of hazardous waste sites. The term *bioenhancement* has been coined to describe the use of foreign microbes to deliver specific gene sequences via conjugation to a target indigenous bacterial population.[7] In contrast, the dispersal of plasmids carrying antibiotic resistance genes can be detrimental and is surely undesirable. These two different scenarios give rise to following question: What environmental parameters affect the frequency of exchange of genetic information in biofilms? To find answers to this question, a reliable means for the *in situ* quantification of gene transfer events in undisturbed biofilms must be available. Until now, gene transfer frequencies have mostly been evaluated by plating transconjugants (recipient cells that have received a plasmid from a donor cell) or transformants (recipient cells that have taken up free DNA) on selective agar plates. This method is thus dependent on the subsequent cell division and colony formation by transconjugants or transformants under selective pressure. However, the discovery of the green fluorescent protein (GFP), which provides a label for living bacteria and can be used as a reporter molecule for the investigation of gene expression or protein localization,[8] enabled the *in situ* microscopic detection of gene transfer events on a single-cell level. This approach has until now been limited to conjugal gene transfer.[9–14] Natural genetic transformation[15–19] has mostly been determined on the basis of transformant growth on selective agar media. A more promising approach is the concurrent use of GFP-labeled plasmids, automated confocal laser scanning microscopy (CLSM), and semiautomated quantitative image processing to determine conjugation[20] or transformation[21] frequencies in biofilms. These studies have shown that gene transfer frequencies detected *in situ* in biofilms by CLSM are higher than those quantified on selective agar media. In addition, this approach delivers information concerning the location and distribution of transformants and transconjugants within the biofilm.

[7] L. J. Ehlers and E. J. Bouwer, *Wat. Sci. Technol.* **39**, 163 (1999).

[8] M. Chalfie, Y. Tu, G. Euskirchen, W. W. Ward, and D. C. Prasher, *Science* **263**, 802 (1994).

[9] B. B. Christensen, C. Sternberg, and S. Molin, *Gene* **173**, 59 (1996).

[10] B. B. Christensen, C. Sternberg, J. B. Andersen, and S. Molin, *APMIS Suppl.* **84**, 25 (1998).

[11] B. B. Christensen, C. Sternberg, J. B. Andersen, L. Eberl, S. Moller, M. Givskov, and S. Molin, *Appl. Environ. Microbiol.* **64**, 2247 (1998).

[12] C. Dahlberg, M. Bergstrom, and M. Hermansson, *Appl. Environ. Microbiol.* **64**, 2670 (1998).

[13] O. Geisenberger, A. Ammendola, B. B. Christensen, S. Molin, K.-H. Schleifer, and L. Eberl, *FEMS Microbiol. Lett.* **174**, 9 (1999).

[14] B. Normander, B. B. Christensen, S. Molin, and N. Kroer, *Appl. Environ. Microbiol.* **64**, 1902 (1998).

[15] M. G. Lorenz and W. Wackernagel, *Microbiol. Rev.* **58**, 563 (1994).

[16] F. Gebhard and K. Smalla, *Appl. Environ. Microbiol.* **64**, 1550 (1998).

[17] K. M. Nielsen, A. M. Bones, and J. D. van Elsas, *Appl. Environ. Microbiol.* **63**, 3972 (1997).

[18] G. J. Stewart and C. D. Sinigalliano, *Appl. Environ. Microbiol.* **56**, 1818 (1990).

[19] H. G. Williams, M. J. Day, J. C. Fry, and G. J. Stewart, *Appl. Environ. Microbiol.* **62**, 2994 (1996).

[20] M. Hausner and S. Wuertz, *Appl. Environ. Microbiol.* **65**, 3710 (1999).

[21] L. Hendrickx, M. Hausner, and S. Wuertz, *Wat. Sci. Technol.* **41**, 155 (2000).

Green Fluorescent Protein as Labeling Tool

The green fluorescent protein is a biologically formed fluorophore responsible for the green emitted light of the bioluminescent system of the jellyfish *Aequorea victoria*.[8] Since its discovery, GFP has proved to have many advantages as a detection marker for living eukaryotic as well as prokaryotic cells. Unlike other commonly used reporter genes, no additional substrate is needed to allow direct *in vivo* monitoring of GFP-labeled cells. Labeled cells can be detected simply by visual inspection on irradiation.[22] With a size of only 238 amino acids, GFP is easily amenable to mutation, resulting in the creation of GFP variants with different excitation or emission spectra, improved brightness,[23,24] destabilized fluorescence expression,[25] or temperature-desensitized fluorescence expression.[26]

The use of GFP as a biological marker has many advantages. Most importantly, there is no selective pressure needed to observe *gfp*-expressing bacteria. Bacterial communities can be investigated without any outside mechanical, chemical, or biological disturbance. This allows simultaneous localization and detection of gene expression. It even allows detection of nonculturable cells.[20,27] However, the influence on the metabolism of the labeled cells due to the expression of this artificial protein remains unclear. The choice of using GFP to investigate *in situ* horizontal gene transfer is an obvious one, considering the simple detection of GFP expression after genes encoding GFP have been received by bacterial cells via horizontal gene transfer.

Preparation of Samples

To investigate horizontal gene transfer, GFP can be combined with fluorochrome tags such as general nucleic acid stains[21] or oligonucleotide probes.[13,20] General nucleic acid dyes are available with differing cell permeability, fluorescence enhancement on binding nucleic acids, excitation and emission spectra, DNA/RNA selectivity, and binding activity. Whereas some nucleic acid stains differentiate between live and dead cells, others do not. Fluorescently labeled oligonucleotide probes can be obtained with varying degrees of selectivity and can be conjugated to any desired fluorochrome. When biofilm samples are to be stored for future investigation or hybridized with gene probes, the biofilm needs to undergo

---

[22] D. C. Prasher, *Trends Genet.* **11,** 320 (1995).

[23] R. Heim and R. Y. Tsien, *Curr. Biol.* **6,** 178 (1996).

[24] T.-T. Yang, P. Sinai, G. Green, P. A. Kitts, Y.-T. Chen, L. Lybarger, R. Chervenak, G. H. Peterson, D. W. Piston, and R. Kain, *J. Biol. Chem.* **273,** 8212 (1998).

[25] J. B. Andersen, C. Sternberg, L. K. Poulson, S. P. Bjorn, M. Givskov, and S. Molin, *Appl. Environ. Microbiol.* **64,** 2240 (1998).

[26] Y. Kimata, M. Iwak, C. R. Lim, and K. Kohno, *Biochem. Biophys. Res. Commun.* **232,** 69 (1997).

[27] J. C. Cho and S. J. Kim, *FEMS Microbiol. Lett.* **170,** 257 (1999).

paraformaldehyde fixation. If the biofilm will be investigated immediately without hybridization, no fixation step is required.

Growth of Biofilms on Slides

Grow a biofilm on a glass microscopic slide in a petri dish filled with the appropriate medium and inoculate with one colony of a bacterial strain on a slowly tilting table until the desired state of the biofilm is reached.

Growth of Biofilms in Flow Cells

1. Construct a stainless steel[28] or Plexiglas[29] continuous flow cell system consisting of four separate flowthrough channels (40 mm long, 4 mm wide, 8 mm in height), sealed with 0.2-mm-thick coverslips (24 × 50 mm) and supplied with medium via inlets and outlets connected with Tygon or Teflon tubes. The entire flow cell, including tubing, can be sterilized by autoclaving or by rinsing with 0.5% (w/v) sodium hypochlorite.

2. Under sterile conditions, introduce medium into the flowthrough channels, using a peristaltic pump. Disconnect the medium flow and clamp the tubing connecting the inlets of the flow cell with the medium reservoir.

3. Sterilize the surface of the tubing near the outlet of the flow cell with ethanol, insert a sterile syringe tip connected to a sterile syringe containing an overnight culture of the recipient strain through the sterilized tubing into the flow channel, and inoculate the channels with 200 μl of the culture.

4. Cell attachment is allowed to take place during the following 2–3 hr.

5. Reconnect the medium flow and allow the biofilm to develop on the coverslips for the desired time period, depending on the strain and medium used.

Conjugation

1. Introduce donor and, if needed, helper cells to the recipient biofilm by pumping the cell solution into the channels.

2. Disengage the pump, and allow 2 hr for plasmid transfer between donor and recipient cells.

3. Wash for an additional 12 hr with phosphate-buffered saline [PBS; NaCl (8 g/liter), KCl (0.2 g/liter), Na_2HPO_4 (0.2 g/liter); pH 7.0] to allow GFP expression in recipient cells.

[28] M. Kuehn, M. Hausner, H.-J. Bungartz, M. Wagner, P. A. Wilderer, and S. Wuertz, *Appl. Environ. Microbiol.* **64,** 4115 (1998).
[29] G. M. Wolfaardt, J. R. Lawrence, R. D. Robarts, S. J. Caldwell, and D. E. Caldwell, *Appl. Environ. Microbiol.* **60,** 434 (1994).

Transformation

1. Pump medium containing a desired amount of specific DNA prepared by a standard alkaline DNA extraction method[30] into the flow channel.

2. Wash the channels with mineral medium (or 0.01 M MgSO$_4$) without DNA for an additional 12 hr to allow GFP expression in transformant cells.

Fixation of Biofilm on Slides

1. Gently rinse the biofilm-covered slide with PBS.

2. The biofilm is submerged in a 4% (w/v) paraformaldehyde (PFA) solution in a fresh petri plate for 1 hr at room temperature. The PFA solution is prepared as described.[31]

3. Wash the slide with PBS.

4. Drying the slide for 10 min at 65° will, in addition, fix the biofilm.

5. When necessary, dehydration can be carried out by incubating the slides for 3 min in ethanol solutions of increasing purity (50, 75, and 100%). After air drying of the microscopic slides, the samples can be stored in an air-tight vessel at −20° or stained immediately.

Fixation of Biofilm in Flow Channels

1. Rinse the biofilm by pumping PBS for 30 min through the flow channels.

2. Replace PBS in the flow channels with 4% (w/v) PFA, using the pump. Fix the biofilms for 1 hr at room temperature.

3. Wash the biofilms by pumping PBS through the channels for 1 hr.

Staining Biofilms Grown on Slides with Syto 17 or Syto 60 Nucleic Acid Stains

Syto 17 and Syto 60 are provided by Molecular Probes (Eugene, OR).

1. Gently rinse the slides with 0.01 M MgSO$_4$. The manufacturer recommends that buffers containing phosphate be avoided, as they may increase nonspecific background signals.

2. Incubate 1 cm^2 of the microscopic slide for 1–30 min at room temperature with 100 μl of a 50 nM–20 μM Syto solution diluted with 0.01 M MgSO$_4$, as recommended by the manufacturer.

3. Rinse the biofilm with 0.01 M MgSO$_4$.

[30] J. Sambrook, E. F. Fritsch, and T. Maniatis, "Molecular Cloning: A Laboratory Manual," 2nd Ed. Cold Spring Harbor Laboratory, Cold Spring Harbor, New York, 1989.

[31] R. I. Amann, *in* "Molecular Microbial Ecology Manual," p. 1. Kluwer Academic Publishers, Dordrecht, The Netherlands, 1995.

4. Unfixed biofilms can be directly monitored after covering them with a microscope coverslip. An unfixed biofilm cannot be stored after microscopic monitoring. Fixed biofilms need to be air dried. Dried, fixed, and Syto-stained biofilms can be stored for several months in a sealed 50-ml disposable plastic tube at $-20°$. For microscopy investigations, apply a drop of antifading agent (e.g., Citifluor, London, UK) to the slide. Note that antifading agents may contain toxic or carcinogenic components and should be handled and applied in a fume hood while wearing gloves. After monitoring by microscopy the antifading agent can be gently rinsed off the biofilm-covered slide with distilled water. The waste is collected according to laboratory safety rules for hazardous compounds. After drying, the biofilm can be stored in a disposable plastic 50-ml tube at $-20°$.

Staining Biofilms in Flow Channels with Nucleic Acid Stains

1. Pump 10 ml of a 50 nM–20 μM Syto solution diluted with 0.01 M MgSO$_4$ into the flow channels.
2. Rinse overnight with 0.01 M MgSO$_4$ by pumping at the same rate.

Hybridization of Biofilms on Slides with rRNA-Directed Oligonucleotide Probes

Fixed biofilms on slides are dotted with 16 μl of hybridization buffer[32] mixed with 2 μl (100 ng) of oligonucleotide probe. Biofilms are hybridized in a moisture chamber at 46° for 1.5 hr, washed, and prepared for microscopy as described previously.[32]

Hybridization of Biofilms in Flow Cells with rRNA-Directed
Oligonucleotide Probes

1. To precondition the fixed biofilms, hybridization buffer[32] containing the appropriate formamide concentration is gently pumped into the flow channels.
2. Desired oligonucleotide gene probes are added to hybridization buffer in a separate vial to obtain a final concentration of 6.25 ng of gene probe per microliter of hybridization buffer. The preconditioning hybridization buffer is pumped out of the flow channels.
3. The buffer/probe mix is pumped into the flow channels so that the biofilms on the lower coverslip are covered.
4. The inlet and outlet tubes are clamped with metal clamps. The flow cell including tubing is transferred to a sealable plastic dish. Hybridization is carried out for 1.5 hr at 46°.

---

[32] W. Manz, R. Amann, W. Ludwig, M. Wagner, and K.-H. Schleifer, *System. Appl. Microbiol.* **15**, 593 (1992).

5. After hybridization, the flow cell is reconnected to the pump. Appropriate washing buffer,[32] prewarmed to 48°, is pumped through the channels to displace the hybridization buffer/probe solution.

6. The flow cell is again transferred to a sealable plastic dish and incubated for 30 min at 46°.

7. The washing buffer is displaced with distilled water and the hybridized biofilm can be observed with a microscope.

Monitoring of *in Situ* Gene Transfer

Transformation in a Monoculture Biofilm

To investigate natural transformation in a monoculture biofilm, discrimination between two cell types is needed: recipients and transformants. Whereas recipients can be detected with a general nucleic acid dye that helps visualize the total biofilm, transformants will be detected only after receiving and expressing genes encoding a GFP variant. Hendrickx *et al.*[21] used plasmid pGAR2, containing the enhanced yellow fluorescent protein (*eyfp*) insert from pEYFP (Clontech, Palo Alto, CA), for transformation of the highly naturally competent *Acinetobacter* sp. BD413. The monoculture BD413 biofilm was dyed with the Syto 60 general nucleic acid stain. Syto 60-stained cells could be detected with an LSM410 confocal laser scanning microscope (Zeiss, Jena, Germany), using the 633-nm laser line and a 665-nm longpass emission filter. A 488-nm laser line and a 515- to 540-nm bandpass emission filter were used for the detection of cells expressing *eyfp*. Transformed cells were detected in superimposed optical sections in the *xy* plane of the biofilm as Syto 60-stained cells expressing *eyfp* (Fig. 1A; see color insert).

Conjugation in Defined Biofilm

Investigation of conjugation in a biofilm implies discrimination between at least three cell types: donors, recipients and transconjugants. Whereas donors carry and express the transferable genes encoding a GFP variant, recipients can be labeled with an oligonucleotide probe specific for the recipient strain. Recipients that have accepted and expressed the GFP variant will be detected as labeled cells expressing GFP. Note that triparental conjugation involves a fourth player, the helper strain. In the conjugation process it acts as a silent partner, providing the donor with the necessary genes encoding the cognate conjugation system to transfer the plasmid carrying a GFP variant to the recipient strain. When the GFP variant-carrying plasmid is transferred to the helper strain, cells will express fluorescence and be rightfully counted as donor cells.

Hausner and Wuertz[20] investigated triparental conjugation between *Escherichia coli* donor strain GM16(pRK415::*gfp*), *Alcaligenes eutrophus* recipient strain AE104, and *E. coli* helper strain CM404(pRK2013). To visualize the recipient cells the biofilm was hybridized with the rRNA-directed oligonucleotide probe BET42a,[32] labeled with the fluorescent dye tetramethylrhodamine-5-isothiocyanate (TRITC), which is specific for the β subgroup of proteobacteria. The hybridized biofilms were investigated by CLSM, using the 488- and 543-nm laser lines, in combination with the 515- to 540-nm bandpass and 590-nm longpass emission filters to detect GFP-expressing donors and TRITC-labeled recipients, respectively. Transconjugants were recognized as TRITC-labeled cells expressing GFP fluorescence (Fig. 1B; see color insert).

Monitoring of in Situ Gene Transfer on Line with Dual Fluorescence Labeling Using Green Fluorescent Protein Variants

The creation of various GFP variants has opened up the possibility of using fluorescent proteins for dual/multiple labeling. With dual fluorescence labeling using GFP variants, gene transfer can be monitored over time without interruption of the process. Until now dual labeling in prokaryotic cells was established with GFP mutants excitable in the UV range.[24,33] But considering the mutagenic character of UV radiation, it is more desirable to use GFP variants excitable at a longer wavelength. Promising variants with separable fluorescence spectra and excitable in the non-UV range have already been engineered. EYFP has an excitation peak at 513 nm and an emission spectrum in the yellow–green region (527 nm). Enhanced cyan fluorescent protein (ECFP; Clontech) has a fluorescence excitation (major peak at 433 nm and a minor peak at 453 nm) and emission (major peak at 475 nm and a minor peak at 501 nm) spectrum compatible with EYFP for the purpose of dual fluorescence labeling.

Discoveries of biological tags with features similar to GFP are reported to be useful for multilabeling in combination with GFP variants. The GFP homolog drFP583[34] from a *Discosoma* species, with an absorption maximum at 558 nm and emission maximum at 583 nm, could be used as a reporter protein in combination with the red-shifted GFP variant P11,[35] which has an excitation peak at 489 nm and an emission peak at 511 nm.

Automated Image Acquisition

A predefined protocol is needed to obtain reliable data for the quantification of gene transfer in biofilms. Obviously, intact biofilms are three-dimensional (3D) structures. Thus, gene transfer must be quantified in a biofilm volume unit. Such

[33] P. J. Lewis and A. L. Marston, *Gene* **227**, 101 (1999).

[34] M. V. Matz, A. F. Fradkov, Y. A. Labas, A. P. Savitsky, A. G. Zaraisky, M. L. Markelov, and S. A. Lukyanov, *Nature Biotechnol.* **17**, 969 (1999).

[35] R. Heim, D. C. Prasher, and R. Y. Tsien, *Proc. Natl. Acad. Sci. U.S.A.* **91**, 12501 (1994).

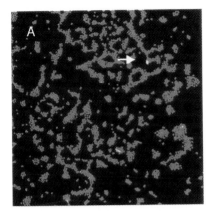 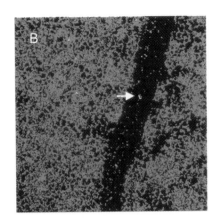

Fig. 1. *Acinetobacter calcoaceticus* (A) and *Alcaligenes eutrophus* (B) recipient biofilms depicting transformant (A) and transconjugant (B) cells. *Acinetobacter calcoaceticus* biofilms were stained with the general nucleic acid Syto 60. *Alcaligenes eutrophus* biofilms were hybridized with a Cy5-labeled oligonucleotide probe. Transformants and transconjugants express EYFP fluorescence. Both images are superpositions of two single *xy* optical sections, collected with the CLSM. Recipients were assigned the color red; transformants [white arrow (A)] and transconjugants [white arrow (B)] emit both EYFP and probe-conferred fluorescence. The length of the horizontal image edge is 318 μm.

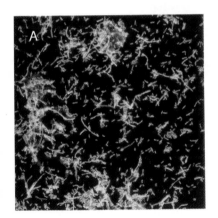 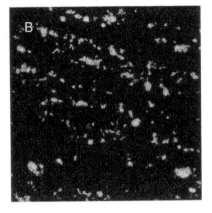

Fig. 2. Two examples depicting different outcomes of *in situ* hybridization of YFP-expressing cells. Both images are superpositions of two single *xy* optical sections, collected with a CLSM. Oligonucleotide probe-labeled cells were assigned the color red; cells expressing *gfp* or *eyfp*, which could not be hybridized by the probe, are depicted in green; *gfp*- or *eyfp*-expressing, oligonucelotide probe-labeled cells appear yellow. (A) *Sphingomonas* sp. LB126[42a] transconjugant cells (yellow) emitting GFP-conferred fluorescence, and hybridized with a Cy5-labeled oligonucleotide probe. Note GFP-expressing elongated cells (green), which could not be hybridized by the probe. (B) *Alcaligenes eutrophus* transconjugant cells (yellow), emitting EYFP-conferred fluorescence, and hybridized with a Cy5-labeled oligonucelotide probe. The length of the horizontal image edge is 318 μm.

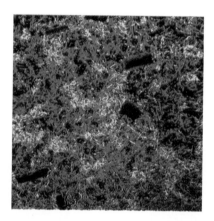

Fig. 3. *Acinetobacter calcoaceticus* cells, expressing *gfp* (shown in green), and counterstained with the nucleic acid stain Syto 60 (depicted in red). Note red cells, which do not show GFP fluorescence, *gfp*-expressing green cells, which could not be counterstained with the nucleic acid stain, and a few Syto 60-stained cells that emit GFP-conferred fluorescence (yellow). The length of the horizontal image edge is 128 μm.

an approach necessitates the use of CLSM,[36] which allows for the collection of confocal *xy* optical sections obtained in stacks along the *z* axis. To minimize the subjective choice of a microscopic field to be scanned (i.e., repeatedly finding a position in the biofilm where a high number of transconjugants or transformants is apparent), an automated scanning procedure[28] should be employed. This quantitative optical method for routine measurements of biofilm structures under *in situ* conditions is based on macro routines designed to perform automated investigations of biofilms, using image acquisition and image analysis techniques. The key components of this method are the on-line collection of confocal two-dimensional (2D) optical images in the *xy* plane from a 3D domain of interest followed by the off-line analysis of these 2D images. A 3D reconstruction of the relevant biological information can be generated with the quantitative data extracted from each 2D image.

The method was developed for the LSM410 confocal laser scanning microscope coupled to an Axiovert 135TV inverse fluorescence microscope (both instruments from Zeiss). Before quantification, the flow cells with the stained or hybridized biofilm to be investigated must be firmly mounted to the specimen support frame incorporated into the motorized stage. Next, the user checks the biofilm parameters (the staining or hybridization quality, transconjugant or transformant frequency, *gfp* expression, and thickness and density of the biofilm) with the microsope in the fluorescence mode and finds a suitable starting position. A prescan of the selected area must be performed to locate the biofilm–glass interface, the $z = 0$ starting plane, and to optimize the scan parameters. These include contrast, brightness, pinhole size, laser intensity, and possibly the use of the average filter during image collection. The use of this filter, which eliminates background noise, is recommended for weak signals such as in the case of low GFP fluorescence. However, it must be kept in mind that an increased number of scans will have a photobleaching effect on the probes used. The thickness of the biofilm can be easily determined by scanning through the entire depth of the biofilm. The decision must be made as to whether the GFP and probe (or nucleic acid stain)- conferred fluorescence will be collected simultaneously, using multichannel imaging, or separately with two subsequent scan runs (see Problems and Solutions, below). Next, the image collection macro is activated and the user is prompted to input information on the image edge length, the number of stacks in the domain of interest, the vertical step interval (Δz) between consecutive vertical *xy* sections (which depends on the choice of the objective lens and fluorescence properties of the scanned probes; see Problems and Solutions), and the file name for image storage.

The decision on the total number of stacks to be scanned in order to obtain representative data is dependent on the heterogeneity of the biofilm. A method for determining statistically representative areas of microbial biofilms is discussed

[36] D. E. Caldwell, D. R. Korber, and J. R. Lawrence, *Adv. Microbial Ecol.* **12**, 1 (1992).

by Korber et al.[37] It is also advisable to obtain data from different locations in the flow channel (i.e., near the wall, in the middle, near the outlet or inlet, etc.). Choosing the location for the scan depends on the question to be answered by the experiment and should be considered during experimental design.

Semiautomated Digital Image Processing and Analysis

The collected gray images can be processed with Quantimet 570 image processing software (Leica, Cambridge, UK). The macro routines used allow for semiautomated processing of image data.[28] The user input involves pixel calibration according to the objective used for image collection with the CLSM, and specifying the size and position of the area of interest for measurements. The use of gray image transformation filters and appropriate thresholds for binarization of transformed images is also determined by the user. In most cases, good-quality CLSM images do not require filtering. The automatic macro routines involve image acquisition, image transformation with the user-specified filters, image binarization using the user-selected thresholds, colocalization (i.e., detection of common pixels in two corresponding images), quantification of areal fraction (the ratio of detected pixels to the total pixel number in the measured frame) in binarized images, and data output into a user-specified table, which can be read into a Microsoft (Redmond, WA) Excel file. Biovolumes are calculated by the trapezoidal rule. A more detailed discussion of the applied image processing is presented by Kuehn et al.[28] To find a possible relationship between the size and location of recipient clusters, the distribution of transconjugant volume, and conjugation frequency, 3D image-processing methods [Interactive Data Language (IDL) software; Research Systems, Boulder, CO] can be applied.[38]

Problems and Solutions

Inability to Detect Green Fluorescent Protein

GFP detection in cells containing the *gfp* gene is not always guaranteed. Tsien[39] summarized the factors that affect the detectability of green fluorescent protein. At the DNA level low expression may not provide a sufficient amount of GFP molecules for cells to be detected. Posttranslational fluorophore formation can be hindered after satisfactory expression levels of *gfp* with problems such as solubility, folding hindrance, or availability of O_2.[39] This may result in inter- and

[37] D. R. Korber, J. R. Lawrence, M. J. Hendry, and D. E. Caldwell, *Binary* **4**, 204 (1992).

[38] K. Rodenacker, A. Brühl, M. Hausner, M. Kuehn, V. Liebscher, M. Wagner, G. Winkler, and S. Wuertz, *Image Anal. Stereol.* **19**, 151 (2000).

[39] R. Y. Tsien, *Annu. Rev. Biochem.* **67**, 509 (1998).

intraspecies variation in GFP expression (see below). Also, the molecular properties of perfectly folded GFP variants cannot always offer the expected fluorescence regarding excitation or emission wavelength, extinction coefficient, or quantum yield.[39] Control experiments with labeled pure culture species are necessary to obtain optimized CLSM detection settings.

Another problem is the occurrence of photoisomerization or photobleaching and dimerization. In some instances background signals could compromise the reliability of GFP detection. Tsien[39] also noted that the location of GFP itself could be confined to small subregions of cells, which could lead to an underestimation of the biovolume of *gfp*-expressing cells. On the other hand, the location of GFP could be diffuse, as if cells leak GFP out. This can lead to an overestimation of GFP detection.

Even strong *gfp* expression and good GFP maturation with desired molecular properties in an environment with no disturbing background signals may not be without problems. When *gfp* expression is too strong and GFP production too high, bacteria could suffer from toxicity effects. There have been (mostly unpublished) reports regarding toxicity associated with the overexpression of *gfp* in bacteria.[40] For example, *eyfp*-expressing *Acinetobacter* sp. BD413 cells could not be maintained on selective medium.[41]

Inter- and Intraspecies Variation in Green Fluorescent Protein Expression

Experiments with different recipient strains have shown that the intensity of GFP-mediated fluorescence may vary between different strains, GFP variants, and between cells of the same population. For example, *Acinetobacter* sp. BD413 cells labeled with the pRK415::*ecfp* plasmid homogeneously expressed strong cyan fluorescence.[42] In contrast, *Sphingomonas* sp. LB126 cells labeled with the same plasmid expressed weak ECFP-mediated fluorescence.[42] However, when the same strain was labeled with pRK415::*gfp*, elongated *Sphingomonas* sp. LB126 cells expressed strong fluorescence, whereas weak GFP fluorescence emanated only from shorter, oval cells in the same population.[42] Differences in intensities in GFP fluorescence among different transconjugants have also been reported by Geisenberger *et al.*[13] To optimize detection of even weakly fluorescent cells, the CLSM settings must be carefully selected. This is especially problematic when variations within the same population are noted. Increasing the brightness and contrast parameters to detect weakly fluorescent cells may lead to overestimation of the signal from brightly fluorescent cells.

[40] S. A. Endow and D. W. Piston, *in* "Green Fluorescent Protein: Properties, Applications and Protocols" (M. Chalfie and S. Kain, eds.), p. 272. John Wiley & Sons, Chichester, 1998.

[41] L. Hendrickx, M. Hausner, and S. Wuertz, submitted (2001).

[42] L. Hendrickx and M. Hausner, unpublished data (1999).

Inability to Stain gfp-Expressing Cells with Oligonucleotide Probes or with Nucleic Acid Stains

Previous studies have revealed that transconjugants with high GFP-mediated fluorescence intensities cannot be detected by fluorescence *in situ* hybridization (FISH).[13] This phenomenon is observed, for example, with *Sphingomonas* species transconjugant cells exhibiting high GFP fluorescence (Fig. 2A[43]; see color insert). In this case, elongated transconjugant cells, which emanate strong GFP fluorescence, cannot be detected by the oligonucleotide probe. On the other hand, experiments with other recipient strains, such as *Alcaligenes eutrophus*, have shown that the majority of the transconjugants, showing moderate levels of GFP fluorescence, can be detected with the BET42a probe directed against the beta subclass of proteobacteria (Fig. 2B; see color insert). The same effect is observed with nucleic acid stains to an alarming extent. For example, when a *gfp*-expressing *Acinetobacter* sp. biofilm is stained with the Syto 60 nucleic acid stain it is observed that less than one-third of the cells cannot be visualized with the nucleic acid counterstain (Fig. 3; see color insert). At present, it is not known what causes this unfavorable side effect of GFP expression. However, it can be speculated that the high protein levels in the cell may pose a permeability barrier to the fluorescent probes.

Spectral Overlap

It has been observed that enhanced green fluorescent protein (EGFP), EYFP, and ECFP have long emission tails[42] that extend into the excitation spectra of fluorophores excitable in the green spectral range. For this reason, dual labeling of cells with these GFP variants and dyes such as TRITC or the cyanine derivative Cy3 may result in spectral overlap of the two fluorophores. We have observed this phenomenon in *egfp*-expressing cells labeled with a TRITC-conjugated lectin.[44] In this case both GFP and TRITC emission overlapped to produce an emission in the orange–red spectral range.

Molecular proximity and insufficient separation of emission and excitation ranges can result in fluorescence resonance energy transfer (FRET). This is a quantum mechanical phenomenon that occurs when two fluorophores are less than 100 Å apart. Under these conditions, the emission of one fluorophore, the donor, overlaps the excitation spectrum of the second fluorophore, the acceptor, and produces emission of the acceptor. The emission of the donor, which would normally occur in the absence of the acceptor, cannot be detected.[39] Theoretically, FRET could pose an unexpected problem when using a combination of fluorophores or GFP variants with overlapping emission and excitation spectra.

[43] L. Bastiaens, D. Springael, P. Wattiau, H. Harms, R. deWachter, H. Verachtert, and L. Diels, *Appl. Environ. Microbiol.* **66,** 1834 (2000).
[44] A. Johnsen, unpublished data (1998).

When spectral overlaps are to be avoided, it is advisable to use a green fluorescent protein in combination with a dye that has an excitation maximum in the red spectral range. In our experience EGFP or EYFP works best in combination with Cy5-labeled gene probes or the nucleic acid stain Syto 60 (Molecular Probes). The emission and excitation spectra are sufficiently separated so that the emissions of both species can be independently detected. Such a combination of fluorophores is also suitable for multichannel imaging.

Optimizing Thickness of Optical Sections Obtained with Confocal Laser Scanning Microscope

To avoid the overestimation of biovolumes it is important to minimize the thickness of the optical sections obtained with a CLSM. Failure to do this could result in greater signal intensity, which will lead to overestimation of the scanned area and eventually to overestimation of biovolumes. In addition, should a donor cell and a recipient cell be positioned vertically above each other, both donor and recipient signals would be collected from the same xy location if the thickness of the optical section were too large. During semiautomated image processing, such overlapping signals could be falsely interpreted as transconjugant cells. The vertical dimension of the optical section is described by the half-maximum width of the optical section (full width, half maximum, FWHM), which takes into account the axial intensity weighting function $I(z)$ of the confocal microscope. The FWHM is dependent on the excitation and emission wavelengths of the fluorochrome used, the numerical aperture and magnification of the objective, the refractive index of the immersion medium, and the CLSM digital pinhole parameter.[45] In general, decreasing the pinhole size will greatly decrease the FWHM. Similarly, using a higher magnification objective and decreasing the refractive index of immersion medium will reduce the FWHM. In our experience, a ×63, 1.2NA water immersion objective is the best choice when investigating hydrated biofilms. Approximating the refractive index of the immersion medium to that of the investigated sample will also diminish the spreading of the intensity weighting function caused by spherical aberration. Similarly, a ×100, 1.3NA oil immersion objective is a good choice for the investigation of biofilms on slides to which an antifading agent has been applied.

Mechanically and Optically Induced Image Offset

Offset in dual-channel images as a result of two corresponding single-channel images not being congruent can occur when the motorized specimen stage, which shifts during automatic image acquisition in user-determined increments, does not move exactly the specified distance. Even a difference of 0.25 μm may result in

---

[45] H. Nasse, personal communication (1999).

an incorrect overlap of, for example, GFP- and oligonucleotide probe-conferred fluorescence. Underestimation of the number of transconjugants or transformants could result if the mechanical misalignment causes incomplete overlap of GFP- and probe-conferred fluorescence. This effect is reduced by a 1-pixel dilation of the area covered by recipients (i.e., oligonucleotide probe- or nucleic acid stain-positive cells[38]). In contrast, overestimation could be caused if a *gfp*-expressing donor cell and a probe-labeled recipient cell overlap as a result of the mechanical misalignment. The superposition would be incorrectly interpreted as a transconjugant cell. For this reason reference images must be taken at the beginning of each single-channel scanning procedure. When the collection of images of one fluorescence type (e.g., oligonucleotide probe) is complete, a new scan at the original starting position should be made and compared in all three directions with the reference scan. If necessary, the realignment of the motorized stage must be performed manually. To avoid the possibility of an unwanted mechanical misalignment and to minimize photobleaching of sensitive probes, multiple channel imaging is recommended. Several factors must be optimized in order to minimize optical offsets resulting from multichannel imaging. These include alignment of the optical pathway and choosing the most suitable lens and immersion medium (e.g., water immersion objective for hydrated biofilms). These and other related factors are discussed in more detail by Lawrence *et al.*[46]

Control Experiments

In light of this information, it is obvious that control experiments must be carried out in order to minimize false-positive or false-negative results. Recipient biofilms should be hybridized or stained and scanned in order to check the efficiency of the hybridization or staining procedure. These biofilms should also be scanned, using the CLSM settings for the detection of GFP signals, in order to check for autofluorescence of recipients. Spectral overlaps (between GFP and probes) should be minimized by the appropriate choice of the counterstaining probe and by optimization of CLSM settings, including the choice of beamsplitters and emission filters. Next, single-species transconjugant or transformant biofilms should also be hybridized or stained and scanned to evaluate the efficiency of probe binding to *gfp*-expressing cells. As discussed above, some *gfp*-expressing cells prevent probe entry. This may in fact result in an underestimation of transconjugant or transformant biovolumes. Further negative control experiments for conjugation in biofilms would include setups in which no gene transfer could take place (i.e., matings without helper strain or fixation of the biofilm immediately after the

---

[46] J. R. Lawrence, G. M. Wolfaardt, and T. R. Neu, *in* "Digital Image Analysis of Microbes" (M. H. F. Wilkinson and F. Schut, eds.), p. 431. John Wiley & Sons, New York, 1998.

addition of donor cells). The data obtained from these experiments provide information on the false detection of transconjugants as a result of mechanical or optical shifts or of choosing inappropriate CLSM settings, which result in a large FWHM and thus possible overlap of donor and recipient cells (see discussion above). Overlaps detected and quantified as transconjugants in these negative controls should be used for background correction of data obtained from actual experiments. Prior to semiautomated image processing and quantification, visual inspection of the collected data using programs such as IDL is highly recommended. This allows the user to spot noncellular material, such as autofluorescing crystals, which could lead to false-positive results. Positive controls should be carried out on agar plates, using classic microbiology techniques, to ensure that conjugation or transformation with the strains used does take place. Where possible, nonfixed biofilms from parallel setups should be scraped off coverslips at the end of the experiment, suspended in PBS buffer, homogenized, and plated out on selective agar plates. Conjugation and transformation frequencies determined by these methods should be compared with the *in situ* frequencies obtained with the CLSM.

Conclusions and Perspectives

The use of the green fluorescent protein in combination with the CLSM makes quantitative *in situ* investigation of gene transfer possible. So far, the frequency of horizontal gene transfer has been quantitatively established in model biofilms, using defined donor and recipient strains. The importance of appropriate controls has been clearly shown. It remains to be seen how well transfer of genetic information to unknown recipients in natural biofilms or sludge flocs can be quantified. In previous studies, *gfp*-expressing transconjugants have been microscopically detected in marine bacterial communities[12] or in activated sludge.[13] In these systems, however, autofluorescence may pose a serious hindrance to the automated detection of GFP-conferred fluorescence in unknown recipient strains. Nevertheless, with carefully designed experiments and controls, it may be possible to overcome this problem. The automated quantitative detection of GFP-tagged transformants or transconjugants in natural biofilms or aggregates should be explored.

[14] Transcriptional Analysis of Genes Involved in *Pseudomonas aeruginosa* Biofilms

By TIMNA J. O. WYCKOFF and DANIEL J. WOZNIAK

Introduction

Pseudomonas aeruginosa is a common gram-negative bacterium found mainly in soil and water that poses minimal threat to healthy individuals. However, *P. aeruginosa* is the major cause of death in patients suffering from cystic fibrosis. In these patients, *P. aeruginosa* causes increasingly serious chronic lung infections by forming biofilm communities on the lung epithelium. Thus, biofilm formation by *P. aeruginosa* is a topic of significant medical importance.

Our laboratory studies the transcriptional control of genes involved in the biosynthesis of alginate, an extracellular polysaccharide produced by clinical isolates of *P. aeruginosa,* and believed to be involved in biofilm formation.[1,2] Transcriptional control of a gene is often studied by utilizing a fusion of the gene's promoter with a reporter gene [chloramphenicol acetyltransferase (CAT), green fluorescent protein (GFP), etc.]. The fusion can be introduced into cells either on a replicating plasmid, or by homologous recombination at the promoter site in the genome. Both of these methods have limitations. The plasmid-based system requires antibiotic selection that may have physiological effects interfering with data interpretation. In addition, having multiple copies of the promoter in the cell may titrate important regulators. The insertion of a gene fusion into the genome at the wild-type locus may interrupt an operon, causing polar effects, and removes the possibility of studying autoregulation. In addition, the absence of the wild-type gene may disrupt temporal control of a cascade of regulatory events.

This chapter describes a new method of transcriptional analysis developed by T. T. Hoang *et al.* in the laboratory of H. Schweizer.[3] This method avoids many of the above-described pitfalls by placing a single copy of the gene fusion at a defined, nonessential, location in the genome. Hoang *et al.*[3] have also engineered a method of removing unwanted plasmid sequence, thus avoiding the need for antibiotic selection in the final strain. We describe two specific applications of this method, but it should be applicable to a variety of transcriptional questions in *P. aeruginosa.*

[1] B. C. Hoyle, L. J. Williams, and J. W. Costerton, *Infect. Immun.* **61,** 777 (1993).

[2] D. G. Davies, A. M. Chakrabarty, and G. G. Geesey, *Appl. Environ. Microbiol.* **59,** 1181 (1993).

[3] T. T. Hoang, A. J. Kutchma, A. Becher, and H. P. Schweizer, *Plasmid* **43,** 59 (2000).

Construction of Strains

Plasmid Mini-CTX1

The plasmid mini-CTX1 was developed by Hoang *et al.*[3] (Fig. 1A). It contains an *Escherichia coli* origin of replication (*ori*) and tetracycline resistance gene (*tet*) for selection. In addition, it contains the φCTX integrase gene (*int*) and *attP* sequence for insertion of the plasmid at the *attB* sequence in the *P. aeruginosa* chromosome. Finally, Flp recombinase target (FRT) sequences are appropriately

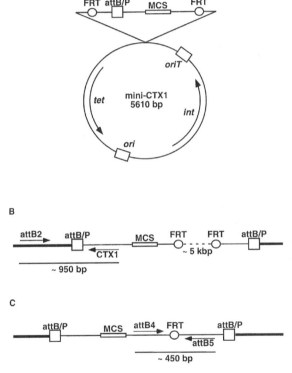

FIG. 1. Mini-CTX1 system (not drawn to scale). (A) Map of plasmid mini-CTX1. FRT, Flp recombinase target; MCS, multiple cloning site; *int*, φCTX integrase gene. (B) Diagram of mini-CTX1 integrated into the *P. aeruginosa* chromosome at *attB*. Thin lines designate plasmid sequence, whereas thick lines designate the chromosome. The positions of primers attB2 and CTX1 and the expected PCR product are shown. (C) Diagram of *P. aeruginosa* chromosome after removal of unwanted plasmid sequence by Flp recombinase. The positions of primers attB4 and attB5 and the expected PCR product are shown. [Adapted from T. T. Hoang, A. J. Kutchma, A. Becher, and H. P. Schweizer, *Plasmid* **43**, 59 (2000).]

located for removal from the chromosome subsequent to integration of all but the area directly adjacent to the convenient multiple cloning site (MCS).

Introduction of Mini-CTX1 into Pseudomonas aeruginosa

We use a triparental mating protocol to introduce mini-CTX1 plasmids containing various gene fusion inserts into *P. aeruginosa* PA01 and FRD1 as follows.

1. Grow overnight cultures of the *P. aeruginosa* acceptor strain in LBNS [tryptone (10 g/liter), yeast extract (5 g/liter)], an *E. coli* donor strain carrying the mini-CTX1-based plasmid in LB [tryptone (10 g/liter), yeast extract (5 g/liter), NaCl (10 g/liter)] containing tetracycline (15 μg/ml), and a helper strain, HB101/pRK2013,[4] in LB containing kanamycin (30 μg/ml).

2. Combine 500 μl of acceptor, 300 μl of donor, and 200 μl of helper with 2 ml of LBNS. Pass this mixture through a Swinex filter holder (Millipore, Bedford, MA) containing a 0.45-μm pore size nitrocellulose filter (Millipore) and allow filter to incubate on a nonselective LANS plate [LBNS, agar (15 g/liter)] for 6–24 hr.

3. Suspend the cells on the filter in 2 ml of LBNS and plate various amounts on LANS plates containing irgasan (25 μg/ml), to eliminate donor and helper *E. coli* strains and to select for *P. aeruginosa,* and tetracycline (100 μg/ml) to select for mini-CTX1-containing cells.

4. Verify recombinants by streaking individual colonies on LANS plates containing tetracycline (100 μg/ml).

At this point in the protocol, tetracycline resistance could be due to integration of the mini-CTX1-based plasmid at *attB,* integration of the plasmid by homologous recombination at another locus, or a spontaneous mutation. We use the polymerase chain reaction (PCR) to confirm integration of the mini-CTX1-based plasmid into the chromosome at *attB,* using a primer (attB2, 5′-GTCGCCGCCGGCGATGC-3′) that anneals upstream of the *attB* site in the *P. aeruginosa* chromosome, and a primer (CTX1, 5′-CCTCGTTCCCAGTTTGTTCC-3′) that anneals downstream of the *attP* site in plasmid mini-CTX1 (Fig. 1B). When a mini-CTX1-based plasmid is integrated into the *P. aeruginosa* chromosome at *attB,* a PCR with genomic DNA and primers attB2 and CTX1 yields an approximately 950-bp fragment. *Pseudomonas aeruginosa* PA01 yields no PCR fragment under these conditions. We prepare genomic DNA from isolated colonies as follows.

1. Grow an overnight culture of the strain to be tested in appropriate medium and antibiotic selection.

2. Centrifuge 100 μl of the culture for 1 min at maximum speed in a microcentrifuge tube at room temperature.

---

[4] D. Figurski and D. R. Helinski, *Proc. Natl. Acad. Sci. U.S.A.* **76,** 1648 (1979).

3. Discard the supernatant and suspend in 25 μl of sterile water.

4. Boil for 20 min.

5. Centrifuge again for 1 min and use 1 μl as template for a 10-μl PCR. We obtain best results by using *Taq* polymerase (Perkin-Elmer, Norwalk, CT) as specified by the manufacturer.

Removal of Unwanted Plasmid DNA Sequence by Flp Recombinase

Hoang *et al.*[3] constructed mini-CTX1 to have FRT sites flanking the MCS and *attP* site. This means that the rest of mini-CTX1 can be removed from the chromosome after integration by the action of Flp recombinase, contained on plasmid pFLP2.[5] In addition, pFLP2 can be selected against by sucrose, as it contains the *sacB* gene. This yields a strain containing only the sequence within and directly adjacent to the MCS of mini-CTX1, and requiring no antibiotic selection. We introduce Flp recombinase into the integrants by a biparental mating protocol as follows.

1. Grow overnight cultures of the *P. aeruginosa* integrant in LBNS containing tetracycline (100 μg/ml) and *E. coli* SM10/pFLP2 in LB with ampicillin (100 μg/ml). SM10[6] contains the necessary "helper" genes to allow for mobilization of pFLP2,[5] which contains the Flp recombinase.

2. Combine 500 μl of *P. aeruginosa* and 300 μl of SM10/pFLP2 with 2 ml of LBNS. Pass this mixture through a Swinex filter as described above and allow the filter to incubate on a nonselective LANS plate for 6 hr. This mating is efficient and overnight incubation is unnecessary.

3. Suspend the cells on the filter in 2 ml of LBNS and plate various dilutions on LANS plates containing irgasan (25 μg/ml) to select for *P. aeruginosa* and carbenicillin (250 μg/ml) to select for pFLP2-containing cells.

4. Colony purify on nonselective LANS plates, as the pFLP2 plasmid is unstable.

5. Streak isolated colonies to LANS plates containing 5% (w/v) sucrose to select against the pFLP2 plasmid.

6. Colony purify on nonselective LANS plates. Strains in which the unwanted mini-CTX1 sequence and pFLP2 have been removed should be tetracycline and carbenicillin sensitive.

We use PCR to confirm removal of unwanted mini-CTX1 sequence, using a primer (attB4, 5'-CGCCCTATAGTGAGTCG-3') that anneals downstream of the MCS in mini-CTX1, and a primer (attB5, 5'-CGCCCCAACCTCGCTGG-3') that anneals upsteam of the *attP* site in mini-CTX1 (Fig. 1C). When unwanted mini-CTX1 sequence has been removed, a PCR with genomic DNA and primers attB4 and attB5 yields an approximately 450-bp fragment. In genomic DNA from

[5] T. T. Hoang, R. R. Karkhoff-Schweizer, A. J. Kutchma, and H. P. Schweizer, *Gene* **212,** 77 (1998).
[6] R. Simon, U. Priefer, and A. Puhler, *Bio/Technology* **1,** 784 (1983).

strains containing the entire mini-CTX1, these primers anneal more than 5 kbp away from each other and, under short extension times, no PCR fragment is formed. We prepare genomic DNA from isolated colonies as described above.

Applications

We have used the mini-CTX1 system for two specific biofilm applications to date. First, we inserted a p*fliC–xylE* gene fusion at the *attB* site to test the feasibility of using this system to assess temporal changes in gene expression during the development of a *P. aeruginosa* biofilm. Second, we inserted a p*tac–gfp* gene fusion to test the feasibility of using single-copy, chromosomal GFP to visualize spatial differentiation of gene expression in a biofilm.

Assaying Promoter Activity: pfliC–xylE Gene Fusion

Formation of *P. aeruginosa* biofilms requires the presence of flagella.[7] The gene *fliC* encodes flagellin, the structural protein of a *P. aeruginosa* flagellum. Its promoter is thus a good choice for tracking expression of flagellar genes. The area from -260 to $+22$ contains 5′ regulatory sequences and the start of transcription.[8] We constructed a fusion of this region with the *xylE* reporter gene. This common reporter encodes catechol-2,3-dioxygenase, an enzyme for which there is a simple, spectrophotometric assay.[9] We cloned the p*fliC–xylE* fusion into mini-CTX1, integrated it into *P. aeruginosa* PA01, and removed unwanted mini-CTX1 sequences as outlined above. We were then able to assay the strain for *fliC* expression under various conditions. The Xyl E assay is performed as follows (adapted from Karkhoff-Schweizer and Schweizer[10]).

1. This assay can be performed with planktonic cultures in any stage of growth or with cells scraped from a growing biofilm. The number of cells needed depends on the strength of the promoter, but the lower limit in our hands is approximately the equivalent of 1 ml with an A_{540} of 0.04.

2. *For planktonic cells:* Centrifuge an appropriate amount of cell culture in a microcentrifuge at maximum speed for 1 min at room temperature. Discard the supernatant and suspend in 1 ml of Xyl E buffer [50 mM potassium phosphate (pH 7.5), 10% (v/v) acetone].

For biofilm cells: Scrape biofilm cells from the surface, using a rubber policeman, into 1 ml of Xyl E buffer. To obtain enough cells, we grow biofilms

[7] G. A. O'Toole and R. Kolter, *Mol. Microbiol.* **30**, 295 (1998).

[8] P. A. Totten and S. Lory, *J. Bacteriol.* **172**, 7188 (1990).

[9] M. M. Zukowski, D. R. Gaffney, D. Speck, M. Kauffman, A. Findell, A. Wisecup, and J. Lecocq, *Proc. Natl. Acad. Sci. U.S.A.* **80**, 1101 (1983).

[10] R. R. Karkhoff-Schweizer and H. P. Schweizer, *Gene* **140**, 7 (1994).

in 15×100 mm petri dishes containing 15 ml of innoculated LBNS. This is a modification of the method described by O'Toole and Kolter,[11] in which a biofilm forms at the liquid–air interface. After 3 hr, we typically recover 1 ml of cells with an A_{540} of 0.08. This corresponds to approximately 8×10^7 viable cells.

3. In a quartz cuvette combine 10 μl of 100 mM catechol freshly made in water with 990 μl of cell sample. Mix well, place in a spectrophotometer, and record the A_{375} for 2 min. If highly expressed promoters are being analyzed, it may be necessary to assay a dilution (in Xyl E buffer) of the sample.

4. The enzyme activity of the sample is calculated by using the extinction coefficient of the reaction product, 2-hydroxymuconic semialdehyde ($\epsilon_{375} = 4.4 \times 10^4 \, M^{-1}$), and corrected for the number of cells by dividing by the A_{540} of the sample. Activities are thus expressed as nanomoles product liberated per minute per 10^9 cells (1 ml with an A_{540} of 1).

Using the above described method with overnight cultures of planktonic cells, motile *P. aeruginosa* PA01 has no detectable Xyl E activity. PA01 with a promoterless *xylE* integrated at *attB* has a background level of 3.1 nmol/min per 10^9 cells. PA01 containing p*fliC–xylE* at *attB* has Xyl E activity of 65 nmol/min per 10^9 cells. When Alg T, a negative regulator of *fliC*,[12] is introduced on a plasmid, the Xyl E activity is reduced to background levels (5.6 nmol/min per 10^9 cells). Alg T has no effect on the Xyl E activity of wild-type PA01 or the strain containing the promoterless *xylE*. These results show that a single-copy *fliC* promoter at *attB* is expressed and regulated as previously shown.

Visualizing Promoter Activity: ptac–gfp Gene Fusion

To test the feasibility of using a single copy of *gfp* at *attB* as a way of visualizing spatial differences in gene expression in a developing biofilm, we constructed a gene fusion of p*tac* from pKK223-3 (Promega, Madison, WI) with GFPmut3* (kindly provided by J. B. Anderson, Technical University of Denmark, Lyngby, Denmark).[13] We cloned this fusion into mini-CTX1 and integrated it into the chromosome of PA01 and removed unwanted sequence as outlined above. We then prepared a biofilm of this strain for confocal microscopy as follows.

1. Place a sterile no. 1.5 coverslip in a petri dish of inoculated LBNS such that the medium comes about halfway up the coverslip. Allow biofilm to form for the desired amount of time. The biofilms shown in Fig. 2 were grown for 20 hr.

[11] G. A. O'Toole and R. Kolter, *Mol. Microbiol.* **28,** 449 (1998).

[12] E. S. Garrett, D. Perlegas, and D. J. Wozniak, *J. Bacteriol.* **181,** 7401 (1999).

[13] J. B. Andersen, C. Sternberg, L. K. Poulsen, S. P. Bjorn, M. Givskov, and S. Molin, *Appl. Environ. Microbiol.* **64,** 2240 (1998).

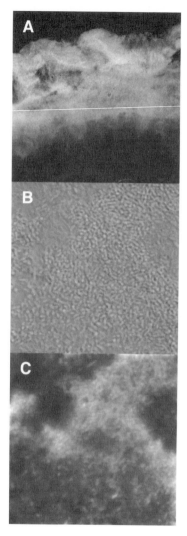

FIG. 2. Confocal laser scanning microscope images of a *P. aeruginosa* PA01/*ptac–gfp* biofilm. (A) Image acquired using ×10 objective and detecting GFP fluorescence. A section of biofilm formed at the liquid–air interface of a glass coverslip half-immersed in inoculated medium is clearly visible. (B) Image acquired using ×63 objective and detecting transmitted light. This is the same section of biofilm shown in (C). (C) Image acquired using. ×63 objective and detecting GFP fluorescence. GFP fluorescence seen here matches the pattern of cells seen in (B).

2. Using sterile forceps, remove the coverslip from the medium, rinse gently by submerging in a beaker of water, and fix the biofilm by placing the coverslip biofilm side up in a petri dish of 4% (w/v) *p*-formaldehyde for 10 min.

3. Rinse with water and mount on a microscope slide (biofilm side down) in a drop of sterile 90% (v/v) glycerol. Seal the edges of the coverslip to the slide with clear fingernail polish.

Confocal miscroscopy of the PA01/p*tac–gfp* biofilm shows that all cells express levels of GFP adequate for visualization (Fig. 2). This suggests that gene fusions of *gfp* at *attB* may be a useful method for visualizing and monitoring gene expression during biofilm development.

Concluding Remarks

The mini-CTX1 system, developed by Hoang *et al.*,[3] should help enhance our knowledge of the regulation of biofilm development at the transcriptional level. We have established the correct functioning at the *attB* site of the *fliC* promoter, which may show differential expression in a developing biofilm. In addition, it was shown that a single copy of p*tac–gfp* at the *attB* site is sufficient for visualization by confocal microscopy, suggesting a role for the mini-CTX1 system with the GFP reporter system.

While the mini-CTX1 system has many strengths—the absence of antibiotic selection and the placement of a single copy of a gene fusion in a defined location of the chromosome—it does have its limitations. The detection of a reporter signal in single copy will be limited by promoter strength and assay sensitivity. At this time, the system will be most useful with highly expressed promoters and/or sensitive reporter assays. In addition, the study of transcriptional regulation of a specific gene in a biofilm population has its own limitations. Biofilms are complex populations; scraping the whole population and assaying for reporter activity may not reveal the nuances of changes in a subset of that population. Nonetheless, the mini-CTX1 system promises to be a powerful new tool for transcriptional analysis of genes involved in *P. aeruginosa* biofilms.

Acknowledgments

We would like to thank Dr. Herb Schweizer (Colorado State University) for allowing us access to mini-CTX1 prior to its publication. We also thank Ken Grant (Wake Forest University Baptist Medical Center) for help with the confocal microscopy. Finally, this work is supported by Public Health Service Grant 1RO1 AI45779 to D.J.W.

[15] First Stages of Biofilm Formation: Characterization and Quantification of Bacterial Functions Involved in Colonization Process

By Thanh-Thuy Le Thi, Claire Prigent-Combaret, Corinne Dorel, and Philippe Lejeune

Introduction

Bacterial colonization of abiotic materials and biofilm formation have important detrimental consequences in medicine (contamination of catheters, prostheses, indwelling devices, artificial organs) and in many economic fields (biofouling of marine materials, wall growth in fermentation processes, contamination of food production lines). There is therefore a strong need to design surface coating methods able to interfere with the colonization process in order to prevent, or at least to delay, biofilm development. To reach this objective, increasing attention is being paid to the physiology and genetics of the initial stages of adhesion. Bacterial appendages and adhesins responsible for the linkage of the first pioneering cells to the surface (or the conditioning film) are potential targets for antiadhesion molecules grafted or smeared on the surface. Likewise, the determination of the mechanisms resulting in the movement of bacteria toward surfaces, the identification of the functions involved in the sensing of the particular microenvironments encountered at interfaces, and the description of the regulatory networks allowing the developmental processes necessary for the structural development of biofilms would help find nontoxic surface treatments able to lead the microorganisms away from the locations they usually contaminate. This would be particularly important in the field of indwelling medical devices. The development of infections on foreign bodies, such as prostheses and artificial organs, frequently results from bacteria introduced at the same time as the implanted device. Any delay (even a few hours) in the colonization process could successfully increase the capacities of the antibiotic therapy and the immunological defenses to eradicate the infection.

This chapter describes methods to obtain and analyze bacterial mutants with altered adhesion properties. It also presents some techniques used to characterize cellular functions involved in the first stages of biofilm development by model bacteria. This should provide a basic outline for studying the formation of biofilms by other organisms.

Rapid Methods for Quantification of Adhesion

Bacterial adhesion to abiotic surfaces is studied in a wide variety of experimental systems. The choice of a particular system is dictated by the objectives of the experiment. Careful measurements of adhesion kinetics can be obtained, under various experimental conditions, in flow devices such as parallel plate flow chambers equipped with microscopic techniques.[1] However, the ability to screen large amounts of mutants and rapidly identify altered phenotypes is critical to procedures such as random transposon mutagenesis. Rapid methods, largely based on biofilm formation in the wells of microtiter plates, have therefore been developed.

Screening Methods for Detection of Mutants with Altered Adhesion Abilities

To screen a library of *Pseudomonas fluorescens* clones obtained by transposon insertion, Cowan and Fletcher[2] utilized 96-well microtiter plates. Fifty-microliter samples of cell suspensions (obtained after purification of individual mutants, culture in test tubes, and standardization of the optical density) are placed in the wells. The cells are allowed to adsorb for 2 hr at 15°. At the end of that period, unbound cells are removed by pouring water over the surface of the entire microtiter plate. The stream is directed toward a few empty wells that were not used in the assay, and the water then floods into adjacent wells containing the bacteria. The dish is then inverted to drain water off, and the process is repeated twice. The entire dish is then immersed in a glass dish containing Bouin's fixative, and after approximately 4 min fixation is stopped by flushing the microtiter plate. Congo red is used to stain cells for subsequent measurement. Staining is accomplished as follows: water is aspirated from each well and replaced with 45 μl of cetyl pyridinium chloride (0.34%, v/v). Congo red (10 μl) is added to each well and the microtiter plate is rotated for 5 min. Stain is rinsed away by exhaustive washing with water poured over the plate, and the plate is allowed to dry. Just before reading, 50 μl of 95% (v/v) ethanol is added to each well to elute stain from the cells, and the plate is covered and rotated for 10 min. Absorption of eluted stain is then read at 490 nm on a microtiter plate reader.

Several groups developed even faster similar techniques. For instance, Dorel *et al.*[3] screened a mini-Mu insertion library of an adherent *Escherichia coli* K-12 strain. A sample of each clone is taken from the selection plate with a toothpick and the cells are suspended and directly grown at 30° in a 96-well polystyrene plate containing 0.2 ml of M63 minimal medium. Biofilm formation is visualized after 24 hr of culture as follows: planktonic cells are transferred to another plate (in order to recover the clones of interest) and the biofilm that has developed on the bottom of the well is washed twice by gentle pipetting with M63 medium. The plate is then dried for 1 hr at 80°. A drop of crystal violet is then added to each well, followed, after a few minutes, by extensive washes with M63 medium. A first screening

[1] B. Gottenbos, H. C. Van Der Mei, and H. J. Busscher, *Methods Enzymol.* **310,** 523 (1999).
[2] M. M. Cowan and M. Fletcher, *J. Microbiol. Methods* **7,** 241 (1987).
[3] C. Dorel, O. Vidal, C. Prigent-Combaret, I. Vallet, and P. Lejeune, *FEMS Microbiol. Lett.* **178,** 169 (1999).

of noncolored wells allows the isolation of obvious nonadherent mutants. Their colonization properties are then more carefully assayed in 24-well polystyrene plates, where the putative mutants are cultivated in 2 ml of M63 medium for 24 hr at 30°. The thickness of the biofilm is rapidly quantified as follows: for each well, the liquid medium and two washes (with M63 medium) are pooled and referred to as swimming cells. The biofilm is recovered in 1 ml of M63 medium by scraping the bottom of the well and pipetting up and down. The number of surface-attached and swimming bacteria is estimated from the optical density at 600 nm and compared to give the adhesion percentage corresponding to each clone.

More Precise Methods for Quantification of Adhesion

Several methods using slides or beads of various materials to quantify adhesion have been described. Their level of accuracy obviously cannot compete with sophisticated flow devices (for a critical review, see Bos et al.[4]), but some of these methods are simple to use and efficient when a large number of strains or culture conditions must be evaluated.

To quantify the first stages of adhesion of various bacterial species or the efficiency of coating treatments to delay colonization, Le Thi and Lejeune[5] have developed a microscopy technique with the aim of minimizing subjective bias during observation. Strips of various materials, such as glass, Plexiglas, polystyrene, Teflon, or stainless steel, are glued to microscope slides and incubated without shaking in petri dishes containing microbial suspensions or cultures. After various incubation times, the liquid containing the free-living bacteria is removed and the dishes are filled with water or 10 mM MgSO$_4$. After a few minutes, a second, similar wash is performed and the dishes are filled with a solution (2 μg ml^{-1}) of 4′,6-diamidino-2-phenylindole (DAPI). Thirty minutes later, the slides are washed twice in water and glued (by their bottom face) on a second microscope slide, and the strips are covered with a coverslip and a drop of glycerol (80%, v/v). The samples are examined under oil immersion, using a Leitz (Leica Microsystems, Wetzlas, Germany) Laborlux S epifluorescence microscope with Leitz filter block A2 (excitation, 340 to 380 nm; diachronic mirror, 455 nm; suppression filter, 470 nm; magnification, ×625, corresponding to fields of 1.38 mm^2). For each sample, 25 fields are randomly selected (in 5 areas situated in the middle and in the 4 quadrants of the sample) and rapidly analyzed in order to classify the sample according to the following scale (Fig. 1):

Stage 1: From zero to five bacteria in each field
Stage 2: Only isolated bacteria (no microcolonies)

[4] R. Bos, H. C. Van Der Mei, and H. J. Busscher, *FEMS Microbiol. Rev.* **23,** 179 (1999).
[5] T. T. Le Thi and P. Lejeune, unpublished results (1999).

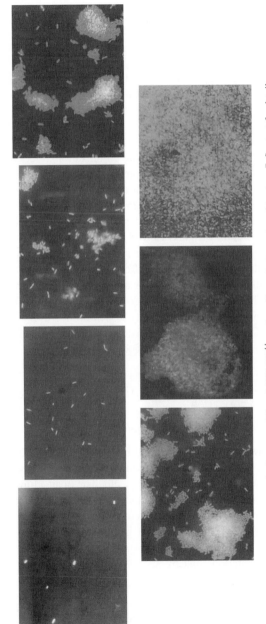

Fig. 1. First stages of adherence of *E. coli* PHL628[11] to glass. Top to bottom: stage 1 to stage 7. See text for details.

Stage 3: Countable isolated bacteria and small microcolonies (fewer than 100 bacteria)

Stage 4: Large microcolonies (more than 100 bacteria), but not confluent

Stage 5: Confluent microcolonies and isolated bacteria

Stage 6: Microcolonies occupying one-quarter of at least one field

Stage 7: At least one of the fields totally covered by the biofilm

Stage 8: The majority of the fields totally covered, but portions of the substrate remain visible

Stage 9: All fields totally covered

This method gives precise and reproducible colonization kinetics and allows fine discrimination of the adhesion-delaying properties of surface-coating treatments, especially when low initial bacterial densities are used.

Another way to assay adhesion is to use packed beds. For instance, Jucker *et al.*[6] utilized such a system to quantify the adhesion of *Stenotrophomonas maltophilia* and *Pseudomonas putida* on glass or Teflon at various ionic strengths. In their system, microbial suspensions are passed through glass columns (1.0-cm i.d., length, 9.5 cm) filled with either glass or Teflon beads. The influent is supplied to the vertical downflow columns by a peristaltic pump at a flow rate varying from 70 to 80 cm hr^{-1}. The influent bacterial density remains constant during the whole experiment (90 min). The effluent density is determined every 5 min by optical density and the ratio between the two densities is calculated. This technique gives reproducible curves that can be used to model adhesion.

Methods for Isolation of Mutants with Modified Colonization Properties

Isolation of Adherent Mutants from Nonadherent Strains

It is well known that biofilm formation on the walls of culture vessels can markedly perturb industrial processes and long-term experiments based on chemostat or cyclic flow culture conditions. This mechanism of attachment and growth of organisms on the surface of continuous-culture vessels was named "wall growth"[7] and has been reported for numerous bacterial species having an innate tendency to stick to surfaces.[8,9] Their attachment generally results in a thin, nonvisible film continually discharging cells into the medium. Another type of wall growth involves "sticky" mutations that frequently arise during the course of long-term

[6] B. A. Jucker, H. Harms, and A. J. B. Zehnder, *J. Bacteriol.* **178,** 5472 (1996).

[7] D. E. Dykhuizen and D. L. Hartl, *Microbiol. Rev.* **47,** 150 (1983).

[8] D. H. Larsen and R. L. Dimmick, *J. Bacteriol.* **88,** 1380 (1964).

[9] G. A. Murgel, L. W. Lion, C. Acheson, M. L. Shuler, D. Emerson, and W. C. Ghiorse, *Appl. Environ. Microbiol.* **57,** 1987 (1991).

experiments. Adhesion of these types of mutants usually results in a thick biofilm visible with the naked eye on the wall of the culture vessel.[7] Cyclic flow cultures are based on the periodic replacement of a large part (even the totality) of the exhausted medium by an equal volume of fresh medium. At each new cycle, the fresh medium is seeded with the bacteria remaining on the walls of the vessel at the end of the previous cycle. This method of culture intrinsically exercises, therefore, a strong positive selective pressure in favor of adherent mutants. Moreover, the reservoir of potential mutations able to confer the adherent phenotype and the efficiency of the selective pressure are so formidable that cyclic flow experiments must frequently be prematurely interrupted because of the total confinement of bacterial growth to the walls of the culture vessel.[10]

This technique of selection of spontaneous mutations can be used to activate cryptic structures of adhesion and is particularly relevant when applied to a well-characterized strain. Such a procedure has been applied to a classic laboratory strain of E. coli, K-12, which is unable to colonize inert surfaces. An adherent mutant was isolated and, because of the large number of tools and knowledge available, the mutation responsible for this phenotype was easily located by classical genetics, cloned, and sequenced.[11] The adhesion structures derepressed by this mutation are curli, a particular class of thin and flexible fimbria previously described[12] as determinants of pathogenicity in some natural isolates of E. coli. Because of their potential synthesis by K-12 strains, Vidal et al.[11] searched for the presence of curli on clinical strains of E. coli isolated from patients with catheter-related bacteremia and demonstrated their constitutive expression and their essential role in adhesion to biomaterials.

The enrichment in spontaneous mutants able to adhere to the wall of cyclic flow culture devices is also a method by which to select suppressor mutations restoring adhesion in nonadherent derivatives.

Inactivation of Functions Required for Adhesion by Transposon Insertion

As shown in Table I,[13–20] transposon mutagenesis followed by screening of nonadherent derivatives has been successfully used to identify bacterial genes

[10] R. Longin, unpublished observation (1988).

[11] O. Vidal, R. Longin, C. Prigent-Combaret, C. Dorel, M. Hooreman, and P. Lejeune, *J. Bacteriol.* **180,** 2442 (1998).

[12] A. Olsén, A. Jonsson, and S. Normark, *Nature (London)* **338,** 652 (1989).

[13] C. Yun, B. Ely, and J. Smit, *J. Bacteriol.* **176,** 796 (1994).

[14] L. A. Pratt and R. Kolter, *Mol. Microbiol.* **30,** 285 (1998).

[15] P. Genevaux, P. Bauda, M. S. DuBow, and B. Oudega, *FEMS Microbiol. Lett.* **173,** 403 (1999).

[16] P. Genevaux, P. Bauda, M. S. DuBow, and B. Oudega, *Arch. Microbiol.* **172,** 1 (1999).

[17] G. A. O'Toole and R. Kolter, *Mol. Microbiol.* **30,** 295 (1998).

[18] G. A. O'Toole and R. Kolter, *Mol. Microbiol.* **28,** 449 (1998).

[19] C. Heilmann, M. Hussain, G. Peters, and F. Götz, *Mol. Microbiol.* **24,** 1013 (1997).

[20] P. I. Watnick and R. Kolter, *Mol. Microbiol.* **34,** 586 (1999).

TABLE I
TRANSPOSON INSERTIONS CONFERRING A NONADHERENT PHENOTYPE

Organism	Gene	Function	Ref.
Caulobacter sp. MCS6		Holdfast synthesis	Yun *et al.*[13]
Escherichia coli 2K1056	*fliC, flhD, mot*	Motility	Pratt and Kolter[14]
	fim	Type I fimbriae synthesis	Pratt and Kolter[14]
Escherichia coli W3110	*dsbA*	Folding of periplasmic proteins	Genevaux *et al.*[15]
	rfaG, rfaP, galU	Lipopolysaccharide core biosynthesis	Genevaux *et al.*[16]
Escherichia coli PHL644	*ompR*	Regulation of curli synthesis	Vidal *et al.*[11]
	csgA	Curli synthesis	Vidal *et al.*[11]
	cpxA	Regulation of curli synthesis	Dorel *et al.*[3]
Pseudomonas aeruginosa PA14	*sad*	Motility	O'Toole and Kolter[17]
	pilBCD	Type IV fimbriae synthesis	O'Toole and Kolter[17]
Pseudomonas fluorescens WCS365	*sad*	Motility	O'Toole and Kolter[18]
	clpP	Clp protease synthesis	O'Toole and Kolter[18]
Staphylococcus epidermidis 0-47	*atlE*	Autolysin synthesis	Heilmann *et al.*[19]
Vibrio cholerae El Tor	*mot*	Motility	Watnick and Kolter[20]
	msh	MSHA pili synthesis	Watnick and Kolter[20]
	eps	Exopolysaccharide synthesis	Watnick and Kolter[20]

and functions required for adhesion. The delivery vehicle for the transposons was different in each case; most of these genetic tools have been previously reviewed.[21]

The majority of the functions revealed by these studies are structural components of the bacterial envelope: holdfast in the stalked *Caulobacter*[13]; type I fimbriae,[14] lipopolysaccharide,[16] and curli[11] in various adherent strains of *E. coli;* type IV fimbriae in *Pseudomonas aeruginosa*[17]; flagella in *E. coli*,[14] *P. aeruginosa,*[17] *P. fluorescens,*[18] and *Vibrio cholerae*[20]; an autolysin in *Staphylococcus epidermidis*[19]; and MSHA pili (also type IV pili) in *V. cholerae.*[20] Interestingly, regulatory functions have also been identified: the two-component sensor systems Cpx A/Cpx R and Env Z/Omp R (which are clearly involved in the regulation of curli synthesis in *E. coli*[3,11]), and two proteins acting at a posttranslational level (Dsb A in *E. coli*[15] and Clp P in *P. fluorescens*[18]).

Quantification of Gene Expression in Biofilms

Measurement of Promoter Activity

Various methods to quantify gene expression (fusions with reporter genes and assays of transcripts) in biofilms have been reviewed in this series.[22] Using

---

[21] J. H. Miller, ed. *Methods Enzymol.* **204** (1991).
[22] C. Prigent-Combaret and P. Lejeune, *Methods Enzymol.* **310,** 56 (1999).

one of these techniques, Prigent-Combaret *et al.*[23] have shown that the level of transcription of 38% of 446 random insertions of a transposon carrying the reporter gene *lacZ* differed in planktonic and biofilm cells of *E. coli* K-12. Comparison of the levels of expression of several *lacZ* fusions in the two populations indicated that the cells were not surrounded by the same microenvironment. Differences in cellular density, oxygen concentration, and osmolarity were probably partially responsible for the establishment of the different physiological states. As it is known that in *E. coli* there is a quasilinear relationship between the external osmolarity and the internal concentration of potassium,[24] these authors assayed the intracellular K^+ concentration of both populations.[23] A significant difference confirmed the hypothesis that the biofilm bacteria were confronted with a microcosmos of higher osmolarity than were the planktonic cells. This higher osmolarity could be a result of the attraction of ionic molecules within the biofilm to counterbalance the electric charges existing at the solid surface, the bacterial appendages, and the various exopolymers of the matrix.

Proteome Analysis

A sensitive method to compare the physiological states of two bacterial populations is two-dimensional electrophoresis of the cellular proteins. Differentially expressed proteins can then be extracted from the gels and identified by partial sequencing. In the field of biofilm study, few data have been collected in this way. Perrot *et al.*[25] have compared the protein patterns from gel-entrapped *E. coli* cells (a situation similar to a biofilm) with those of free-living bacteria and identified a first set of differentially synthesized proteins. There is no doubt that this methodology will rapidly reveal numerous functions critical for biofilm formation.

Acknowledgments

Research in the authors' laboratory was funded by grants from the French Defense Ministry (96/048 DRET) and the Centre National de la Recherche Scientifique (Réseau Infections Nosocomiales).

[23] C. Prigent-Combaret, O. Vidal, C. Dorel, and P. Lejeune, *J. Bacteriol.* **181,** 5993 (1999).

[24] W. Epstein and S. G. Schultz, *J. Gen. Physiol.* **49,** 221 (1965).

[25] F. Perrot, M. Hebraud, R. Charlionet, G. A. Junter, and T. Jouenne, *Electrophoresis* **21,** 645 (2000).

Section III

Growth of Bacteriophage in Bacteria in Biofilms

[16] Phenotype Characterization of Genetically Defined Microorganisms and Growth of Bacteriophage in Biofilms

By ROBERT J. C. MCLEAN, BRIAN D. CORBIN, GRANT J. BALZER, and GARY M. ARON

Introduction

Many investigators have shown biofilms to be a dominant mode of growth of bacteria and other microorganisms in nature (reviewed in Costerton *et al.*[1]). Indeed, the Centers for Disease Control (Atlanta, GA) now estimate that at least 62% of all bacterial infectious diseases, particularly chronic infections, involve biofilms.[2] In contrast to their planktonic counterparts, biofilm organisms are quite resistant to antibiotics[3] and disinfectants.[4] Biofilms are also important in other areas including wastewater treatment, biofouling of pipes, and microbially induced corrosion (reviewed in Costerton *et al.*[1]). As a consequence, there has been a tremendous upsurge of interest in studying the physiology and genetics of microorganisms within biofilms.

The early studies of biofilms (reviewed in Costerton *et al.*[1,2]) were largely descriptive. These studies employed conventional scanning (SEM) and transmission electron microscopy (TEM) to observe biofilms with a high degree of resolution. Biofilms are highly hydrated, typically >90% H_2O in most environments. As a result, the dehydration employed for conventional SEM and TEM causes hydrated structures within biofilms to collapse, giving a mistaken impression that individual cells were simply thrown together. Other drawbacks to conventional SEM and TEM are that the chemicals (including glutaraldehyde, OsO_4, and other metal-containing stains such as uranyl acetate and lead citrate) and high vacuum cause a loss of cell viability. Although environmental SEM permits the examination of hydrated biofilms, its major drawback is that organic molecules cannot be resolved unless stained with heavy metals.[5] Overall, SEM and TEM allow the observation of biofilms at fixed points of time. Although scanning confocal laser microscopy (SCLM) lacks the resolving power of SEM and TEM, it permits the observation of fully hydrated and living biofilms. Consequently, we now realize

[1] J. W. Costerton, Z. Lewandowski, D. E. Caldwell, D. R. Korber, and H. M. Lappin-Scott, *Annu. Rev. Microbiol.* **49,** 711 (1995).

[2] J. W. Costerton, P. S. Stewart, and E. P. Greenberg, *Science* **284,** 1318 (1999).

[3] J. C. Nickel, I. Ruseska, J. B. Wright, and J. W. Costerton, *Antimicrob. Agents Chemother.* **27,** 619 (1985).

[4] M. W. LeChevallier, C. D. Cawthon, and R. G. Lee, *Appl. Environ. Microbiol.* **54,** 649 (1988).

[5] B. Little and P. Wagner, *Can. J. Microbiol.* **42,** 367 (1996).

FIG. 1. SCLM image of a mixed community wastewater biofilm. As first described by Lawrence *et al.*,[5a] regions of high cell density (microcolonies) are surrounded by regions of low cell density (water channels). Scale bar represents 0 to 50 μm. (R. J. C. McLean, unpublished data, 2000.)

that biofilms are dynamic and heterogeneous microbial communities (Fig. 1[5a]), capable of changing over time and in response to nutrients and other environmental factors.[6] SCLM and other microscopy techniques are described in detail elsewhere.[7-9]

While the early descriptive studies were important in establishing the importance and ubiquitous nature of biofilms, we now need to address the many fundamental issues of biofilm biology. Current techniques for growing biofilms and studying biofilm-induced gene expression and physiology are presented throughout this volume and in *Methods in Enzymology,* Volume 310. Here, we address two major issues: (1) identifying and testing genes for their importance in biofilm physiology, and (2) studying the interaction of bacteriophage with bacteria in their natural environment (i.e., biofilms).

To date, several genes and gene products have been identified as being important for biofilm formation and growth. These include the quorum-sensing genes *lasI* and *lasR* and the alginate (capsule) biosynthesis gene *algC* in

[5a] J. R. Lawrence, D. R. Korber, B. D. Hoyle, J. W. Costerton, and D. E. Caldwell, *J. Bacteriol.* **173,** 6558 (1991).

[6] G. A. James, D. R. Korber, D. E. Caldwell, and J. W. Costerton, *J. Bacteriol.* **177,** 907 (1995).

[7] J. R. Lawrence and T. R. Neu, *Methods Enzymol.* **310,** 131 (1999).

[8] D. Phipps, G. Rodriguez, and H. Ridgway, *Methods Enzymol.* **310,** 178 (1999).

[9] T. A. Fassel and C. E. Edmiston, Jr., *Methods Enzymol.* **310,** 194 (1999).

Pseudomonas aeruginosa.[10,11] In *Escherichia coli,* identified genes to date include the alternate sigma factor *rpoS,*[12] colanic acid locus *wca,* tripeptidase T (*pepT*), and *proU.*[13] In *Streptococcus mutans, gbpA* involved in glycosyltransferase has been identified.[14] Investigators may wish to use these genes as positive controls when adapting techniques for screening biofilm-activated genes. Other chapters in this volume (from the laboratories of Ceri [8], Kolter [2], Lejeune [15], and Schoolnik [1][14a]) describe several elegant strategies that can be employed for the study of biofilm-specific gene expression. These approaches include the use of gene microarrays, *in vivo* expression technology, random transposon mutagenesis with promotorless reporter genes, subtractive hybridization, and the use of two-dimensional gel electrophoresis to identify patterns of protein expression from the genome (proteome analysis). Once a candidate gene has been identified, several experimental strategies can be employed. If the gene is fully or partially sequenced, then a computer search for other genes (i.e., bioinformatics) can be used to estimate function and distribution.[15] In the first part of this chapter, we present several experimental strategies that can be used to test the function and importance of candidate genes.

One of the myths in the scientific community is that biofilms are indestructible. While quite resistant to most antimicrobial agents, the organisms within biofilms have coevolved with their predators and parasites. Consequently, it would be predicted that biofilm organisms would be susceptible to predation and parasitism. Whereas predation by larger organisms such as protozoa or invertebrates is expected,[16] parasitism by smaller entities such as bacteriophage has been described only more recently[17] (B. D. Corbin, R. J. C. McLean, and G. M. Aron, unpublished data, 2000). Yet, these interactions may play an important role in microbial ecology and possible gene transfer via transduction. In the latter part of this chapter, we describe some techniques and strategies that can be used to study phage–biofilm interactions.

[10] D. G. Davies, A. M. Chakrabarty, and G. G. Geesey, *Appl. Environ. Microbiol.* **59,** 1181 (1993).

[11] D. G. Davies, M. R. Parsek, J. P. Pearson, B. H. Iglewski, J. W. Costerton, and E. P. Greenberg, *Science* **280,** 295 (1998).

[12] J. L. Adams and R. J. C. McLean, *Appl. Environ. Microbiol.* **65,** 4285 (1999).

[13] C. Prigent-Combaret, O. Vidal, C. Dorel, and P. Lejeune, *J. Bacteriol.* **181,** 5993 (1999).

[14] K. R. O. Hazlett, J. E. Mazurkiewicz, and J. E. Banas, *Infect. Immun.* **67,** 3909 (1999).

[14a] G. K. Schoolnik, F. H. Yildiz, N. A. Dolganov, K. Meibom, D. Schnappinger, M. I. Voskuil, M. A. Wilson, and K. H. Chong, *Methods Enzymol.* **336,** Chap. 1, 2001 (this volume); P. N. Danese, L. A. Pratt, and R. Kolter, *Methods Enzymol.* **336,** Chap. 2, 2001 (this volume); M. D. Parkins, M. Altebaeumer, H. Ceri, and D. G. Storey, *Methods Enzymol.* **336,** Chap. 8, 2001 (this volume); T. T. Le Thi, C. Prigent-Combaret, C. Dorel, and P. Lejeune, *Methods Enzymol.* **336,** Chap. 15, 2001 (this volume).

[15] M. S. Gelfand, E. V. Koonin, and A. A. Mironov, *Nucleic Acids Res.* **28,** 695 (2000).

[16] I. Sibille, T. Sime-Ngando, L. Mathieu, and J. C. Block, *Appl. Environ. Microbiol.* **64,** 197 (1998).

[17] M. M. Doolittle, J. J. Cooney, and D. E. Caldwell, *Can. J. Microbiol.* **41,** 12 (1995).

General Experimental Strategies for Biofilm Growth

As evidenced in the literature, there are a number of experimental devices and media that can be employed for biofilm growth. Although the plethora of devices may be confusing to newcomers to this field, there are a number of commonly employed techniques for biofilm growth that represent a good starting point. These are listed in Dibdin and Wimpenny.[18] In the context of phenotypic characterization, the choice of a biofilm growth apparatus will be influenced by (1) the gene(s) being investigated, (2) the type of information being sought (e.g., morphological, biochemical, genetic analysis), and (3) the financial and physical resources available to the investigator. With respect to the latter issue, there are a number of inexpensive devices that can be purchased or constructed. Another issue will be the choice of organism(s) to be investigated (addressed below).

As an integral component of experimental design, we strongly emphasize the importance of performing background work. Among other things, the investigator should address (1) the reliability of equipment and media, (2) the growth characteristics and stability of the strain(s) under investigation, (3) the ease, reliability, and reproducibility of any techniques used to assay biofilm growth and/or gene expression, (4) the reproducibility of experimentation, (5) the number of replicates needed for accurate data analysis, and (6) the ease of cleaning any equipment. With respect to the first point, some equipment or tubing may leak with repeated use. For example, the Robbins Device (RD),[19] commonly employed in many biofilm studies, does have a tendency to leak. This problem can be anticipated by running water through the system to test for leaks. Teflon tape can be used to seal the screws and connections in the RD, and the apparatus can also be placed in a container capable of containing a spill. Sometimes, it may be necessary to modify or design a piece of equipment. For example, the RD, originally constructed of brass, was developed by J. Robbins[20] as a means to investigate biofouling and corrosion-causing biofilms from pipe surfaces. Before its development, sample collection of pipe-fouling organisms consisted of removing sections from a pipe with a saw. Nickel et al.[3] developed the smaller version of the RD, commonly used today. This modified RD (available commercially from Tyler Research, Edmonton, AB, Canada) is constructed of acrylic and is capable of testing various compounds for biofilm formation.

Environmental selection may strongly influence the stability of the strains being tested. While this may not be a significant issue in short-term mutant screening protocols, such as the microtiter assay for adhesion defective mutants,[21] it

[18] G. Dibdin and J. Wimpenny, *Methods Enzymol.* **310**, 296 (1999).

[19] R. J. C. McLean, M. Whiteley, B. C. Hoskins, P. D. Majors, and M. M. Sharma, *Methods Enzymol.* **310**, 248 (1999).

[20] W. F. McCoy, J. D. Bryers, J. Robbins, and J. W. Costerton, *Can. J. Microbiol.* **27**, 910 (1981).

[21] G. A. O'Toole and R. Kolter, *Mol. Microbiol.* **28**, 449 (1998).

may become an issue during longer term studies, particularly those employing continuous culture (G. J. Balzer, D. A. Siegele and R. J. C. McLean, unpublished data, 2000). Detection of revertants or other mutations is always an issue that should be addressed. Often these may be identified on the basis of phenotypic characteristics, when the organisms are grown on plates. Investigators may employ continual antibiotic selection to ensure the stability of a plasmid. However, it must be remembered that biofilm growth itself can enhance bacterial survival in the presence of an antibiotic, even if the strain lacks traditionally recognized R factors.[3]

Organism and Strain Selection

Although biofilm formation is quite a common phenomenon with most microorganisms, extensive research has been conducted only in a few organisms. *Pseudomonas aeruginosa* is by far the most widely studied biofilm-forming organism, as it readily forms biofilms that are important in medical and industrial environments.[1] Biofilm studies involving genetically defined strains have also been conducted in other gram-negative organisms such as *Pseudomonas fluorescens, E. coli,* and Vibrio cholerae.[12,13,21] Because of their medical importance, the majority of biofilm studies conducted with gram-positive organisms involve *Staphylococcus aureus, Staphylococcus epidermidis,* and *Streptococcus mutans.*[14,22,23] One of the main advantages of working with these aforementioned organisms is that the culture conditions and media for biofilm growth have been well established. With other organisms, the investigator will need to establish culture conditions that will permit reproducible biofilm growth and formation. As stated earlier, other chapters in this volume,[14a] and [25] in Volume 337 of this series,[14b] address various strategies that can be used to identify biofilm-defective mutants. Once isolated, these mutant strains should be stored at −80°, using 10% (v/v) glycerol as a cryoprotectant.[19] When testing these, the investigator may wish to classify these mutants into the following general categories.

1. Genes involved in initial adhesion and colonization, such as those encoding pili, capsule, chemotaxis, flagella, or capsule production. Such genes may encode for initial adhesion to the substratum[21] or else adhesion to other cells[14]

2. Genes downregulated for biofilm growth (an example would be *E. coli* flagella genes[13])

3. Genes upregulated for biofilm growth (an example is the *wca* locus for colanic acid production in *E. coli*[13])

4. Housekeeping genes (an example would be those involved in glycolysis)

[22] C. Heilmann, C. Gerke, F. Perdreau Remington, and F. Goetz, *Infect. Immun.* **64**, 277 (1996).
[23] S. E. Cramton, C. Gerke, N. F. Schnell, W. W. Nichols, and F. Götz, *Infect. Immun.* **67**, 5427 (1999).

5. Genes of importance in mature biofilms

6. Genes of importance for the return to the planktonic state

7. Genes involved in mixed population biofilms. These genes could either facilitate cell–cell synergistic interactions or else facilitate competition[24]

8. Regulatory genes

9. Genes involved in protection of the biofilm organisms against antimicrobial agents or predation

10. Genes of indeterminant function

The above list is an estimation of the classes of genes involved in bacterial growth as biofilms. Undoubtedly, future work will result in a modification of this list. Nevertheless, classifying genes in such a manner will assist the investigator in adopting a strategy for phenotypic characterization.

Strategies for Biofilm Phenotype Characterization

Genes in category 1 (initial biofilm formation) will be primarily important in the colonization of a surface and may have a limited role after that point. Microtiter assays[21] or else short-term (0- to 48-hr) experiments using flow cells or the RD are suitable for evaluating null mutations of category 1 genes. In comparison with an isogenic parent, the mutant strain would be expected to lack the ability to form a biofilm or else to form a biofilm with an altered structure. This latter characteristic may be typical when a gene, necessary for cell–cell adhesion, has been removed. Characterization of strains carrying null mutations can be used for all other categories; however, the experimental design will differ. For example, genes involved in biofilm maturation (class 5) will become evident only when biofilms have been grown for some time (typically >48 hr). The Calgary biofilm device, which is capable of growing biofilms for extended periods of time,[25] has potential applications for screening large numbers of mutants in classes 5 and 9.

Genes involved in class 7 (mixed population biofilms) or class 9 (protection against predation or the immune system) will require more complex testing protocols that involve the use of multiple organisms[19] or even animal model experiments.[26] In such instances, it is of key importance to be able to track and identify the organisms of interest throughout the experiment.

For studies of genes that are either upregulated or downregulated during biofilm growth (categories 2 and 3), the investigator may wish to insert the gene of interest

[24] J. D. Hillman, B. I. Yaphe, and K. P. Johnson, *J. Dent. Res.* **64**, 1272 (1985).

[25] H. Ceri, M. E. Olson, C. Stremick, R. R. Read, D. W. Morck, and A. Buret, *J. Clin. Microbiol.* **37**, 1771 (1999).

[26] K. P. McDermid, D. W. Morck, M. E. Olson, N. D. Boyd, A. E. Khoury, M. K. Dasgupta, and J. W. Costerton, *J. Infect. Dis.* **168**, 897 (1993).

into plasmids of varying copy number. Here, an assumption is made that gene expression will be proportional to copy number. An alternative strategy for categories 2 and 3 is to move a candidate gene into a tightly regulated operon (such as the arabinose operon in *E. coli*[27]). This strategy will enable the investigator to activate or deactivate a gene at will. The use of reporter gene fusions is also popular in studies of gene regulation in biofilms, in that levels of gene expression typically vary considerably from cell to cell throughout a biofilm.[28] Several commonly used reporter systems are available including *lacZ*, *gfp*, and *lux*. The advantage of *gfp*, a eukaryotic gene encoding a green fluorescent protein from *Aequorea victoria*, is that it does not require exogenously added substrate in order to be visualized.[29] Several precautions should be addressed when using reporter gene fusions. Investigators should perform background work to ensure that the reporter gene will work in their experimental setup, is expressed only when desired, and that its level of expression correlates with that of the gene under investigation. A second consideration is the stability or longevity of the fusion gene product. Stable reporter genes can be used to identify bacteria that have expressed a gene at some point during growth. In contrast, unstable reporter genes with a short half-life will give information about genes that are currently being expressed. Reporter systems for studying gene expression and cell physiology are addressed in detail in Korber *et al.*[30] and Christensen *et al.*[31]

Two-dimensional gel electrophoresis[32] can be used to enhance studies of biofilm and planktonic cell gene expression. For example, when investigating class 1 mutations (initial adhesion), an investigator may use transposon mutagenesis to identify five candidate genes. Although mutations for these genes can all be screened, a question can arise concerning whether all potential genes for this function (in this case adhesion) have been identified. Initially, two-dimensional gel electrophoresis studies of wild-type bacteria would be used to compare patterns of *de novo* gene expression in planktonic and newly adherent organisms. Similar studies would be carried out with the various null mutants. In theory, each of the mutant strains would lack the ability to produce one or more proteins, seen on the wild-type gel. Through similar studies with the remaining mutants, the investigator can see which proteins remain to be accounted for.

[27] D. A. Siegele and J. C. Hu, *Proc. Natl. Acad. Sci. U.S.A.* **94**, 8168 (1997).

[28] B. B. Christensen, C. Sternberg, J. B. Andersen, L. Eberl, S. Møller, M. Givskov, and S. Molin, *Appl. Environ. Microbiol.* **64**, 2247 (1998).

[29] G. V. Bloemberg, G. A. O'Toole, B. J. Lugtenberg, and R. Kolter, *Appl. Environ. Microbiol.* **63**, 4543 (1997).

[30] D. R. Korber, G. M. Wolfaardt, V. Brözel, R. MacDonald, and T. Niepel, *Methods Enzymol.* **310**, 3 (1999).

[31] B. B. Christensen, C. Sternberg, J. B. Andersen, R. J. Palmer, Jr., A. Toftgaard Nielsen, M. Givskov, and S. Molin, *Methods Enzymol.* **310**, 20 (1999).

[32] D. Blankenhorn, J. Phillips, and J. L. Slonczewski, *J. Bacteriol.* **181**, 2209 (1999).

TABLE I

DEFINED CARBON-LIMITED MEDIUM SUITABLE FOR CHEMOSTAT GROWTH
OF GRAM-NEGATIVE BACTERIA INCLUDING *Escherichia coli, Aeromonas hydrophila,*
AND *Pseudomonas aeruginosa*[a] AS AMENDED[b]

Trace mineral solution	
Nitrilotriacetic acid	1.5 g
$MgSO_4$ (anhydrous)	1.47 g
$CoCl_2 \cdot 6H_2O$	0.1 g
$MnSO_4$	0.5 g
NaCl	1.0 g
$FeSO_4 \cdot 7H_2O$	0.1 g
$ZnSO_4 \cdot 7H_2O$	0.1 g
$CuSO_4 \cdot 5H_2O$	0.01 g
$NaBO_4$	0.01 g
$Na_2MoO_4 \cdot 2H_2O$	0.01 g
Preparation of trace mineral solution: Add nitrilotriacetic acid to 500 ml of H_2O, adjust to pH 6.5, add remaining components, and fill to 1 liter final volume. Filter sterilize through a 0.2-μm pore size filter	
Calcium chloride stock solution: Add 1.7 g of $CaCl_2 \cdot 2H_2O$ to 1 liter of H_2O and sterilize by autoclaving	
Glucose stock solution: Dissolve 12.5 g of glucose in 100 ml of H_2O and filter sterilize through a 0.2-μm pore size filter	
Carbon-Limited Defined medium	
NH_4Cl	0.24 g
K_2HPO_4	1.0 g
H_2O	1 liter
Dissolve compounds together and adjust to pH 7.2. Autoclave for 15 min at $121°$ and allow to cool to room temperature. Then aseptically add	
Trace mineral solution	0.6 ml
$CaCl_2$ solution	1.0 ml
Glucose solution	20 ml

[a] See Ref. 32a.
[b] See Ref. 33.

Formulations for Media

Although there are many different growth regimens and formulations for media, we highly recommend the use of defined media and continuous culture. In this fashion, the investigator can control the growth rate and physiology of the organisms within the biofilm. A recipe for a carbon-limited defined medium, useful for growing many strains of *E. coli* and other heterotrophs, is shown in Table I.[32a,33] When auxotrophs are grown, this medium can be supplemented with

[32a] F. C. Niedhardt, P. L. Bloch, and D. F. Smith, *J. Bacteriol.* **119,** 736 (1974).
[33] M. Whiteley, E. Brown, and R. J. C. McLean, *J. Microbiol. Methods* **30,** 125 (1997).

the appropriate nutrient(s), or alternatively with a complex supplement such as 0.4% (w/v) casamino acids. If necessary, antibiotics can also be added to the medium, As we cautioned previously, the biofilm mode of growth does confer resistance to many antibiotics. It is possible that cells within a microcolony may lose genes that are coupled to an antibiotic resistance marker. Always monitor cells during and after a biofilm experiment to ensure that they retain the original phenotype (e.g., resistance to a particular antibiotic).

Before experimentation, the cultures are streaked onto an appropriate solid medium [such as Luria–Bertani (LB) agar] from frozen storage, checked for purity, and then subcultured at least once in the defined medium. In this fashion, the investigator ensures the purity of the inoculum and minimizes the loss of unstable plasmids.

Chemostat Growth of Biofilm Cultures

Once mutants have been screened and identified, detailed studies of their role(s) in biofilms should be performed. We find that chemostat culture offers one of the most rigorous approaches for this task. Because of the setup time and equipment involved, it is not as well suited for large-scale screening. It has been our experience that reproducibility of experimental data between experimental runs (a characteristic concern with batch-grown biofilms) is enhanced with continuous culture biofilms grown on defined media. Many commercially available chemostats and fermentors are available. The expense of these commercially available systems can be quite formidable. As described earlier, we designed a simple chemostat from a laboratory flask, which can be constructed inexpensively with appropriate glass-blowing facilities.[33] For continuous culture experiments, we recommend reviving the culture(s) from frozen stock and streaking on solid medium to check for purity. In this fashion, the investigator reduces the difficulties of working with genetically unstable strains. The plate cultures are transferred to a broth culture, which is then used to inoculate the chemostat. We allow the chemostat culture to grow in batch mode (i.e., fresh medium is not added) overnight, after which fresh medium is added. The planktonic culture is grown for a minimum of one to three dilutions (i.e., 100–300 hr at a dilution rate of $0.01 \ hr^{-1}$) to allow the culture to become equilibrated, after which the chemostat is connected to the RD. This procedure also works well with mixed culture experiments. Details of this procedure are addressed in McLean et al.[19]

Phage–Biofilm Studies

As bacteriophage and bacteria have coevolved over many millions of years, it would be expected that phage would be able to infect bacteria within biofilms. To date, relatively few studies of this aspect of phage ecology have been conducted.

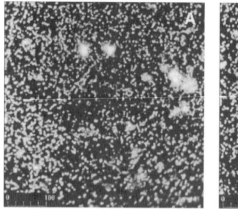
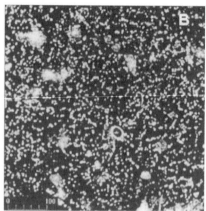

FIG. 2. SCLM image of an *E. coli* biofilm before infection with T4 phage (A) and after 4 hr of infection (B) with T4 phage added at a concentration of 100 MOI. Note that the biofilm structure remains relatively intact after phage infection. Scale bars represent 0 to 100 μm.

Doolittle *et al.*[17] have shown that T4 phage has the ability to infect *E. coli* biofilms and that these lytic infections can be observed by scanning confocal laser microscopy[34] (Fig. 2). Sutherland and co-workers[35,36] have shown that phage, capable of infecting biofilms, possess depolymerase enzymes that enable the bacteriophage to penetrate the biofilm matrix. Although these four descriptive studies establish the ability of phage to infect biofilms, virtually nothing else is known about phage–biofilm interactions.

On the basis of studies with planktonic bacteria, bacteriophage infections require first that the host bacterium and phage come into contact, and that the phage then binds to a suitable receptor on the bacterial cell surface.[37] Phage replication is dependent on the presence of receptors on the bacterial cell surface, host physiology,[38] and virus-specific gene activation and expression. Traditionally, phage replication has been divided into four separate stages: (1) adsorption of the phage to the host, (2) penetration of the phage particle into the bacterium, (3) intracellular multiplication of the virus, and (4) lysis of the host cell with release of progeny phage.[39] In relation to biofilms, it would be anticipated that the biofilm matrix would prevent adsorption of the phage to the host or the spread of progeny

[34] M. M. Doolittle, J. J. Cooney, and D. E. Caldwell, *J. Indust. Microbiol.* **16**, 331 (1996).

[35] K. A. Hughes, I. W. Sutherland, and M. V. Jones, *Microbiology* **144**, 3039 (1998).

[36] K. A. Hughes, I. W. Sutherland, J. Clark, and M. V. Jones, *J. Appl. Microbiol.* **85**, 583 (1998).

[37] H. W. Ackermann, *Adv. Virus Res.* **51**, 135 (1997).

[38] H. Hadas, M. Einav, I. Fishov, and A. Zaritsky, *Microbiology* **143**, 179 (1997).

[39] M. H. Adams, "Methods of Study of Bacterial Viruses: Bacteriophages." Interscience Publications, New York, 1959.

phage. As well, the altered physiology of biofilm bacteria may also affect phage replication. While a detailed review of phage biology is beyond the scope of this chapter, investigators would be well advised to become familiar with this area.

On the basis of our experience with T4 and other phage replication in biofilms (B. D. Corbin, E. A. Weaver, M. Whiteley, R. J. C. McLean, and G. M. Aron, unpublished data, 2000), we recommend the following background work be performed with the phage.

1. When a phage is to be isolated from the environment, we recommend first making a lawn of the target organism on a petri dish and then spreading a liquid with the suspected bacteriophage on the lawn. For example, we isolated a bacteriophage, specific for *Chromobacterium violaceum,* by spreading 100 μl of a suspension containing 10^9 CFU/ml on a plate of R2A (Difco, Detroit, MI) agar. After this suspension had dried, we obtained wastewater from a sewage treatment plant, filtered it through a 0.2-μm pore size filter, and spread 100 μl of this filtrate on the plate. In this particular case, the plaques were quite small (1–2 mm in diameter). Alternatively, phage can be obtained from other investigators or from culture collections such as the American Type Culture Collection (Manassas, VA; *http://www.atcc.org*). As a general strategy for growth, purification, and storage of phage, we recommend the protocols described in Sambrook *et al.*[40]

2. Perform a one-step growth experiment, using the phage and a planktonic culture of the organism to be investigated. In the case of *E. coli* and T4 phage, inoculate 10 ml of LB broth with 100 μl of culture and grow overnight (18 hr) at 37° with shaking. Use 1 ml of this culture to inoculate a flask with 50 ml of LB broth. The flask culture is then placed in a shaker incubator for 3 hr which should yield a cell density of approximately 5×10^8 CFU/ml. After this 3-hr growth, the suspension is diluted 1 : 10 and mixed with T4 at a multiplicity of infection (MOI) of 10. The MOI is defined as the ratio of plaque-forming units (bacteriophage concentration) to colony-forming units (bacterial density). After a 10-min absorption period, the suspension is diluted 1 : 1000 to stop phage adsorption, and the infected cell suspension is immediately assayed for infectious centers and for an increase in plaque-forming units at 15-min intervals. This information will enable the investigator to determine the duration of the phage latent period as well as the burst size (number of progeny phage produced per infected cell).

The majority of our experiments to date (B. D. Corbin and G. M. Aron, unpublished data, 2000) have employed a chemostat RD[33] to grow biofilms on disks of silicone rubber. From our experience, *E. coli* biofilms grown at a dilution rate of 0.028 hr^{-1} will colonize silicone disks at a density of ca. 8×10^7 CFU/cm^2

[40] J. Sambrook, E. F. Fritsch, and T. Maniatis, "Molecular Cloning," 2nd Ed., Cold Spring Harbor Laboratory Press, Cold Spring Harbor, New York, 1989.

after 48 hr. These colonized disks are then removed from the RD and placed in scintillation vials containing 10 ml of a phage suspension at varying phage concentrations (MOIs). Alternatively, *E. coli* biofilms can be grown in a chemostat coupled to a flow cell.[12] The flow cell is then disconnected from the chemostat and a phage suspension is added. The flow cell enables the study of phage–biofilm interactions by SCLM (Fig. 2). In this fashion, investigators are able to control the physiology (growth rate and nutrient limitation) of the biofilm organism, and the concentration of the phage. Longer term experiments will involve mixing the phage with the chemostat culture. Here, it would be anticipated that the phage, planktonic cultures, and biofilm cultures would achieve a steady state.

Summary and Conclusions

Phenotypic characterization will be a pivotal aspect of future research in understanding the biofilm mode of growth. We hope that the concepts and techniques presented in this chapter will benefit other investigators in this field. Although initial studies will necessarily involve monocultures, eventually mixed culture work will have to be performed to understand biofilm growth in the natural environment. As the study of biofilm–phage interactions is new, there is considerable fundamental work that needs to be addressed. Here, we anticipate that some phage are better adapted to growth in biofilms, some are adept in growing in mixed culture biofilms, and others are better adapted to infecting planktonic organisms. Whereas biofilms are now widely accepted as a fundamental aspect of microbial growth in nature, the field of phage ecology is quite new and an exciting challenge for the future.

Acknowledgments

Work in the authors' laboratories has been funded by the Biology Department and by Faculty Research Grants from Southwest Texas State University (R. J. C. M. and G. M. A.), the Texas Higher Education Coordinating Board (R. J. C. M.), and the Environmental Protection Agency (R. J. C. M.).

Section IV

Biofilms of Staphylococci

[17] Methods for Studying Biofilms Produced by *Staphylococcus epidermidis*

By MARGARET A. DEIGHTON, JILLIAN CAPSTICK, EWA DOMALEWSKI, and TRUNG VAN NGUYEN

Introduction

Coagulase-negative staphylococci, in particular *Staphylococcus epidermidis*, once regarded as harmless skin commensals or contaminants of clinical material, are now accepted as major nosocomial pathogens. These bacteria are the most common cause of medical device-associated infection, especially if host immunity is depressed because of underlying disease or treatment with immunosuppressive drugs. Common types of device-associated infection include prosthetic valve endocarditis, infections of intravascular devices, peritonitis associated with continuous ambulatory peritoneal dialysis, central venous shunt infections, and bacteremia (reviewed in Boyce[1]). Bloodstream infection with coagulase-negative staphylococci is also common in low birth weight newborns requiring intensive care.[2,3]

Bayston and Penny[4] were the first to describe medical devices coated with a heavy mass of bacteria and extracellular material, now referred to as *biofilm*. The ability to grow on implanted devices as an adherent multilayered biofilm is now considered to be an important virulence factor of coagulase-negative staphylococci, in particular *S. epidermidis*.[5]

Biofilm production by *S. epidermidis* is a complex process that is not completely understood. The initial, reversible phase is mediated by surface proteins, which probably interact via hydrophobic reactions.[6–8] During the subsequent phase, bacteria become enmeshed in a mass of extracellular material that is firmly attached to the underlying surface. Several groups have described polymers

[1] J. M. Boyce, *in* "The Staphylococci in Human Disease" (K. B. Crossley and G. L. Archer, eds.), p. 309. Churchill Livingstone, New York, 1997.

[2] D. G. Sidebottom, J. Freeman, R. Platt, M. F. Epstein, and D. A. Goldmann, *J. Clin. Microbiol.* **26**, 713 (1988).

[3] C. C. Patrick, S. L. Kaplan, C. J. Baker, J. T. Parisi, and E. O. Mason, *Pediatrics* **84**, 977 (1989).

[4] R. Bayston and S. R. Penny, *Dev. Med. Child. Neurol.* **14**(Suppl. 27), 25 (1972).

[5] G. D. Christensen, L. Baldassarri, and W. A. Simpson, *in* "Infections Associated with Indwelling Medical Devices" (A. L. Bisno and F. A. Waldvogel, eds.), p. 45. American Society for Microbiology, Washington, D.C., 1994.

[6] C. Heilmann, C. Gerke, F. Perdreau-Remington, and F. Götz, *Infect. Immun.* **64**, 277 (1996).

[7] C. Heilmann, M. Hussain, G. Peters, and F. Götz, *Mol. Microbiol.* **24**, 1013 (1997).

[8] C. Heilmann and F. Götz, *Zentralbl. Bakteriol.* **287**, 69 (1998).

associated with this second accumulation phase of biofilm formation.[9–13] Polysaccharide adhesin (PS/A) is a β-1,6-linked polyglucosamine that is essential for cell accumulation, resistance to phagocytosis, and virulence in animal models.[9,14–16] Slime-associated antigen (SAA) was initially described by Christensen et al.[10] and later shown to consist mainly of N-acetylglucosamine.[12] Polysaccharide intercellular adhesin (PIA) is a β-1,6-linked glucosaminoglycan[11,17] necessary for biofilm formation[6,8] and experimental device-related infection.[18,19] Thus preparations of PS/A, SAA, and PIA have a common polyglucosamine backbone but apparently differ in molecular size, substitutions on the glucosamine, and associated protein material.[12,13,16] Both PS/A and PIA are synthesized by a protein encoded by the *ica* locus of *S. epidermidis*.[16,20] Biofilm formation has been described in *Staphylococcus aureus* grown in tryptone soy broth (TSB) supplemented with glucose.[21,22] The process involves PIA synthesis by proteins encoded by the *ica* locus.[23]

In an earlier volume of this series, Christensen et al.[24] listed four questions asked by investigators interested in the colonization of medical devices. In essence they are as follows:

1. How do microorganisms differ in their ability to form biofilms on standard biomaterials under standard conditions? This question is applied to clinical isolates

[9] M. Tojo, N. Yamashita, D. A. Goldmann, and G. B. Pier, *J. Infect. Dis.* **157**, 713 (1988).

[10] G. D. Christensen, L. P. Barker, T. P. Mawhinney, L. M. Baddour, and W. A. Simpson, *Infect. Immun.* **58**, 2906 (1990).

[11] D. Mack, M. Nedelmann, A. Krokotsch, A. Schwarzkopf, J. Heesemann, and R. Laufs, *Infect. Immun.* **62**, 3244 (1994).

[12] L. Baldassarri, F. Donelli, A. Gelosia, M. C. Voglino, W. A. Simpson, and G. Christensen, *Infect. Immun.* **64**, 3410 (1996).

[13] M. Hussain, M. Herrmann, C. von Eiff, F. Perdreau-Remington, and G. Peters, *Infect. Immun.* **65**, 519 (1997).

[14] Y. Kojima, M. Tojo, D. A. Goldmann, T. D. Tosteson, and G. B. Pier, *J. Infect. Dis.* **162**, 435 (1990).

[15] H. Shiro, G. Meluleni, A. Groll, E. Muller, T. D. Tosteson, D. A. Goldmann, and G. B. Pier, *Circulation* **92**, 2715 (1995).

[16] D. McKenney, J. Hübner, E. Muller, Y. Wang, D. A. Goldmann, and G. B. Pier, *Infect. Immun.* **66**, 4711 (1998).

[17] D. Mack, W. Fischer, A. Krokotsch, K. Leopold, R. Hartmann, H. Egge, and R. Laufs, *J. Bacteriol.* **178**, 175 (1996).

[18] M. E. Rupp, J. S. Ulphani, P. D. Fey, K. Bartscht, and D. Mack, *Infect. Immun.* **67**, 2627 (1999).

[19] M. E. Rupp, J. S. Ulphani, P. D. Fey, and D. Mack, *Infect. Immun.* **67**, 2656 (1999).

[20] C. Heilmann, O. Schweitzer, C. Gerke, N. Vanittanakom, D. Mack, and F. Götz, *Mol. Microbiol.* **20**, 1083 (1996).

[21] R. Baselga, I. Albizu, E. De La Cruz, E. Del Casho, M. Barberan, and B. Amorena, *Infect. Immun.* **61**, 4857 (1993).

[22] M. G. Ammendolia, R. Di Rosa, L. Montanaro, C. R. Arciola, and L. Baldassarri, *J. Clin. Microbiol.* **37**, 3235 (1999).

[23] S. E. Cramton, C. Gerke, N. F. Schnell, W. W. Nichols, and F. Götz, *Infect. Immun.* **67**, 5427 (1999).

[24] G. D. Christensen, L. Baldassarri, and W. A. Simpson, *Methods Enzymol.* **253**, 477 (1995).

and, more recently, to evaluate mutants deficient in different aspects of biofilm formation.

2. How do environmental conditions affect biofilm density of reference strains on standard biomaterials?

3. How do biomaterials differ in their ability to support biofilms of reference strains of bacteria under standard conditions?

4. How do antimicrobial agents and host defenses affect the development and maintenance of biofilms?

One of the problems encountered when examining the literature on biofilm production by *S. epidermidis* is the plethora of methods and approaches that have been used to investigate the interaction between the bacteria and glass or plastic surfaces. This chapter begins with an overview of methods that have been used to study biofilm formation by staphylococci. Next, methods for assaying early attachment and mature biofilm are described. Although deceptively simple, these methods require careful attention to detail in order to obtain meaningful results. An attempt is made to address the problems associated with differences in methodology between research groups, by presenting methodology that is accepted by a majority of investigators.

Overview of Methods Used for Studying *Staphylococcus epidermidis* Biofilm

The initial attachment of staphylococci to polystyrene or biomaterial surfaces is studied by incubating cell suspensions in contact with a plastic surface for a brief period, usually 1 hr. The adherent cells are counted microscopically[6] or by performing viable counts after vortexing or sonication.[11,16,25–27] The number of attached cells can also be estimated by growing the bacteria in the presence of radioactively labeled substrates that become incorporated into the cell[28–31] or by assaying for released bacterial products such as urease[32] or ATP, using bioluminescence protocols.[25,33,34] All these methods are suitable for estimating the numbers of bacteria that attach during the first hour or two of the process. Most investigators

[25] K. G. Kristinsson, *J. Med. Microbiol.* **28**, 249 (1989).

[26] D. J. Siverhus, D. D. Schmitt, C. E. Edmiston, D. F. Bandyk, G. R. Seabrook, M. P. Goheen, and J. B. Towne, *Surgery* **107**, 613 (1990).

[27] G. Giridhar, A. S. Kreger, Q. N. Myrvik, and A. G. Gristina, *J. Biomed. Mater. Res.* **28**, 1289 (1994).

[28] A. Pascual, A. Fleer, N. A. C. Westerdaal, and J. Verhoef, *Eur. J. Clin. Microbiol.* **5**, 518 (1986).

[29] C. P. Timmerman, A. Fleer, J. M. Besnier, L. deGraaf, F. Cremers, and J. Verhoef, *Infect. Immun.* **59**, 4187 (1991).

[30] E. Muller, S. Takeda, D. A. Goldmann, and G. B. Pier, 1991. *Infect. Immun.* **59**, 3323 (1991).

[31] G. J. C. Veenstra, F. F. M. Cremers, H van Dijk, and A. Fleer, *J. Bacteriol.* **178**, 537 (1996).

[32] W. M. Dunne and E. M. Burd, *Appl. Environ. Microbiol.* **57**, 863 (1991).

prefer to use viable counting techniques, because they are simple to perform and inexpensive.

Biofilm formation on the sides of glass test tubes is detected visually[35–41] or assayed spectroscopically.[42] Biofilm density on polystyrene tissue culture plates is measured by direct spectroscopic analysis of stained material after drying[6,39,43–50] or after elution of stained material.[51] Radiometric assays have also been used to measure biofilm on tissue culture plates.[45] Modifications that have been made to this method to answer specific questions are discussed later in this chapter.

The recognition that the ability to form biofilm on polystyrene tissue culture plates may not reflect the situation *in vivo* has led investigators to develop methods to examine biofilms or estimate their density on medically relevant polymers. In general, these methods are less reproducible than the standard quantitative assay because of inherent properties of medical biomaterials. These materials may not be transparent, they have rough surfaces, are irregular in shape, and difficult to cut without creating rough edges that support denser biofilm.

[33] A. Ludwicka, L. M. Switalski, A. Lundin, G. Pulverer, and T. Wadström, *J. Microbiol. Methods* **4,** 169 (1985).

[34] F. Schumacher-Perdreau, C. Heilmann, G. Peters, F. Götz, and C. Pulverer, *FEMS Microbiol. Lett.* **117,** 74 (1994).

[35] G. D. Christensen, W. A. Simpson, A. L. Bisno, and E. H. Beachey, *Infect. Immun.* **37,** 318 (1982).

[36] D. S. Davenport, R. M. Massanari, M. A. Pfaller, M. J. Bale, S. A. Streed, and W. J. Hierholzer, *J. Infect. Dis.* **153,** 332 (1986).

[37] F. Diaz-Mitoma, G. K. M. Harding, D. J. Hoban, R. S. Roberts, and D. E. Low, *J. Infect. Dis.* **156,** 555 (1987).

[38] M. A. Deighton, J. C. Franklin, J. W. Spicer, and B. Balkau, *Epidemiol. Infect.* **101,** 99 (1988).

[39] M. Pfaller, D. Davenport, M. Bale, M. Barrett, F. Koontz, and R. M. Massanari, *Eur. J. Clin. Microbiol. Infect. Dis.* **7,** 30 (1988).

[40] D. J. Freeman, F. R. Falkiner, and C. T. Keane, *J. Clin. Pathol.* **42,** 872 (1989).

[41] M. Hussain, J. G. M. Hastings, and P. J. White, *J. Med. Microbiol.* **34,** 143 (1991).

[42] C.-L. Tsai, D. J. Schurman, and R. Lane Smith, *J. Orthopaed. Res.* **6,** 666 (1988).

[43] G. D. Christensen, W. A. Simpson, J. J. Younger, L. M. Baddour, F. F. Barrett, D. M. Melton, and E. H. Beachey, *J. Clin. Microbiol.* **22,** 996 (1985).

[44] M. A. Deighton and B. Balkau, *J. Clin. Microbiol.* **28,** 2442 (1990).

[45] M. Hussain, C. Collins, J. G. M. Hastings, and P. J. White, *J. Med. Microbiol.* **37,** 62 (1992).

[46] D. Mack, N. Siemssen, and R. Laufs, *Infect. Immun.* **60,** 2048 (1992).

[47] L. Baldassarri, W. A. Simpson, G. Donelli, and G. D. Christensen, *Eur. J. Clin. Microbiol. Infect. Dis.* **12,** 866 (1993).

[48] E. Muller, J. Hübner, N. Gutierrez, S. Takeda, D. A. Goldmann, and G. B. Pier, *Infect. Immun.* **61,** 551 (1993).

[49] W. Zeibuhr, V. Krimmer, S. Rachid, I. Lößner, F. Götz, and J. Hacker, *Mol. Microbiol.* **32,** 345 (1999).

[50] M. H. Wilcox, *J. Clin. Pathol.* **47,** 1044 (1994).

[51] R. Bayston and J. Rodgers, *J. Clin. Pathol.* **43,** 866 (1990).

Methods involving direct examination of the biofilm on the surfaces of biomedical polymers include fluorescence microscopy,[52-54] scanning electron microscopy[55-58] and confocal laser scanning microscopy. The latter displays microcolonies embedded in the deeper layers of the biofilm.[58] Indirect methods that are applicable to the estimation of biofilm density on biomedical polymers include viable counting after sonication, bioluminescence, isotopic labeling, and dye elution. These are described briefly below.

Biofilm density on biopolymer surfaces can be estimated indirectly by sonication followed by direct plate counting.[26] Silverhus *et al.*[26] obtained reliable results after incubation of various vascular graft materials in a suspension of *S. epidermidis* for periods of 2 hr (initial attachment) to 24 hr (mature biofilm). After sonication at 20 kHz for 10 min, the number of bacteria released from the surface was determined by viable count. This method reliably removes 95% of bacteria at the initial attachment phase[11] and is applicable to a wide variety of biomaterials of different shapes; however, in our hands, it suffered from poor reproducibility when applied to mature biofilms of *S. epidermidis* on biomaterial surfaces. On the basis of scanning electron microscopy after sonication, we attributed poor reproducibility to the variation in the size of released bacterial aggregates and to the uneven, irregular structure of some biomaterial surfaces. Another problem with the method is that biomaterials differ in the power and frequency of oscillation required to release a standard percentage of bacteria while retaining viability.[59]

Bioluminescence assays overcome the problem of inconsistent release of bacteria from biofilm on the biomaterial surface because they measure bacterial ATP released during growth. A disadvantage of this method for estimating biofilm density is the decreasing level of bacterial ATP as growth rate diminishes.[33] However, Hussain *et al.*,[41] in a study of five biofilm-positive and two biofilm-negative *S. epidermidis* strains, obtained general agreement between a bioluminescence assay and a modified spectroscopic plate assay. The procedure is described in detail by Christensen *et al.*[24] in a previous volume of this series.

The use of radioactively labeled substrates is less reliable for estimating biofilm density than for assaying initial attachment because the ratio of colony-forming

[52] T. I. Ladd, D. Schmiel, J. C. Nickel, and J. W. Costerton, *J. Clin. Microbiol.* **21**, 1004 (1995).
[53] J. Zufferey, B. Rime, P. Francioli, and J. Bille, *J. Clin. Microbiol.* **26**, 175 (1998).
[54] S. L. Fessia and M. J. Griffin, *Perit. Dial. Int.* **11**, 144 (1991).
[55] G. Peters, R. Locci, and G. Pulverer, *J. Infect. Dis.* **146**, 479 (1982).
[56] T. R. Franson, N. K. Sheth, H. D. Rose, and P. G. Sohnle, *J. Clin. Microbiol.* **20**, 500 (1984).
[57] L. Baldassarri, A. Gelosia, G. Donelli, M. Mignozzi, E. Fiscarelli, and G. Rizzoni, *J. Mater. Sci. Mater. Med.* **5**, 601 (1994).
[58] S. P. Gorman, C. G. Adair, and W. M. Mawhinney, *Epidemiol. Infect.* **112**, 551 (1994).
[59] M. Deighton and J. Capstick, unpublished data (1990).

units (CFU) to counts per minute becomes progressively unstable after the first few hours of incubation.[24] Moreover, most investigators favor methods that avoid the use of radiolabeled substrates because of the hazards associated with the use and disposal of radioactive products.

A simple dye elution technique, based on the Christensen quantitative biofilm assay, was described.[60] The method appears to have promise for comparing biofilms on different biopolymers; however, reproducibility was less than optimal, as indicated by the high standard deviations reported. This variability was likely to be due to the inherent uneven nature of the surfaces of the biomaterials examined.

Several investigators have established flow systems to examine biomaterials for their ability to support initial attachment or biofilm formation under conditions of shear stress similar to those in the intravascular environment.[61–66] Olson et al.[61] used a modified Robbins device (MRD) to examine attachment and biofilm formation on disks coated with a cyanoacrylate tissue adhesive. The MRD exposes sample disks to a circulating bacterial suspension. After aseptic removal of the disks at specific times, attachment and biofilm formation may be assessed by (1) scraping followed by sonication to disperse clumps, and then counting the number of released bacteria; (2) epifluorescence; or (3) scanning electron microscopy. The same group[67] later developed a chemostat-coupled MRD that was used successfully by Linton et al.[65] to examine the initial attachment of S. epidermidis RP62A to various biomaterials. Khardori et al.[63] used a modified Robbins device to examine catheter segments under flow conditions in the presence and absence of antimicrobial agents. Another approach was adopted by Ceri et al.[66] The Calgary biofilm device, based on a 96-well microtiter tray format, is suited for antimicrobial susceptibility testing in a clinical laboratory. The device produces equivalent biofilms on 96 pegs exposed to bacterial cultures under conditions of shear stress. After incubation in the presence of antimicrobial agents, the pegs are sonicated and the number of released bacteria is determined by plate counts or turbidity measurement to determine minimum biofilm eradication concentrations (MBECs).[66]

[60] K. Merritt, A. Gaind, and J. M. Anderson, J. Biomed. Mater. Res. **39,** 415 (1998).

[61] M. E. Olson, I. Ruseska, and J. W. Costerton, J. Biomed. Mater. Res. **22,** 485 (1988).

[62] J. M. Higashi, I. Wang, D. M. Shales, J. M. Anderson, and R. E. Marchant, J. Biomed. Mater. Res. **39,** 341 (1998).

[63] N. Khardori, E. Wong, H. Nguyen, C. Jeffery-Wiseman, E. Wallin, R. P. Tewari, and G. P. Bodey, J. Infect. Dis. **164,** 108 (1991).

[64] I. Wang, J. M. Anderson, M. R. Jacobs, and R. E. Marchant, J. Biomed. Mater. Res. **29,** 485 (1995).

[65] C. J. Linton, A. Sherriff, and M. R. Millar, J. Appl. Microbiol. **86,** 194, (1999).

[66] H. Ceri, M. E. Olson, C. Stremick, R. R. Read, D. Morck, and A. Buret, J. Clin. Microbiol. **37,** 1771 (1999).

[67] J. Jass, J. W. Costerton, and H. M. Lappin-Scott, J. Indust. Microbiol. **15,** 283 (1995).

Finally, the polymers involved in biofilm formation may be assayed by immunological procedures for estimating *N*-acetylglucosamine,[68] SAA,[12] PS/A,[16,48] or PIA.[46]

This chapter focuses on conventional methods for measuring the initial attachment phase of biofilm formation and the density of mature biofilms.

Storage and Handling of Staphylococci

Staphylococci are stored as freeze-dried cultures or at −70°, preferably on storage beads such as Protect (Technical Service Consultants Lancashire, UK). When retrieving stocks from storage, it is important to be aware that pure cultures of staphylococci, in particular *S. epidermidis,* consist of a mixture of phenotypes that differ in attachment and accumulation capabilities. Phenotypes may also differ in their ability to survive after storage at −70°, and therefore the phenotypic mix retrieved after storage is unlikely to be identical to that of the original parent. To minimize phenotypic change, cultures retrieved from storage are examined to ensure that the characteristics of the parent culture are retained. Working cultures are stored on agar plates at 4°. Because phenotypic shifts can also occur after serial subculture,[69] the number of subcultures made before examining for biofilm is kept to an absolute minimum. Reference strains (Table I) are stored as freeze-dried cultures.

Quantitation of Initial Attachment Stage of Biofilm Formation

The number of bacteria attached to a surface depends both on the incubation time and the initial inoculum. As a significant proportion of the inoculum attaches within 5 min and saturation occurs after 30–60 min, short exposure times are appropriate.[33] Most investigators use dense inocula suspended in a nonnutrient solution and short exposure periods. The importance of using a standard suspending fluid was discussed in a previous volume of this series.[24] Protocols that avoid the use of radioisotopes are described below.

Method Based on Total Bacterial Counting

This method (adapted from Heilmann *et al.*[6]) measures the attachment of staphylococci to polystyrene by staining and counting adherent cells. It is used mainly to compare mutants for their attachment capabilities. The test strain is grown in TSB for 18 hr, centrifuged, and suspended in phosphate-buffered saline (PBS), to provide a suspension of 10^8 CFU/ml on the basis of the OD_{600}. Suspension is performed gently, using a syringe fitted with a 23-gauge needle to

[68] V. L. Thomas, B. A. Sanford, R. Moreno, and M. A. Ramsay, *Curr. Microbiol.* **35**, 249 (1997).
[69] G. D. Christensen, L. M. Baddour, J. T. Parisi, S. N. Abraham, D. L. Hasty, J. H. Lowrance, J. A. Josephs, and W. A. Simpson, *J. Infect. Dis.* **161**, 1153 (1990).

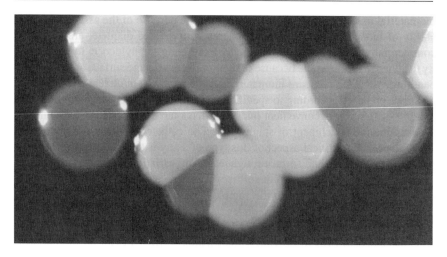

FIG. 1. Colonial variants of *S. epidermidis* on Congo red agar [M. Deighton, S. Pearson, J. Capstick, D. Spelman, and R. Borland, *J. Clin. Microbiol.* **30**, 2385 (1992); M. A. Deighton, J. Capstick, and R. Borland, *Epidemiol. Infect.* **108**, 423 (1992)]. The dark colonies are the major phenotype, and the pale colonies are variants.

the concentration of iron[73]; and the concentration of other divalent cations in the growth medium.[74] The density of biofilm generated is also affected by the size of the initial inoculum used in the assay.[73]

In addition, *S. epidermidis*[75-77] and *S. aureus*[21] frequently undergo reversible phenotypic changes (phase variation) involving biofilm formation as well as several other properties. Major phase shifts may occur after serial subculture of *S. epidermidis*[75] and *S. aureus*[21] or after retrieval from storage at $-70°$.[37,44] Phenotypic variation of biofilm production involves reversible insertion of IS*256* into one of the four genes required for the accumulation phase of biofilm formation (*icaC* gene).[49] Thus it is important to examine the colonial morphology of strains before testing, especially after retrieval from storage and to examine the phenotypic mixture of cultures at the conclusion of quantitative and qualitative biofilm assays. This may be achieved by direct plating on Memphis agar[69] or Congo red agar[8,76,77] and assessing cultures for the proportion of phase variants present (Fig. 1, Table II).

[73] M. A. Deighton and R. Borland, *Infect. Immun.* **61**, 4473 (1993).

[74] W. M. Dunne and E. M. Burd, *Microbiol. Immunol.* **36**, 1019 (1992).

[75] G. D. Christensen, L. M. Baddour, and W. A. Simpson, *Infect. Immun.* **55**, 2870 (1987).

[76] M. Deighton, S. Pearson, J. Capstick, D. Spelman, and R. Borland, *J. Clin. Microbiol.* **30**, 2385 (1992).

[77] M. A. Deighton, J. Capstick, and R. Borland, *Epidemiol. Infect.* **108**, 423 (1992).

TABLE II
FORMULATIONS OF CONGO RED AGAR

Ref.	Agar base	Congo red (%, w/v)	Carbohydrate supplement
Freeman et al. (1989)[a]	Agar no. 2 (Oxoid); brain–heart infusion broth (Oxoid)	0.08	Glucose or sucrose (5%, w/v)
Deighton et al. (1992)[b,c]	Blood agar base no. 2 (Oxoid)	0.003	None
Heilmann et al. (1996)[d]		0.08	Glucose (1%, w/v)

[a] D. J. Freeman, F. R. Falkiner, and C. T. Keane, *J. Clin. Pathol.* **42**, 872 (1989).
[b] M. Deighton, S. Pearson, J. Capstick, D. Spelman, and R. Borland, *J. Clin. Microbiol.* **30**, 2385 (1992).
[c] M. A. Deighton, J. Capstick, and R. Borland, *Epidemiol. Infect.* **108**, 423 (1992).
[d] C. Heilmann and F. Götz, *Zentralbl. Bakteriol.* **287**, 69 (1998).

Colonial Morphology on Congo Red Agar

Three different formulations of Congo red agar (CRA) have been described.[8,40,76,77] These differ in the amount of Congo red, agar base, and carbohydrate supplementation (Table II). On the CRA described by Freeman et al.,[40] biofilm-positive strains produce black crystalline colonies, whereas biofilm-negative strains have pink colonies with black centers or specks. Strains assessed as moderate or weak biofilm producers by the quantitative assay may give discrepant results. The media described by Deighton et al.[76,77] (Fig. 1) and Heilmann and Götz[8] were designed to display phenotypic variants and biofilm mutants, respectively.

Detection of Biofilm on Sides of Glass Test Tubes

The earliest method for examining *S. epidermidis* biofilms was the qualitative test described by Christensen et al.[35] Although referred to as the *tube adherence test* in early publications, this test displays the outcome of both stages of biofilm formation, i.e., the multilayered structure consisting of microcolonies embedded in an amorphous extracellular matrix (referred to as *slime*). Test strains are incubated for 24 hr at 37° on nutrient or blood agar plates. Cultures are examined for colonial variation. If more than one colonial morphotype is present, the major phenotype is selected. Alternatively, major and minor morphotypes are examined separately. Three or four well-isolated colonies with identical morphology are touched with a straight wire, which is used to inoculate 2 ml of TSB in a test tube with a loosely fitting cap. If the medium is opalescent or cloudy, the inoculum is too heavy. Larger volumes of medium or tight caps are not appropriate because anaerobic conditions[72] and dissolved CO_2 reduce the amount of biofilm produced by some strains of *S. epidermidis*.[70,71] Cultures are incubated at 37° for 24 hr without shaking. Tilting the tubes at an angle of 45° from the vertical position enhances oxygenation and facilitates reading.[38] After incubation, the growth medium is decanted and the biofilm is washed gently three times with water, to remove nonadherent bacteria, taking care not to dislodge the biofilm. The biofilm is fixed at 60° for 1 hr and then stained with safranin (0.1%, w/v) by rolling the stain around the sides of the tube for 1 min. This procedure estimates biofilm density by staining entrapped bacteria. Alcian blue, which stains extracellular polysaccharide, is equally effective.[35] Excess stain is removed by washing three times in clean water, and then the test tube is inverted until dry. Biofilm is considered present if a stained film lines the sides of the test tube. Ring formation at the air–liquid interface without a stained film needs to be interpreted with caution. It may indicate a strain that produces biofilm only under aerobic conditions[72] or a strongly positive strain that has lost biofilm during early stationary phase or during incubation.[38] Suitable controls are *S. epidermidis* RP62A (biofilm positive) and *Staphylococcus hominis* SP2 (biofilm negative) (Table I). The amount of biofilm is estimated visually and recorded as strong (+++), moderate (++), weak (+), or negative (–) (Fig. 2). All tests are repeated at least twice.

The tube assay is simple to perform. It is also reproducible, provided protocols are followed with careful attention to detail, fresh cultures are tested after a minimum number of subcultures, and cultures showing significant phase variation are avoided. Any discrepant results can usually be resolved by repeating the assay with a new culture prepared from the original stock culture. In early studies, poor reproducibility of the tube assay and discrepancies between the tube and quantitative assays[37,43,44] were reported. In hindsight, it is likely that these discrepancies were caused by differences in oxygenation of cultures in test tubes compared with microtiter plates and phenotypic variation of cultures. It is also the case that the initial step in biofilm formation depends on hydrophobic interactions, and that glass presents a hydrophilic surface, whereas polystyrene is hydrophobic. Mutants that lack certain surface proteins associated with cell surface hydrophobicity fail to produce biofilm on polystyrene but are still able to do so on glass.[6]

Quantitation of Mature Biofilm on Polystyrene under Static Conditions

The quantitative biofilm assay, for measuring the density of a mature biofilm of *S. epidermidis* on plastic surfaces, was first described by Christensen *et al.*[43] The assay has been used extensively to examine collections of pathogenic and commensal coagulase-negative staphylococci for their potential to form biofilm on polystyrene surfaces. Now that molecular techniques for manipulation of staphylococci are more widely available, this method provides a simple, reliable, and inexpensive test for examining mutants. Many modifications to the original assay

FIG. 2. Tube biofilm assay. Strong (+++), moderate (++), and weak biofilm displayed on the sides of glass test tubes.

have been described in the literature (Table III). Although these modifications are unlikely to alter the major conclusions drawn, they make direct comparison of the work of different research groups difficult. The following method is based on Christensen's method with modifications described by Baldassarri *et al.*[47]

Preparation of Inoculum. A 5-ml volume of TSB is inoculated with three or four colonies from a 24-hr culture on nutrient agar or blood agar and incubated without shaking for approximately 18 hr ($\pm$30 min). If colonial variants are present it is important to ensure that all colonies picked belong to the predominant phenotype. Alternatively, major and minor morphotypes are examined separately. The broth culture is centrifuged, and then the cells are suspended in fresh TSB to a density of approximately 10^7 CFU/ml, on the basis of the OD_{600}. Any cell aggregates are broken up with a syringe fitted with a 23-gauge needle, and then the suspension is vortexed briefly at high speed to obtain a smooth suspension. A standard inoculum is important because biofilm density increases with increasing initial inoculum.[73]

Inoculation of Tissue Culture Plates. Two hundred microliters of the diluted culture is inoculated into the wells of 96-well flat-bottom tissue culture plates. It is

TABLE III
Protocols for Quantitative Biofilm Assay

Ref.	Growth medium	Inoculum	Microtiter plate	Culture conditions	Washing procedure	Fixation	Stain	Stain wavelength (nm)	Cutoff OD
Christensen et al. (1985)[a]	TSB	1 : 100 dilution of overnight broth culture	Falcon and Corning	18 hr, 37°	Four washes in PBS	Bouin's	Hucker crystal violet	570	Low: 0.12 High: 0.24
Pfaller et al. (1988)[b]	TSB	1 × 10⁵ CFU/ml	Corning	48 hr, 35°	Two washes in PBS		Safranin	490	0.415
Deighton and Balkau (1990)[c]	TSB	1 : 100 dilution of overnight broth	Linbro (Nunc)	18 hr, 37°	Four washes in PBS	Bouin's	Hucker crystal violet	600	Low: 0.3 High: 0.6
Dunne (1990)[d]	TSB	5 × 10⁵ CFU/ml	GIBCO	18 hr, 37°	Four washes in Hanks' balanced salt solution	Methanol	Hucker crystal violet	570	
Hussain et al. (1991)[e]	TSB, HHW (chemically defined)	10⁷ CFU/ml		18 hr, 37°	Two washes in saline	Bouin's	Safranin (0.1%, w/v)	490	
Mack et al. (1992)[f]	TSB	1:100 dilution of overnight broth	Nunc	20–24 hr, 37°	Four washes in PBS	Bouin's	Gentian violet	570	0.1
Baldassarri et al.(1993)[g]	TSB + 1% (w/v) glucose	1:2 dilution of overnight broth	Nunc	24 hr, 37°	Four washes in PBS	60°, 1 hr	Hucker crystal violet	570	Low: 0.12 High: 0.24
Muller et al. (1993)[h]	TSB	5 × 10⁵–5 × 10⁶/ml	Costar	18 hr, 37°	Four washes in PBS	Bouin's	Safranin	490	
Wilcox (1994)[i]	TSB	10⁶ CFU/ml	Dynatech	37°	Three washes in saline	Formalin (10%, v/v)	Crystal violet Bouin's crystal violet	546 600	

Heilman et al. (1996)[j]	TSB + 0.25% (w/v) glucose	1:200 dilution of overnight broth	Greiner	24 hr, 37°	Two washes in PBS	Air	Safranin (0.1%, w/v)	490	
Ziebuhr et al. (1997)[k]	TSB	1:200 dilution of overnight broth	Greiner	24 hr, 37°	Four washes in PBS	Bouin's	Gentian violet (0.4%, w/v)	490	0.12
McKenney et al. (1998)[l]	TSB + 0.25% (w/v) glucose	1:200 dilution of overnight broth	Costar	24 hr, 37°	Two washes in PBS	Air dried	Safranin (0.025%, w/v)	490	

Abbreviations: TSB, tryptone soy broth.

[a] G. D. Christensen, W. A. Simpson, J. J. Younger, L. M. Baddour, F. F. Barrett, D. M. Melton, and E. H. Beachey, *J. Clin. Microbiol.* **22**, 996 (1985).

[b] M. Pfaller, D. Davenport, M. Bale, M. Barrett, F. Koontz, and R. M. Massanari, *Eur. J. Clin. Microbiol. Infect. Dis.* **7**, 30 (1988).

[c] M. A. Deighton and B. Balkau, *J. Clin. Microbiol.* **28**, 2442 (1990).

[d] W. M. Dunne, *Antimicrob. Agents Chemother.* **34**, 390 (1990).

[e] M. Hussain, J. G. M. Hastings, and P. J. White, *J. Med. Microbiol.* **34**, 143 (1991).

[f] D. Mack, N. Siemssen, and R. Laufs, *Infect. Immun.* **60**, 2048 (1992).

[g] L. Baldassarri, W. A. Simpson, G. Donelli, and G. D. Christensen, *Eur. J. Clin. Microbiol. Infect. Dis.* **12**, 866 (1993).

[h] E. Muller, J. Hübner, N. Gutierrez, S. Takeda, D. A. Goldmann, and G. B. Pier, *Infect. Immun.* **61**, 551 (1993).

[i] H. Wilcox, *J. Clin. Pathol.* **47**, 1044 (1994).

[j] C. Heilmann, C. Gerke, F. Perdreau-Remington, and F. Götz, *Infect. Immun.* **64**, 277 (1996).

[k] W. Ziebuhr, C. Heilmann, F. Götz, P. Meyer, K. Wilms, E. Strauße, and J. Hacker, *Infect. Immun.* **85**, 890 (1997).

[l] D. McKenney, J. Hübner, E. Muller, Y. Wang, D. A. Goldmann, and G. B. Pier, *Infect. Immun.* **66**, 4711 (1998).

important to use the same brand of tissue culture plates throughout a set of experiments. Polystyrene tissue culture plates are sterilized by γ irradiation and treated to reduce the hydrophobicity. Because the initial attachment phase of biofilm formation is mediated by hydrophobic interactions, differences in hydrophobicity could affect biofilm density. Microtiter plates used for enzyme-linked immunosorbent assays (ELISAs) assays have different surface characteristics and give different results in biofilm assays.[43] The plates are set up so that each strain is tested in quadruplicate. Each plate includes a medium control and reference strains such as *S. epidermidis* RP62A (biofilm positive) and *S. hominis* SP2 (biofilm negative) (Table I).

Incubation Conditions. Microtiter plates are incubated in a wet box for 24 hr (±30 min). It is important to adhere to a standard time of incubation, because the density of biofilm is dependent on incubation time.[72]

Washing and Staining Biofilm. First, the OD_{600} is read. This step is used as an additional control to detect poor growth or contamination. Growth from selected wells is subcultured onto Congo red agar (Fig. 1, Table II) to detect possible phenotypic shifts that could cause unexpected results. The contents of all wells are then carefully decanted into a discard container for contaminated material. Each plate is washed three times in clean water, taking care to remove nonadherent cells, while preserving the structure of the biofilm. With care, this can be achieved aseptically with gloved hands, large containers of water, and a stack of paper towels. All contaminated material is autoclaved. We have found that alternative washing procedures, such as using a manual pipette or mechanical plate washer, are difficult to standardize because some disruption to the biofilm is inevitable during the process. After draining, the biofilm is fixed for 1 hr in a hot air oven at 60°. This method of heat fixing is comparable to Bouin's fixative and avoids using explosive chemicals.[47] In our own experience heat-fixed biofilms are more stable and give more consistent readings than those fixed with Bouin's fixative. The biofilm is stained with Hucker crystal violet (Table IV) for 1 min, and then excess stain is decanted and the plate is washed several times by dunking in a large container of clean tap water. Washing with fresh water is continued until the washings are free of stain. At all stages of the process, careful handling is essential to preserve the integrity of the biofilm. Biofilm integrity is carefully monitored during washing and any wells that lose visible sheets of material are excluded from subsequent calculations. The plate is drained and allowed to dry in air for 30 min in an inverted position at 37° or at room temperature overnight. It is important to ensure that plates are completely dry before reading.

Measurement of Biofilm. The density of the biofilm (OD_{570}) is measured with a spectrophotometer, with wells containing uninoculated medium to blank the instrument. The instrument should have a maximum OD reading of at least 2.5 to accommodate strong biofilm producers. The means and standard deviations

TABLE IV
HUCKER CRYSTAL VIOLET FOR STAINING BIOFILMS

Ingredient	Amount
Crystal violet	2 g
Alcohol, 95% (v/v)	20 ml
Ammonium oxalate	0.8 g
Distilled water	80 ml

Procedure

The crystal violet is dissolved in alcohol. The ammonium oxalate is dissolved in distilled water. The two solutions are mixed, left to stand at room temperature for 24 hr, and then filtered. Before use the stain is examined for precipitate and filtered again if necessary

of quadruplicate readings are calculated. The assay is repeated on at least two occasions.

Interpretation of Results. The mean OD_{570} of the test wells is interpreted as follows:

Positive	≥ 0.24
Weak	≥ 0.12 and < 0.24
Negative	< 0.12

In the original description of the method by Christensen *et al.*[43] the low cutoff value was chosen on the basis of 3 standard deviations above that of uninoculated medium. The high cutoff value was obtained by doubling the low value. Pfaller *et al.*[39] defined the cutoff value for a biofilm-positive isolate as 3 standard deviations above the mean optical density for a collection of tube test-negative isolates. Using the method described above, *S. epidermidis* RP62A is strongly positive, with optical density values of 2.5, and *S. hominis* SP2 is negative. With respect to clinical isolates, strong biofilm producers have optical density values of 1.5 or higher, nonbiofilm producers have values of about 0.06, and many strains are in the weak or low positive range. In general, the assay is reproducible in the classification of strains into three categories (positive, weak, and negative). However, it is not unusual to observe some variation in the optical density values, particularly in the intermediate range. This variability is due to the random insertion of a transposon into the *ica* gene cluster.[49]

Measurement of Biofilm Production by Staphylococcus aureus. Biofilm production by *S. aureus* is measured by a slight modification of the standard quantitative assay. Instead of using standard TSB [containing 1% (w/v) glucose], TSB is supplemented with an additional amount of filter-sterilized glucose. Cramton *et al.*[23] suggest adding 0.25% (w/v) glucose to standard. TSB (final concentration, 1.25%, w/v) whereas 2% (final concentration) is suggested by Ammendolia *et al.*[22]

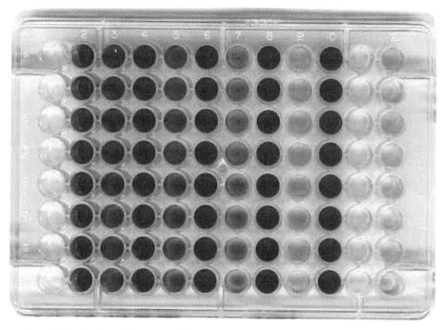

FIG. 3. Quantitative biofilm assay with mixed cultures of biofilm-positive and biofilm-negative isolates of *S. epidermidis*. Column 1, medium blank; columns 2, 4, 6, 8, and 10, strain RP62A (biofilm positive); columns 3, 5, 7, and 9, mixtures of strains RP62A and A211 (biofilm negative, rows A–D) or A250 (biofilm negative, rows E–H) in ratios of 80 : 20, 60 : 40, 40 : 60, 20 : 80; column 11, rows A–D, A211; rows E–H, A250.

Measurement of Biofilm Produced by Mixed Staphylococcal Cultures

We have developed a method for measuring the biofilm produced by staphylococci in mixed culture.[78] Inocula are prepared by centrifuging 18-hr TSB cultures and suspending them in fresh TSB as in the standard assay (approximately 10^7 CFU/ml). Mixed inocula are prepared in sterile Wassermann tubes in the following proportions: strain A:strain B = 100 : 0, 80 : 20, 60 : 40, 40 : 60, 20 : 80, 0 : 100. Matching suspensions in which strain B is replaced by TSB are also prepared. Two hundred-microliter aliquots of the suspensions are transferred to the wells of 96-well flat-bottom tissue culture trays, such that each is tested in quadruplicate and cultures are processed as described for the standard quantitative adherence assay. Positive, negative, and uninoculated controls are prepared as for the standard assay (Fig. 3). To confirm that mixtures consisting of two strains are retained after incubation, the growth from selected wells is plated onto Congo red agar.

---

[78] M. A. Deighton, E. Domalewski, and T. Nguyen, unpublished data (2000).

Modification of Quantitative Assay to Reflect Conditions Found in Human Host

The standard methods for measuring biofilm production may be modified to provide conditions that reflect those encountered by staphylococci in the human host. The knowledge that higher levels of CO_2 are found in most body fluids than in air led Denyer *et al.*[70] to develop a modified biofilm assay in which microtiter plates are incubated in an atmosphere of 5% CO_2–95% air. Because coagulase-negative staphylococci are the leading cause of peritonitis in patients receiving continuous ambulatory peritoneal dialysis, the use of pooled spent peritoneal dialysate in place of standard culture medium and incubation in 5% CO_2 may provide relevant information on biofilm formation on Tenckhoff catheters placed in the peritoneal cavity.[79] Biofilm assays have also been performed under conditions of iron limitation, in order to reflect conditions that bacteria encounter in the mammalian host.[73] Finally, several investigators have adapted the quantitative assay to examine the effect of subinhibitory concentrations of antimicrobial agents on biofilm formation.[80–83]

[79] M. H. Wilcox, D. G. E. Smith, J. A. Evans, S. P. Denyer, R. G. Finch, and P. Williams, *J. Clin. Microbiol.* **28,** 2183 (1990).

[80] K. H. Schadow, W. A. Simpson, and G. D. Christensen, *J. Infect. Dis.* **157,** 71 (1988).

[81] W. M. Dunne, *Antimicrob. Agents Chemother.* **34,** 390 (1990).

[82] M. H. Wilcox, R. G. Finch, D. G. E. Smith, P. Williams, and S. P. Denyer, *J. Antimicrob. Chemother.* **27,** 577 (1991).

[83] M. E. Rupp and K. E. Hamer, *J. Antimicrob. Chemother.* **41,** 155 (1998).

[18] Methods to Detect and Analyze Phenotypic Variation in Biofilm-Forming Staphylococci

By WILMA ZIEBUHR, ISABEL LOESSNER, VANESSA KRIMMER, and JÖRG HACKER

Introduction.

Staphylococci are the most common cause of medical device-associated infections, especially in immunocompromised patients.[1] The pathogenesis of these infections is favored by the ability of certain *Staphylococcus epidermidis* and *Staphylococcus aureus* strains to express a slimy matrix in which the bacteria become embedded and form thick, multilayered biofilms on smooth surfaces. Numerous proteins and polysaccharides involved in this process have been

[1] M. E. Rupp and G. L. Archer, *Clin. Infect. Dis.* **19,** 231 (1994).

described.[2-5] One of these factors is polysaccharide intercellular adhesin (PIA), which contributes essentially to staphylococcal biofilm accumulation.[6] The molecular basis of PIA expression has been determined and is described in detail elsewhere in this volume.[6a] The enzymes involved in PIA synthesis are encoded by the *ica* operon comprising the *icaA*, *icaD*, *icaB*, and *icaC* genes.[7,8] Epidemiological studies of the distribution of this genetic information among staphylococci revealed remarkable differences. Thus, the *ica* operon was found to be widespread in *S. epidermidis* isolates causing polymer-associated infections. However, *S. epidermidis* saprophytic isolates from the skin or mucosa of healthy persons rarely contain *ica*-specific DNA and the majority of these strains are, therefore, biofilm negative.[9,10] The *ica* operon has also been detected in *S. aureus* and a range of other staphylococcal species.[11,12] In contrast to the situation in *S. epidermidis*, all *S. aureus* strains analyzed so far contain the entire *ica* gene cluster, but only a few express the operon and produce the polysaccharide intercellular adhesin *in vitro*.

Staphylococcus aureus and *S. epidermidis* clinical isolates are known to vary in a wide range of phenotypic properties. Thus, it has been shown that cells from a single parental clone can differ in terms of growth rate, exoprotein production, antibiotic susceptibility, and adhesion.[10,13-16] It has been suggested that this

[2] G. D. Christensen, L. P. Barker, T. P. Mawhinney, L. M. Baddour, and W. A. Simpson, *Infect. Immun.* **58**, 2906 (1990).

[3] M. Hussain, M. Herrmann, C. von Eiff, F. Perdreau-Remington, and G. Peters, *Infect. Immun.* **65**, 519 (1997).

[4] M. Tojo, N. Yamashita, D. A. Goldmann, and G. B. Pier, *J. Infect. Dis.* **157**, 713 (1988).

[5] D. Mack, M. Nedelmann, A. Krokotsch, A. Schwarzkopf, J. Heesemann, and R. Laufs, *Infect. Immun.* **62**, 3244 (1994).

[6] D. Mack, W. Fischer, A. Krokotsch, K. Leopold, R. Hartmann, H. Egge, and R. Laufs, *J. Bacteriol.* **178**, 175 (1996).

[6a] D. Mack, K. Bartscht, C. Fischer, H. Rohde, C. de Grahl, S. Dobinsky, M. A. Horstkotte, K. Kiel, and J. K.-M. Knobloch, *Methods Enzymol.* **336**, Chap. 20, 2001 (this volume).

[7] C. Gerke, A. Kraft, R. Süssmuth, O. Schweitzer, and F. Götz, *J. Biol. Chem.* **273**, 18586 (1998).

[8] C. Heilmann, O. Schweitzer, C. Gerke, N. Vanittanakom, D. Mack, and F. Götz, *Mol. Microbiol.* **20**, 1083 (1996).

[9] N. B. Frebourg, S. Lefebvre, S. Baert, and J. F. Lemeland, *J. Clin. Microbiol.* **38**, 877 (2000).

[10] W. Ziebuhr, C. Heilmann, F. Götz, P. Meyer, K. Wilms, E. Straube, and J. Hacker, *Infect. Immun.* **65**, 890 (1997).

[11] D. McKenney, K. L. Pouliot, Y. Wang, V. Murthy, M. Ulrich, G. Doring, J. C. Lee, D. A. Goldmann, and G. B. Pier, *Science* **284**, 1523 (1999).

[12] S. E. Cramton, C. Gerke, N. F. Schnell, W. W. Nichols, and F. Gotz, *Infect. Immun.* **67**, 5427 (1999).

[13] R. Baselga, I. Albizu, M. De La Cruz, E. Del Cacho, M. Barberan, and B. Amorena, *Infect. Immun.* **61**, 4857 (1993).

[14] G. D. Christensen, L. M. Baddour, B. M. Madison, J. T. Parisi, S. N. Abraham, D. L. Hasty, J. H. Lowrance, J. A. Josephs, and W. A. Simpson, *J. Infect. Dis.* **161**, 1153 (1990).

[15] M. Deighton, S. Pearson, J. Capstick, D. Spelman, and R. Borland, *J. Clin. Microbiol.* **30**, 2385 (1992).

[16] M. Mempel, H. Feucht, W. Ziebuhr, M. Endres, R. Laufs, and L. Gruter, *Antimicrob. Agents Chemother.* **38**, 1251 (1994).

phenotypic variation might contribute to the successful adaptation of the bacteria to environmental changes. The organization of bacteria in biofilms has been recognized as a crucial mechanism by which to withstand unfavorable external conditions.[1,17] With respect to *S. epidermidis* it was shown that the majority of the clinical *S. epidermidis* isolates grow in multicellular clusters and that the removal of these biofilms (e.g., by antibiotics) is difficult. Therefore, it is assumed that staphylococcal biofilm formation represents an advantage for the bacteria that contributes considerably to pathogenesis. At the same time, however, numerous studies give evidence that *S. epidermidis* has a strong capacity to vary biofilm expression.[13,15,18] Thus, it was observed that within a *S. epidermidis* biofilm population biofilm-negative variants are currently released.[10,19] In this chapter we describe the detection and genetic analysis of these biofilm-negative variants. In addition, we show the application of the 16S rRNA *in situ* hybridization technique for the detection of staphylococci grown in biofilms and in infected tissues.[20]

Detection and Analysis of Biofilm-Negative *Staphylococcus epidermidis* Variants

Our investigations of biofilm formation in *S. epidermidis* have demonstrated that the expression of the *ica* operon is highly variable among *ica*-positive *S. epidermidis*.[10,16] In a PIA-expressing population from a single bacterial colony, *S. epidermidis* biofilm variants occur that are hampered in PIA production. These PIA-negative colonies can be easily detected by their color on Congo red agar (CRA), and the frequency of their occurrence can be determined. Subsequently, the biofilm-negative phenotype of the clones is confirmed by a quantitative adhesion assay in polystyrene tissue culture plates. In the course of our studies we have observed that in *ica*-positive *S. epidermidis* isolates various genetic mechanisms contribute to the varying PIA expression. They include the reversible on and off switch of polysaccharide intercellular adhesin production by various phase variation mechanisms and the irreversible block of biofilm generation by chromosomal rearrangements. In addition, regulatory influences play a role that leads in turn to a downregulation of *ica* expression. To differentiate between these mechanisms the biofilm-negative variants are analyzed by pulsed-field gel electrophoresis, polymerase chain reaction (PCR), Southern hybridization, and nucleotide sequencing.

[17] J. W. Costerton, P. S. Stewart, and E. P. Greenberg, *Science* **284,** 1318 (1999).
[18] L. M. Baddour, L. P. Barker, G. D. Christensen, J. T. Parisi, and W. A. Simpson, *J. Clin. Microbiol.* **28,** 676 (1990).
[19] W. Ziebuhr, V. Krimmer, S. Rachid, I. Lößner, F. Götz, and J. Hacker, *Mol. Microbiol.* **32,** 345 (1999).
[20] V. Krimmer, H. Merkert, C. von Eiff, M. Frosch, J. Eulert, J. F. Lohr, J. Hacker, and W. Ziebuhr, *J. Clin. Microbiol.* **37,** 2667 (1999).

Detection of Polysaccharide Intercellular Adhesin-Negative Variants on Congo Red Agar and Determination of Their Frequencies

Congo red agar (CRA) has been described as a suitable solid medium for the detection of slime-producing staphylococcal isolates.[21] On this medium, PIA-expressing isolates grow as black-colored colonies with a rough surface. In contrast, PIA-negative colonies exhibit a red color and have a smooth surface. The CRA is prepared by mixing commercially available tryptic soy broth (TSB, 30 g/liter; Difco, Detroit, MI) with glucose (10 g/liter) and agar (15 g/liter) in distilled water. The pH is adjusted to pH 7.0 and the solution is autoclaved. For 1 liter of agar solution 800 mg of Congo red dye (Merck, Darmstadt, Germany) is dissolved in distilled water, filter sterilized, and added to the TSB–glucose–agar solution after cooling.

For the frequency determination of biofilm-negative variants a single bacterial colony is picked and diluted in phosphate-buffered saline. Fifty milliliters of tryptic soy broth is inoculated with an appropriate aliquot of this dilution and incubated at 37° in a shaker. The bacterial cell number in the inoculum is determined by plating the corresponding aliquot on tryptic soy agar. During the incubation period the bacterial growth curve of the liquid culture is recorded by taking samples for optical density measurement. Samples taken at the beginning, at the middle, and at the end of the exponential growth phase are diluted and plated onto Congo red agar. After incubation overnight at 37° and additionally for 24 hr at room temperature the colony morphology is analyzed by plate stereomicroscopy. The number of red-colored colonies and the total number of colonies are counted and the putative biofilm-negative, red-colored colonies are picked for further analysis.

The probability (P) of the occurrence of a biofilm-negative variant from a biofilm-forming inoculum (per cell and generation) is determined by using the equation described by Gally *et al.*[22]:

$$P = 1 - (1 - x)^{1/n}$$

where n is the number of generations and x is the number of biofilm-negative colonies divided by the total number of colonies. The number of generations (n) is determined as

$$n = (\log N - \log N_0)/\log 2$$

where N is the total number of bacterial cells in the culture and N_0 is the number of bacterial cells in the inoculum.

---

[21] D. J. Freeman, F. R. Falkiner, and C. T. Keane, *J. Clin. Pathol.* **42**, 872 (1989).

[22] D. L. Gally, J. A. Bogan, B. I. Eisenstein, and I. C. Blomfield, *J. Bacteriol.* **175**, 6186 (1993).

Determination of Biofilm Formation in Quantitative Biofilm Assay

To confirm that the red-colored colonies picked from the CRA represent biofilm-negative variants, a biofilm assay is performed by using polystyrene tissue culture plates.[23] The bacteria are grown overnight in tryptic soy broth, diluted 1 : 100 in fresh medium, and distributed in 96-well tissue culture plates. After overnight incubation at 37° the bacterial cultures are poured out. Plates are washed three times with phosphate-buffered saline (PBS) and remaining bacteria are fixed by air drying. After staining with a 0.4% (w/v) crystal violet solution, the optical density of the adherent biofilm is determined at 490 nm in an enzyme-linked immunosorbent assay (ELISA) reader. Values >0.12 are regarded as biofilm positive.

As mentioned above, *ica* expression can be influenced by regulatory mechanisms that have not been characterized in detail to date. However, studies of *ica* expression under various environmental influences give evidence that *S. epidermidis* biofilm formation is enhanced by high salt and other external stress conditions.[24] To separate downregulated variants from those that carry the genetic alterations described below, the growth medium can be supplemented with 1 to 3% (w/v) sodium chloride. From our experience, these conditions will overcome the block of *ica* expression and induce biofilm production.

Selection of Biofilm-Positive Revertant Strains by Serial Passages

A typical feature of biofilm-forming *S. epidermidis* is the phase variable expression of the *ica* operon.[10,14] It means that the biofilm-forming phenotype of the parent strain can be restored after repeated passages of a PIA-negative variant. To select for these revertants, a single colony of a biofilm-negative variant is isolated and incubated at 37° in tryptic soy broth in a tissue culture flask. After 24 hr the medium is replaced. This procedure is repeated until a biofilm of adhering bacteria becomes visible on the bottom of the tissue culture flask (maximum of 15 days). After washing with PBS, the adhering bacterial cells are scratched from the bottom and streaked out on Congo red agar. After incubation at 37° overnight and an additional 24 hr at room temperature, single, black-colored colonies are isolated and again analyzed by the quantitative adhesion assay.

Genetic Analyses of Variants

In the *S. epidermidis* strains analyzed so far, the frequency of occurrence of biofilm-negative variants ranges from 10^{-4} to 10^{-5}. In general, three different

---

[23] G. D. Christensen, W. A. Simpson, J. J. Younger, L. M. Baddour, F. F. Barrett, D. M. Melton, and E. H. Beachey, *J. Clin. Microbiol.* **22,** 996 (1985).
[24] S. Rachid, S. Cho, K. Ohlsen, J. Hacker, and W. Ziebuhr, *Adv. Exp. Med. Biol.* **485,** 159 (2000).

groups of biofilm-negative variants occur. The first group consists of variants in which *ica* expression is downregulated. These variants return to the PIA-expressing phenotype within the first two passages in tryptic soy broth or can be easily recognized by the addition of sodium chloride to the growth medium, which will induce PIA production.

The second group includes phase variants. They exhibit a relatively stable biofilm-negative phenotype, but can revert to the PIA-producing phenotype of the parent strain after 9 to 12 passages in liquid medium. The genetic background of phase variation in *ica* expression is known only in part. One mechanism has been shown to be mediated by the insertion and precise excision of the naturally occurring insertion sequence IS256 into and from the *ica* genes.[19] It seems likely that the *ica* operon represents a hot spot for the integration of the element and IS256 insertional mutants have been observed *in vitro* as well as in *S. epidermidis* strains obtained from polymer-associated infections (W. Ziebuhr, I. Loessner, V. Krimmer, and J. Hacker, unpublished observations, 2000). Because IS256 can be precisely excised from the *ica* genes (including an initially duplicated 8-bp target site) the element can function as a genetic switch mediating phase variation of *ica* expression. However, this mechanism plays a role only in 25 to 30% of the phase variants and phase variation of *ica* expression has also been observed in IS256-negative strains. Clearly, other phase variation mechanisms independent of the action of IS256 do exist, and remain to be determined.

Finally, a third group of PIA-negative variants will occur, which do not revert to the biofilm-forming phenotype of the parent strain even after extensive passages in liquid medium. These variants have usually undergone large chromosomal rearrangements including complete or incomplete deletions of the *ica* operon. These genome rearrangements affect *ica* expression and sometimes also influence the expression of antibiotic resistance genes. Obviously, they play a role *in vivo* during an infection, because chromosomal rearrangements have been identified in a clinical course of shunt-associated *S. epidermidis* meningitis.[25]

To differentiate the genetic mechanisms underlying varying PIA expression the variants will be analyzed genetically by different methods. First, the wild-type strain, the biofilm-negative variants, and any revertants are investigated by comparison of their *Sma*I- and/or *Sst*II-specific macrorestriction patterns after pulsed-field gel electrophoresis (PFGE). For PFGE, well-established standard protocols are available.[26] This method will exclude any possible contamination by other staphylococcal strains, but more important, it will identify larger rearrangements of the staphylococcal chromosome. Figure 1A gives an example of rearrangements in PIA-negative variants resulting in several fragment shifts (indicated by arrows)

[25] W. Ziebuhr, K. Dietrich, M. Trautmann, and M. Wilhelm, *Int. J. Med. Microbiol.* **290**, 115 (2000).
[26] F. Linhardt, W. Ziebuhr, P. Meyer, W. Witte, and J. Hacker, *FEMS Microbiol. Lett.* **74**, 181 (1992).

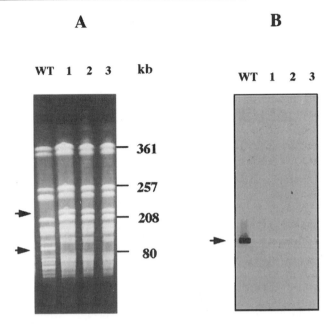

FIG. 1. Pulsed-field gel electrophoresis of *Sma*I-restricted genomic DNA (A) and Southern hybridization with an *ica*-specific DNA probe (B). WT, PIA-positive *S. epidermidis* isolate; 1–3, PIA-negative variants.

in comparison with the wild-type pattern. To identify *ica* deletion mutants the gels are blotted onto Nylon membranes and hybridized with *ica*-specific DNA probes. Figure 1B demonstrates the results of such an experiment. The PIA-negative variants (lanes 1 to 3) have completely lost the genetic information for biofilm formation. In addition, this Southern hybridization experiment indicates that the deleted *ica* gene cluster was originally situated on one of the fragments that had undergone a band shift.

The genetic analysis of phase variants includes the investigation of the biofilm-forming wild type, the PIA-negative variants, and the corresponding revertants. Because the IS256-mediated phase variation is the only mechanism characterized so far, the analysis will be aimed at the detection of *ica*::IS256 insertional mutants. This can be done by Southern hybridization of *Eco*RI-restricted chromosomal DNA with *icaAD*-, *icaB*-, and *icaC*-specific DNA. Insertional mutagenesis of any of these genes will result in a band shift and subsequent hybridization of the blot with an IS256-specific probe will visualize those fragments cohybridizing with the element and the *ica*-specific DNA. The exact IS256 insertion sites and the target site duplications are then identified by PCR amplification of the corresponding *ica* fragments by using the following oligonucleotide primers:

icaA1: 5'-GACCTCGAAGTCAATAGAGGT-3'
icaA2: 5'-CCCAGTATAACGTTGGATACC-3'
icaAD1: 5'-TATGCTACATGGTGAAGCCC-3'
icaAD2: 5'-GATAGTTTAACGCGAGTGCGC-3'
icaB1: 5'-ATGGCTTAAAGCACACGACGC-3'
icaB2: 5'-TATCGGCATCTGGTGTGACAG-3'
icaC1: 5'-ATAAACTTGAATTAGTGTATT-3'
icaC2: 5'-ATATATAAAACTCTCTTAACA-3'

Enlarged PCR fragments containing the IS element are analyzed by direct nucleotide sequencing, using the IS256-specific outward reading sequencing primers IS256-forward (5'-GGACCTACATGATGAATG-3') and IS256-reverse (5'-CTTGGGTCATGTAAAAGT-3'). Likewise, the excision of the element and the restoration of the *ica* wild-type sequence are confirmed by nucleotide sequencing of the corresponding PCR fragments of the revertants.

Taken together, the studies indicate that biofilm-forming *S. epidermidis* use different genetic mechanisms to switch off PIA production. At the present stage of experimental work the question arises as to why biofilm expression is highly variable in *S. epidermidis*. In a sense, bacteria organized in biofilms can be considered to resemble multicellular organisms, and this point of view is supported by the fact that bacterial cells in biofilms differ considerably from their planktonic counterparts in terms of metabolic activity, gene expression, and an inherent higher resistance to antibiotics.[27] It is conceivable that in the course of a polymer-associated infection by biofilm-forming staphylococci planktonic cells are currently released from the sessile population, which cannot be efficiently eliminated by host defense mechanisms. In this respect, it has been hypothesized that, in addition to mechanic detachment, specific genetic programs might exist abolishing extracellular matrix substance expression.[17] It is tempting to speculate that the genetic mechanisms described here might contribute to this process. However, more experimental work is needed to substantiate this hypothesis.

Investigation of Staphylococci in Biofilms and in Infected Tissue by 16S rRNA-Directed *in Situ* Hybridization

The ability of clinically relevant staphylococci to generate biofilms and to vary their phenotypic properties is suggested to contribute to staphylococcal virulence. In addition to the variation of biofilm formation, *S. epidermidis* as well as *S. aureus*

---

[27] J. A. Shapiro, *Annu. Rev. Microbiol.* **52,** 81 (1998).

clinical isolates are able to reduce considerably their growth rate. These strains exhibit a small colony variant (SCV) phenotype that is characterized by altered growth under aerobic conditions, diminished exoprotein production, and increased resistance to aminoglycosides. Moreover, these bacteria enter epithelial cells and persist intracellularly.[28] Staphylococcal SCVs are associated with infections of prosthetic joints and other implanted biomaterials and are often not detectable by conventional microbiological techniques.[29] One possibility to overcome this problem is the use of diagnostic tools that allow the direct detection and investigation of the bacteria *in situ* in infected tissues or in biofilms on the artificial implants.[20]

The use of the 16S rRNA-directed *in situ* hybridization technique with fluorescently labeled oligonucleotides is a suitable method with many advantages. First, each bacterial cell contains multiple copies of 16S rRNA in its ribosomes. Hence, the technique is sensitive enough to detect single bacterial cells. Second, 16S rRNA genes are highly conserved through bacterial evolution. They consist of regions that are common to all eubacteria, and other regions that are extremely species specific. By using the appropriate gene probes, it is possible to detect either any bacterial pathogen or, when highly specific probes are used, single bacterial species. The technique allows the identification of microorganisms independent of bacterial growth rates and metabolic activities. This is a special advantage for the detection of dormant and metabolically inactive bacteria, because the number of ribosomes is not significantly affected in such organisms. We use this method for the investigation of *S. epidermidis* and *S. aureus* grown in biofilms and for the detection of these species in infected tissues obtained from patients with septicemic loosenings of orthopedic implants.

Protocol for Investigation of Staphylococcus epidermidis and Staphylococcus aureus in Biofilms and in Infected Tissues

For the detection of *S. epidermidis,* the 16S rRNA-targeted oligonucleotide probe SEP1 (5'-CTG ACC CCT CTA GAG ATA GAG T-3') described by Zakrzewska-Czerwinska *et al.,*[30] is used. *Staphylococcus aureus* cells are visualized by using the probe SA-P1 (5'-TTG ACA ACT CTA GAG ATA GAG C-3').[31] The DNA oligonucleotide probes are synthesized and labeled with either Cy3 or fluorescein by MWG Biotech (Ebersberg, Germany). The biofilm-forming staphylococcal strains are grown overnight in chamber slides (Lab-Tek II chamberslide,

[28] R. A. Proctor, B. Kahl, C. von Eiff, P. E. Vaudaux, D. P. Lew, and G. Peters, *Clin. Infect. Dis.* **27** (Suppl. 1), S68 (1998).

[29] R. A. Proctor, P. van Langevelde, M. Kristjansson, J. N. Maslow, and R. D. Arbeit, *Clin. Infect. Dis.* **20**, 95 (1995).

[30] J. Zakrzewska-Czerwinska, A. Gaszewska-Mastalarz, G. Pulverer, and M. Mordarski, *FEMS Microbiol. Lett.* **79**, 51 (1992).

[31] R. W. Bentley, N. M. Harland, J. A. Leigh, and M. D. Collins, *Lett. Appl. Microbiol.* **16**, 203 (1993).

Nalge Nunc International, Chicago, IL) in tryptic soy broth at 37°. After decanting, the supernatant slides are washed three times with phosphate-buffered saline to remove all nonadhering bacteria. Bacterial biofilms are fixed by air drying. For permeabilization the cells are treated for 10 min with 0.25% (w/v) saponin, 0.25% (w/v) Triton X-100 in phosphate-buffered saline, washed in 5 mM sodium hydrogen phosphate, 5 mM disodium hydrogen phosphate (pH 7.5), and dehydrated for 5 min in 50, 70, and 100% ethanol. The slides are then exposed to an enzyme mix [lysostaphin (750 µg/ml) lysozyme (5 mg/ml) 5 mM sodium phosphate, 0.05% (w/v) saponin] for 1 hr at 37° and are washed again in 5 mM sodium hydrogen phosphate, 5 mM disodium hydrogen phosphate (pH 7.5) and dehydrated for 5 min in 50, 70, and 100% ethanol.

Hybridization is performed by applying 15 µl of hybridization solution [0.9 M NaCl, 20 mM Tris-HCl (pH 7.5), 0.01% (w/v) sodium dodecyl sulfate (SDS), 43% (v/v) formamide, 50 ng of probe] to the microscope slides, which are then incubated for 3 hr at 43° in an isotonically equilibrated humid chamber. Removal of the unbound probe and washing are performed at 43° for 20 min in a wash buffer containing 40 mM sodium chloride, 5 mM EDTA, 0.01% (w/v) SDS and 20 mM Tris-HCl (pH 7.5).[32] The slides are rinsed briefly with distilled water, air dried, and mounted with Citifluor solution (Citifluor, London, UK). Fluorescence is detected with a Zeiss (Oberkochen, Germany) Axioplan microscope equipped with an epifluorescence unit. Previous studies have given evidence that the gene probes SA-P1 and SEP1 are highly specific for the identification of *S. aureus* and *S. epidermidis,* respectively, and that the technique is suitable for the detection of staphylococcal cells, even when they are embedded in a thick matrix of extracellular polysaccharides.[20] Figure 2 (see color insert) shows an example from the investigation of a bacterial biofilm consisting of *S. aureus* and *S. epidermidis* cells grown in a mixed culture in chamber slides. *Staphylococcus aureus* cells are imaged in red by the Cy3-labeled SA-P1 probe and *S. epidermidis* cells are visualized in green by using the fluorescein-labeled SEP1 gene probe. The image indicates that cells of both species can be clearly differentiated by this technique and their distribution within the biofilm can be visualized. The 16S rRNA-directed *in situ* hybridization represents a basic method for the investigation of bacterial biofilms and a combination with laser scanning microscopy will extend its usefulness for structural analyses.

The method can also be adapted for the *in situ* detection of *S. aureus* and *S. epidermidis* in infected tissue. This application is especially useful in infections of orthopedic implants, which are often caused by small colony variants. For a successful detection of the bacteria, the handling and correct processing of the material are crucial. Therefore, the tissue samples should be fixed in PBS–formaldehyde buffer at 4° immediately after obtaining. The buffer is prepared by dissolving 0.8 g

---

[32] R. I. Amann, L. Krumholz, and D. A. Stahl, *J. Bacteriol.* **172,** 762 (1990).

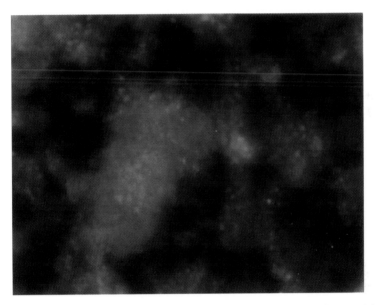

Fig. 2. 16S rRNA *in situ* hybridization analysis of a mixed staphylococcal biofilm consisting of *S. aureus* cells (red) and *S. epidermidis* cells (green).

A   **B**

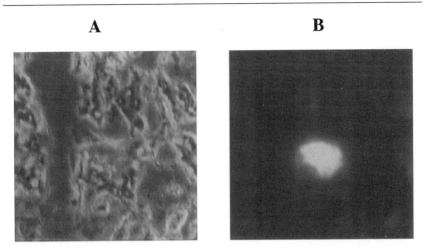

FIG. 3. *In situ* hybridization of connective tissue obtained from a progressive loosening of an artificial hip arthroplasty. (A) Phase-contrast and (B) fluorescence micrograph taken after 16S rRNA-directed hybridization with the *S. aureus*-specific Cy3-labeled SA-P1 oligonucleotide probe.

of paraformaldehyde in 20 ml of prewarmed phosphate-buffered saline (maximum of 70°). After cooling the solution is filter sterilized and stored at 4°. Fixation of the tissue in this buffer will stabilize the RNA and the samples can be stored at 4° up to 1 year. The tissue samples are then embedded in paraffin for thin-layer dissection. For the hybridization procedure, the paraffin must be completely removed from the tissue sections. For this purpose, the thin layers are treated twice with Rotihistol (Roth, Karlsruhe, Germany) for 10 min and then for 10 min with a Rotihistol–ethanol (1 : 1, v/v) mixture. For permeabilization, hybridization, and detection, the samples are then processed in the same manner as described above.

Figure 3 demonstrates the detection of *S. aureus* cells in a tissue sample obtained from a patient with septic loosening of a hip arthroplasty. Conventional microbiological diagnostics had failed to identify the infectious agent, but use of the *in situ* hybridization method revealed the detection of a cell cluster of *S. aureus* cells in the thin-layer dissection.

In summary, the method has proved to be suitable for the investigation of staphylococcal infections associated with medical devices. It can be used for the structural analysis of biofilms generated by staphylococci on polymer surfaces and represents a useful additional diagnostic tool for the detection of these bacteria in infected tissues.

Acknowledgments

We thank Katja Dietrich for excellent technical assistance. The authors' work is supported by BMBF grant 01K19608, the Graduiertenkolleg Infektiologie, and the Fond der Chemischen Industrie.

[19] *In Vivo* Models to Evaluate Adhesion and Biofilm Formation by *Staphylococcus epidermidis*

By MARK E. RUPP and PAUL D. FEY

Introduction

Adhesion of *Staphylococcus epidermidis* to biomaterials is a pivotal event in the pathogenesis of prosthetic device infections. Adhesion is a complex multistep event that is governed by various adhesins at different stages of the process. The early stages of adhesion are characterized by initial attachment of organisms to the surface of the biomaterial, whereas later stages of adhesion involve cellular aggregation and elaboration of biofilm. In addition, the pathogenesis of biomaterial-based infections involves interactions between the host, the microbe, and the biomaterial that are impossible to duplicate in *in vitro* systems. Therefore, to better study the adhesion of *S. epidermidis* to biomaterials and the pathogenesis of prosthetic device infections, investigators have developed *in vivo* models to better mimic the human condition.

Background

A number of investigators have developed *in vivo* models to study staphylococcal pathogenesis. To better characterize the influence of host serum constituents and inflammatory cells on staphylococcal disease, Zimmerli and colleagues developed a tissue cage model in guinea pigs.[1] Other investigators have used *in vivo* models to study specific adhesins and biofilm. For example, Christensen and colleagues utilized a mouse foreign body infection model, in which intravascular catheter segments were implanted under the skin to examine the role of biofilm production in the pathogenesis of *S. epidermidis* infections.[2] A number of investigators have used similar models in several animal species to investigate pathogenesis of *S. epidermidis* and *Staphylococcus aureus* infections.[3-5] Gallimore *et al.* implanted catheters in the mouse peritoneal cavity to mimic peritoneal dialysis-associated infection.[6] To better study intravascular infections, several investigators have used

[1] W. Zimmerli, F. A. Waldvogel, P. Vaudaux, and U. E. Nydegger, *J. Infect. Dis.* **146,** 487 (1982).
[2] G. D. Christensen, W. A. Simpson, A. L. Bisno, and E. H. Beachey, *Infect. Immun.* **40,** 407 (1982).
[3] C. C. Patrick, M. R. Plaunt, S. V. Hetherington, and S. M. May, *Infect. Immun.* **60,** 1363 (1992).
[4] D. W. Lambe, K. P. Ferguson, J. L. Keplinger, C. G. Gemmell, and J. H. Kalbfleisch, *Can. J. Microbiol.* **36,** 455 (1990).
[5] R. J. Sherertz, W. A. Carruth, A. A. Hampton, M. P. Byron, and D. D. Solomon, *J. Infect. Dis.* **167,** 98 (1993).
[6] B. Gallimore, R. F. Gagnon, R. Subang, and G. K. Richards, *J. Infect. Dis.* **164,** 1220 (1991).

a rat or rabbit endocarditis model.[7–9] Kojima *et al.* utilized a rabbit central venous catheter-associated infection model to study the effect of antibody directed against *S. epidermidis* polysaccharide adhesin.[10] In this chapter we describe two *in vivo* models used in our laboratory to study the pathogenic significance of adhesins and biofilm production of *S. epidermidis.*

It should be emphasized, however, that *in vivo* models should be used only after *in vitro* methods have been explored. Bacterial strains to be utilized in *in vivo* models must be selected carefully. Whenever possible, only genetically characterized strains should be used when examining the importance of putative virulence determinants. Careful study design is necessary to limit the number of animals used to the minimum that will yield meaningful data.

Mouse Foreign Body Infection Model

The mouse foreign body infection model is an attractive model with which to study bacterial adhesion and to assess the significance of biofilm formation by *S. epidermidis.* It reflects the complex milieu associated with biomaterial-based infection and it is relatively simple to use from a technical standpoint. The model mimicks the conditions found in the subcutaneous space of humans and thus serves as an *in vivo* model of prosthetic device infection such as intravascular catheter infections.

Animals

Male Swiss-Albino mice (25–30 g) are used in this model.

Bacteria

Staphylococcus epidermidis is harvested from broth culture by centrifugation, suspended in phosphate-buffered saline (PBS), and diluted to appropriate density. Attention should be paid to the growth phase of the bacteria because there are increasing data regarding the regulation of cell components of *S. epidermidis* in relationship to growth phase.[11] In addition, the composition of the nutrient broth should be defined because the amount of nutrients has been shown to influence the expression of adhesins and biofilm.[12]

[7] L. M. Baddour, G. D. Christensen, J. H. Lowrance, and W. A. Simpson, *J. Med. Microbiol.* **41,** 259 (1994).

[8] H. Shiro, E. Muller, N. Gutierrez, S. Boisot, M. Grout, T. D. Tosteson, D. Goldman, and G. B. Pier, *J. Infect. Dis.* **169,** 1042 (1994).

[9] F. Perdreau-Remington, M. A. Sande, G. Peters, and H. F. Chambers, *Infect. Immun.* **66,** 2778 (1998).

[10] Y. Kojima, M. Tojo, D. A. Goldmann, T. D. Toseteson, and G. B. Pier, *J. Infect. Dis.* **162,** 435 (1990).

[11] C. Vuong, F. Götz, and M. Otto, *Infect. Immun.* **68,** 1048 (2000).

[12] D. Mack, N. Siemessen, and R. Laufs, *Infect. Immun.* **60,** 2048 (1992).

Biomaterial

One-centimeter segments of 14-gauge Teflon intravenous catheters (Quik Cath; Baxter, Morton Grove, IL) serve as the foreign body. The Teflon catheter segments can easily be interchanged with segments of other catheter materials (polyurethane, silastic, latex, etc.) or materials found in other prosthetic devices (Dacron, Gortex, metal alloys, etc.) to study the effect of alternative biomaterials.

Preparative Regimen

All surgical instruments and drapes are steam sterilized. Surgery is conducted under a laminar airflow hood that is disinfected with isopropyl alcohol or dilute bleach. To reduce the risk of regurgitation and aspiration, animals are fasted for 12 hr before the procedure.

Anesthesia

The mice are anesthetized with ketamine and xylazine administered via the intraperitoneal route. The dose should be individualized to achieve deep anesthesia and not simply sedation, and ranges from 100 to 200 mg/kg for ketamine and 5 to 16 mg/kg for xylazine.

Surgical Procedure

After the induction of anesthesia, the flank of the mouse is shaved with surgical clippers. The animal's eyes are coated with boric acid ophthalmic ointment to prevent exposure keratitis. The skin is cleansed with povidone-iodine and a small incision is made through the skin. A 1-cm segment of catheter or other biomaterial sample is placed in the subcutaneous space. The biomaterial sample is inoculated with bacteria by dipping the sample in a suspension of bacteria or inoculating a suspension of bacteria (quantified by plate count) into the biomaterial bed. The wound is then closed with a single monofilament 3-0 Nylon suture.

Abscess Formation and Burden of Infection

On postoperative day 7 the mice are killed by inhalation of carbon dioxide. Abscess formation is scored visually from 0 to 4 (0, no tissue reaction; 1, slight vascularization; 2, inflammation with increased vascularization; 3, capsule formation; 4, capsule formation and purulence). The catheter segments are aseptically removed, placed in sterile microcentrifuge tubes with 1 ml of PBS, and vortexed at high speed for 1 min. The wash fluid is quantitatively cultured on Mueller–Hinton agar plates. In addition, the tissue surrounding the catheter can be removed, weighed, homogenized, and quantitatively cultured.

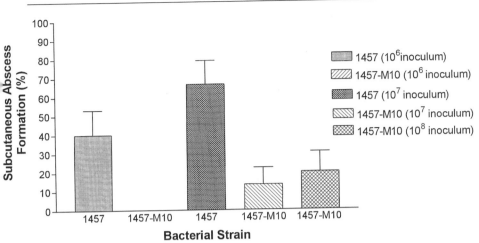

Fig. 1. Subcutaneous abscess formation by *S. epidermidis* 1457 and syngeneic PIA-negative mutant 1457-M10 in the mouse foreign body infection model. Visually apparent abscesses developed in 40 and 67% of mice inoculated with 10^6 and 10^7 CFU, respectively, of wild-type *S. epidermidis* 1457 compared with 0 and 13% of mice inoculated with equal numbers of *S. epidermidis* 1457-M10 ($p < 0.05$). Only 20% of animals inoculated with 10^8 CFU of *S. epidermidis* 1457-M10 developed a subcutaneous abscess. Columns represent mean abscess formation and error bars represent the standard error of the mean. [Reproduced with permission from M. E. Rupp, J. S. Ulphani, P. D. Fey, K. Bortscht, and D. Mack, *Infect. Immun.* **67**, 2627 (1999).]

Notes

1. The adequacy of the wash process in removing adherent bacteria has been confirmed by electron microscopy.[13]

2. Because it would be difficult to differentiate between possible contaminants and the strain of *S. epidermidis* experimentally inoculated, molecular epidemiologic testing can be performed to demonstrate that the strain recovered from the catheter or abscess is the same as the experimental strain. We generally analyze restriction fragment length polymorphism patterns by pulsed-field gel electrophoresis to accomplish this task. Standard procedures are followed.[14]

Key Results

Figure 1 illustrates the incidence of abscess formation associated with the inoculation of varying numbers of either wild-type *S. epidermidis* 1457 or the syngeneic polysaccharide intracellular adhesin (PIA)-negative mutant 1457-M10.[13]

[13] M. E. Rupp, J. S. Ulphani, P. D. Fey, K. Bartscht, and D. Mack, *Infect. Immun.* **67**, 2627 (1999).
[14] M. E. Rupp, J. Han, and R. V. Goering, *Infect. Dis. Obstet. Gynecol.* **2**, 218 (1995).

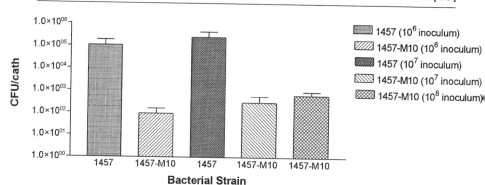

FIG. 2. Recovery of *S. epidermidis* 1457 and its syngeneic PIA-negative mutant 1457-M10 from implanted subcutaneous catheter segments in the mouse foreign body infection model. There were significantly greater numbers of strain 1457 adherent to the catheters compared to strain 1457-M10 at both the 10^6 and 10^7 CFU inoculum levels ($p < 0.05$). Bars represent the mean of log-transformed adhesion values, and the lines represent the standard error of the mean. Although there were fewer bacteria recovered from the catheters of mice inoculated with 10^8 CFU of *S. epidermidis* 1457-M10 than from the catheters inoculated with either 10^7 or 10^6 CFU of *S. epidermidis* 1457, these differences did not reach statistical significance. [Reproduced with permission from M. E. Rupp, J. S. Ulphani, P. D. Fey, K. Bartscht, and D. Mack, *Infect. Immun.* **67**, 2627 (1999).]

Figure 2 demonstrates a significant difference in the number of organisms recovered from the explanted catheters in the mouse foreign body model associated with the various levels of inocula of *S. epidermidis* 1457 and the PIA-negative mutant 1457-M10.[13]

Rat Central Venous Catheter-Associated Infection Model

Although we have successfully employed the mouse foreign body infection model described above to study the importance of staphylococcal adhesins and biofilm formation in the pathogenesis of biomaterial-based infection, the mouse model has a number of limitations. The model does not involve the vascular space and does not duplicate the conditions found in an intravascular catheter-associated environment. Consequently, features such as blood flow dynamics, serum proteins and other blood components, and humoral immunity are not accurately reflected. Therefore, we developed the rat central venous catheter-associated infection model.[15]

Animals

Male Sprague-Dawley rats (400 g) are used in this model. Rats are acclimated for 7 days before surgery. They are housed in individual polypropylene shoe-box

---

[15] M. E. Rupp, J. S. Ulphani, P. D. Fey, and D. Mack, *Infect. Immun.* **67**, 2656 (1999).

cages on cob bedding. Cages and bedding are changed twice per week. Water and feed (standard laboratory rat diet; Harlan, Indianapolis, IN) are given *ad libitum*. Animals are maintained under the following environmental conditions: lighting, 12 hr on/off cycle; temperature, 22 to 25°; air changes, 10/hr; humidity, 50%.

Bacteria

Staphylococcus epidermidis are grown in broth culture and are harvested and suspended as described for the mouse model above.

Intravascular Catheter

Intravascular catheters are constructed from silicone rubber tubing (Specialty Manufacturing, Saginaw, MI).[16] A 17-cm length of 0.064-cm intravenous tubing is slid onto a 20-gauge blunt needle with a Luer lock hub. A 10-cm length of 0.15-cm i.d. tubing encases the cannula and is attached to the hub with heat-shrinkable tubing (Cole-Parmer Instrument, Vernon Hills, IL). The outer tubing serves to protect the subcutaneous and external portions of the inner cannula from stress encountered in manipulation of the catheter after placement. An injection adapter is attached to the Luer lock of the needle.

Restraint Jacket

Animal restraint jackets (Fig. 3) are used to secure the catheters and prevent the rats from disrupting the surgical site or chewing on the external portion of the catheter.[17] The jacket is constructed from elastic material and is secured with a Velcro closure. A flap covers the catheter hub, which is secured to the jacket with elastic bands.

Insertion of Central Venous Catheter

Preparative Regimen. All surgical instruments and drapes are steam sterilized. Surgery is conducted under a laminar airflow hood that is disinfected with isopropyl alcohol or dilute bleach. To reduce the risk of regurgitation and aspiration, animals are fasted for 12 hr before the procedure.

Anesthesia. Rats are anesthetized with ketamine and xylazine administered via the intraperitoneal route. The dose is individualized to achieve deep anesthesia and not simply sedation and ranges from 40 to 90 mg/kg for ketamine and from 5 to 13 mg/kg for xylazine.

Surgical Procedure. After the induction of anesthesia, the back and neck are shaved with surgical clippers. The animal's eyes are coated with boric acid ophthalmic ointment to prevent exposure keratitis. The back and neck are cleansed

---

[16] M. E. Rupp and J. S. Ulphani, U.S. Patent 5,762,636 (1998).
[17] M. E. Rupp and J. S. Ulphani, U.S. Patent 5,839,393 (1998).

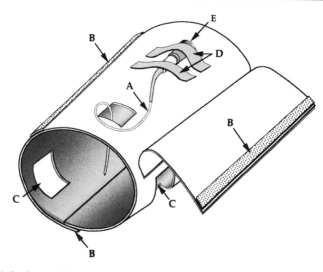

FIG. 3. Animal restraint jacket. (A) catheter, (B) Velcro closure, (C) limb holes, (D) elastic straps, (E) injection adapter. [Reproduced with permission from J. S. Ulphani and M. E. Rupp, *Lab. Anim. Sci.* **49**, 283 (1999).]

with povidone-iodine and the surgical site is draped with sterile drapes. An approximately 2-cm skin incision is made in the right lateral neck. A 16-gauge needle is used to create a subcutaneous tract from the neck incision to the scapular region and the catheter is advanced through the needle. The needle is removed, leaving 4 to 5 cm of catheter in the subcutaneous tract with the exit site in the midscapular region. Blunt dissection is carried out to locate and isolate the jugular vein. Two 3.0 silk sutures are placed around the vein. A small incision is made in the vein and the catheter is advanced into the sinus venosus. Another technique that we have used to introduce the catheter into the vein was described by Harms and Ojeda, and utilizes an insertion needle.[18] The catheter is secured at the insertion site with the 3.0 silk suture and to overlying fascia and muscle with a second silk suture. The skin is then closed with monofilament suture. The catheter is flushed with heparin solution [20 units of heparin/ml in 0.9% (w/v) NaCl] to maintain patency.

Central Venous Catheter-Associated Infection. Twenty-four hours after placement of the catheters, blood is obtained from the central venous catheter and cultured to ensure sterility. The bacterial strain of interest is next injected into the catheter and allowed to dwell for 15 min. The catheter is then flushed with heparin solution. Preliminary dose-ranging experiments should be conducted to determine inoculum size, as it varies from strain to strain. For example, an inoculum of

[18] P. G. Harms and S. R. Ojeda, *J. Appl. Physiol.* **36**, 391 (1974).

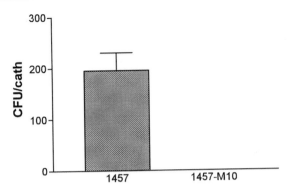

FIG. 4. Recovery of wild-type *S. epidermidis* 1457 and its PIA-negative syngeneic mutant 1457-M10 from implanted central venous catheters in the rat central venous catheter-associated infection model. A mean of 195 CFU was recovered from the catheters of rats inoculated with 10^4 CFU of strain 1457 compared with none from the catheters of rats inoculated with 10^4 CFU of strain 1457-M10 ($p < 0.05$).

10^4 CFU of *S. epidermidis* 1457 reliably results in a central venous catheter-associated infection, whereas 10^7 CFU of *S. epidermidis* 0-47 is required.[15,19] Experiments with *S. aureus* reveal that inoculation with as few as 100 CFU results in central venous catheter-associated infection.[20]

On day 8, the rats are killed. Blood from the central venous catheter and the periphery is obtained and quantitatively cultured by plating 0.1 ml on trypticase soy agar (TSA) plates (Remel, Lenexa, KS). The location of the distal portion of the catheter in the superior vena cava is confirmed and the catheter and surrounding venous tissue are aseptically removed, placed in phosphate-buffered saline, and washed by vortexing at maximum speed for 1 min. The wash fluid is then quantitatively cultured. To ascertain the extent of metastatic disease, the heart, lungs, liver, and kidneys are aseptically removed, weighed, homogenized in 1 ml of PBS, and quantitatively cultured on TSA.

Key Results

Figure 4 illustrates the number of organisms recovered from the explanted catheters of rats challenged with PIA-positive wild-type strain *S. epidermidis* 1457 (mean ± standard deviation, 195 ± 92 CFU) or PIA-negative *S. epidermidis* 1457-M10 (no bacteria recovered from catheters) (seven rats in each group).[15] Similarly, using this model we have observed significant differences in the pathogenicity of

[19] M. E. Rupp, P. D. Fey, C. Heilmann, and F. Götz, *J. Infect. Dis.*, in press.
[20] J. S. Ulphani and M. E. Rupp, *Lab. Anim. Sci.* **49**, 283 (1999).

S. epidermidis O-47 and isogenic mutants (mut 1 and mut 2) that are deficient in PIA or the autolysin (Atl E).[19] There are also differences in the propensity of *S. epidermidis* strains to cause metastatic disease. In general, strains deficient in the production of PIA or autolysin are less likely to cause an infection associated with metastatic disease of peripheral organs.[15,19]

Discussion and Conclusion

Staphylococcus epidermidis is the most common cause of nosocomial bacteremia and infections of implanted prosthetic medical devices.[21,22] It has long been thought that bacterial adhesion and biofilm production are important in the pathogenesis of these infections. Efforts to study *S. epidermidis* have been hampered by a lack of genetic techniques to create syngeneic mutants. These hurdles have been cleared and a number of investigators have described putative virulence determinants. Bacterial adhesion and the pathogenesis of prosthetic device infections is a complex process that involves interactions between the host, the device, and the microbe. Because of its complexity it is difficult to duplicate this milieu in *in vitro* systems. Therefore, we have employed two *in vivo* models to study the adhesins of *S. epidermidis*. In these models we have demonstrated the importance of PIA and Atl E in the pathogenesis of *S. epidermidis* biomaterial-based infection. PIA, which is synthesized by gene products of the *ica* locus, is a polysaccharide whose expression is dependent on glucose in the growth medium.[12,23–25] Syngeneic, PIA-negative mutants are capable of initial attachment to plastic but are unable to form multilayer colonies and produce biofilm.[23,26] Phenotypic and genotypic studies have shown that PIA and the hemagglutinin of *S. epidermidis* are identical or closely related.[27,28] In both the mouse foreign body infection model and the rat central venous catheter-associated infection model, we demonstrated the critical importance of PIA in the pathogenesis of experimental biomaterial-based infection.[13,15,19]

[21] M. E. Rupp and G. L. Archer, *Clin. Infect. Dis.* **19**, 231 (1994).

[22] Centers for Disease Control and Prevention, *Am. J. Infect. Control.* **27**, 520 (1999).

[23] D. Mack, M. Nedelmann, A. Krokotsch, A. Schwarzkopf, J. Heesemann, and R. Laufs, *Infect. Immun.* **62**, 3244 (1994).

[24] C. Heilmann, C. Gerke, F. Perdreau-Remington, and F. Götz, *Infect. Immun.* **64**, 277 (1996).

[25] C. Heilmann, O. Schweitzer, C. Gerke, N. Vanittanakom, D. Mack, and F. Götz, *Mol. Microbiol.* **20**, 1013 (1996).

[26] D. Mack, W. Fischer, A. Krokotsch, K. Leopold, R. Hartmann, H. Egge, and R. Laufs, *J. Bacteriol.* **178**, 175 (1996).

[27] P. D. Fey, J. S. Ulphani, F. Götz, C. Heilmann, D. Mack, and M. E. Rupp, *J. Infect. Dis.* **179**, 1561 (1999).

[28] D. Mack, J. Riedewald, H. Rohde, T. Magnus, R. H. Feucht, H. A. Elsner, R. Laufs, and M. E. Rupp, *Infect. Immun.* **67**, 1004 (1999).

The autolysin of *S. epidermidis,* Atl E, is a proteinaceous adhesin described by Heilmann and colleagues.[24,25] Syngeneic Atl E-deficient mutants are unable to initially attach to plastic but are able to form multicell clusters, produce biofilm, and mediate hemagglutination.[24,25] In the rat central venous catheter-associated infection model, we demonstrated that initial adhesion (mediated by the autolysin Atl E) and aggregation/biofilm production (mediated by PIA) are essentially equally important in the pathogenesis of experimental intravascular catheter infection. It is hoped that, through improved understanding of *S. epidermidis* biomaterial-based infection, the production of less infection-prone devices or the development of novel methods for prevention or treatment of prosthetic device infection will be realized.

Acknowledgments

This work was supported in part by grants 96006810 and 9951167Z from the American Heart Association. We thank Joseph Ulphani for expert technical assistance.

[20] Genetic and Biochemical Analysis of *Staphylococcus epidermidis* Biofilm Accumulation

By Dietrich Mack, Katrin Bartscht, Claudia Fischer, Holger Rohde, Clemens De Grahl, Sabine Dobinsky, Matthias A. Horstkotte, Kathrin Kiel, and Johannes K.-M. Knobloch

Introduction

Staphylococcus epidermidis is the most prominent coagulase-negative staphylococcal species that colonizes human skin and mucous membranes. It exhibits a low pathogenic potential in the normal human host. Today coagulase-negative staphylococci rank among the five most frequent causative organisms of nosocomial infections. The reason is a tremendous increase in the incidence of nosocomial sepsis and infections related to implanted foreign materials such as intravascular and peritoneal dialysis catheters, cerebrospinal fluid shunts, prosthetic heart valves and prosthetic joints, vascular grafts, cardiac pacemakers, and intraocular lenses.[1-3]

---

[1] A. L. Bisno and F. A. Waldvogel, "Infections Associated with Indwelling Medical Devices," 2nd Ed. American Society of Microbiology, Washington, D.C., 1994.

[2] M. E. Rupp and G. L. Archer, *Clin. Infect. Dis.* **19,** 231 (1994).

It is believed that the specific virulence of *S. epidermidis* in medical device-related infections is linked to an unusual ability to colonize polymer surfaces in multilayered biofilms. *In vitro,* a proportion of *S. epidermidis* strains produce a macroscopically visible, adherent biofilm on test tubes or tissue culture plates with a morphology in scanning electron micrographs similar to that of infected intravascular catheters.[4–6] By now, this phenotype is regularly referred to as *biofilm formation,* whereas the partially ambiguous term *slime production* was used earlier.[7–9]

Biofilm formation may be divided into two phases.[8,10,11] The first step, primary attachment of staphylococcal cells to a polymer surface, is a complex process involving various physicochemical factors, including participation of staphylococcal surface proteins Ssp 1 and Ssp 2, the major autolysin AtlE, and binding via matrix proteins as mediated by the fibrinogen-binding protein Fbe and other unknown receptors, or polysaccharide factors akin to the capsular polysaccharide adhesin PS/A (reviewed in Refs. 8 and 9).

In the second phase the attached bacteria proliferate and accumulate in a multilayered biofilm. Using transposon mutagenesis, our group identified a linear homoglycan composed primarily of N-acetylglucosamine in β-1,6-glycosidic linkages containing deacetylated amino groups, succinate, and phosphate.[12,13] The polysaccharide is separated into two differentially charged polysaccharide species I and II by anion-exchange chromatography.[13] It is referred to as polysaccharide intercellular adhesin (PIA), which is functional in cell-to-cell adhesion and essential for biofilm accumulation of most clinical *S. epidermidis* strains.[12–16] In addition, PIA

[3] J. D. Thylefors, S. Harbarth, and D. Pittet, *Infect. Control Hosp. Epidemiol.* **19,** 581 (1998).

[4] G. D. Christensen, W. A. Simpson, A. L. Bisno, and E. H. Beachey, *Infect. Immun.* **37,** 318 (1982).

[5] G. D. Christensen, W. A. Simpson, J. J. Younger, L. M. Baddour, F. F. Barrett, D. M. Melton, and E. H. Beachey, *J. Clin. Microbiol.* **22,** 996 (1985).

[6] M. Hussain, M. H. Wilcox, P. J. White, M. K. Faulkner, and R. C. Spencer, *J. Hosp. Infect.* **20,** 173 (1992).

[7] M. Hussain, M. H. Wilcox, and P. J. White, *FEMS Microbiol. Rev.* **10,** 191 (1993).

[8] D. Mack, *J. Hosp. Infect.* **43**(Suppl.), S113 (1999).

[9] D. Mack, K. Bartscht, S. Dobinsky, M. A. Horstkotte, K. Kiel, J. K. M. Knobloch, and P. Schäfer, *in* "Handbook for Studying Bacterial Adhesion: Principles, Methods, and Applications" (Y. H. An and R. J. Friedman, eds.), p. 307. Humana Press, Totowa, New Jersey, 2000.

[10] C. Heilmann, C. Gerke, F. Perdreau-Remington, and F. Götz, *Infect. Immun.* **64,** 277 (1996).

[11] C. Heilmann, M. Hussain, G. Peters, and F. Götz, *Mol. Microbiol.* **24,** 1013 (1997).

[12] D. Mack, M. Nedelmann, A. Krokotsch, A. Schwarzkopf, J. Heesemann, and R. Laufs, *Infect. Immun.* **62,** 3244 (1994).

[13] D. Mack, W. Fischer, A. Krokotsch, K. Leopold, R. Hartmann, H. Egge, and R. Laufs, *J. Bacteriol.* **178,** 175 (1996).

[14] D. Mack, N. Siemssen, and R. Laufs, *Infect. Immun.* **60,** 2048 (1992).

[15] D. Mack, N. Siemssen, and R. Laufs, *Zentralbl. Bakteriol. Suppl.* **26,** 411 (1994).

[16] D. Mack, M. Haeder, N. Siemssen, and R. Laufs, *J. Infect. Dis.* **174,** 881 (1996).

is essential for hemagglutination of erythrocytes by *S. epidermidis.*[17–20] Synthesis of PIA *in vivo* and *in vitro* depends on the enzymatic activity of the gene products of the *icaADBC* locus of *S. epidermidis.*[10,19,21,22] It was demonstrated in two relevant animal foreign body infection models, using a well-characterized syngeneic biofilm-negative transposon mutant 1457-M10, that a functional *icaADBC* locus and the ability to produce PIA are essential for the pathogenesis of *S. epidermidis* biomaterial-related infections.[12,23–25] By transposon mutagenesis four unlinked genetic loci were identified, the inactivation of which leads to a biofilm-negative phenotype.[26] All class I mutants have insertions in the *icaADBC* locus, whereas in mutants of classes II, III, and IV, apparently regulatory genetic loci are inactivated by the Tn*917* insertions. These genetic loci control expression of PIA synthesis and biofilm formation by directly or indirectly influencing expression of *icaADBC* at the level of transcription.[26] It has been reported that, in contrast to previous reports,[27] PS/A has a structure closely related if not identical to that of PIA and that the synthesis of PS/A also depends on the *icaADBC* locus.[28] As detailed analytical data were not reported the determination of the relation of PIA and PS/A requires further study. A surface protein referred to as accumulation-associated protein (AAP) was described also to be required for *S. epidermidis* biofilm formation.[29]

Considerable progress has been made in deciphering the molecular mechanisms responsible for *S. epidermidis* biofilm formation because of the establishment of methods for the genetic manipulation of *S. epidermidis,* facilitating the analysis of factors involved in the course of biofilm development.[8,9] These techniques include transposon mutagenesis and gene cloning, and generation of specific antisera reactive with factors functional in biofilm accumulation, which

[17] M. E. Rupp and G. L. Archer, *Infect. Immun.* **60**, 4322 (1992).

[18] M. E. Rupp, N. Sloot, H. G. Meyer, J. Han, and S. Gatermann, *J. Infect. Dis.* **172**, 1509 (1995).

[19] D. Mack, J. Riedewald, H. Rohde, T. Magnus, H. H. Feucht, H. A. Elsner, R. Laufs, and M. E. Rupp, *Infect. Immun.* **67**, 1004 (1999).

[20] P. D. Fey, J. S. Ulphani, F. Götz, C. Heilmann, D. Mack, and M. E. Rupp, *J. Infect. Dis.* **179**, 1561 (1999).

[21] C. Heilmann, O. Schweitzer, C. Gerke, N. Vanittanakom, D. Mack, and F. Götz, *Mol. Microbiol.* **20**, 1083 (1996).

[22] C. Gerke, A. Kraft, R. Süssmuth, O. Schweitzer, and F. Götz, *J. Biol. Chem.* **273**, 18586 (1998).

[23] M. Nedelmann, A. Sabottke, R. Laufs, and D. Mack, *Zentralbl. Bakteriol.* **287**, 85 (1998).

[24] M. E. Rupp, J. S. Ulphani, P. D. Fey, K. Bartscht, and D. Mack, *Infect. Immun.* **67**, 2627 (1999).

[25] M. E. Rupp, J. S. Ulphani, P. D. Fey, and D. Mack, *Infect. Immun.* **67**, 2656 (1999).

[26] D. Mack, H. Rohde, S. Dobinsky, J. Riedewald, M. Nedelmann, J. K. M. Knobloch, H.-A. Elsner, and H. H. Feucht, *Infect. Immun.* **68**, 3799 (2000).

[27] M. Tojo, N. Yamashita, D. A. Goldmann, and G. B. Pier, *J. Infect. Dis.* **157**, 713 (1988).

[28] D. McKenney, J. Hübner, E. Muller, Y. Wang, D. A. Goldmann, and G. B. Pier, *Infect. Immun.* **66**, 4711 (1998).

[29] M. Hussain, M. Herrmann, C. von Eiff, F. Perdreau-Remington, and G. Peters, *Infect. Immun.* **65**, 519 (1997).

allow the characterization of polysaccharides of the glycocalyx of the organism functional for biofilm development. These methods may also be useful for the analysis of alternative mechanisms leading to biofilm formation of *S. epidermidis* or to determine the mechanisms of biofilm formation of other organisms.

Assay of Biofilm Formation *in Vitro*

For genetic analysis of *S. epidermidis* biofilm formation it is necessary to have a simple, reproducible *in vitro* assay, which allows the detection of quantitative differences in biofilm formation of various strains or mutants. Most researchers use the semiquantitative microtiter plate assay first described by Christensen *et al.*[5] for quantitation of biofilm formation. In our laboratory we use the following procedure, which is reproducible when all variables such as growth medium, incubation time, performance of washing steps, and fixation and staining procedures are carefully controlled.[14]

1. Test strains are grown in tryptic soy broth lacking glucose [TSBαGlc$_{Oxoid}$ prepared from tryptone (Oxoid, Basingstoke, England), neutralized soya peptone (Oxoid), NaCl, and dipotassium phosphate as indicated by the manufacturer] from 6 hr to overnight with shaking at 37°.

2. Cultures are diluted 1 : 100 into trypticase soy broth (TSB$_{BBL}$; Becton Dickinson, Cockeysville, MD) and volumes of 200 μl are inoculated into each well of 96-well tissue culture plates (Nunclon Delta; Nunc, Roskilde, Denmark) and incubated for 20–24 hr at 37°. Tests are performed in quadruplicate.

3. The plates are washed four times with 200 μl of phosphate-buffered saline (PBS) per well, using a multichannel pipette. The wells are emptied by flicking the plates.

4. Adherent bacterial cells are fixed for 15 min with, per well 150 μl of Bouin's fixative [0.5 g of picric acid is solubilized in 37.5 ml of distilled water; 12.5 ml of 37% (v/v) formaldehyde and 2.5 ml of concentrated acetic acid are added]. Fixed cells are washed once with PBS. As an alternative adherent biofilm cells are air dried before staining.[30]

5. Adherent bacterial cells are stained with 150 μl of gentian violet per well for 5 min. The staining solution is flicked away and the stained bacterial biofilms are washed five times with running tap water.

6. After air drying the optical density of stained adherent bacterial biofilms is measured at 570 nm (using 405 nm as the reference wavelength) in an automatic spectrophotometer (Behring, Marburg, Germany). A value of 0.1 OD$_{570}$ is used as a break point to differentiate biofilm-producing from biofilm-negative strains.

[30] L. Baldassarri, W. A. Simpson, G. Donelli, and G. D. Christensen, *Eur. J. Clin. Microbiol. Infect. Dis.* **12**, 866 (1993).

TABLE I
DIFFERENTIAL BIOFILM PRODUCTION BY CLINICAL
Staphylococcus epidermidis ISOLATES GROWN IN DIFFERENT
TRYPTIC SOY BROTH MEDIA

Strain	Biofilm $(OD_{570})^a$	
	TSB_{BBL}	TSB_{Oxoid}
8188	2.40	2.40
7874	2.26	2.30
10730	2.30	0.67
10590	2.20	0.66
1815	2.17	0.06
1516	2.10	0.08

[a] Strains were grown in the respective medium in 96-well tissue culture plates, the plates were washed with PBS, cells were fixed with Bouin's fixative, and adherent bacterial biofilms were stained with gentian violet. Biofilm-producing strains were defined as having a mean OD_{570} greater than 0.1.

This assay has been modified by various researchers, who have used different staining solutions, different brands of microtiter plates, and tryptic soy broth manufactured by various companies. All these variables may influence the test results of individual strains under study. The most significant differences result from the growth media and the type of microtiter plates used. In our experience coagulase-negative staphylococci most firmly attach to polystyrene microtiter plates modified for tissue culture use (e.g., Nunclon Delta; Nunc) as compared with plates used for microtitration or bacteriologic use. Significant differences in adhesion values are obtained under identical conditions when tryptic soy broth containing glucose obtained from different manufacturers is used, although the principal ingredients are similar (Table I). We have also observed, when using TSB_{BBL} produced by Becton Dickinson, significant changes in biofilm formation by certain reference strains and transposon mutants depending on the lot of the respective medium.[26] It is therefore advisable to obtain tryptic soy broth of the same lot for at least 1 year of work to keep assay conditions constant.

Preparation of Antisera Specific for Factors Functional in Biofilm Accumulation

We raised antisera against whole cells of strong biofilm-producing *S. epidermidis* strains, which were grown as a biofilm in tissue culture dishes. These antisera were then absorbed with biofilm-negative *S. epidermidis* strains in such a way as to allow identification of antigens specific for biofilm-producing strains. As determined later, these absorbed antisera specifically detected PIA.[13,14] This approach is potentially useful to detect factors specifically associated with biofilm formation of other organisms.

1. Overnight cultures of biofilm-producing *S. epidermidis* 1457 in TSBα-Glc$_{Oxoid}$ are diluted 1 : 100 into 10 ml of TSB$_{BBL}$ per 9-cm tissue culture dish (Nunclon Delta; Nunc), which are incubated for 20 hr at 37°. The biofilm-negative *S. epidermidis* 5179 is used in parallel as a control. The medium is aspirated and the bacteria are scraped off from the plastic surface, using a disposable cell scraper (Nunc), into 4 ml of PBS per dish. Formaldehyde is added to a final concentration of 0.5% (v/v) and the suspension is gently shaken (100 rpm) for 20 hr. Cells are collected by centrifugation (1500g for 10 min), suspended in PBS at an OD$_{578}$ of 1.0, and stored in small samples at −20°. Samples are plated onto blood agar to control sterility of the antigen preparations.

2. After obtaining preimmune sera, rabbits are immunized three times a week at regular intervals by intravenous injection of 0.2 ml of the bacterial suspension for the first week, 0.4 ml for the second week, 0.8 ml for the third week, and 1.0 ml thereafter for about 3 months. Rabbits are exsanguinated and serum is stored at −20°.

3. The antiserum is absorbed with two biofilm-negative *S. epidermidis*-strains, 5179 and 9896.[14] As preliminary experiments revealed that the expression of antigens specific for biofilm-producing *S. epidermidis* is extremely glucose dependent the antiserum is further absorbed with *S. epidermidis* 1457 grown in tryptic soy broth lacking glucose [TSBαGlc$_{Difco}$ prepared with tryptone (Difco, Detroit, MI), neutralized soya peptone (Oxoid), NaCl, and dipotassium phosphate as indicated by the manufacturer].[14] Later, the syngeneic biofilm-negative transposon mutant *S. epidermidis* 1457-M11 alone is used for absorption of sera with similar results.[13]

Overnight cultures of the strains in TSBαGlc$_{Difco}$ are diluted 1 : 100 into 800 ml of TSB$_{BBL}$ or TSBαGlc$_{Difco}$, respectively, and incubated with shaking for 20 to 24 hr. Bacteria are harvested and washed two times in PBS containing 0.05% (w/v) NaN$_3$. The wet weight of the cell pellets is recorded and the cells are suspended at 10% (w/v) in the same buffer. Formaldehyde is added to a final concentration of 1.5% (v/v) and the cell suspension is gently mixed, using a magnetic stirring bar, for 90 min at room temperature. The cells are then sedimented and washed twice in PBS containing 0.05% (w/v) NaN$_3$. Finally the cells are suspended in 9 ml of the same buffer for each 800-ml culture equivalent.

4. For absorption of serum 1 ml of each of the three bacterial preparations prepared in step 3 is sedimented sequentially into one Eppendorf microcentrifuge tube (2.2 ml) and the bacteria are suspended in 1 ml of rabbit antiserum. For absorption with the syngeneic biofilm-negative mutant 1457-M11, 2 ml of the bacterial preparation is used per milliliter of antiserum. Absorption is carried out

for 6–8 hr at room temperature by gently agitating the bacterial suspension on a rotating wheel. Bacterial cells are sedimented twice at $16,000g$ for 15 min in a minicentrifuge and the absorbed antiserum is stored at $-20°$. Reactivities of antisera absorbed with the two different bacterial preparations are similar.

5. In addition, an antiserum is raised against purified PIA.[21] Rabbits are immunized with 0.5 mg of purified polysaccharide I of PIA[13] emulsified in Freund's complete adjuvant (Calbiochem, Fankfurt, Germany) followed by two additional boosters at 4-week intervals. The reactivity of this antiserum is similar to that of the absorbed antisera.

Detection of Polysaccharide Intercellular Adhesin by Various Immunochemical Methods

Specific detection of PIA is essential for any question regarding the physiology of *S. epidermidis* in biofilms. Several immunochemical methods were developed for qualitative and quantitative detection of PIA *in situ* on the bacterial cell surface[14,15] or in bacterial extracts at different stages of purification.[12–14,16] Indirect immunofluorescence is the method of choice for detection of PIA on bacterial cells in biofilm communities.

Indirect Immunofluorescence Test

Standard Immunofluorescence Test. Overnight cultures of bacteria grown in $TSB\alpha Glc_{Oxoid}$ are diluted $1:100$ into TSB_{BBL} and incubated for 22 hr at $37°$ with shaking. As an alternative, the bacteria under study are grown in 5-cm tissue culture dishes (Nunclon Delta; Nunc). In this case the medium is aspirated and bacteria are scraped into 4 ml of PBS. Bacterial suspensions are directly diluted in PBS to an OD_{578} of 0.3 to 0.5 and volumes of 20 µl are applied per field of immunofluorescence slides with 10 individual sample fields (BioMerieux, Marcy l'Etoile, France). Slides are dried, fixed with cold acetone for 2 min, and stored at $4°$ until use.

Volumes (20 µl) of a PIA-specific rabbit antiserum (absorbed or raised against purified PIA) diluted $1:50$ in PBS are applied per field. After 30 min of incubation in a wet chamber at $37°$ slides are washed three times with PBS and dried. Fluorescein-conjugated anti-rabbit IgG (Sigma, Deisenhofen, Germany) diluted in PBS ($1:80$) is applied for 30 min at $37°$ in the dark. After washing, slides are mounted in 90% (v/v) glycerol in PBS, pH 8.6, containing 0.1% (w/v) *p*-diphenyleneamine[31] and viewed with an Orthoplan fluorescence microscope (E. Leitz, Wetzlar, Germany).

---

[31] G. D. Johnson, R. S. Davidson, K. C. McNamee, G. Russell, D. Goodwin, and E. J. Holborow, *J. Immunol. Methods* **55**, 231 (1982).

Double Immunofluorescence Staining. In a second approach superior demonstration of the relation of PIA expression and presence of cells in clusters of *S. epidermidis* biofilms can be observed.[15] Bacteria grown as a biofilm are scraped from tissue culture plates and are spread at an appropriate dilution onto standard microscope slides. The immunofluorescence test is performed as described above, using slightly larger volumes of the anti-PIA and fluorescein-conjugated antisera (about 100 µl/slide). However, after drying, the slides are not mounted for microscopy but are again fixed in cold acetone for 2 min. After drying an anti-staphylococcal antiserum, raised against the biofilm-negative *S. epidermidis* 5179, diluted 1 : 50 in PBS, is applied. After washing and drying of the slides rhodamine-conjugated anti-rabbit IgG (Sigma) diluted in PBS (1 : 80) is applied. The slides are then washed, dried, and mounted for microscopy as described above. Using the appropriate filter set, PIA-expressing cells are labeled green, whereas all staphylococci are labeled red independent of PIA expression (Fig. 1).

Detection of Polysaccharide Intercellular Adhesin by Coagglutination in Bacterial Extracts

The assay allows the quantitation of PIA in bacterial extracts using protein A-positive *Staphylococcus aureus* Cowan I as a carrier of specific anti-PIA antibodies. The sensitivity of the assay is in the range of other coagglutination assays, detecting as little as 0.05 µg of PIA per milliliter, using reagents prepared in our laboratory.[14]

Preparation of Bacterial Extracts. Antigen extracts of *S. epidermidis* grown on tissue culture dishes in TSB_{BBL} are prepared by sonication.[14] Two 9-cm tissue culture dishes (Nunclon Delta; Nunc) are each inoculated with an overnight culture of the respective strain grown in $TSB\alpha Glc_{Oxoid}$ diluted 1 : 100 into 10 ml of TSB_{BBL} or other suitable growth medium and incubated at 37° for 20–24 hr. Bacterial cells are scraped off the surface into the growth medium to avoid loss of cells of biofilm-negative strains when aspirating the growth medium. The bacteria are then collected by centrifugation, the medium is removed, and the cells are suspended in 5 ml of PBS. The cell-associated PIA is quantitatively extracted by sonicating twice for 30 sec on ice, using a 3-mm microtip of a sonicator disintegrator (Branson sonifier 250-D; Branson Ultrasonics, Danbury, CT) at 70% of maximal amplitude. Samples are removed for the determination of cell density by measuring the OD_{578} of appropriate dilutions. Bacterial cells are removed by centrifugation at 1500 g for 15 min. Extracts are clarified by centrifugation at 16,000g for 1 hr at 4° and stored at −20°.

Preparation of Staphylococcus aureus Cowan I for Coagglutination. A procedure described by Kessler[32] is followed with modifications. *Staphylococcus aureus* Cowan I is grown in 800 ml of TSB_{BBL} for 20 hr at 37° with shaking. Cells are

---

[32] S. W. Kessler, *Methods Enzymol.* **73**, 442 (1981).

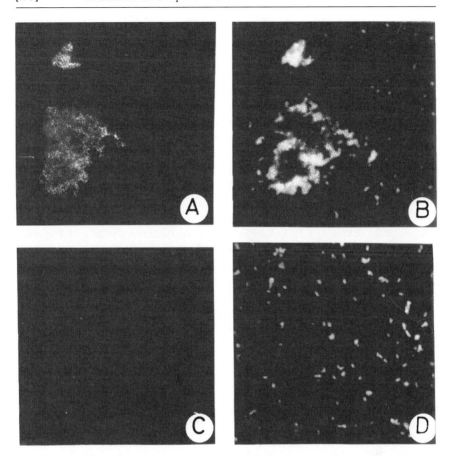

FIG. 1. Immunofluorescence detection of PIA. Analysis of biofilm-producing wild-type *S. epidermidis* 13-1 (A and B) and its syngeneic biofilm-negative transposon mutant M10 (C and D) in a double-immunofluorescence assay using PIA-specific absorbed antiserum (A and C) or an antistaphylococcal antiserum raised against biofilm-negative *S. epidermidis* 5179 (B and D). PIA is expressed predominantly with cells located in large clusters, whereas it is almost not detected with cells in pairs or tetrads.

sedimented and washed twice in PBS containing 0.05% (w/v) NaN_3. The wet weight of the cells is recorded and cells are suspended at a density of 10% (w/v) in the same buffer. Formaldehyde is added to a final concentration of 1.5% (v/v) and cells are gently agitated on a magnetic stirrer for 90 min at room temperature. Cells are sedimented, suspended in the same volume of buffer as described above, and then transferred to a glass Erlenmeyer beaker. The cell suspension is heated in a water bath at 80° for 5 min. Cells are cooled rapidly in an ice bath, washed twice in PBS containing 0.05% (w/v) NaN_3, and suspended in the same buffer at a concentration of 10% (w/v) as determined by the wet cell weight. One-milliliter volumes are frozen at −80° until use.

Preparation of Polysaccharide Intercellular Adhesin-Specific Coagglutination Reagent. The *S. aureus* Cowan I suspension (1 ml) is washed twice in PBS containing 0.05% (w/v) NaN$_3$ and is resuspended in 900 μl of the same buffer. Rabbit anti-PIA antiserum (100 μl), either absorbed or raised against purified PIA, is added. The mixture is incubated at room temperature for 15 min with occasional mixing. The sensitized *S. aureus* Cowan I cells are sedimented in a microcentrifuge (10,000*g* for 1 min) and washed three times with PBS containing 0.05% (w/v) NaN$_3$. Finally, the sensitized cells are resuspended in 10 ml of the same buffer and stored at 4°. The coagglutination reagent can be used for about 2 weeks without significant deterioration of performance.

Coagglutination for Detection of Polysaccharide Intercellular Adhesin. For coagglutination assay of PIA, samples (5 μl) of bacterial extracts or fractions during purification of PIA are mixed with 15 μl of coagglutination reagent on microscope slides. On a standard microscope slide three different samples can be tested simultaneously. By using an appropriate carrier, up to four microscope slides can be processed at the same time. Agglutination is evaluated after 3 min in bright light against a dark background. When using a new batch of coagglutination reagent, and especially when a new preparation of *S. aureus* Cowan I is used, controls including bacterial extracts of PIA-negative strains and PBS should be performed to evaluate the cutoff point for positive coagglutination. When a new antiserum is introduced control experiments using unsensitized *S. aureus* Cowan I and carrier cells sensitized with the respective preimmune sera should be evaluated with antigen preparations of a panel of *S. epidermidis* strains of interest. Also, the medium used for growth of the bacteria should be evaluated for eventual cross-reactivity when using this coagglutination assay.

Quantitation of antigen in bacterial extracts is performed by analysis of serial 2-fold dilutions of the respective extracts in PBS. The antigen titers are defined as the highest dilutions displaying a positive coagglutination test.

Detection of Polysaccharide Intercellular Adhesin by Enzyme-Linked Immunosorbent Assay Inhibition

Although the coagglutination assay can be used to screen fractions during purification of PIA it is desirable to have a specific assay that can dectect and quantitate PIA in multiple samples simultaneously. To this end an assay was developed that allows the detection of PIA in bacterial extracts by enzyme-linked immunosorbent assay (ELISA) inhibition, using whole bacterial cells to sensitize the plates.[13]

Preparation of Enzyme-Linked Immunosorbent Assay Plates. The biofilm-producing *S. epidermidis* 1457 is used as antigen. Cells are grown on four 9-cm tissue culture plates in TSB$_{BBL}$ for 24 hr at 37°. The medium is aspirated and adherent cells are scraped into a total volume of 16 ml of distilled water. Cell clusters are disintegrated by 20 to 30 strokes with a Dounce homogenizer with a loosely fitting pestle. An aliquot is removed, diluted to a total volume of 5 ml with PBS, and sonicated twice for 30 sec, using a 3-mm microtip of a sonicator disintegrator (Branson sonifier 250-D) at 70% of maximal amplitude for determination of cell numbers. The cell density of the main sample is then adjusted to an OD_{578} of 1.0 in 0.3% (w/v) methylglyoxal (Sigma), pH 8.0, and the 96-well plates (Greiner, Nürtingen, Germany) are coated with aliquots (100 μl/well) for 90 min at 37°.[33] A stock solution of 2% (w/v) methylglyoxal is prepared by mixing 5 ml of 40% (w/v) methylglyoxal with 15 ml of distilled water. The pH is adjusted to pH 8.0 by adding 10% (w/v) $NaHCO_3$. The solution is made up to 100 ml by adding distilled water. This stock solution is stable at 4° for several weeks. Plates are washed two times with washing buffer [PBS plus 0.5% (v/v) Tween 20 (Merck, Darmstadt, Germany)] and blocked overnight at 4° with washing buffer containing 0.5% (w/v) bovine serum albumin (BSA; Serva, Heidelberg, Germany). After washing with buffer the plates are stored at 4° until use (plates can be stored for several days while filled with wash buffer).

As an alternative for measuring PIA concentration by ELISA inhibition, Maxi-Sorb ELISA plates (Nunc) are coated with purified polysaccharide I of PIA[13] diluted in PBS at a concentration of 0.125 μg/well overnight at room temperature.[16]

Enzyme-Linked Immunosorbent Assay-Inhibition Procedure. A calibration curve is established by analyzing 2-fold dilutions of a crude PIA-containing bacterial extract of biofilm-producing *S. epidermidis* 9142. Antigen-containing samples to be tested are diluted in PBS in order to reach an extinction within the linear range of the reference curve. For convenience, several dilutions of a sample can be investigated in parallel. All samples are analyzed in duplicate. The absorbed anti-PIA antiserum is diluted in washing buffer as required to reach a half-maximal extinction in the ELISA as determined in prior control experiments. Samples (20 μl) are mixed with 60 μl of the diluted absorbed antiserum in a microtiter plate and incubated at 37° for 60 min. Samples are then transferred to ELISA plates (50 μl/well) and incubated at 37° for 90 min. After washing four times with buffer ELISA plates are incubated with alkaline phosphatase-conjugated anti-rabbit monoclonal antibody (Sigma) diluted 1 : 5000 in washing buffer (50 μl/well) at 37° for 60 min. Plates are washed again four times and substrate buffer [100 μl/well; phosphatase substrate no. 104 (1.41 mg/ml; Sigma) in 0.5 mM $MgCl_2$, 0.02% (w/v) NaN_3, 1 M diethanolamine, pH 9.6] is added. After incubation at 37° for 30 min 2 M NaOH (50 μl/well) is added and the absorbance is measured at 405 nm with an automatic spectrophotometer (Behring). Depending on the inhibition of the ELISA signal, the antigen concentration of samples is established from the reference curve in arbitrary ELISA units (Fig. 2). If a purified sample of PIA is used to establish

---

[33] S. J. Challacombe, *in* "ELISA and Other Solid Phase Immunoassays" (D. M. Kemeny and S. J. Challacombe, eds.), p. 319. John Wiley & Sons, Chichester, UK, 1988.

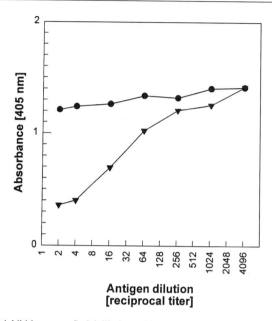

FIG. 2. ELISA inhibition assay. Serial dilutions of bacterial extracts of the biofilm-producing PIA-positive *S. epidermidis* 13-1 (▼) and the syngeneic biofilm-negative, PIA-negative mutant M11 (●) were mixed with absorbed PIA-specific antiserum for 1 hr at 37°. The ELISA plates coated with biofilm-producing *S. epidermidis* 1457 were then developed as described in text. From the almost linear part of the inhibition curve of the reference extract of *S. epidermidis* 13-1 the concentration of PIA can be estimated in arbitrary ELISA units.

the reference curve the concentration of PIA can be expressed as micrograms of hexosamine per milliliter.[16]

Detection of Polysaccharide Intercellular Assay by Immuno-Dot-Blot Assay

Antigens reactive with PIA-specific antisera are detected in bacterial extracts or fractions obtained during purification by immuno-dot-blot assay. An Immobilon polyvinylidene difluoride (PVDF) membrane (Millipore, Bedford, MA) cut to the appropriate dimensions is sequentially equilibrated with methanol followed by 50 mM sodium phosphate buffer, pH 7.5. The membrane is placed onto a filter paper wetted by the same buffer and samples are then applied in 5-μl volumes. The membrane is blocked for 1 hr in 3% (w/v) milk powder in water and then washed in TTSB [125 mM NaCl, 20 mM Tris-HCl, 0.1% (v/v) Tween 20, pH 7.5]. The primary PIA-specific antiserum is diluted 1 : 800 in TTSB and incubated with the blocked membrane for 60 min. After washing three times for 15 min in TTSB the membrane is incubated for 60 min with second antibody (biotinylated anti-rabbit-IgG; Sigma) diluted 1 : 5000 in TTSB. After three additional washing steps the membrane

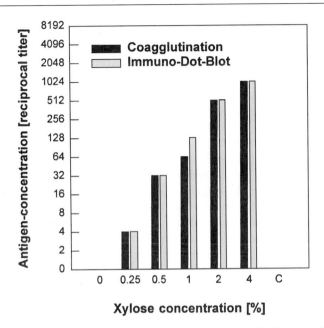

FIG. 3. Comparison of performance of coagglutination and immuno-dot-blot assays for detection of PIA. *Staphylococcus carnosus* × pTX*icaADBC* containing the synthetic genes for PIA under the control of a xylose-inducible promoter were incubated in the presence of various xylose concentrations (given as percentages) and bacterial extracts were prepared. *Staphylococcus carnosus* × pTX16 without the synthetic genes for PIA synthesis was used as control (C). PIA concentration in the various extracts was determined by coagglutination and immuno-dot-blot assay. Similar titers were obtained by both tests.

is incubated for an additional 60 min with streptavidin–horseradish peroxidase conjugate (Amersham Life Science, Little Chalfont, England) diluted 1 : 1500 in TTSB. The membrane is developed after three final washings in buffer according to instructions given by the manufacturer, using ECL Western blotting detection reagents (Amersham Life Science). After 1 min the membrane is covered with Saran Wrap and exposed to Hyperfilm ECL high-performance chemiluminescence film (Amersham Life Science) for 5 to 7 sec. After development, the film gives a permanent record of the assay.

The immuno-dot-blot assay has sensitivity similar to that of the coagglutination assay when serial dilutions of PIA-containing extracts are investigated (Fig. 3). However, with the coagglutination assay the determination of the respective end points is more convenient. Therefore, we use the immuno-dot-blot assay routinely for primary screening of column fractions during purification of PIA, whereas the concentration of polysaccharides in the respective reactive fractions is determined by coagglutination.

Purification of Polysaccharide Intercellular Adhesin

The composition of the *S. epidermidis* glycocalyx, or slime, has attracted major interest. In 1990, it became apparent that slime preparations prepared from *S. epidermidis* strains grown in complex medium and/or on agar are regularly contaminated by carbohydrate compounds derived from agar or the complex growth medium used.[34,35] Chemically defined medium and solidification of medium with silica gel instead of agar or parallel analysis of polysaccharide preparations derived from pairs of strains, which differentially express the polysaccharides of interest, is mandatory to obtain meaningful analytical results.

Therefore, to purify PIA, bacteria were grown in dialyzed TSB_{BBL} and the syngeneic biofilm-negative, PIA-negative transposon mutant 1457-M11 was analyzed in parallel.[13] In addition, when the chemically defined medium HHW^{36} was used to grow these strains similar compositional data were obtained for PIA prepared from these cells compared with cells grown in TSB_{BBL}.[37]

Preparative extraction of polysaccharide intercellular adhesin. *Staphylococcus epidermidis* strains are grown for 20–24 hr at 37° with shaking at 100 rpm/min in beakers, each containing about 300 ml of dialyzed TSB_{BBL}, which is prepared by dialysis of 100 ml of 10-fold concentrated TSB_{BBL} against 900 ml of water. Bacterial cells from 2 to 4 liters of dialyzed TSB_{BBL} are collected by centrifugation and are suspended in 20 ml of PBS. The antigen is extracted by sonicating (Branson sonifier 250-D) four times for 30 sec on ice, using a 5-mm microtip at maximal amplitude. Bursts of sonication are separated by intervals of 30 sec to keep the bacterial suspension cool. Cells are removed by centrifugation at 6000 rpm in a Beckman (Fullerton, CA) JA 17 rotor at 4° and extracts are clarified by centrifuging for 60 min at 12,000 rpm in the same rotor. The extracts are filter sterilized, dialyzed overnight against 2 liters of PBS or 50 m*M* Tris-HCl, pH 7.5, depending on which type of column is chosen for fractionation, and concentrated to a volume of about 3 ml, using Centriprep 10 (Amicon Millipore, Beverley, MA).

Gel-filtration chromatography. The polysaccharide in PBS was originally fractionated by gel filtration on a 1.6 × 100 cm Sephadex G-200 (Pharmacia, Freiburg, Germany) column equilibrated with PBS.[12] However, we now use Sephacryl S-200 or Sephacryl S-300 (Pharmacia) with similar results. The columns are eluted at a flow rate of about 0.3 ml/min and fractions of 1.5 ml are collected. The void and inclusion volumes of the columns are determined with 2-MDa blue dextran, 20-kDa yellow dextran, and vitamin B_{12} (dye kit; Pharmacia). PIA is detected near the void volume of the columns, using one of the assays described above as appropriate, and antigen-containing fractions are concentrated with Centriprep 10.

[34] D. T. Drewry, L. Galbraith, B. J. Wilkinson, and S. G. Wilkinson, *J. Clin. Microbiol.* **28,** 1292 (1990).
[35] M. Hussain, J. G. Hastings, and P. J. White, *J. Infect. Dis.* **163,** 534 (1991).
[36] M. Hussain, J. G. Hastings, and P. J. White, *J. Med. Microbiol.* **34,** 143 (1991).
[37] D. Mack and A. Krokotsch, unpublished results (1999).

Anion-exchange chromatography. Concentrated antigen preparations purified by gel filtration are dialyzed against 2 liters of 50 mM Tris-HCl, pH 7.5. The concentrated antigen in a volume of about 2.5 ml is applied to a 1.6 × 11 cm Q-Sepharose (Pharmacia) column equilibrated with the same buffer. The column is first eluted with 40 ml of 50 mM Tris-HCl, pH 7.5, followed by a 200-ml continuous gradient from 0 to 1000 mM NaCl in the same buffer. The column is eluted at a flow rate of 0.3 ml/min and fractions of 2.4 ml are collected. The polysaccharide antigen is detected by specific immunochemical assay as described above. Antigen-containing fractions are pooled, concentrated, and dialyzed against 2 liters of 50 mM sodium acetate, pH 5.0. Hexosamine content of the preparations is determined by colorimetric assay, which requires buffers free of Tris-HCl, which interferes with the assay.[12,38]

As the purification scheme contains several dialysis steps, which are time-consuming and are prone to severe losses of the polysaccharide, we have modified our purification scheme. We now use 50 mM sodium phosphate buffer, pH 7.5, to equilibrate the gel filtration and the Q-Sepharose columns. Therefore, the antigenic material extracted from the bacteria is concentrated and applied directly to the Sephacryl S-200 or S-300 column. Antigen-reactive fractions are simply concentrated and used directly in the following anion-exchange chromatography step. Similar elution profiles of PIA are observed when using this buffer system. However, at the present time we investigate whether subtle differences in the chromatographic behavior of polysaccharides I and II of PIA exist when using the two different buffer systems.[13]

The concentrated PIA preparations exhibit a tendency to spontaneous aggregation of the polysaccharide, leading to almost quantitative precipitation. Although not reproducibly observed, there are several situations that seem to enhance precipitation of PIA, including freeze-drying, dialysis against distilled water, and repeated freezing and thawing of concentrated polysaccharide preparations. To avoid these bouts we store concentrated PIA preparations at 4° in buffers of at least 50 mM ionic strength.

Genetic Manipulation of *Staphylococcus epidermidis* and Transposon Mutagenesis

The genetic analysis of *S. epidermidis* biofilm formation is difficult because these organisms are apt to resist genetic manipulation methods. A standard transposon mutagenesis experiment requires (1) introduction of the carrier plasmid of the transposon into the selected host strain, (2) proof of genetic linkage of transposon insertions and observed phenotypic changes, (3) reintroduction into the respective mutant of the cloned wild-type allele of the genes inactivated by the transposon

---

[38] R. Lane-Smith and E. Gilkerson, *Anal. Biochem.* **98**, 478 (1979).

for phenotypic complementation, and eventually, (4) transfer of the transposon insertions into other host strains with different genetic backgrounds. Although the standard techniques of electroporation and protoplast transformation normally used to perform all these manipulations work in principle with a small number of selected *S. epidermidis* strains,[10,39–42] they turn out to be completely inefficient with the clinical *S. epidermidis* isolates used in various other studies.[12,19,43] For this reason we make use of plasmid mobilization by coconjugation and have developed a phage transduction system for transduction of plasmids and chromosomal genetic markers using *S. epidermidis* phages.[12,19,23,26]

Mobilization of Plasmids by Coconjugation

Staphylococcus aureus WBG4883 containing the conjugative plasmid pWBG636 is used to mobilize plasmids into desired recipient strains (properties of strains and plasmids used are listed in Table II.[44–46] This is accomplished in two successive steps. First, pWBG636 is conjugated into the host strain carrying the plasmid to be mobilized. In the second step this host strain, now containing pWBG636 and the respective plasmid to be mobilized, is mated with *S. epidermidis* 9142 as the recipient. Plasmids successfully mobilized by this strategy are pTV1ts containing the *Enterococcus faecalis* transposon Tn917 and pTX*icaADBC* containing the cloned *icaADBC* locus encoding the synthetic genes for PIA synthesis.[22,47] Donor strains of the mobilized plasmids are *S. epidermidis* 23[12] and *Staphylococcus carnosus* TM300.[26]

Donor and recipient strains are mated on membrane filters as described.[48] To this end strains are grown in brain–heart infusion (BHI) broth overnight at 30° (pTV1ts) or 37° with shaking, using the appropriate antibiotics for selection. Volumes of recipient (3 ml) and of donor cultures (1 ml) are filtered onto 0.45-μm pore size nitrocellulose filters by vacuum suction followed by washing with BHI broth. The filters are incubated on BHI agar plates (bacterial side up) at 30 or 37° as appropriate for 6 to 20 hr. Bacterial growth eluted from the filters is plated on peptone yeast (PY) agar [1.0% (w/v) peptone, 0.5% (w/v) yeast

[39] E. Muller, J. Hübner, N. Gutierrez, S. Takeda, D. A. Goldmann, and G. B. Pier, *Infect. Immun.* **61,** 551 (1993).

[40] L. Grüter, O. König, and R. Laufs, *FEMS Microbiol. Lett.* **66,** 215 (1991).

[41] J. Augustin and F. Götz, *FEMS Microbiol. Lett.* **54,** 203 (1990).

[42] F. Götz and B. Schumacher, *FEMS Microbiol. Lett.* **40,** 285 (1987).

[43] P. D. Fey, M. W. Climo, and G. L. Archer, *Antimicrob. Agents Chemother.* **42,** 306 (1998).

[44] E. E. Udo, D. E. Townsend, and W. B. Grubb, *FEMS Microbiol. Lett.* **40,** 279 (1987).

[45] E. E. Udo and W. B. Grubb, *J. Med. Microbiol.* **31,** 207 (1990).

[46] E. E. Udo and W. B. Grubb, *FEMS Microbiol. Lett.* **60,** 183 (1990).

[47] P. J. Youngman, in *"Bacillus subtilis* and Other Gram-Positive Bacteria" (A. L. Sonenshein, J. A. Hoch, and R. Losick, eds.), p. 585. American Society for Microbiology, Washington, D.C., 1993.

[48] D. E. Townsend, S. Bolton, N. Ashdown, and W. B. Grubb, *J. Med. Microbiol.* **20,** 169 (1985).

TABLE II
STRAINS AND PLASMIDS

Staphylococcal strain	Plasmid	Antibiotic resistance	Properties	Ref.
S. epidermidis 1457	p1457	—	Biofilm-positive wild type	14,23
1457c × pTV1ts	pTV1ts	Ery[R], Cm[R]	Biofilm-positive Tn mutagenesis host	23
23	pTV1ts	Ery[R], Cm[R]	Donor of pTV1ts in conjugation	40
9142	Multiple	Cip[R]	Biofilm-positive wild type	14
13-1	pWBG636, pTV1ts	Cm[R], Ery[R], Cip[R], Gm[R]	Biofilm-positive Tn mutagenesis host	12
9142-M10	—	Ery[R], Cip[R]	Biofilm-negative transductant	19
1457-M10	—	Ery[R]	Biofilm-negative transductant	12
M10	pWBG636	Ery[R], Cip[R], Gm[R]	Biofilm-negative Tn*917* mutant	12
M11	pWBG636	Ery[R], Cip[R], Gm[R]	Biofilm-negative Tn*917* mutant	12
8400, 8188, 7874, 10730, 10590, 1815, 1516			Biofilm-positive clinical isolates	16
5179, 9896			Absorption of PIA-specific antisera	14
PS48, PS71			Propagation of *S. epidermidis* phages	50,51
S. aureus WBG4883	pWBG636	Gm[R]	Conjugative mobilization of plasmids	46
S. carnosus × pTX*icaADBC*	pTX*icaADBC*	Tet[R]	Biofilm-positive in presence of xylose	22
S. carnosus × pTX16	pTX16	Tet[R]	*icaADBC*-negative control strain	22
	pTV1ts	Ery[R], Cm[R]	Temperature-sensitive replicon, Tn*917*	47
	pTX*icaADBC*	Tet[R]	*icaADBC* under control of xylose-inducible promoter	22

Abbreviations: Ery, erythromycin; Cm, chloramphenicol; Cip, ciprofloxacin; Gm, gentamicin; Tet, tetracycline.

extract, 0.5% (w/v) NaCl, 0.1% (w/v) glucose, 1.5% (w/v) agar, pH 7.5] while selecting for pTV1ts (erythromycin and chloramphenicol, 10 μg/ml), pTX*icaADBC* (tetracycline, 10 μg/ml), and/or for pWBG636 (gentamicin, 8 μg/ml) as required. Transconjugants are purified on selective PY agar plates and are mated with the recipient strain *S. epidermidis* 9142 on membrane filters as described above. Bacteria are plated on PY agar, selecting with the antibiotics appropriate for the respective plasmid and for *S. epidermidis* 9142 with ciprofloxacin (2 μg/ml). Transconjugant clones are selected and further characterized. Thomas and Archer[49] used the conjugative *S. aureus* plasmid pGO1 in a similar approach to mobilize recombinant derivatives of plasmid pC221 into different *S. epidermidis* strains; however, mobilization of other functional plasmids was partly unsuccessful because of rearrangements. It should be noted that we observed transposition- and temperature-sensitive elimination of pTV1ts only with 3 of about 20 dif-

[49] W. D. Thomas, Jr., and G. L. Archer, *Plasmid* **27,** 164 (1992).

ferent transconjugant clones obtained.[12] Similar results were obtained with pTX*icaADBC*, where also clones were observed that had lost the xylose-inducible expression phenotype of biofilm formation.[26] Although this approach solves the problems of genetic manipulation of *S. epidermidis* only partially, it is a valuable alternative for plasmid transfer into selected *S. epidermidis* strains.[43,49] It is even more valuable as *S. carnosus* TM300, the prototype coagulase-negative staphylococcal cloning host, apparently can also serve as a donor in conjugative mobilization of plasmids into *S. epidermidis* by pWBG636.[26]

Transduction of Plasmids and Chromosomal Markers by Staphylococcus epidermidis Phages

We make use of *S. epidermidis* typing phages 48 and 71 for transduction experiments.[12,23,50,51] These phages are propagated on the homologous propagating strains *S. epidermidis* PS48 and PS71. For propagation of phages NB2+ broth [nutrient broth no. 2 (Oxoid) containing 0.04% (w/v) $CaCl_2$] is used for broth cultures and Staphylococcus typing (ST) agar [2% (v/v) nutrient broth no. 2, 0.5% (w/v) NaCl, 0.04% (w/v) $CaCl_2$, 1.2% (w/v) agar] for plaque titrations and growth of high-titer phage stocks.[52] For soft agar overlays the respective agar medium is prepared with only 0.7% (w/v) agar.[12]

Adaptation of Phages to Clinical Staphylococcus epidermidis Strains. Staphylococcus epidermidis phages 48 and 71 were adapted to *S. epidermidis* 9142, 8400, and 1457, respectively, by plaque purification.[12,23] To accomplish this the respective strains are grown in NB2+ broth overnight at 37° and the bacteria are spread onto ST-agar plates with cotton applicators. To increase the possibility of adapting the phage to a new host strain the bacterial suspension may be heated at 56° for 2 min before spreading on ST-agar plates. After drying for 15 to 30 min, volumes (10–20 μl) of 10-fold dilutions of phage lysates are spotted onto the agar plates. After drying for an additional 30–60 min at room temperature plates are incubated at 30° for 18–22 hr. Phage plaques are picked with a micropipettor and diluted into 300 μl of NB2+ broth. After vigorous shaking for 5 min to extract the phage particles appropriate dilutions are used in plaque titrations or for preparations of high-titer phage stocks of the plaque-purified virus. At high phage concentration inhibition of bacterial growth may occur, which is reminiscent of bacterial lysis due to infection. However, if lysis or plaques are not formed at higher dilutions of the inoculum attempts to isolate infectious phage from these inhibition zones are mostly unsuccessful.

[50] B. A. Dean, R. E. Williams, F. Hall, and J. Corse, *J. Hyg. (Lond.)* **71**, 261 (1973).

[51] V. T. Rosdahl, B. Gahrn-Hansen, J. K. Moller, and P. Kjaeldgaard, *APMIS* **98**, 299 (1990).

[52] M. J. de Saxe and C. M. Notley, *Zentralbl. Bakteriol. [Orig. A]* **241**, 46 (1978).

Preparation and Titration of High-Titer Phage Stocks. High-titer phage stocks [approximately 5×10^9 plaque-forming units (PFU) per milliliter] for transduction experiments are grown according to the soft agar technique.[53] The respective host strain is grown in NB2+ broth to a density of about 0.3 OD_{578} at 37°. All other incubations are performed at 30°. Bacterial suspensions (0.5 ml each) are mixed in triplicate with the phage inoculum (approximately 4×10^5 PFU/ml in 0.5 ml of NB2+). Three milliliters of ST-soft agar cooled to 48° is added per tube and mixed thoroughly on a vortex mixer. Extensive formation of bubbles should be avoided. The mixture is spread onto ST-agar plates. After incubation for 18 to 20 hr 5 ml of NB2+ broth is added per plate and the soft agar is transferred with a glass spatula into 50-ml polypropylene centrifuge tubes. Phage is extracted by vigorous shaking for 5 min. The agar is sedimented twice by centrifugation at $1500 \times g$ for 15 min. Phage lysates are sterilized by passing through a 0.2-μm pore size filter (Sartorius, Göttingen, Germany).

Phage titers are determined in a procedure similar to growth of high-titer phage stocks. The respective permissive *S. epidermidis* strain is grown in NB2+ broth to a density of 0.3 OD_{578}. Ten-fold dilutions of the phage lysate are prepared in the same medium. Bacteria (0.5 ml/tube) are dispensed into two glass tubes and 450 μl of the respective phage dilutions is added per tube. After mixing 3 ml of ST-soft agar is added. After vortexing, the mixture is spread onto ST-agar plates, which are incubated at 30° for 18 hr. The plaque titer is calculated from the numbers of plaques and the respective dilution. Routinely, titers of 10^9 to 10^{10} PFU/ml are obtained.

Transduction Procedure. For phage transduction a procedure described by Kayser *et al.*[54] is followed with modifications.[12,23,54] *Staphylococcus epidermidis* recipients in transduction experiments are subcultured on blood agar at 37° for 18 hr. Bacteria from one or two blood agar plates are suspended to a density of 0.5–1.0×10^{10} CFU/ml in NB-2+ broth. To overcome possible restriction barriers, the recipient strains can be heated to 56° for 3 min before transduction.[23]

For transduction of chromosomal markers, e.g., transposon Tn917 insertions, lysates grown on the individual transposon mutants are UV irradiated until the number of plaque-forming units is reduced by 90%. We use a UV source of a sterile work bench at a distance of about 40 to 50 cm. The phage lysate (2.5 ml) is spread into a 9-cm bacteriological polystyrene petri dish, the lid is removed, and the lysate is irradiated for 90 sec just before transduction is initiated. However, each UV source needs to be calibrated in advance. UV irradiation increases the number of transductants about 5- to 10-fold. For transduction of plasmid markers UV irradiation is not necessary.

The recipient cells (1 ml) are infected with the phage lysate (in 1 ml of NB2+ broth) at a multiplicity of infection of less than 1 PFU per bacterial cell and

[53] M. Swanstrom and M. H. Adams, *Proc. Soc. Exp. Biol. Med.* **78**, 372 (1953).
[54] F. H. Kayser, J. Wüst, and P. Corrodi, *Antimicrob. Agents Chemother.* **2**, 217 (1972).

incubated for 30 min at 37°. Phage adsorption is stopped by adding 1 volume of ice-cold BHI broth containing 40 mM sodium citrate. After centrifugation at 1500g at 4° for 10 min bacterial cells are additionally washed twice with 2 ml of ice-cold BHI broth containing 20 mM sodium citrate. Finally, recipient cells are suspended in 3 ml of the same medium and are incubated with shaking for 3 hr at 37°. Recipient cells are then mixed with BHI soft agar containing 20 mM sodium citrate and the appropriate antibiotics and are plated on BHI agar containing antibiotics [erythromycin (10 μg/ml), chloramphenicol (10 μg/ml), or tetracycline (10 μg/ml) as appropriate]. Transductants are isolated after incubation at 37° for 36 to 48 hr. When transducing plasmids, such as pTV1ts, with temperature-sensitive replicons recipient cells are plated in BHI soft agar immediately after washing the cells and incubation is at 30°. Transduction frequencies are expressed as transductants per PFU of the non-UV-irradiated phage lysate. In all experiments a sample of the recipient strain not infected with the phage lysate is processed in parallel as a control. Similarly, a sample (1 ml) of the phage lysate is plated onto the BHI agar similar to the transduction recipients for control of sterility of the lysate.

Transposon Mutagenesis

An *S. epidermidis* strain is usually selected for a transposon mutagenesis experiment because (1) it is a clinical isolate that can be genetically manipulated by one or another of the methods described above and (2) it has the phenotypic properties of interest. There are several prerequisites that increase the yield of a transposon mutagenesis experiment in *S. epidermidis*. First, the host strain should be free of additional plasmids, as Tn917 or its derivatives have a high predilection for insertion into plasmids, resulting in 90 to 95% nonchromosomal transposon insertions if other plasmids are present.[23,55] If the host strain of interest is not free of plasmids, it should be cured, if at all possible, of its respective plasmids before introducing the transposon carrying plasmid pTV1ts.[23,56,57]

For transposon mutagenesis *S. epidermidis* 1457c × pTV1ts, cured of the endogenous plasmid p1457,[23] is grown in PY broth [1.0% (w/v) peptone, 0.5% (w/v) yeast extract, 0.5% (w/v) NaCl, 0.1% (w/v) glucose, pH 7.5] containing 10 μg/ml concentrations of both chloramphenicol and erythromycin at 30° overnight with shaking.[23,26] The culture is diluted 1 : 100 into PY broth containing erythromycin at 2.5 μg/ml. After 5 hr of incubation at 30° dilutions are plated on BHI agar containing erythromycin (20 μg/ml), so that about 20 to 100 transposon mutants grow per plate. Plates are incubated for 36 hr at 45.5°, transposon mutants

[55] K. E. Weaver and D. B. Clewell, *in* "Streptococcal Genetics" (J. J. Ferretti and R. Curtiss III, eds.), p. 17. American Society for Microbiology, Washington, D.C., 1987.

[56] W. C. Olson, Jr., J. T. Parisi, P. A. Totten, and J. N. Baldwin, *Can. J. Microbiol.* **25**, 508 (1979).

[57] W. M. Shafer and J. J. Landolo, *Infect. Immun.* **25**, 902 (1979).

are subcultured onto BHI agar containing erythromycin (20 μg/ml) at 45.5°, and the resulting clones are tested for a biofilm-negative phenotype by the biofilm assay.

It is absolutely necessary to use an incubator that holds the temperature within a narrow range. The optimal temperature for elimination of pTV1ts and transposition must be determined experimentally for each host strain and a given incubator. For *S. epidermidis* 9142 and 1457 a temperature of 44.5 to 45.5° is optimal. Increasing this optimal temperature for only 0.5 to 1.0° in most cases is lethal to the bacterial cells. This range is so narrow that even the position of the plates of a transposon mutagenesis experiment within the incubator must be checked in advance in preliminary experiments.

Inoculating individual clones by insertion of sterile wooden tooth picks directly into microtiter tissue culture plates for the biofilm assay resulted in isolation of 9 independent isogenic biofilm-negative mutants out of 4500 clones tested, using the plasmid-free host strain *S. epidermidis* 1457c. Only 2 syngeneic mutants, M10 and M11, out of 6000 clones were isolated when using the plasmid-containing host strain *S. epidermidis* 13-1.[12,23,26] An alternative approach is selection for the transposition of the transposon and elimination of the temperature-sensitive plasmid pTV1ts under antibiotic selection in liquid medium followed by spreading the resulting transposon mutants after several rounds of enrichment for non adherent mutants onto an indicator agar, such as Congo red,[39,58] for biofilm-producing *S. epidermidis* strains.

Quantitation of Primary Attachment of *Staphylococcus epidermidis* to Polystyrene and Extracellular Matrix Proteins

An ELISA-based method was developed to evaluate attachment of *S. epidermidis* to polystyrene modified for cell culture (Nunclon Delta; Nunc), which is also used in the biofilm assay. In contrast to other published methods this assay compares attachment of *S. epidermidis* with the same surface used to evaluate biofilm formation *in vitro*.[10,12,27,59] In addition, the assay can be modified for screening large numbers of clones with impaired attachment properties from a transposon mutagenesis experiment.

Attachment assay. Staphylococcus epidermidis strains are grown in TSB$_{BBL}$ with shaking and 2-fold serial dilutions made in PBS. Samples of these dilutions (100 μl) are applied in triplicate to wells of 96-well tissue culture plates (Nunclon Delta; Nunc) and incubated for 1 hr at 37°. Bacterial cell densities are determined

[58] D. J. Freeman, F. R. Falkiner, and C. T. Keane, *J. Clin. Pathol.* **42**, 872 (1989).

[59] C. P. Timmerman, A. Fleer, J. M. Besnier, L. De Graaf, F. Cremers, and J. Verhoef, *Infect. Immun.* **59**, 4187 (1991).

by plating of appropriate dilutions. All wells are washed five times with buffer [PBS containing 0.05% (v/v) Tween 20 and 0.05% (w/v) NaN$_3$, 200 μl/well], using a multichannel pipette, and adherent cells are fixed *in situ* for 10 min at room temperature with a solution of 1.0% (v/v) formaldehyde in PBS. After an additional washing with buffer, quenching is performed with 0.1 M glycine for 20 min at 37°. Residual binding sites are blocked by incubation with 3% (w/v) bovine serum albumin in PBS containing 0.05% (w/v) NaN$_3$ for 2 hr at 37° or overnight at 4°. After washing three times with buffer, attached cells are detected by ELISA, using rabbit antiserum raised against the biofilm-negative *S. epidermidis* 5179 and alkaline phosphatase-coupled anti-rabbit IgG (Sigma). Anti-staphylococcal antiserum (100 μl/well) diluted 1 : 2500 in washing buffer containing 0.5% (w/v) bovine serum albumin is applied for 90 min at 37°. After washing three times with buffer alkaline phosphatase-conjugated anti-rabbit monoclonal antibody (Sigma) diluted 1 : 2500 in buffer (100 μl/well) is applied for 1 hr at 37°. Plates are washed three times and substrate buffer [100 μl/well; phosphatase substrate no. 104 (1.45 mg/ml; Sigma) in 0.5 mM MgCl$_2$, 0.02% (w/v) NaN$_3$, 1 M diethanolamine, pH 9.6] is added. After incubation at 37° for 30 min 2 M NaOH (50 μl/well) is added and the absorbance is measured at 405 nm with an automatic spectrophotometer (Behring). Differences in the attachment capacity are indicated by a shift of the response curve to higher inocula as observed by comparison of attachment of *S. epidermidis* 1457 to the unmodified surface and the identical surface preincubated for 60 min with fetal calf serum (Fig. 4A).

For screening a transposon library for mutants impaired in attachment the clones are replicated into microtiter plates (PS U-96; Greiner, Nürtingen, Germany) and grown overnight in TSB$_{BBL}$. The Greiner microtiter plates are used because *S. epidermidis* strains do not attach to this surface. The bacteria in the microtiter plates are resuspended by pipetting to ensure even suspension and the individual clones are diluted into PBS into a microtiter plate (Nunclon Delta; Nunc) with a surface modified for tissue culture. A multichannel pipette is used for all steps. The cell density is chosen in such a way that the wild-type strain exhibits an A$_{405}$ of 1.5 to 2.0 in the attachment ELISA as described above. All clones indicating impaired attachment with an A$_{405}$ < 0.5 are retested in the assay as described above.

Binding to matrix proteins. This assay was also modified to allow the comparison of attachment of *S. epidermidis* 1457 and its syngeneic mutant 1457-M10 to immobilized fibronectin and fibrinogen.[24] Bacteria are grown in TSB$_{BBL}$ for 18 hr at 37° with agitation. Cells are harvested by centrifugation and suspended and diluted in PBS. Bacterial cell densities are determined by plating of appropriate dilutions. Microtiter plates (PS U-96; Greiner) are coated at 150 μl/well with human fibronectin (Boehringer, Mannheim, Germany) or human fibrinogen (Sigma) in PBS for 16 hr at 4° at concentrations of 0.1, 1, and 10 μg/ml. Plates are washed three times with buffer and blocked as described above. After

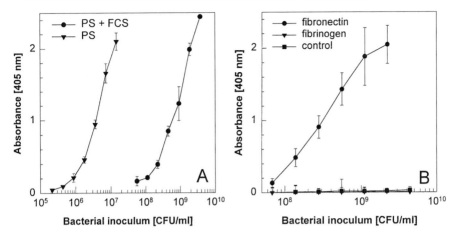

FIG. 4. (A) Primary attachment of *S. epidermidis* 1457 to polystyrene cell culture plates (Nunclon Delta; Nunc). Plates were either unmodified (▼) or modified by preincubation with 10% (v/v) fetal calf serum in PBS (●). Bacteria were inoculated into the plates at various densities for 1 hr at 37°. Attached bacterial cells were detected by ELISA as described. (B) Attachment of *S. epidermidis* 1457 to polystyrene microtiter plates (PS U-96; Greiner, Nürtingen, Germany) coated with human fibronectin (1 μg/ml; ●) or human fibrinogen (1 μg/ml; ▼), or uncoated (■) as a control. Bacteria were inoculated into the plates at various densities and incubated for 1 hr at 37°. Attached bacterial cells were detected by ELISA as described.

washing, 100-μl samples of bacterial suspension at varying densities are added in triplicate to the wells of coated microtiter plates for 1 hr at 37°. After washing attached cells are detected by ELISA as described above.

Using this assay, specific density-dependent attachment to fibronectin-coated surfaces is observed whereas binding to a fibrinogen-coated surface cannot be demonstrated with the strain used (Fig. 4B).[24] This assay probably can also be modified for screening of a transposon library for mutants impaired in binding to fibronectin.

Detection of *icaADBC* Locus in Clinical *Staphylococcus epidermidis* Isolates by Polymerase Chain Reaction

As expression of biofilm formation is variable for a given strain, depending on the type of medium used and the type of surface used to attach the biofilm, it is necessary to develop reliable genotypic methods to detect the *icaADBC* locus in different clinical *S. epidermidis* isolates. To reach this goal we evaluated three primer pairs specific for *icaA, icaB,* and *icaC*[21,22,60] for specific detection

---

[60] W. Ziebuhr, V. Krimmer, S. Rachid, I. Lößner, F. Götz, and J. Hacker, *Mol. Microbiol.* **32**, 345 (1999).

TABLE III

OLIGONUCLEOTIDE PRIMERS FOR DETECTION OF *icaaADBC* GENES BY
POLYMERASE CHAIN REACTION FOR *Staphylococcus epidermidis*

Primer	Sequence
*ica*Aforward (1446–1463)[a]	5′-TCG ATG CGA TTT GTT CAA ACA T-3′
*ica*Areverse (1661–1640)	5′-CTG TTT CAT GGA AAC TCC-3′
*ica*Bforward (2629–2649)	5′-TGG ATC AAA CGA TTT ATG ACA-3′
*ica*Breverse (2949–2932)	5′-ATG GGT AAG CAA GTG CGC-3′
*ica*Cforward (4087–4106)	5′-GGC GTC GGA ATG ATG TTA AG-3′
*ica*Creverse (4186–4165)	5′-AAT TCC AGT TAG GCT GGT ATT G-3′

[a] Position of the respective primers in the sequence of the *icaR-icaADBC* operon as given in the GenBank database (accession number U43366).[21,22,60]

of the *icaADBC* locus in clinical *S. epidermidis* isolates resulting in amplified fragments of 215, 320, and 99 bp, respectively (Table III). Polymerase chain reactions (PCRs) were performed with all three primer pairs in separate reactions and these results were compared with detection of the *icaADBC* locus with a cloned *icaADBC*-specific probe in Southern blot hybridization. Seventy *S. epidermidis* clinical isolates including several reference strains phenotypically representing a continuum from biofilm negative to strongly biofilm producing were evaluated.[16,19] All three primer pairs revealed a positive signal with all strains detected as positive by hybridization with the *icaADBC*-specific probe. Moreover, all three primer pairs revealed a negative PCR result with all strains detected as *icaADBC* negative by Southern hybridization.

For PCR, 50-μl samples of PCR mix containing final concentrations of 0.2 μ*M* for each primer, 0.2 m*M* dNTPs, 1.5 m*M* MgCl$_2$, and 1.0 U/50 μl for *Thermus brockianus* DNA polymerase (DyNAzyme DNA polymerase kit; Finnzymes, Espoo, Finland) are dispensed into reaction vials. Two drops of mineral oil are added to prevent evaporation. Using a sterile pipette tip, part of a colony of the respective *S. epidermidis* strain is transferred into the reaction vial. The vial is thoroughly mixed and after 5 min of denaturation at 94° 35 cycles are performed [30 sec at 94° (denaturation), 30 sec at 55° (annealing), 45 sec at 72° (elongation)], followed by a final elongation step for 7 min at 72°. The resulting PCR products are analyzed with 2% (w/v) NuSieve (3 : 1) agarose gels in TBE buffer (0.045 *M* Tris–borate, 0.001 *M* EDTA).[61]

[61] J. Sambrook, E. F. Fritsch, and T. Maniatis, "Molecular Cloning: A Laboratory Manual," 2nd Ed. Cold Spring Harbor Laboratory Press, Cold Spring Harbor, New York, 1989.

Acknowledgments

We thank Rainer Laufs for continuous support. We thank Vibeke T. Rosdahl (Statens Serum Institute, Copenhagen, Denmark) for providing phages and propagating strains, and W. B. Grubb (Curtin University of Technology, Perth, Australia) and Friedrich Götz (Molekulare Genetik, University of Tübingen, Germany) for bacterial strains and plasmids. The photographic work of C. Schlüter is acknowledged. Work in the laboratory of the authors is supported by grants from the Deutsche Forschungsgemeinschaft; the Bundesministerium für Bildung, Wissenschaft, Forschung, und Technologie; and the Paul Gerson Unna-Forschungszentrum für Experimentelle Dermatologie (Beiersdorf, Hamburg, Germany) given to D.M.

[21] *In Vitro* Methods to Study Staphylococcal Biofilm Formation

By Sarah E. Cramton, Christiane Gerke, and Friedrich Götz

Introduction

Chronic nosocomial infections by gram-positive bacteria have become more prevalent with the increased use of prosthetic biomedical implants. Two of the most commonly isolated organisms associated with biomedical implant infections are *Staphylococcus epidermidis,* a member of the normal skin microbiota, and *Staphylococcus aureus,* a resident of aural and nasal cavities in a large proportion of healthy individuals. Staphylococci opportunistically colonize intravenous catheters, artificial heart valves, replacement hip joints, and many other prosthetic devices. Chronic infection of a prosthetic implant can serve as a septic focus that can lead to osteomyelitis, acute sepsis, and death, particularly in immunocompromised patients.[1] Bacteria colonize these artificial substrates as a biofilm, multiple layers of cells enclosed in a self-produced polymer matrix. Cells within a biofilm are less sensitive to antibiotic agents and host immune defenses and therefore are difficult to treat clinically.[2] Often, the only means of eliminating the infection is to remove the prosthetic implant. By studying the process of biofilm formation in these organisms, we hope to find means by which to combat the establishment of biomedical implant-associated colonization and to eradicate established biofilm infections.

---

[1] G. D. Christensen, L. Baldassarri, and W. A. Simpson, *in* "Infections Associated with Indwelling Medical Devices" (A. L. Bisno and F. A. Waldvogel, eds.), p. 45. ASM Press, Washington, D.C., 1994.

[2] B. D. Hoyle and J. W. Costerton, *Prog. Drug Res.* **37,** 91 (1991).

A number of methods have been developed that are useful for the *in vitro* cultivation of staphylococcal biofilms. These methods enable us to identify and study the molecular mechanisms that lead to biofilm formation. Syngeneic mutant strains for genes important for the formation or maintenance of a biofilm *in vitro* can then be constructed and tested in animal models for *in vivo* virulence and biofilm formation.

In Vitro Biofilm Assay

In vitro biofilm formation by staphylococci is still frequently assayed according to the method developed by one of the pioneers in the field.[3,4] Bacteria are grown overnight in rich media, for example, tryptic soy broth (TSB) (17 g of tryptone, 3 g of soy meal, 5 g of NaCl, 2.5 g of $K_2HPO_4 \cdot 3H_2O$, and 2.5 g of glucose per liter), supplemented with an additional 0.25% (w/v) glucose (TSB+Glc). Overnight cultures are then diluted 1 : 200 to 1 : 1000 in fresh TSB+Glc. The TSB medium should be autoclaved as briefly as possible (e.g., 12 min) to prevent damage to the glucose. The diluted staphylococci can then be pipetted in 200-μl portions into the wells of a sterile U-bottomed polystyrene microtiter plate. Duplicate or triplicate sample wells are recommended because of the sensitivity of some strains to the subsequent washing procedures. The microtiter plate should then be incubated overnight at 37°.

The "phenotype" of cells growing in the microtiter wells should be noted carefully before removing nonadherent cells and medium from the microtiter wells by inverting the plate and vigorously banging the plate on an absorbent surface (e.g., paper towels). The wells should then be *gently* washed twice with phosphate-buffered saline (PBS) (7 mM Na_2HPO_4, 3 mM NaH_2PO_4, 130 mM NaCl, pH 7.4), and the plate vigorously emptied onto the absorbant surface and dried in an inverted position for at least 1 hr. Particularly for weak biofilm-forming strains, washing must be done quickly and carefully, and the plate must remain inverted to avoid residual medium or PBS from running back into the well and "washing" the attached cells again. After drying, the wells can be stained by pipetting 0.1% (w/v) safranin into the wells, removing it vigorously as described for washing, and maintaining the plate in an inverted position until dry. Because of the inconsistency of this method, attributable to individual or day-to-day variations in washing intensity, it is important to have the appropriate controls, both positive and negative, present on each plate so that a standard can be established. In addition, some strains seem to lose their ability to form strong *in vitro* biofilms after several passages

[3] G. D. Christensen, W. A. Simpson, A. L. Bisno, and E. H. Beachey, *Infect. Immun.* **37,** 318 (1982).
[4] G. D. Christensen, W. A. Simpson, J. J. Younger, L. M. Baddour, F. F. Barrett, D. M. Melton, and E. H. Beachey, *J. Clin. Microbiol.* **22,** 996 (1985).

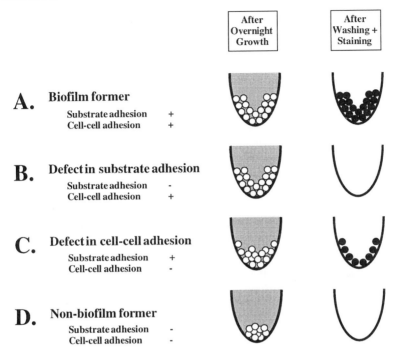

FIG. 1. Biofilm phenotypes in *in vitro* microtiter plate assay. (A) Wild-type strain adheres to the solid substrate and is able to mediate cell–cell adhesion. (B) Mutant defective for substrate adhesion forms confluent layer on bottom of microtiter plate well, but cell mass is removed on washing. (C) Mutant defective for cell–cell adhesion binds to substrate surface, but cells sediment in bottom of well after overnight growth and unattached cells are washed away, leaving a single layer of cells attached to the solid substrate. (D) Strain able to mediate neither substrate adhesion nor cell–cell adhesion accumulates at bottom of microtiter plate well and all cells are removed on washing.

on plates, and therefore a fresh batch from frozen stocks is recommended if a particular strain seems to be weakening in its biofilm phenotype.

Attached cells in microtiter plate wells can be "quantitated" in an ELISA reader (450 nm), which essentially measures the number of adhering cells in each well relative to other wells on the same plate. A qualitative analysis by eye is often more useful, however, to distinguish the different "phenotypes" in each well. Figure 1 shows the four biofilm phenotypes that can be expected on the basis of a model in which biofilm formation *in vitro* requires two genetically distinguishable steps: primary adhesion to a solid substrate followed by cell–cell adhesion that allows bacteria to accumulate the multiple layers of the biofilm. Analysis of the phenotype requires a combination of the appearance of bacteria in the microtiter plate well after overnight growth as well as the final phenotype after washing and staining. A wild-type strain that can mediate both substrate and cell–cell adhesion appears as a confluent layer on the bottom of the microtiter well, both in culture and after washing and staining. A strain that can mediate cell–cell adhesion but not substrate adhesion appears much like a wild-type strain after overnight incubation; however, because of the inability to adhere to the substrate, the cells are removed with the medium and the well is blank after washing and staining. Cells that are unable to mediate cell–cell adhesion often appear to have accumulated a sediment at the bottom of the microtiter plate well after overnight incubation. Sometimes, a thin halo of adherent cells can be seen surrounding the pile-up at the bottom of the well. After washing and staining, a single layer of cells is left attached to the bottom of the microtiter plate well, but the rest of the cells are gone. A strain that can mediate neither cell–substrate nor cell–cell adhesion tends to accumulate as a pellet of sedimented bacteria in the bottom of the microtiter well after overnight growth and the well is blank after washing and staining.

Biofilm assays can also be performed on glass surfaces, which are hydrophilic, in contrast to hydrophobic polystyrene surfaces, and may produce different results with some strains, depending on the affinity for different substrate materials. The 1- or 2-ml glass vials that are often used for freezing and storing bacterial stocks are useful for such assays, although these vials usually have flat bottoms. Alternatively, glass test tubes may be used; however, some effort should be exerted to ensure that the tubes are relatively new and without scratches caused by dishwashing brushes, because this may give the bacteria an advantage in adhering to the crevices on the surface of the glass. Crystal glass microtiter plates exist and are made by manufacturers of spectrophotometer cuvettes. Unfortunately, the cost is prohibitive for large-scale tests and cleaning the plates for reuse again introduces the risk of scratches on the glass surfaces, which may affect results.

Molecular Genetic Manipulation

Transposon Mutagenesis

Transposon mutagenesis in staphylococci can be performed with transposon Tn917, originally isolated from *Enterococcus faecalis*,[5] and carried on temperature-sensitive plasmid pTV1ts. The transposon carries an erythromycin resistance cassette and the plasmid is chloramphenicol resistant. The plasmid can best be transformed into the staphylococcal strain of interest either by electroporation[6] or

[5] J. H. Shaw and D. B. Clewell, *J. Bacteriol.* **164,** 782 (1985).

[6] J. C. Lee, *in* "Methods in Molecular Biology" (J. A. Nickoloff, ed.), p. 209. Humana Press, Totowa, New Jersey, 1995.

protoplast transformation,[7,8] with incubations carried out at 30°. The mutagenesis and curing of plasmid pTV1ts are performed as follows.

Grow an overnight culture at 30° with shaking in 5 ml of basic medium (BM) (10 g of casein-hydrolyzed peptone, 5 g of yeast extract, 5 g of NaCl, 1 g of glucose monohydrate, 1 g of $K_2HPO_4 \cdot 3H_2O$, pH 7.2, per liter) and erythromycin (10 μg/ml). Transfer 100 μl of overnight culture to 25 ml of prewarmed fresh BM and incubate with shaking at 42° (S. aureus and Staphylococcus carnosus; S. epidermidis may have to be incubated at a slightly lower temperature to ensure growth) without antibiotic selection overnight. This incubation is best carried out in a shaking water bath to ensure that the medium is warm enough to prevent plasmid replication. Plate 100 μl of a 1 : 100 dilution of the overnight culture on prewarmed BM–agar plates containing erythromycin (10 μg/ml) and incubate at 41° overnight. Approximately 100 colonies per plate can be expected under these conditions. A sampling of colonies should be picked to BM–agar plates containing erythromycin (10 g/ml) or chloramphenicol (10 μg/ml) to confirm that the plasmid cure rate is at least 95%. Colonies can then be picked to selective plates or otherwise assayed to screen for mutants with the desired phenotype.

Construction of Deletion/Replacement Mutants

The gene(s) of interest and a minimum of 500 bp upstream and downstream should be amplified by long-range polymerase chain reaction (PCR), chromosomal DNA from the wild-type strain of interest, and primers that contain appropriate restriction endonuclease recognition sites. The homologous recombination conditions described below are optimized for flanking DNA sequence of about 2 kb both upstream and downstream; less flanking sequence may lead to less efficient recovery of recombinants. This fragment can then be cloned into an appropriate vector, preferably temperature sensitive. Shuttle vectors of the pBT series[9] and derivatives are convenient because they allow cloning in Escherichia coli and are temperature sensitive in staphylococci. This construct can be used both to complement mutant phenotypes (assays must be performed at 30°) and to construct a deletion mutant.

The deletion mutant plasmid can be constructed by using the above constructed plasmid, inverse PCR, and primers that flank the gene(s) to be deleted, oriented toward the upstream and downstream flanking regions. The primers should again contain appropriate restriction enzyme recognition sites to allow the cloning of an antibiotic resistance marker in place of the deleted gene(s) of interest.

[7] F. Götz, B. Kreutz, and K. H. Schleifer, Mol. Gen. Genet. **198**, 340 (1983).
[8] F. Götz and B. Schumacher, FEMS Microbiol. Lett. **40**, 285 (1987).
[9] R. Brückner, FEMS Microbiol. Lett. **151**, 1 (1997).

A chromosomal deletion/replacement mutant can be created by homologous recombination with the deletion construct above. Particularly when ample flanking DNA sequence is present, deletion/replacement mutants can be constructed in multiple strains of the same species if desired. After transformation of the knockout construct into the strain of interest, the homologous recombination can be performed as follows:

Grow the wild-type strain of interest containing the deletion/replacement construct overnight in BM at 30° with selection for the antibiotic resistance marker on the plasmid. The overnight culture should then be diluted 1 : 1000 in fresh BM and grown again overnight at 30° with antibiotic selection. In most cases, about 15 cell division generations are sufficient, but this step can be repeated several times if the flanking DNA sequences are short and/or if the efficiency of recombination seems to be low. The culture should then be diluted 1 : 1000 and grown overnight at 42° without antibiotic selection twice to allow for curing of the plasmid carrying the deletion/replacement construct. Incubation in a shaking water bath is recommended to ensure that the medium actually has the desired temperature. *Staphylococcus epidermidis* may have to be grown at a slightly lower temperature to ensure growth. The culture should then be diluted 1 : 100 and plated on TSB plates selective for the marker replacing the gene(s) of interest. Positive colonies should be picked to plates containing the marker of interest as well as plates containing the antibiotic to which the plasmid had resistance. Colonies carrying both markers represent either (1) noncured plasmid, in which case incubation at 42° without selection should be repeated, or (2) single cross-over events in which the plasmid has integrated into the chromosome. This problem may be solved by repeating the incubation at 30° with antibiotic selection several times. Homologous recombination and plasmid curing should always be confirmed by amplifying chromosomal DNA by PCR and/or Southern blotting to show that the deletion/replacement region of the chromosome has the expected size and by testing the mutations for the expected phenotype, if known.

Molecular Basis of Adhesion and Accumulation

Biofilm formation is thought to be a two-step process that requires primary adhesion to a solid substrate followed by cell–cell adhesion. These two processes have been shown to be genetically distinct in *S. epidermidis* by the isolation of biofilm-negative transposon mutations in genes that effect each of these processes.[10]

---

[10] C. Heilmann, C. Gerke, F. Perdreau-Remington, and F. Götz, *Infect. Immun.* **64**, 277 (1996).

Adhesion

Staphylococcus aureus binds many plasma components such as fibrinogen, fibrin, collagen, vitronectin, and laminin, among others.[1,11,12] *Staphylococcus epidermidis,* on the other hand, seems to bind to serum components with a much weaker affinity, if at all.[12] *Staphylococcus epidermidis* shows a preference for polymer components of biomedical implants and may bind directly to the surface, while *S. aureus* seems to have a greater affinity for metal and tissue–implant boundaries, and is probably dependent on precoating of the surface(s) with serum proteins.

A gene with similarity to the *S. aureus* major autolysin gene (*atl*) was identified in *S. epidermidis* via a transposon mutant with a biofilm-negative phenotype *in vitro.*[10,13] The function of Atl E has been inferred by similarity to the *S. aureus* gene product. Atl is found at the cell septum and is involved in hydrolysis of the cell wall during cell division and separation.[14] The biofilm-negative phenotype of the *atlE* mutant is probably an indirect effect resulting from a change in cell surface hydrophobicity due to (1) the absence of the protein itself on surface or (2) a disruption of normal cell separation during cell division.

Because the *atlE* mutant is unable to adhere to the surface of a polystyrene microtiter plate, the entire accumulation of cells is removed with washing of the wells, using the procedure described above and diagrammed in Fig. 1. A primary adhesion assay was developed to show that the *atlE* mutant binds to polystyrene at a much lower rate than does a syngeneic wild-type strain, as follows.[10]

Grow an overnight culture to early stationary phase in BM. Alternatively, actively dividing cells can be assayed by diluting an overnight culture to an OD_{578} of 0.1 in BM and allowing the culture to grow to an OD_{578} between 1.0 and 2.0 before proceeding. Dilute the bacteria to an OD_{578} of 0.1 in PBS. Incubate 10 ml of the cell suspension in a sterile petri dish for 30 min at 37°. Remove the cell suspension and wash the floor of the petri plate five times with PBS. Place a microscope coverslip over a section of the petri plate floor and examine the arrangement and number of cells with a phase-contrast microscope. By counting the number of cells adhering to the petri plate within a given surface area, it is possible to quantitate primary adhesion to the polystyrene surface. It is also possible to measure adhesion to glass (or other surfaces, including coated substrates) by incubating a glass microscope slide on the bottom of the plate.

[11] P. E. Vaudaux, D. P. Lew, and F. A. Waldvogel, *in* "Infections Associated with Indwelling Medical Devices" (A. L. Bisno and F. A. Waldvogel, eds.), p. 1. ASM Press, Washington, D.C., 1994.

[12] T. J. Foster and D. McDevitt, *in* "Infections Associated with Indwelling Medical Devices" (A. L. Bisno and F. A. Waldvogel, eds.), p. 31. ASM Press, Washington, D.C., 1994.

[13] C. Heilmann, M. Hussain, G. Peters, and F. Götz, *Mol. Microbiol.* **24,** 1013 (1997).

[14] S. Yamada, M. Sugai, H. Komatsuzawa, S. Nakashima, T. Oshida, A. Matsumoto, and H. Suginaka, *J. Bacteriol.* **178,** 1565 (1996).

Unfortunately, this method, used together with the *atlE* mutant, must be taken circumspectly. Because of the cell separation defect, under the microscope the mutant appears to have small clusters of 8–10 cells. Another autolysin apparently partially compensates for the Atl E defect such that the culture does not consist of one large unseparated cluster of cells, but rather as numerous smaller clusters. The small cell groups, however, create several practical problems. First, optical density measurements with this strain are inaccurate. Second, the mere weight of several cells together may allow the clusters to be more easily washed away than single cells, and finally, quantitation of these small clusters is not accurate when comparing with 8–10 single cells that together have much more surface area exposed to the solid substrate. In summary, this method is useful for evaluating primary adhesion providing that a cell surface component differs between two strains, but may not be as useful when other phenotypes disturb the assay too severely.

Accumulation

A second gene locus that was identified by its biofilm-negative phenotype affects cell–cell adhesion and is mediated by the intercellular adhesion (*ica*) genes. The gene cluster is present in both *S. aureus* and *S. epidermidis* and its presence has been inferred in some other staphylococcal species as well[15]; however, it has never been detected in *S. carnosus*. Staphylococci that are able to mediate adhesion to a solid substrate are able to make a biofilm when complemented with the *ica* genes from *S. epidermidis*[16,17] or *S. aureus*.[15] Mutants that lack *ica* gene product function adhere as a single layer of cells to the surface of a microtiter plate well, but fail to form the multiple layers of a mature biofilm because of the loss of cell–cell adhesion. This adhesive property is conferred by the extracellular polysaccharide produced by the Ica proteins.

Isolation and Chemical Composition of Extracellular Polysaccharide

The extracellular polysaccharide from *S. epidermidis* has been investigated by a number of laboratories, but early results were often contaminated with components derived from the medium or agar. Mack and co-workers[18] were eventually able to

---

[15] S. E. Cramton, C. Gerke, N. F. Schnell, W. W. Nichols, and F. Götz, *Infect. Immun.* **67**, 5427 (1999).

[16] C. Heilmann, O. Schweitzer, C. Gerke, N. Vanittanakom, D. Mack, and F. Götz, *Mol. Microbiol.* **20**, 1083 (1996).

[17] D. McKenney, K. L. Pouliot, Y. Wang, V. Murthy, M. Ulrich, G. Döring, J. C. Lee, D. A. Goldmann, and G. B. Pier, *Science* **284**, 1523 (1999).

[18] D. Mack, W. Fischer, A. Krokotsch, K. Leopold, R. Hartmann, H. Egge, and R. Laufs, *J. Bacteriol.* **178**, 175 (1996).

show that preparations isolated from two different *S. epidermidis* strains at neutral pH consisted of two principal fractions. The majority of the substance isolated comprised linear β-1,6-linked glucosaminoglycans of approximately 130 repeating units. It was estimated that 80% of the glucosamine residues were N-acetylated and the rest were nonacetylated and positively charged. The minor fraction (15% of total) consisted of a similar polysaccharide containing negative charges due to phosphate- or ester-linked succinate groups. Because of its function in cellular aggregation, the polysaccharide was called the polysaccharide intercellular adhesin (PIA).

The isolation procedure separated PIA from the cell surface by sonication and from other cell surface-associated factors by gel filtration. Analysis of a wild-type strain showed the presence of 75% hexosamine after the gel-filtration step, which was absent in a preparation from a syngeneic biofilm-negative strain, while hexose, phosphate, and protein were present in substantial amounts in both preparations. The uncharged major fraction was separated from charged components via anion- and cation-exchange chromatography. The procedure yielded 2–3 mg of purified PIA per liter. Chemical analyses were performed via colorimetric assays for sugars, high-pressure liquid chromatography (HPLC), gas–liquid chromatography (GLC), and nuclear magnetic resonance (NMR) spectroscopy. It was determined that both the β-linkage and the acetylation of backbone residues are important for antibody specificity, with molecules with each of these properties able to compete for antibody binding.

Baldassarri and co-workers[19] isolated a compound that they named slime-associated antigen (SAA), also using sonication and gel filtration, and reported yields of 0.2 mg/liter. They determined that at least 70% of the dry weight consisted of N-acetylglucosamine and was negatively charged. The composition of the remaining 30% could not be identified, but DNA, protein, teichoic acid, long-chain fatty acids, cell wall teichoic acid, peptidoglycan, and lipoteichoic acid were ruled out. Similar results were obtained with two wild-type *S. epidermidis* strains. The authors concluded that SAA has the same structure as PIA.

McKenney and co-workers[20] isolated a substance that they called polysaccharide/adhesin (PS/A) from *S. carnosus* carrying a copy of the *S. epidermidis ica* genes on an introduced plasmid (pCN27),[16] using molecular sieve chromatography and pH 5.0. The preparation was treated with DNase, RNase, trypsin, and hydrofluoric acid to remove teichoic acids and yielded 0.5–2 mg/liter insoluble PS/A. Chemical analyses included detection of amino sugars by colorimetry, GLC–mass spectrometry (MS), and NMR. Their analyses indicated that

[19] L. Baldassarri, G. Donnelli, A. Gelosia, M. C. Voglino, A. W. Simpson, and G. D. Christensen, *Infect. Immun.* **64,** 3410 (1996).
[20] D. McKenney, J. Hubner, E. Muller, Y. Wang, D. A. Goldmann, and G. B. Pier, *Infect. Immun.* **66,** 4711 (1998).

their adhesin, which they renamed poly-N-succinyl β-1,6-glucosamine (PNSG),[17] has the same β-1,6-linked linear polyglucosamine backbone described for PIA, but that in contrast to the uncharged fraction of PIA,[18] the amino groups are substituted primarily with succinate and occasionally acetate. They also reported that antibodies raised against PIA and PNSG cross-react, confirming the similarity, if not identity, of the two compounds.

The difference(s) in the described chemical structure between PIA and PNSG may be merely semantic or the result of slightly different isolation methods; for example, PNSG may represent the minor charged fraction reported for PIA. The differences may, however, be real. It is possible that *S. epidermidis* and *S. aureus* produce an identical polysaccharide backbone but that the amino groups are modified (e.g., acetylated, succinylated) differently, or that the *ica* genes from either organism expressed in *S. carnosus* may produce product(s) not identical with that produced by the natural host. The groups of both Mack and Christensen isolated PIA/SAA from more than one *S. epidermidis* strain and observed identical results, suggesting that strain-to-strain differences do not exist when comparing products from the same organism. Until PIA/PNSG is isolated from representatives of both species as well as *S. carnosus* expressing the *S. epidermidis* or *S. aureus ica* genes, and the products are chemically analyzed side by side, it will be difficult to resolve this issue.

Isolation of Polysaccharide Intercellular Adhesin/Poly-N-Succinyl β-1,6-Glucosamine

We have developed a relatively crude method for preparing PIA/PNSG from cultures of *S. carnosus* carrying the *ica* genes from either *S. epidermidis* or *S. aureus* on a xylose-inducible expression plasmid (pTX15).[21] The method is based largely on the method of Mack and co-workers[18] described above, and should be applicable to wild-type *S. epidermidis* or *S. aureus* strains as well, although the yield will be reduced compared with the overexpression of the *ica* genes in *S. carnosus*.

Grow an overnight culture of *S. carnosus* carrying plasmid pTX15/SE*icaADBC* or pTX15/S*AicaADBC* in the presence of tetracycline (10 mg/ml). Dilute the cells to an OD_{578} of 0.1 in LB (10 g of tryptone, 5 g of yeast extract, and 5 g of NaCl per liter, pH 7.0). When the cells have reached an OD_{578} of 0.4, add xylose to a final concentration of 0.5% (w/v), and allow the culture to grow for an additional 4–6 hr. Harvest the cells by centrifugation for 10 min at 4° and 5000 g. Suspend the cell pellet in a 1 : 50 volume of 0.5 M EDTA, pH 8.0. This may not be entirely possible because the overexpression of PIA/PNSG makes the cells so adherent that the pellet may resemble a piece of chewing gum. For unknown reasons, this clumping

[21] A. Peschel and F. Götz, *FEMS Microbiol. Lett.* **40**, 285 (1996).

effect is more dramatic after overexpression of the *ica* genes from *S. epidermidis* than from *S. aureus*. The bacterial suspension should then be sonicated twice for 20 sec with incubation on ice in between. For larger preparations, this step may need to be repeated until the PIA/PNSG seems to have largely separated from its association with the cell surface. This can be recognized by the dissolution of the chewing gum-like cell pellet. Unlike gram-negative bacteria, sonication does not lyse staphylococci. The entire sonicated preparation should then be incubated twice for 5 min at 100° with incubation for 5 min on ice in between. This step allows the further dissociation of PIA/PNSG from the cell surface without lysis of the bacteria. The cells and cell debris should then be removed by centrifugation twice for 15 min at 4° and 7000 *g*. The supernatant can then be concentrated to approximately 2 ml and applied to an anion-exchange column (e.g., Q-Sepharose). The majority of uncharged PIA/PNSG does not bind to the charged matrix and is found in the flowthrough or the first 2 ml of elution with 50 mM Tris, pH 7.5. Additional charged PIA/PNSG can be eluted with increasing salt concentrations (e.g., 150, 300, 600 mM), but is usually found in the first fraction(s) of each step.

Quantification of Extracellular Polysaccharide

Until the exact structure of PIA/PNSG has been satisfactorily described, including differences (if any) between that produced by *S. epidermidis, S. aureus,* the *ica* genes from either organism expressed in heterologous host *S. carnosus,* and between strains within each species, it will remain difficult to develop a method for objectively quantifying the amount of polysaccharide produced by a particular strain. So far, semiquantitative methods have been used that rely on serial dilution of antibodies raised against *S. epidermidis* PIA or on visual comparison of cell surface extracts from strains prepared in parallel.[22] Both these methods have the drawback that they are dependent on the cross-reactivity of the antibody. Because it is not clear whether the structure of PIA/PNSG is exactly the same from species to species or between strains of the same species, comparisons of nonsyngeneic output remain approximate.

We tested a number of different buffers and conditions for the isolation of PIA/PNSG from the cell surface and into solution so that it could be semiquantitatively assayed, as outlined in Table I. The method and conditions described below proved to be an effective small-scale means of comparing relative PIA/PNSG production in parallel cultures.

Grow bacteria overnight in 5 ml of TSB+Glc medium [or TSB without glucose but supplemented with 0.1% (w/v) glucosamine], with antibiotic selection where appropriate. On the basis of the OD$_{578}$ values, take an equal number of cells from

[22] C. Gerke, A. Kraft, R. Süßmuth, O. Schweitzer, and F. Götz, *J. Biol. Chem.* **273**, 18586 (1998).

TABLE I

ISOLATION OF POLYSACCHARIDE INTERCELLULAR ADHESIN FROM *Staphylococcus epidermidis* RP62A, O-47, AND *Staphylococcus carnosus* (pCN27) UNDER VARIOUS CONDITIONS

Solution	15 min of shaking at room temperature[a]	5 min at 100°[a]	Surface-associated PIA after heating[b]
PBS[c]	−	++	Positive
TBS[d]	−	++	Positive
TBS + Tween 20 (0.5%, v/v)	−	−	Positive
SDS buffer[e]	−	++	Positive
LiCl			
5 M	++	++++	Negative
2.5 M	+	+++	Positive
NaCl (5 M)	+	+++	Positive
EDTA[f]			
0.5 M	++	++++	Negative
0.1 M	++	+++	Positive

[a] PIA in supernatant detected after cell sedimentation.

[b] PIA associated with suspended cells detected with PIA-specific antibody.

[c] PBS, phosphate-buffered saline: 7 mM Na$_2$HPO$_4$, 3 mM NaH$_2$PO$_4$, 130 mM NaCl, pH 7.4.

[d] TBS, Tris-buffered saline: 20 mM Tris-HCl (pH 7.5), 150 mM NaCl.

[e] SDS buffer: 67 mM Tris-HCl, 2% (w/v) SDS, 10% (v/v) glycerol, pH 6.8.

[f] pH 8.0.

each culture. (Strains that form large aggregates may be difficult to quantitate in this manner.) Harvest cells in a 2-ml Eppendorf tube at a relatively slow speed (e.g., 2 min, 6500 rpm) and remove the supernatant as completely as possible. This step may be repeated if more culture is needed than fits in the tube. Suspend the cell pellet in 50–100 µl of 0.5 M EDTA, pH 8.0, and incubate for 5 min at 100°. This step should dissociate the polysaccharide from the cell surface without lysing the cells. Centrifuge at top speed for 5 min and transfer the supernatant to a fresh tube. When working with *S. aureus* in particular, it is necessary to watch out for the cross-reactivity of protein A with all IgGs. This problem can be avoided by treating cell surface extracts with a protease before immunodetection. Forty microliters of cell surface extract can be incubated with 10 µl of proteinase K (20 mg/ml) for 30 min at 37°. One microliter of this mixture can then be spotted (manually or with the help of a dot-blotting apparatus) onto a nitrocellulose filter. The filter should be allowed to air dry for at least 15 min, but can be stored indefinitely before detection.

Block nonspecific binding by incubating the filter for at least 2 hr in TBS [20 mM Tris (pH 7.5), 150 mM NaCl] containing 3% (w/v) bovine serum albumin.

Wash briefly with TBS. Incubate the filter for at least 2 hr in PIA/PNSG-specific antibody diluted in TBS. Wash briefly with TTBS [TBS with 0.5% (v/v) Tween 20]. Wash twice, for 5 min each, with TTBS. Wash for 5 min with TBS. Incubate for at least 1 hr in conjugated secondary antibody diluted in TBS. Repeat the washing procedure. Proceed with a detection method as recommended by the manufacturer of the secondary antibody for Western blots. Detection with either 4-nitroblue tetrazolium chloride/5-bromo-4-chloro-3-indolyl phosphate (NBT/BCIP) or chemiluminescence is possible, with the latter being more sensitive.

In Vitro Assay for Polysaccharide Production

The function of the *ica* genes in *S. epidermidis* and *S. aureus* was determined via an *in vitro* assay for polysaccharide synthesis, using radiolabeled UDP-*N*-acetylglucosamine as a substrate.[15,22] Membrane preparations from *S. carnosus* expressing the *S. epidermidis ica* genes are incubated together with the substrate and radiolabeled sugar polymers are separated on thin-layer chromatography (TLC) plates. With this assay, we were able to show that Ica A is the glucosaminyltransferase, in accordance with its similarity to processive glucosyltransferases.[22] A surprising finding was that the activity of Ica D in the same cell membrane is required to enhance the transferase activity of Ica A. Together, Ica A and Ica D are able to synthesize oligomers of up to 20 units. The presence of Ica C in the same cell membrane causes the synthesis of larger product(s) that do not migrate from the origin on the TLC plate but are able to cross-react with antibodies raised against *S. epidermidis* PIA, suggesting Ica C participates in the synthesis of full-length PIA. The function of Ica B, predicted to be a secreted protein, remains somewhat unclear.[22] The gene sequence shows no similarity to genes of known function and a syngeneic deletion mutant for *icaB* produces some PIA, but the expression may be somewhat delayed (O. Schweitzer and F. Götz, unpublished results, 1997), suggesting that Ica B may be involved in the secretion of PIA to the outside of the cell.

The following method is used to isolate from staphylococcal cells crude membranes that can then be used in the *in vitro* polysaccharide synthesis assay to detect *ica* gene product function.

Crude Membrane Preparation

Grow cultures of bacteria carrying endogenous *ica* genes in LB containing 0.1% (w/v) glucosamine for 14–18 hr. Alternatively, cultures of *S. carnosus* carrying *ica* genes on a xylose-inducible plasmid should be treated as described above for PIA isolation. Cells should be pelleted, the wet weight determined, and the bacteria suspended in 2 μl of buffer A [50 mM Tris-HCl (pH 7.5), 10 mM MgCl$_2$, 4 μM dithiothreitol (DTT)] per milligram wet weight and DNase I added to 10 μg/ml. After addition of 4 mg of glass beads (diameter, 0.15–0.3 mm) per milligram wet weight, bacteria should be vortexed in glass centrifuge tubes three times for 1 min each with incubation on ice in between to disrupt the cells. Intact cells and glass beads can then be collected by centrifugation for 10 min at 2000g. The supernatant should be collected, more buffer added to the cell pellet, and the lysis procedure repeated. Crude cell extracts can then be pooled and the starting volume noted. Membranes can then be separated from the crude extract by ultracentrifugation for 20 min at 200,000g and suspended in 0.5 volume of buffer A (buffer A without DNase I), centrifuged again, and the membranes resuspended in 0.2 volume of buffer A. Triton X-100 extraction of membranes is optional, but enzymatic activity is enriched when this step is included. Add 2% (w/v) Triton X-100 to membranes in buffer A and rotate gently at 4° for 2 hr. Collect the Triton-extracted membranes by ultracentrifugation as described above, wash the membranes by carefully suspending the pellet in 1 or 2 volumes of buffer A, spin again, and resuspend in 1 volume of buffer A. Pelleted cells, crude extracts, and membrane preparations can be stored at −70°.

In Vitro N-Acetyl Glucosaminyltransferase Assay

Membrane preparations (see above) should be incubated together with 2 mM UDP-*N*-acetylglucosamine and 10 μM UDP-*N*-acetyl-D-[U-^{14}C]glucosamine in a reaction volume of 10–50 μl and incubated at 20° for 12 hr. Reaction products can then be separated on NH$_2$-HPTLC plates using acetonitrile–water (65 : 35, v/v) and visualized by exposure to X-ray film for several days to weeks, depending on the strength of the signal(s).

Environmental Factors Influencing the Regulation of Biofilm Formation

The conditions used to study biofilms *in vitro* cannot pretend to mimic the *in vivo* environment, which includes any number of host-produced extracellular matrix and serum components, local salt, pH or oxygen gradients, and variable nutrient concentrations and sources. However, it is possible to find conditions under which biofilms form *in vitro,* and then study how changes in these conditions affect the biofilm or the production of biofilm-related products. The final step would be to find experimental conditions *in vivo* that mimic the differences seen *in vitro.*

Medium and Sugar Source

PIA production can be stimulated *in vitro* by a range of sugar sources, including glucose, sucrose, fructose, and maltose.[23] Because PIA is a glucosaminoglycan, consisting of *N*-acetylglucosamine and glucosamine residues,[18] we

---

[23] D. Mack, N. Siemssen, and R. Laufs, *Infect. Immun.* **60,** 2048 (1992).

TABLE II

POLYSACCHARIDE INTERCELLULAR ADHESIN PRODUCTION BY *Staphylococcus epidermidis* RP62A AND *Staphylococcus carnosus* (pCN27) UNDER VARIOUS GROWTH CONDITIONS

Medium	Sugar	Concentration (%)	Growth	PIA production
TSB_0[a]	None		++	+/−
B_0[b]	None		++	+/−
LB^c	None		+	−
TSB	D-Glucose	0.5	++	++
	D-Glucose	1.0	++	++
LB	D-Glucose	0.5	+	+
	D-Glucose	1.0	+	++
	D-Glucosamine	0.05	+	+++
	D-Glucosamine	0.1	+	++++
	D-Glucosamine	0.25	−	−
	N-Acetyl-D-glucosamine	0.05	+	+++
	N-Acetyl-D-glucosamine	0.1	+/−	++++
	N-Acetyl-D-glucosamine	0.25	−	−

[a] TSB_0, tryptic soy broth: 17 g of tryptone, 3 g of soy meal, 5 g of NaCl, 2.5 g of $K_2HPO_4 \cdot 3H_2O$ per liter.

[b] B_0, basic medium: 10 g of peptone, 5 g of yeast extract, 5 g of NaCl, 1 g of $K_2HPO_4 \cdot 3H_2O$ per liter, pH 7.2.

[c] LB, Luria broth: 10 g of tryptone, 5 g of yeast extract, 5 g of NaCl per liter, pH 7.0.

investigated whether PIA production could be increased by the addition of glucosamine or N-acetylglucosamine to the medium. As seen in Table II, glucosamine or N-acetylglucosamine concentrations of 0.25% seem to be toxic and inhibit cell growth, but concentrations of up to 0.1% of either compound improve PIA production compared with glucose alone or unsupplemented medium.

Anoxic Conditions

Cystic fibrosis (CF) patients are frequently colonized by *S. aureus,* followed by colonization by *Pseudomonas aeruginosa,* which causes the often-lethal lung damage associated with this disease. It has been shown that PIA/PNSG is produced in sputum from the lungs of CF patients suffering from chronic *S. aureus* infections.[17] Because the environment within CF patient sputum is relatively anoxic,[24] we tested the effect of anoxic conditions on PIA/PNSG production *in vitro* and found anaerobiosis to have a stimulatory effect (S. E. Cramton, unpublished,

[24] D. Worlitzsch, K. C. Meyer, P. Birrer, and G. Döring, *Pediatr. Pulmonol.* **17**(Suppl.), A457, 333 (1998).

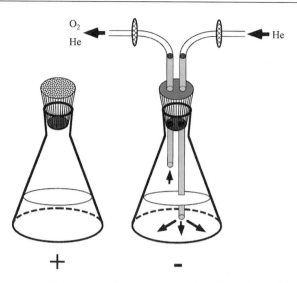

Fig. 2. Apparatuses used to assay cells grown under oxic and anoxic conditions. +, Normal Erlenmeyer flask containing bacteria in medium grown with shaking under ambient oxic conditions; −, Same as flask on left, except that the closure has been replaced by an airtight rubber stopper. Two glass pipes have been bored through the stopper, one allowing filtered helium gas to be bubbled through the medium and the other allowing gas to escape the flask until oxygen in the flask has been replaced by helium. The rubber inlet/outlet tubes are then sealed during growth and the flask is incubated under the same conditions as the flask on the left.

2001).[25] Incubation of cultures in anaerobic jars or in an incubator with a controlled environment (3% O_2, 1% CO_2, 96% N_2) showed some effect, but the increase in PIA/PNSG production was dramatic when oxygen was replaced with helium from the outset of the incubation as diagrammed in Fig. 2. This effect is observed in both *S. aureus* and *S. epidermidis* and is reflected in *ica*-specific RNA transcript levels.

Conclusion and Outlook

In vivo biofilm-forming conditions are difficult to mimic. It is entirely possible that many genes important for biofilm formation *in vivo* are not active or required *in vitro* and have therefore not been identified in the *in vitro* genetic screens performed to date. The genes that have been identified in this manner are clearly involved in biofilm formation *in vitro,* and clinical isolates associated with biomedical implant infections have been shown to produce PIA/PNSG. Preliminary *in vivo* experiments using syngeneic wild-type and *ica* mutant strains show

[25] S. E. Cramton, M. Ulrich, D. Worschlich, F. Götz, and G. Döring, submitted (2001).

mixed results, but the presence and activity of the *ica* gene cluster indicate that it behaves as a virulence factor in at least some animal and cell culture models. It remains to be seen whether other factors can be identified that are involved in the formation or maintenance of staphylococcal biofilms *in vivo,* and that could be potential targets for the prevention or elimination of biomedical implant-associated infections.

Acknowledgments

This project is supported by the German Bundesministerium für Bildung, Wissenschaft, Forschung und Technologie (BMBF) (DLR: 01KI9751/1). The technical assistance of Ulrike Pfitzner is gratefully acknowledged.

[22] Efficient RNA Isolation Method for Analysis of Transcription in Sessile *Staphylococcus epidermidis* Biofilm Cultures

By SABINE DOBINSKY and DIETRICH MACK

Introduction

Staphylococcus epidermidis is a normal inhabitant of human skin and mucous membranes. With the increasing use of foreign biomaterials in medicine these organisms have become one of the most frequently isolated pathogens in nosocomial infections.[1,2] The specific pathogenicity of these skin commensals can be attributed to an unusual ability to colonize polymer surfaces in multilayered communities referred to as *biofilms.*[3,4] Initially, bacteria attach to a polymer surface, followed by accumulation of bacterial cells in multilayered cell aggregates encased in an amorphous glycocalyx.[3,4] A polysaccharide intercellular adhesin (PIA) essential for bacterial accumulation mediates intercellular adhesion in these biofilms.[5,6]

[1] M. E. Rupp and G. L. Archer, *Clin. Infect. Dis.* **19,** 231 (1994).

[2] J. Huebner and D. A. Goldmann, *Annu. Rev. Med.* **50,** 223 (1999).

[3] D. Mack, *J. Hosp. Infect.* **43**(Suppl.), S113–S125 (1999).

[4] D. Mack, K. Bartscht, S. Dobinsky, M. A. Horstkotte, K. Kiel, J. K. M. Knobloch, and P. Schäfer, *in* "Handbook for Studying Bacterial Adhesion: Principles, Methods, and Applications" (Y. H. An and R. J. Friedman, eds.), p. 307. Humana Press, Totowa, New Jersey, 2000.

[5] D. Mack, M. Nedelmann, A. Krokotsch, A. Schwarzkopf, J. Heesemann, and R. Laufs, *Infect. Immun.* **62,** 3244 (1994).

[6] D. Mack, W. Fischer, A. Krokotsch, K. Leopold, R. Hartmann, H. Egge, and R. Laufs, *J. Bacteriol.* **178,** 175 (1996).

Synthesis of PIA requires the expression of the *icaADBC* gene locus of *S. epidermidis.*[7-9] Expression of biofilm formation and PIA by *S. epidermidis* depends significantly on different environmental factors such as type of growth medium used,[10-12] the presence of specific carbohydrates (such as glucose) in the medium,[10,13] and composition of the atmosphere.[11,14] At least three independent regulatory gene loci control expression of the synthetic genes for PIA synthesis on the level of transcription.[12] Apparently, expression of genes relevant for *S. epidermidis* biofilm formation is tightly regulated. Transcriptional activity in biofilms of the *S. epidermidis icaADBC* locus and other gene loci relevant for biofilm formation under different physiologic growth conditions are almost completely unknown at present. Analysis of transcription under these conditions requires the recovery of extremely high-quality mRNA from established *S. epidermidis* biofilms.

Established RNA extraction procedures proved to be unreliable and inefficient for extraction of RNA from staphylococci, because of their extremely stable cell wall.[15-17] In addition, these procedures often require extended periods of enzymatic digestion with proteases or lysostaphin for facilitating lysis of staphylococcal cells, potentially leading to degradation of short-lived mRNA species.[18] Many of these drawbacks have been solved by introduction of a new method for RNA extraction from *Staphylococcus aureus* and mycobacteria, using cell disruption by zirconia/silica beads in a high-speed shaking apparatus (FastPrep system; Bio 101, Vista, CA) and stabilization of extracted RNA by chaotropic reagents containing acid phenol, cetyltrimethylammonium bromide (CTAB), sodium acetate, and dithiothreitol[17] or modifications of this solution (chaotropic RNA stabilization reagent, FastPrep system; Bio 101). By this method mRNA from planktonic

[7] C. Heilmann, O. Schweitzer, C. Gerke, N. Vanittanakom, D. Mack, and F. Götz, *Mol. Microbiol.* **20,** 1083 (1996).

[8] C. Gerke, A. Kraft, R. Süssmuth, O. Schweitzer, and F. Götz, *J. Biol. Chem.* **273,** 18586 (1998).

[9] D. Mack, J. Riedewald, H. Rohde, T. Magnus, H. H. Feucht, H. A. Elsner, R. Laufs, and M. E. Rupp, *Infect. Immun.* **67,** 1004 (1999).

[10] G. D. Christensen, W. A. Simpson, A. L. Bisno, and E. H. Beachey, *Infect. Immun.* **37,** 318 (1982).

[11] M. Hussain, M. H. Wilcox, P. J. White, M. K. Faulkner, and R. C. Spencer, *J. Hosp. Infect.* **20,** 173 (1992).

[12] D. Mack, H. Rohde, S. Dobinsky, J. Riedewald, M. Nedelmann, J. K. M. Knobloch, H.-A. Elsner, and H. H. Feucht, *Infect. Immun.* **68,** 3799 (2000).

[13] D. Mack, N. Siemssen, and R. Laufs, *Infect. Immun.* **60,** 2048 (1992).

[14] L. P. Barker, W. A. Simpson, and G. D. Christensen, *J. Clin. Microbiol.* **28,** 2578 (1990).

[15] K. J. Reddy and M. Gilman, *in* "Current Protocols in Molecular Microbiology" (F. M. Ausubel, R. Brent, R. E. Kingston, D. D. Moore, J. G. Seidman, J. A. Smith, and K. Struhl, eds.), Vol. 1, pp. 4.4.1–4.4.7. John Wiley & Sons, New York, 1993.

[16] J. S. Kornblum, S. J. Projan, S. L. Moghazeh, and R. P. Novick, *Gene* **63,** 75 (1988).

[17] A. L. Cheung, K. J. Eberhardt, and V. A. Fischetti, *Anal. Biochem.* **222,** 511 (1994).

[18] M. Mempel, H. Feucht, W. Ziebuhr, M. Endres, R. Laufs, and L. Grüter, *Antimicrob. Agents Chemother.* **38,** 1251 (1994).

TABLE I
YIELD OF RNA EXTRACTION OF SESSILE *Staphylococcus epidermidis* CULTURES
AT VARIOUS TIME POINTS AFTER BIOFILM INDUCTION

Hours after glucose induction	Amount of biofilm $(OD_{570})^a$	RNA yield/cell density at OD_{578} (μg/OD)	Absorbance ratio (A_{260}/A_{280})
0	0	41.4	2.01
1	0	39.6	2.05
4	1.25	21.4	2.09
6	2.5	3.7	1.90
8	2.5	3.6	1.93

^{a} Biofilm production by *S. epidermidis* was determined by a semiquantitative adhesion assay.[20] Bacteria (200 μl/well) were grown in TSBαGlc$_{Oxoid}$ for 15–17 hr in 96-well tissue culture plates (Nunclon Delta; Nunc, Roskilde, Denmark). Biofilm formation was then induced by adding glucose [5 μl of a 10% (w/v) stock solution per well] at various time points before termination of the experiment. Plates were washed and dried in ambient air, and adherent biofilms were stained with gentian violet. The optical density of stained biofilms was measured at 570 nm (using 405 nm as reference wavelength) in an automatic spectrophotometer (Behring, Marburg, Germany).

cultures of *S. epidermidis* 1457 and RP62A can be successfully prepared for use in Northern blotting experiments demonstrating transcription of *icaADBC*.[12,19]

Extraction of RNA from Sessile *Staphylococcus epidermidis* Biofilm Cells

In tryptic soy broth lacking glucose [TSBαGlc$_{Oxoid}$ prepared from tryptone (Oxoid, Basingstoke, England), neutralized soya peptone (Oxoid), NaCl, and dipotassium phosphate as indicated by the manufacturer] *S. epidermidis* 1457 displays a biofilm-negative phenotype, as biofilm formation depends on the presence of glucose in the growth medium.[10,13] Induction of stationary phase cultures with glucose induces biofilm formation and PIA synthesis, and fully established biofilms are formed 6–8 hr after glucose induction (Table I).[13,20] *Staphylococcus epidermidis* 1457 is precultured in 10 ml of TSBαGlc$_{Oxoid}$ for 6–10 hr with shaking at 160 rpm at 37°. The culture is diluted 1 : 100 in the same medium and 10 ml is inoculated into 9-cm plastic tissue culture plates (Nunc, Roskilde, Denmark) and incubated at 37° for 15–17 hr. Biofilm formation is then induced by the addition of

[19] W. Ziebuhr, V. Krimmer, S. Rachid, I. Lößner, F. Götz, and J. Hacker, *Mol. Microbiol.* **32,** 345 (1999).

[20] D. Mack, K. Bartscht, C. Fischer, H. Rohde, C. de Grahl, S. Dobinsky, M. A. Horstkotte, K. Kiel, and J. K. M. Knobloch, *Methods Enzymol.* **336** [20] (2001) (this volume).

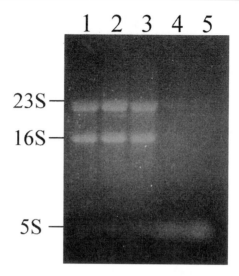

FIG. 1. RNA extraction from *S. epidermidis* biofilms at time points 0 (lane 1), 1 hr (lane 2), 4 hr (lane 3), 6 hr (lane 4), and 8 hr (lane 5) after glucose induction, using zirconia/silica beads and a high-speed shaking apparatus (FastPrep system; Bio 101).[17] Cells were grown and induced as described in text. Cell preparation included only sonication (twice, 30 sec each) to disintegrate the biofilm before RNA extraction. Samples (10 µg) of total cellular RNA were separated on a 1% (w/v) agarose–formaldehyde gel.

glucose [0.25% (w/v) final concentration] to the growth medium at various time points before extraction of RNA.[13]

Yield and composition of total cellular RNA from *S. epidermidis* biofilm cultures at various time points after glucose induction are compared, using the standard extraction protocol with zirconia/silica beads and a high-speed shaking apparatus as described by Cheung *et al.*[17] for extraction of RNA from planktonic *S. aureus* cultures. Before the extraction procedure the biofilms are disintegrated by sonication (twice, 30 sec each). The yield of extracted RNA by this method clearly depends on the age of the biofilm analyzed (Table I). Six hours after glucose induction of biofilm formation the biofilm formed reaches a maximum value ($OD_{570} = 2.5$). Approximately 10 times less RNA is obtained as compared with the yield of a noninduced culture ($OD_{570} = 0$) or a 1-hr biofilm ($OD_{570} = 0$). No significant contamination with DNA or proteins as determined by the A_{260}/A_{280} ratio can be detected in either preparation (Table I). Significant differences in the composition of the extracted RNA are observed depending on the time after glucose induction. From cells harvested 6 or 8 hr after glucose induction the extraction procedure yields predominantly small 5S rRNA (Fig. 1, lanes 4 and 5), indicating incomplete disruption of cells from established *S. epidermidis* biofilms.

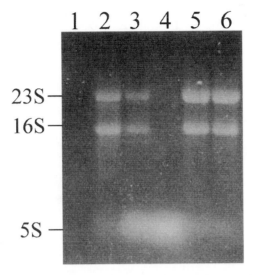

FIG. 2. RNA yield and quality produced by various cell preparation procedures preceding RNA extraction with zirconia/silica beads and a high-speed shaking apparatus (FastPrep system; Bio 101) from *S. epidermidis* biofilms. Samples (10 μg) of total cellular RNA extracted from 8-hr biofilms were analyzed on a 1% (w/v) agarose–formaldehyde gel. Lane 1, simple biofilm disintegration by sonicating twice (30 sec each); lane 2, same preparation method for a 1-hr biofilm, used as positive control; lane 3, cell preparation by additional sonication (five times, 2 min each); lane 4, cell preparation by additional sonication (five times, 1 min each); lane 5, cell wall lysis with lysostaphin, 150 U/ml, 2 min at 37°; lane 6, cell wall lysis with lysostaphin, 150 U/ml, 5 min at 37°.

Several different approaches have been used to overcome these problems. Increasing the processing time in the high-speed shaking apparatus using the zirconia/silica beads does not result in an increase in RNA yield (data not shown). In an attempt to more vigorously destroy the bacterial cell wall, the time of sonication before RNA extraction was prolonged. After sonicating the 8-hr biofilm an additional five times (1 min each), marginal RNA extraction is observed (Fig. 2, lane 4). Sonicating an additional 10 min (five times 2 min each) improves the recovery of 23S and 16S rRNA, but still gives unsatisfactory results as predominantly 5S rRNA is extracted (Fig. 2, lane 3).

Introduction of an enzymatic lysis step with lysostaphin before RNA extraction results in complete disruption of the staphylococcal cell wall. Incubation for 2 min at 37° is already sufficient to allow complete extraction of RNA, including 16S and 23S rRNA species, in high yield and purity (Fig. 2, lane 5). Apparently, the *S. epidermidis* cell wall of biofilm cells is sufficiently disrupted by the zirconia/silica beads only after partial lysis of the peptidoglycan interpeptide bridges by lysostaphin.

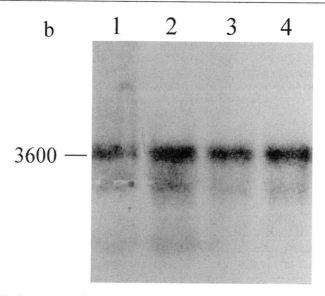

Fig. 3. Northern blot analysis of total cellular RNA extracted from a 4-hr biofilm after various cell preparation procedures. Quantities (10 μg) of each sample were separated on a 1% (w/v) agarose–formaldehyde gel, blotted onto nylon membrane, and hybridized with an *icaC*-specific oligonucleotide. Lane 1, positive control, simple disintegration of the biofilm by sonicating twice, 30 sec each; lane 2, cell preparation by an additional five 1-min sonications; lane 3, cell wall lysis with lysostaphin, 150 U/ml, 2 min at 37°; lane 4, cell wall lysis with lysostaphin, 150 U/ml, 5 min at 37°.

Prolonged incubation times at 37° are a problematic step in RNA extraction procedures.[15,16,18] To verify the functional integrity of the extracted RNA, Northern blot analysis is performed with a ^{32}P-labeled oligonucleotide probe specific for *icaADBC,* the gene locus encoding enzymes for the synthesis of PIA.[7,8] RNA is extracted from a 4-hr biofilm after the cells have been manipulated according to the various protocols. This preparation can serve as positive control, as at this time after biofilm induction bacterial cells are readily disrupted even by the standard method of RNA extraction after sonication (twice, 30 sec each) to disintegrate the biofilm. Similar hybridization patterns are observed with an *icaC*-specific probe with all samples, even when cell wall lysis is performed with lysostaphin for 5 min at 37° (Fig. 3, lane 4). Prolonged sonication (five times, 1 min each) also does no harm to the extracted RNA (Fig. 3, lane 2). Although the *icaADBC* transcript is about 3.6 kb long no indication of degradation of the mRNA species is observed. Because the half-lives of some mRNA species might be even shorter than 2 min it is recommended that a similar control experiment be performed to determine the stability of the specific mRNA under investigation.

Critical for efficient extraction of RNA from established *S. epidermidis* biofilms in high yield and purity is the lysostaphin lysis step before cell disruption, which is also observed with biofilms of other biofilm-producing *S. epidermidis* strains (data not shown). The method probably can be adapted for the analysis of transcription of relevant genes in *S. aureus* biofilms and in biofilms of other gram-positive bacterial species.[21,22]

Optimized RNA Extraction Protocol for Staphylococcus epidermidis Biofilms

Staphylococcus epidermidis 1457 is grown in TSBαGlc$_{Oxoid}$ and induced with glucose as described above. For harvesting *S. epidermidis* biofilms at each time point two 9-cm tissue culture plates are placed on ice. The cells are scraped from the surface with a disposable cell scraper into the growth medium and the suspension is immediately transferred to a prechilled centrifuge tube (Falcon; Becton Dickinson, Heidelberg, Germany) on ice. All further manipulations are carried out at 4°, using precooled solutions. Cells are sedimented by centrifugation (6000g, 10 min) and washed once in an equal volume of ice-cold phosphate-buffered saline. The cell pellet tends to adhere to the inner surface of glass pipettes and therefore suspension of the cells must be performed carefully. Bacteria are then suspended in 10 ml of phosphate-buffered saline and the cell aggregates are disintegrated by sonication (twice, 30 sec each) in an ice bath, using the 3-mm microtip of a sonicator disintegrator (Branson sonifier 250-D; Branson Ultrasonics, Danbury, Connecticut) at 70% of maximal amplitude. Samples are removed to determine the cell density as OD$_{578}$. Cells are sedimented, suspended in 10 ml of ice-cold 0.5 M EDTA, pH 8.0, and washed by repeating sonication (twice, 30 sec each) on ice. After sedimentation, bacteria are suspended in 1 ml of TE buffer (10 mM Tris-HCl, 1 mM EDTA, pH 8.0) containing lysostaphin (150 U/ml; Sigma, Deisenhofen, Germany) and incubated for 2 min at 37°. Immediately thereafter the reaction tube is placed on ice for 5 min. The cells are then collected by centrifugation (13,000g, 5 min) and suspended in 100 μl of ice-cold H$_2$O (aqua ad injectabilia; B. Braun Melsungen, Melsungen, Germany) and 500 μl of chaotropic RNA-stabilizing reagent (CRSR; Bio 101). The suspension is transferred to a 2-ml FastRNA tube (type blue) with zirconia/silica beads (FastPrep system; Bio 101), which already contains 500 μl of acid phenol, pH 4.5 (Amresco, Solon, OH), and 100 μl of chloroform–isoamyl alcohol (24 : 1, v/v; CIA). The tubes are processed in a high-speed shaking apparatus (FP 120 FastPrep cell disruptor; Savant Instruments, Farmingdale, NY), three times (20 sec each) at maximal speed. The phases are separated by centrifuging the samples (13,000g,

[21] S. E. Cramton, C. Gerke, N. F. Schnell, W. W. Nichols, and F. Götz, *Infect. Immun.* **67,** 5427 (1999).
[22] D. McKenney, K. L. Pouliot, Y. Wang, V. Murthy, M. Ulrich, G. Doring, J. C. Lee, D. A. Goldmann, and G. B. Pier, *Science* **284,** 1523 (1999).

30 min). The supernatant is carefully removed and approximately 50–80 μl of the solution over the interphase is left behind to avoid contamination with DNA and proteins. The supernatant is reextracted with 500 μl of CIA and phases are separated again after vortexing for 10 sec. The upper layer is transferred to a new reaction tube and the RNA is precipitated by adding 500 μl of 2-propanol. RNA is sedimented (13,000g, 10 min) and washed in 500 μl of 70% (v/v) ethanol. Residual ethanol is removed with a micropipette and the RNA is dissolved in 100 μl of H_2O. To minimize carbohydrate contamination an additional lithium chloride precipitation is performed by adding 20 μl of 12 M LiCl and 100 μl of 2-propanol per 100 μl of dissolved RNA. After 15 min of incubation on ice the precipitated RNA is sedimented, washed with 70% (v/v) ethanol, and finally dissolved in 50–200 μl of H_2O. The RNA concentration is determined spectrophotometrically at 260 nm. To assess the integrity of the purified RNA, samples (10 μg) are analyzed on 1% (w/v) agarose–formaldehyde gels in morpholinepropanesulfonic acid (MOPS) running buffer (20 mM MOPS, 5 mM sodium acetate, 1 mM EDTA, pH 7.0).

Northern Blot Analysis of Isolated RNA

RNA separated by electrophoresis is blotted onto Zeta-probe membranes (Bio-Rad, Munich, Germany) and fixed by baking at 80° for 30 min. Hybridization is performed with an *icaC*-specific oligonucleotide (5′-GAA ATA GCC ATA CCA TTG TCC-3′) at 52° overnight in modified hybridization buffer followed by two washing steps as suggested by the manufacturer [7% (w/v) sodium dodecyl sulfate (SDS), 20 mM sodium phophate (pH 7.0), 10× Denhardt's solution, 5× standard saline citrate (SSC), 10% (w/v) dextran sulfate, and denatured herring sperm DNA (100 μg/ml)] as described.[12] The membranes are analyzed by autoradiography (Kodak X-Omat X-ray film).

Acknowledgments

We thank Rainer Laufs for continuous support. The photographic work of C. Schlüter is acknowledged. This work was supported by a grant of the Deutsche Forschungsgemeinschaft to D.M.

Section V

Metabolic Potential of Biofilms

[23] Assessment of Metabolic Potential of Biofilm-Associated Bacteria

By Werner Manz, Michael Wagner, and Sibylle Kalmbach

Introduction

In the study of the microbial ecology of bacteria involved in biofilm formation, two major questions concern the metabolic status and phylogenetic identity of community members. For a thorough characterization of microbial biofilms, methods addressing these two questions should be combined. Identification of bacteria has been revolutionized by rRNA-based phylogeny and the development of rRNA-targeted oligonucleotide probes for fluorescent *in situ* hybridization (FISH) in natural microbial communities. For the assessment of metabolic activities, various attempts were made, both on the level of communities and of individual cells.

For determination of general metabolic activities at the community level, determination of the ATP content of biofilm samples is one of the most frequently used methods, allowing, for example, the study of the influence of the substratum on bacterial activity and growth. For an evaluation of the biofilm formation potential and, to a certain extent, the metabolic activities of biofilms under realistic conditions, a biofilm monitor, based on the exposure of inert surfaces and subsequent determination of the ATP content of the biofilms on the test surfaces, has been developed.[1]

At the level of individual cells, a vast number of techniques have been developed. If combined with spectroscopy or flow cytometry, some of them might also be useful tools at the population level. Most measurements of metabolic activities can be roughly grouped into methods detecting activities of specific enzymes, synthesis of macromolecules, or membrane potentials. Some of these methods, however, are not presented here, either because they have not yet been applied to environmental samples of biofilm populations or because they are not widely used in studies of bacterial metabolic activities. They include detection of (1) esterase activity by fluorescent substrates such as fluorescein diacetate or sulfofluorescein diacetate,[2,3] (2) cell division by microcolony formation,[4-6]

---

[1] D. van der Kooij, H. R. Veenendaal, C. Baars-Lorist, D. A. van der Klift, and Y. C. Drost, *Wat. Res.* **29,** 1655 (1995).

[2] T. H. Chrzanowski, R. D. Crotty, J. G. Hubbard, and R. P. Welch, *Microb. Ecol.* **10,** 179 (1984).

[3] S. De Rosa, F. Sconza, and L. Volterra, *Wat. Res.* **32,** 2621 (1998).

[4] H. W. Jannasch, *J. Gen. Microbiol.* **18,** 609 (1958).

[5] J. C. Fry and T. Zia, *J. Appl. Bacteriol.* **53,** 189 (1982).

[6] O. Nybroe, *FEMS Microbiol. Ecol.* **17,** 77 (1995).

(3) membrane potential by rhodamine 123[7,8] or by propidium iodide,[9] (4) membrane integrity by plasmolytic response to osmotic stress,[10] or (5) DNA and RNA degradation by lysozyme treatment and 4',6-diamidino-2-phenylindole (DAPI) staining.[11]

Thus, this chapter starts with the presentation of two of the most widely applied procedures (and important modifications of them) for the detection of bacterial metabolic potentials: the cyanoditolyl tetrazolium chloride (CTC) method[12] and the direct viable count (DVC) method.[13] In situ hybridizations with rRNA-directed oligonucleotide probes, which have been applied only more recently for indirect measurements of bacterial activities,[14–17] are presented in the context of a modification to the DVC method. In addition, the combination of conventional microautoradiography,[18–21] for the assessment of the nutrient utilization of bacterial populations, and FISH is described in detail.[22,23]

Because none of the presented methods is without drawbacks, a combination of different techniques should be applied for a thorough description of the bacterial metabolic potential in biofilm microenvironments.

Direct Viable Count Method and Its Modifications

To determine the potential for growth of single bacterial cells, Kogure and co-workers developed an in situ assay termed the direct viable count method.[13,24] This

[7] A. S. Kaprelyants and D. B. Kell, *J. Appl. Bacteriol.* **72**, 410 (1992).

[8] F. P. Yu and G. A. McFeters, *J. Microbiol. Methods* **20**, 1 (1994).

[9] S. C. Williams, Y. Hong, D. C. A. Danavall, M. H. Howard-Jones, D. Gibson, M. E. Frischer, and P. G. Verity, *J. Microbiol. Methods* **32**, 225 (1998).

[10] D. R. Korber, A. Choi, G. M. Wolfaardt, and D. E. Caldwell, *Appl. Environ. Microbiol.* **62**, 3939 (1996).

[11] D. Weichart, D. McDougald, D. Jacobs, and S. Kjelleberg, *Appl. Environ. Microbiol.* **63**, 2754 (1997).

[12] G. G. Rodriguez, D. Phipps, K. Ishiguro, and H. F. Ridgway, *Appl. Environ. Microbiol.* **58**, 1801 (1992).

[13] K. Kogure, U. Simidu, and N. Taga, *Can. J. Microbiol.* **25**, 415 (1979).

[14] P. F. Kemp, S. Lee, and J. LaRoche, *Appl. Environ. Microbiol.* **59**, 2594 (1993).

[15] S. Lee and P. F. Kemp, *Limnol. Oceanogr.* **39**, 869 (1994).

[16] L. K. Poulsen, G. Ballard, and D. A. Stahl, *Appl. Environ. Microbiol.* **59**, 1354 (1993).

[17] S. Møller, C. S. Kristensen, L. K. Poulsen, J. M. Carstensen, and S. Molin, *Appl. Environ. Microbiol.* **61**, 741 (1995).

[18] A. L. Munro and T. D. Brock, *J. Gen. Microbiol.* **51**, 35 (1968).

[19] H. G. Hoppe, *Mar. Biol.* **36**, 1225 (1976).

[20] L. A. Meyer-Reil, *Appl. Environ. Microbiol.* **36**, 506 (1978).

[21] P. S. Tabor and R. A. Neihof, *Appl. Environ. Microbiol.* **48**, 1012 (1984).

[22] N. Lee, P. H. Nielsen, K. H. Andreasen, S. Juretschko, J. L. Nielsen, K.-H. Schleifer, and M. Wagner, *Appl. Environ. Microbiol.* **65**, 1289 (1999).

[23] C. C. Ouverney and J. A. Fuhrman, *Appl. Environ. Microbiol.* **65**, 1746 (1999).

[24] K. Kogure, U. Simidu, and N. Taga, *Arch. Hydrobiol.* **102**, 117 (1984).

method involves incubation of natural bacterial samples with appropriate nutrients in the presence of the DNA synthesis-inhibiting antibiotic nalidixic acid[13] or a combination of the three antibiotics nalidixic, pipemidic, and piromidic acids, all of which act as gyrase inhibitors.[24] Because cellular functions other than DNA synthesis are not affected by the antibiotics, cells with the metabolic potential for nutrient uptake and synthesis of macromolecules will elongate, but not divide. The enumeration of elongated cells in a sample yields the number of metabolically active bacteria. Bacteria responsive in the DVC assay but unable to form visible colonies under appropriate culture conditions are commonly considered viable but nonculturable (VBNC), and DVC assays have become the most widespread method used to determine the proportion of VBNC cells in natural environments.[25,26]

However, the original DVC method is subject to certain limitations for the application in natural samples; the presence of bacterial species resistant to the antibiotic(s) used being the foremost problem. Other factors possibly affecting the applicability of DVC assays in natural bacterial communities involve differences in the responsiveness of the bacteria to the kinds and concentrations of the added nutrients, to the kinds and concentrations of the antibiotics, and to incubation times. Furthermore, the presence of eukaryotic predators in a sample might pose problems, as may filamentous or sheathed bacteria, for which elongation might be difficult to judge.

Joux and LeBaron[27] proposed an important modification of the DVC procedure with regard to the choice of antibiotics: by using a combination of four quinolones inhibiting DNA synthesis (nalidixic, piromidic, and pipemidic acids, and ciprofloxacin) and one β-lactam (cephalexin) growth of resistant bacteria could be prevented for incubation times of up to 24 hr in a natural marine sample.

Comparison of the effectiveness of nalidixic, piromidic, and pipemidic acids in preventing cell growth in biofilms obtained from drinking water showed that only when using pipemidic acid did total cell counts not increase during the 8 hr incubation period.[28] In the same study, a comparison between yeast extract and R2A medium[29] as nutrient source showed that a higher proportion of cells responded by elongation when R2A was used.

In this chapter we summarize important modifications to the DVC method that help overcome the restrictions of the original procedure. Before presenting the details of the modified DVC assay, however, we would like to stress that some environmental samples (e.g., thick biofilms, where diffusion of antibiotics and

[25] D. B. Roszak and R. R. Colwell, *Microbiol. Rev.* **51**, 365 (1987).

[26] J. D. Oliver, *in* "Starvation in Bacteria," (S. Kjelleberg, ed.), p. 239. Plenum Press, New York, 1993.

[27] F. Joux and P. LeBaron, *Appl. Environ. Microbiol.* **63**, 3643 (1997).

[28] S. Kalmbach, W. Manz, and U. Szewzyk, *FEMS Microbiol. Ecol.* **22**, 265 (1997).

[29] D. J. Reasoner and E. E. Geldreich, *Appl. Environ. Microbiol.* **49**, 1 (1985).

nutrients is hindered) might require protocol adjustments in regard to nutrients, antibiotics, incubation times, and other incubation conditions. Total cell counts must be monitored in all experiments.

Materials and Solutions

Antibiotics: Stock solutions of nalidixic, piromidic, and pipemidic acids (Sigma, Deisenhofen, Germany) are prepared at concentrations of 10 mg/ml in 0.05 M NaOH; stock solutions of cephalexin (Sigma) and ciprofloxacin (Bayer Vital, Leverkusen, Germany) are prepared at concentrations of 10 mg/ml in distilled water. The antibiotic solutions are filter sterilized through 0.2-μm pore size nitrocellulose membranes (Millipore, Eschborn, Germany) and stored at $-20°$

R2A medium,[29] containing per liter of doubly distilled water: 0.5 g of yeast extract, 0.5 g of Difco (Detroit, MI) proteose peptone no. 3, 0.5 g of Cas-amino acids, 0.5 g of glucose, 0.5 g of soluble starch, 0.3 g of sodium pyruvate, 0.3 g of K_2HPO_4, 0.05 g of $MgSO_4 \cdot 7H_2O$; the pH is adjusted to pH 7.2 by addition of K_2HPO_4 or KH_2PO_4

Modified R2A medium[30]: Standard R2A is modified by replacing soluble starch by 0.1% (v/v) Tween 80 (Sigma)

Nutrient broth (0.1%, v/v; Difco)

Formaldehyde solution (3.7%, v/v): 37% (v/v) formaldehyde solution (Merck, Darmstadt, Germany) is freshly diluted 1 : 9 in sterile distilled water

Phosphate-buffered saline (PBS), containing per liter of doubly distilled water: 8 g of NaCl, 0.2 g of KCl, 1.44 g of Na_2HPO_4, 0.24 g of KH_2PO_4; pH 7.2

Procedure

1. Fill the appropriate nutrient solution in sterile glass or plastic tubes or flasks; leave enough headspace for aeration. Tubes or flasks should be closed aseptically with cotton stoppers or other materials allowing aeration. For biofilms from oligotrophic environments, diluted R2A medium or modified R2A medium is appropriate (e.g., 0.5× medium or even lower dilutions); nutrient broth is recommended for biofilms from more nutrient-rich habitats.

2. Add antibiotics at final concentrations of 20 μg/ml (nalidixic acid), 10 μg/ml (piromidic acid), 10 μg/ml (pipemidic acid), 10 μg/ml (cephalexin), and 0.5 μg/ml (ciprofloxacin).

3. Remove substratum with the attached biofilm from its habitat. Take at least duplicate samples for the DVC assay and for direct fixation of the biofilm in formaldehyde solution.

4. Place biofilm in glass or plastic tubes.

[30] S. Kalmbach, W. Manz, J. Wecke, and U. Szewzyk, *Int. J. Syst. Bacteriol.* **49**, 769 (1999).

5. Incubate at room temperature (or at a temperature suitable for the investigated sample) in the dark; usual incubation times vary between 6 and 24 hr. If large numbers of eukaryotic predators are present in a sample or if a sample must be incubated under light and eukaryotic phototrophs are present, suitable inhibitors of eukaryotic cell growth (e.g., cycloheximide) should be added.

6. Remove biofilms and place immediately in 3.7% (v/v) formaldehyde solution, and incubate for 2 hr at 4°. Do not decant the incubation medium, but filter it through a 0.2-μm pore size polycarbonate membrane (Millipore) placed on a 0.45-μm pore size nitrocellulose support membrane (Millipore), using a vacuum filtration unit (Schleicher & Schuell, Dassel, Germany). Fixation of cells on the polycarbonate membrane with formaldehyde solution and subsequent washing with PBS are performed as described for biofilms.

7. Submerse the biofilms gently in PBS solution and air dry at room temperature.

8. If DVC assays are to be combined with *in situ* hybridizations, proceed to step 1 in the protocol for probe active counts (PACs). If DVC assays are not to be combined with *in situ* hybridizations, proceed to step 2 in the protocol for PACs.

Combination of Direct Viable Count and *in Situ* Hybridization: Probe Active Count Assay

To determine the phylogenetic affiliation and abundance of dormant, but potentially active bacteria in biofilm communities, the direct viable count (DVC) method[13] can be coupled with *in situ* hybridization, a combination termed the probe active count (PAC) assay.[28] By using specific probes, it is thus possible to determine activity patterns of defined species or groups of organisms in a biofilm community. In this context rRNA-directed oligonucleotide probes are used both for identification and as indicators of the metabolic status of the detected cells. Because rRNA molecules are integral parts of the protein synthesis machinery the cellular rRNA content, which can be quantified via the signal intensity obtained after *in situ* hybridization, is linked to the metabolic activity.[31,32] However, not only the present metabolic activity but also the physiological history of a cell has strong influence on cellular ribosome content. Thus *in situ* detectability of a bacterial cell with rRNA-targeted probes should be considered the sole indicator of its metabolic potential.

Through the application of PAC assays, sensitivity problems of hybridizations in oligotrophic environments can be overcome. The proportion of cells yielding clear hybridization signals in drinking water biofilms, for example, can be as low as 50% of total cell counts, even when using high-performance fluorochromes such as Cy3. After application of PAC, this percentage could be increased to 80% of all cells.[28]

[31] K. Flärdh, P. S. Cohen, and S. Kjelleberg, *J. Bacteriol.* **174,** 6780 (1992).
[32] M. Fukui, Y. Suwa, and Y. Urushigawa, *FEMS Microbiol. Ecol.* **19,** 17 (1996).

Solutions

DAPI solution: Stock solution of 4',6-diamidino-2-phenylindole (Sigma) is prepared at a concentration of 10 mg/ml in sterile doubly distilled water
Citifluor AF2 (Citifluor, London, UK)
Oligonucleotides, hybridization and washing solutions as described by Manz[33]

Procedure

1. Hybridization of biofilms has been described previously.[33]
2. Total cell counts of native biofilm, of biofilms subjected to the DVC assay, and of detached cells captured on polycarbonate membranes are determined by staining with DAPI at a final concentration of 1 μg/ml for 5 min in the dark. Biofilm-associated cells are stained by placing 50 to 100 μl of DAPI solution onto the substratum and subsequent gentle rinsing with doubly distilled water. Biofilms are air dried and mounted on glass slides, using Citifluor. Cells captured on polycarbonate membranes are stained by placing 20 μl of DAPI solution onto the membrane, followed by immediate mounting on glass slides with Citifluor.
3. The number of detached bacteria per centimeter squared of biofilm surface (D) is calculated by the following equation: $D = NA_m/A_b$, where N is the number of bacteria (per cm^2) on the polycarbonate membrane, A_m is the membrane surface area, and A_b is the surface area of the incubated biofilm.

Cyanoditolyl Tetrazolium Chloride Reduction

The CTC method is based on the reduction of the tetrazolium salt to a fluorescent formazan crystal by the activity of cellular dehydrogenases. Thus, actively respiring cells can be visualized in natural samples of bacteria. CTC has replaced other tetrazolium salts, which had been applied in earlier studies.[34-36] The advantage of the CTC method lies in the simple detectability of the reduced dye, because of its red fluorescence, and CTC has been applied in many aquatic environments.[12,37-40] CTC can be applied to the biofilm sample with or without addition of nutrients, yielding either the number of actively respiring bacteria

[33] W. Manz, *Methods Enzymol.* **310,** 79 (1999).
[34] S. A. Blenkinsopp and M. A. Lock, *Wat. Res.* **24,** 441 (1990).
[35] M. Fukui and S. Takii, *FEMS Microbiol. Ecol.* **62,** 13 (1989).
[36] G. A. McFeters, F. P. Yu, B. H. Pyle, and P. S. Stewart, *J. Microbiol. Methods* **21,** 1 (1995).
[37] L. K. King and B. C. Parker, *Appl. Environ. Microbiol.* **54,** 1630 (1988).
[38] G. Schaule, H.-C. Flemming, and H. F. Ridgway, *Appl. Environ. Microbiol.* **59,** 3850 (1993).
[39] C. E. Heijnen, S. Page, and J. D. van Elsas, *FEMS Microbiol. Ecol.* **18,** 129 (1995).
[40] B. H. Pyle, S. C. Broadway, and G. A. McFeters, *Appl. Environ. Microbiol.* **61,** 4304 (1995).

in situ or the number of cells with the potential for respiratory activity when enough suitable nutrients are present. Some authors, however, reported that they could find no differences between incubation with or without nutrients.[39,40]

Limitations of the CTC assay are (1) the toxicity of CTC itself[41]; (2) the inability of certain bacterial groups to reduce CTC despite being active, which might contribute to low detectability, i.e., only a few percent of all cells in oligotrophic environments[42]; and (3) extracellular reduction of CTC.[43] Because of the toxic effects of CTC, various concentrations of CTC should be tested for the investigated system and the lowest effective concentration should be used.

Solutions

CTC: Stock solutions of 5-cyano-2,3-ditolyl tetrazolium chloride (Polysciences, Eppelheim, Germany) are prepared at concentrations of 25 mM CTC in sterile doubly distilled water and stored at $-20°$

Media: R2A, modified R2A, or nutrient broth as described for DVC assays

Formaldehyde (3.7%, v/v), DAPI solution, and Citifluor as described above for DVC assays

Procedure

1. Fill sterile glass or plastic tubes or flasks with the appropriate medium; leave enough headspace for aeration. Tubes or flasks should be closed aseptically with cotton stoppers or other materials allowing aeration. For incubation without added nutrients, ambient water should be filtered through a 0.2-μm pore size nitrocellulose membrane (Millipore) immediately before the experiment and used as incubation medium. When incubation with nutrients is desired, R2A medium or nutrient broth should be used.

2. Add CTC from the stock solution at a final concentration ranging between 0.5 and 5 mM; ideal concentrations must be determined for the investigated samples and assay conditions (e.g., incubation with or without nutrients).

3. Remove substratum with the attached biofilm from its habitat and place in the glass or plastic tubes. Incubate at room temperature (or at a temperature suitable for the investigated sample) in the dark; usual incubation times vary between 1 and 4 hr. If biofilms are incubated in the presence of nutrients, incubation times should be short in order to prevent cell growth.

4. Remove biofilms and place immediately in 3.7% (v/v) formaldehyde solution, incubate for at least 30 min at $4°$, followed by gentle rinsing with doubly distilled water and air drying.

[41] S. Ullrich, B. Karrasch, H.-G. Hopp, K. Jeskulke, and M. Mehrens, *Appl. Environ. Microbiol.* **62,** 4587 (1996).

[42] T. Schwartz, S. Hoffmann, and U. Obst, *Wat. Res.* **32,** 2787 (1998).

[43] S. Wuertz, P. Pfleiderer, K. Kriebitzsch, R. Späth, T. Griebe, D. Coello-Oviedo, P. A. Wilderer, and H.-C. Flemming, *Wat. Sci. Tech.* **37,** 379 (1998).

5. For total cell counts, stain the biofilms with DAPI at a final concentration of 1 μg/ml by placing 50 to 100 μl of DAPI solution onto each biofilm. Incubate for 5 min in the dark and remove stain by gently rinsing with doubly distilled water. Biofilms are air dried and mounted on glass slides, using Citifluor.

Measurements of *in Situ* Signals by Microscopy

Probe-conferred fluorescence can be detected with an epifluorescence microscope fitted with a high-pressure bulb and appropriate light filter sets, such as a Zeiss (Oberkochen, Germany) Axioplan and Zeiss light filter set no. 01 for DAPI, no. 09 for FLUOS, and no. 15 for CTC detection, and HQ light filter set (AF Analysentechnik, Tübingen, Germany) no. 41007 and no. 41008 for Cy3- and Cy5-labeled probes, respectively. For documentation, color micrographs can be taken with 35-mm microscope camera (MC100; Zeiss) on Kodak EES 1600 color reversal film, or black-and-white micrographs can be produced with Ilford 400 ASA film. Typical exposure times for epifluorescence micrographs are 4–30 sec.

As an alternative to epifluorescence microscopy, probe-conferred fluorescence can be detected by using a confocal laser scanning microscope [e.g., LSM510 (Zeiss); TCS4D (Leica, Heidelberg, Germany); MRC 1000 (Bio-Rad, Hemel Hempstead, UK)]. In this case, the results are obtained by digital image processing, which also offers the basis for quantitative measurements of various parameters[44,45] and subsequent three-dimensional reconstructions of the specimen.

Fluorescence *in Situ* Hybridization Combined with Microautoradiography

The uptake of specific substrates by microorganisms under *in situ* conditions can be studied with radiolabeled substrates in combination with microautoradiography.[46] This method has become widely used in ecological studies and allows conclusions about the metabolic activity of microbial populations to be drawn.[47] A correlation between active substrate uptake and the identity of the organisms, however, is not possible on the basis of microautoradiography alone. A combination with immunofluorescent labeling of specific organisms was developed,[48] but suffers from the drawbacks of immunoassays in natural microbial communities. These consortia are frequently dominated by yet-unculturable microorganisms

[44] H. Daims, A. Brühl, R. Amann, K. H. Schleifer, and M. Wagner, *System. Appl. Microbiol.* **22**, 438 (1999).

[45] M. C. Schmid, U. Twachtmann, M. Klein, M. Strous, S. Juretschko, M. S. M. Jetten, J. W. Metzger, K. H. Schleifer, and M. Wagner, *Syst. Appl. Microbiol.* **23**, 93 (2000).

[46] T. D. Brock and M. L. Brock, *Nature (London)* **209**, 734 (1968).

[47] K. R. Carman, *Microb. Ecol.* **19**, 279 (1990).

[48] C. B. Fliermans and E. L. Schmidt, *Appl. Microbiol.* **30**, 676 (1975).

that cannot be identified by the use of fluorescent antibodies. In biofilms, penetration limitation and unspecific binding of fluorescent antibodies to noncellular compounds cause additional problems.[49] Microautoradiography was successfully combined with rRNA *in situ* hybridization[22] and applied for the characterization of activated sludge communities[22] and seawater picoplankton.[50] For single-cell resolution in sludge samples, flocs were cryosectioned and examined by confocal laser scanning microscopy. Uptake of specific substrates could thus be directly correlated with the phylogenetic identity of the organisms (Fig. 1). The combination of rRNA *in situ* hybridization and microautoradiography can also be used to specifically identify and enumerate members of defined functional bacterial groups in complex environmental samples if appropriate incubation conditions are applied and suitable labeled substrates are used. Probe-identified activated sludge bacteria that take up radioactive fatty acids under anaerobic conditions exclusively in the presence of nitrate almost certainly represent denitrifiers,[22] whereas iron-reducing activated sludge bacteria were identified by anaerobic incubation with labeled substrate in the presence of specific inhibitors of sulfate-reducing and methanogenic prokaryotes.[51]

Here we summarize the details of the method, which allows combining microautoradiography and FISH. The protocol describes the use of labeled acetate as the model compound for the analysis of microbial communities in waste water biofilms. It should be stressed that the applied incubation and washing procedures must be adapted according to which labeled substrate is used and which sample is analyzed. In particular, chemical binding of the labeled substrate to biofilm compounds must be prevented by suitable washing procedures.

Combination of Fluorescence *in Situ* Hybridization with Microautoradiography

Incubation of Biofilm with Radioactive Acetate

Solutions

Sodium acetate, 200 mM

[^{3}H]- or [^{14}C]acetic acid, sodium salt (Amersham, Buckinghamshire, UK)

1× and 3× Phosphate-buffered saline (PBS), pH 7.2, and 4% (v/v) formaldehyde fixative as described by Manz[33]

Ethanol (96%, v/v)

Tissue-Tek (Microm, Walldorf, Germany) embedding solution for cryosectioning

[49] H. Szwerinski, S. Gaiser, and D. Bardtke, *Appl. Microbiol. Biotechnol.* **21**, 125 (1985).

[50] C. C. Ouverney and J. A. Fuhrman, *Appl. Environ. Microbiol.* **65**, 1746 (1999).

[51] J. L. Nielsen, S. Juretschko, M. Wagner, and P. H. Nielsen, submitted (2000).

Procedure

1. Remove substratum with the attached biofilm from its habitat and dilute with sterile-filtered water obtained from the biofilm sampling site to a final dry matter concentration of 1–2 g of suspended solids (SS) per liter in a 9-ml glass vial closed with a butyl rubber stopper. As a control for possible adsorption phenomena pasteurize an extra serum bottle with diluted biofilm for 10 min at 70°.

2. Add acetate to a final concentration of 2 mM together with 10 μCi of sterile [^{3}H]- or [^{14}C]acetate. Incubate the samples at the temperature measured at the biofilm sampling site on a rotary table. If samples are to be incubated anaerobically, evacuate the sample in the serum bottle with "ultra-pure" nitrogen gas, shake gently for 1 hr before addition of the substrate in order to ensure complete oxygen removal, and inject oxygen-free stock solutions of the substrates with a syringe through the rubber stopper, using strict anaerobic techniques. In addition, prepare a control sample without added radioactivity to check for chemography.

3. Incubate the sample for 2 hr. The optimal incubation time might vary depending on the composition and activity of the respective microbial biofilm community.

4. Stop the uptake of radioactively labeled substrate by adding 4% (v/v) formaldehyde fixative to a final concentration of 2% (v/v). If gram-positive bacteria should be targeted in the subsequent rRNA *in situ* hybridization analysis, a parallel sample should be fixed by adding ice-cold 96% (v/v) ethanol [instead of 4% (v/v) formaldehyde fixative] to a final concentration of 50% (v/v). Incubate for 3 hr at 4°.

It should be noted that the addition of ethanol will lead to shrinkage of the biofilm. If preservation of structural details of the architecture is crucial, embedding of the biofilm (e.g., by using agarose or polyacrylamide) prior to fixation should be examined as an amendment to the protocol.

5. Remove biofilm and gently wash it at least three times with 1× PBS for removal of fixative and radioactive substrate in the bulk solution. Carefully transfer a piece of the biofilm into an Eppendorf tube and cover it with at least 0.5 ml of Tissue-Tek embedding solution. Allow the embedding solution to penetrate into the sample for at least 6 hr. Freeze the sample at −20° for at least 6 hr.

6. Quantitative analysis of the absolute uptake of radioactively by labeled substrates by microorganisms can be performed by determination of the total amount of radioactivity in the biofilm after the incubation and washing steps, using liquid scintillation counting.[22]

Fluorescence in Situ Hybridization and 4′,6-Diamidino-2-phenylindole Staining

Solutions

Poly-L-lysine-coated coverslips: Clean coverslips in acid alcohol [1% (v/v) HCl in 70% (v/v) ethanol]. Place the coverslips in a Coplin jar with

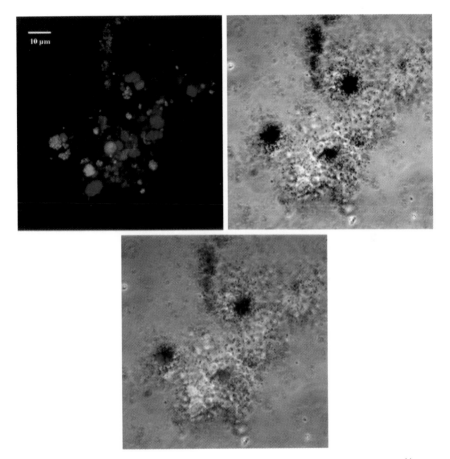

FIG. 1. Combination of FISH and microautoradiography for studying aerobic uptake of [^{14}C]propionate by complex microbial communities in activated sludge of an industrial wastewater treatment plant [S. Juretschko, G. Timmermann, M. Schmid, K. H. Schleifer, A. Pommerening-Röser, H.-P. Koops, and M. Wagner, *Appl. Environ. Microbiol.* **64**, 3042 (1998)]. *Upper left:* FISH analysis: *Nitrosococcus mobilis* is specifically stained red, members of the *Azoarcus/Thauera* cluster are labeled blue, and *Azoarcus tolulyticus* shows lila probe-conferred fluorescence. *Upper right:* Microautoradiography of the identical microscopic field. *Bottom:* Superimposed FISH and microautoradiography pictures demonstrate that other members of the *Azoarcus/Thauera* cluster than *A. tolulyticus* were able to take up propionate.

0.01% (w/v) poly-L-lysine (room temperature) for 5 min. Drain the coverslips and dry for 1 hr at 60° or overnight at room temperature

Ethanol [50, 80, and 96% (v/v)]

PBS (1×), oligonucleotides, hybridization and washing solutions as described by Manz[33]

DAPI solution as described above for DVC assays

Procedure

1. Prepare 1- to 10-μm-thick cryosections of the embedded biofilm, using a cryomicrotome (e.g., MIKROM HM500; Mikrom), and place them on poly-L-lysine-coated coverslips. Dry the sections at 46° for 10 min.

2. Dehydrate the cryosections on the coverslips by successive passages through 50, 80, and 96% (v/v) ethanol (3 min each). Allow to air dry.

3. Soak a strip of Whatman (Clifton, NJ) 3MM paper in 1.5 ml of hybridization buffer and place it in a 50-ml tube. Allow the chamber to equilibrate for 5 min at the hybridization temperature (46°).

4. Mix 8 μl of hybridization buffer with 1 μl of working solution of each fluorescent oligonucleotide probe and spread it onto the sample.

5. Quickly transfer the coverslips to the preheated moisture chamber and hybridize for 2 hr.

6. Remove the coverslips from the moisture chamber and rinse the probe(s) from the coverslips with hybridization buffer preheated to the washing temperature (48°).

7. Transfer the coverslips to a polypropylene screw-top tube filled with 50 ml of preheated washing buffer and incubate for 15 min at 48°.

8. Remove the coverslips from the washing buffer and dip them briefly into double-distilled ice-cold H_2O; shake away the excess water immediately, and allow to air dry.

9. For DAPI stain, spread 50 μl of DAPI solution (1 μg/ml) onto the biofilm sections and allow to react for 5 min at 0° in the dark and remove stain by rinsing the coverslips with ice-cold double-distilled water. Shake away excess water and allow the biofilm sections to air dry.

Microautoradiography

Solutions

Hypercoat Nuclear Emulsion LM-1 for light microscopy (Amersham), Arlington Heights, IL)

Developer (Kodak D19), 40 g/liter

Fixative: 30% (w/v) thiosulfate

Citifluor AF2 (Citifluor, London, UK)

Procedure

Note: All the following steps should be performed in the dark.

1. Allow the film emulsion to melt for 10 min at 45° and pour it into a dipping vessel (e.g., Hypercoat dipping vessel; Amersham).

2. Dip each coverslip carrying a hybridized biofilm section into the vessel for approximately 5–10 sec. Avoid the formation of air bubbles. Transfer the coverslips to tissue paper and allow residual fluid to drip off for 5–10 sec. Carefully dry the backside of each coverslip. Place each slip horizontally on a tray. Allow to air dry for 2–3 hr.

3. Place the coverslips in a slide container together with a small package of dried silica gel. Cover the slide container with aluminum foil. Allow the slips to expose the film for 2–7 days. For each incubation experiment several biofilm sections should be analyzed in parallel. To test for the optimal exposure time, a coverslip should be developed every day and the increase in number of cells covered with silver grains should be determined microscopically (see below).

4. Prepare the developing and fixative solutions freshly in separate jars. Mount the coverslips on a holder. Immerse the holder into the developing solution for 3–5 min. Subsequently transfer the holder to a jar filled with doubly distilled water and incubate for 1 min. Finally, transfer the coverslips to the fixative and incubate for 4 min.

5. Wash the coverslips carefully in a jar under slowly running tap water for 10 min.

6. Incubate the coverslips in doubly distilled water for 2 min; repeat.

7. Allow the coverslips to air dry.

8. Mount the samples with Citifluor.

The coverslips should be examined by inverse epifluorescence microscopy or inverse confocal laser scanning microscopy (e.g., LSM510; Carl Zeiss) to avoid masking of DAPI- and probe-conferred fluorescence of radioactively labeled cells by the overlying silver grains within the autoradiographic film.[22] Comparative estimations of the amount of radioactive labeled substrate taken up by different cell clusters or filaments can be performed by microscopically counting the respective number of silver grains within the film.

Section VI

Extracellular Polymers

[24] Characterization of Extracellular Chitinolytic Activity in Biofilms

By Ace M. Baty III, Zhenjun Diwu, Glen Dunham,
Callie C. Eastburn, Gill G. Geesey, Amanda E. Goodman,
Peter A. Suci, and Somkiet Techkarnjanaruk

Introduction

It is common for bacteria to produce extracellular enzymes having some form of degradative activity. In some cases these enzymes serve to protect cells from antagonistic substances,[1–3] or to convert a large and/or insoluble biopolymer to an assimilable nutrient source.[4,5] In some cases the physiological benefit to the bacterium is not entirely evident.[6,7] Extracellular enzymes may be membrane bound, but in many cases they are released into the surrounding medium.[8] It has been shown that these relatively large molecules become immobilized in the extracellular polymeric matrix in which cells in flocs and biofilms are embedded.[9] Most proteins adsorb irreversibly to substrata having a variety of surface chemistries,[10] and transport by convection is reduced near any solid surface, regardless of the flow regimen in the bulk liquid.[11] Thus, extracellular enzymes have a tendency to become an integral and significant component of the biofilm/substratum microenvironment, influencing cell physiology and biofilm ecology.

In this chapter methods for characterizing extracellular enzyme activity associated with biofilms are presented. There are three components of the methodology. The first is a precipitating fluorogenic enzyme substrate (ELF-97-N-acetyl-β-D-glucosaminide) (ELF-NAG),* used to localize extracellular catalytic activity.

---

[1] G. Harth and M. A. Horwitz, *J. Biol. Chem.* **274**, 4281 (1999).

[2] J. M. Frere, *Mol. Microbiol.* **16**, 385 (1995).

[3] A. Kharazmi, *Immunol. Lett.* **30**, 201 (1991).

[4] E. A. Bayer, L. J. Shimon, Y. Shoham, and R. Lamed, *J. Struct. Biol.* **124**, 221 (1998).

[5] G. P. Hazlewood and H. J. Gilbert, *Prog. Nucleic Acid Res. Mol. Biol.* **61**, 211 (1998).

[6] R. J. Lamont and H. F. Jenkinson, *Microbiol. Mol. Biol. Rev.* **62**, 1244 (1998).

[7] M. J. Benedik and U. Strych, *FEMS Microbiol. Lett.* **165**, 1 (1998).

[8] R. G. Wetzel, *in* "Microbial Enzymes in Aquatic Environments" (R. J. Chrost, ed.), p. 6. Springer-Verlag, New York, 1991.

[9] B. Frolund, T. Griebe, and P. H. Nielsen, *Appl. Microbiol. Biotechnol.* **43**, 755 (1995).

[10] J. D. Andrade, *in* "Surface and Interfacial Aspects of Biomedical Polymers," Vol. 2: "Protein Adsorption" (J. D. Andrade, ed.), p. 1. Plenum Press, New York, 1985.

*ELF-97 is a registered trademark used by the company Molecular Probes (Eugene, OR). ELF is an acronym for enzyme-labeled fluorescence. We use ELF to refer to the precipitating fluorophore produced by enzyme cleavage: 2-(2′-hydroxy-5′-chlorophenyl)-6-chloro-4(3H)-quinazolinone.[11a]

Similar methodologies have been used to probe enzyme activity in biofilms[12] and microbial flocs.[13] ELF-NAG is soluble and nonfluorescent until it is cleaved by the enzyme. Once cleaved, one of the products fluoresces and precipitates onto the substratum. Because the soluble substrate cannot enter the cell, the assay indicates activity of extracellular enzymes exclusively.** Additional information is obtained by using green fluorescent protein (GFP) to report promoter (transcriptional) activity of relevant genes using established techniques.[14]

The second component of the methodology is the use of thin films to fabricate a substratum with properties that mimic a natural surface, but allow cells and fluorescent areas to be detected by conventional light microscopy with micron spatial resolution. Many relevant natural surfaces are either too rough and/or fluoresce too strongly to enable this level of resolution by conventional microscopy. Therefore, elaborate and time-consuming methodologies must be applied to characterize biofilms on these surfaces.[15] Incorporation of thin films into an *in vitro* system allows relevant hypotheses about the natural system to be addressed by relatively simple methods. Spin-casting a substance onto a planer surface produces films that are optically smooth, and so thin (e.g., 50 nm) that the fluorescence from the material is a negligible background. The method is relatively simple, requiring only that a volatile solvent for the substance can be found.

The third component of the methodology is a spin-off of the use of thin films. There are a number of well-established techniques that can be used to characterize thin films under hydrated conditions nondestructively, in real time. These include the quartz crystal microbalance,[16] surface plasmon resonance,[17] total internal reflection fluorescence,[18] and attenuated total reflection Fourier transform

[11] S. Vogel, "Life in Moving Fluids," pp. 127–162. Princeton University Press, Princeton, New Jersey, 1981.

[11a] K. D. Larison, R. BreMiller, S. Wells, I. Clements, and R. P. Haugland, *J. Histochem. Cytochem.* **43**, 77 (1995).

[12] C. T. Huang, K. D. Xu, G. A. McFeters, and P. S. Stewart, *Appl. Environ. Microbiol.* **64**, 1526 (1998).

[13] F. Van Ommen Kloeke and G. G. Geesey, *Microb. Ecol.* **38**, 201 (1999).

[14] B. B. Christensen, C. Sternberg J. B. Andersen, R. J. Palmer, Jr., A. T. Nielsen, M. Givskov, and S. Molin, *Methods Enzymol.* **310**, 20 (1999).

[15] B. Normander, N. B. Hendriksen, and O. Nybroe, *Appl. Environ. Microbiol.* **65**, 4646 (1999).

[16] H. Ebato, C. A. Gentry, J. N. Herron, W. Muller, Y. Okahata, H. Ringsdorf, and P. A. Suci, *Anal. Chem.* **66**, 1683 (1994).

[17] W. Knoll, W. Hickel, M. Sawodny, J. Stumpe, and H. Knobloch, *Fresenius J. Anal. Chem.* **341**, 272 (1990).

[18] Y.-L. Cheng, B. K. Lok, and C. R. Robertson, *in* "Surface and Interfacial Aspects of Biomedical Polymers," Vol. 2: "Protein Adsorption" (J. D. Andrade, ed.), p. 121. Plenum Press, New York, 1985.

**Our results indicate that ELF-NAG does not enter into the cytoplasm of *Pseudoalteromonas* sp. strain S91. However, it may penetrate into other types of bacteria.

infrared spectroscopy (ATR-FTIR).[19] Using one or more of these techniques, it is possible, in principle, to measure degradation of the thin film by an associated biofilm. In the special case where the thin film incorporates a substrate for an extracellular enzyme, the chosen technique can provide a measure of the kinetics of substrate degradation. We have designed an *in vitro* system that enables both microscopic images and ATR-FTIR data to be acquired from a biofilm under flowing conditions.[20]

The success of the methods outlined above is demonstrated here, using results obtained from a chitinolytic marine bacterium, *Pseudoalteromonas* sp. strain S91 (S91) cultured on thin films of chitin.[21–23] Degradation of particulate organic matter (POM) in the ocean is one of the processes that has a major influence on the global ecology. Bacteria are thought to contribute significantly to degradation of POM.[24–26] The associated ecological system has been termed the microbial loop.[27] POM is, of course, too large to enter bacterial cells, and even the component biopolymers are, in general, not cell permeable. Therefore, bacterial extracellular enzymes are an essential component of the microbial loop.[28] Bacterial chitinolytic enzymes are thought to be chiefly responsible for the conversion of the enormous marine chitin reservoir into dissolved organic matter (DOM).[29] Although extracellular enzymes produced by bacterial biofilms undoubtedly make a significant contribution to carbon cycling,[30] there is a paucity of data providing direct information about marine biofilms associated with POM. A significant obstacle to acquiring this data has been methodological, i.e., overcoming the difficulties associated with detection of single cells on relevant natural surfaces.

Bacteria capable of hydrolyzing high molecular weight chitin are widely distributed in sediments, on the surface of plants, in the gut of marine mammals, and

[19] K. Knutzen and D. J. Lyman, *in* "Surface and Interfacial Aspects of Biomedical Polymers," Vol. 1: "Surface Chemistry and Physics" (J. D. Andrade, ed.), p. 197. Plenum Press, New York, 1985.

[20] P. A. Suci, K. J. Siedleki, R. J. Palmer, Jr., D. C. White, and G. G. Geesey, *Appl. Environ. Microb.* **63,** 4600 (1997).

[21] A. M. Baty, Physiological Diversity of an Attached Marine Bacterial Population," pp. 1–142. Ph.D. Thesis, Montana State University, Bozeman, Montana, 1999.

[22] A. M. Baty, C. C. Eastburn, Z. Diwu, S. Techkarnjanaruk, A. E. Goodman, and G. G. Geesey, *Appl. Environ. Microbiol.* **66,** 3566 (2000).

[23] A. M. Baty, C. C. Eastburn, S. Techkarnjanaruk, A. E. Goodman, and G. G. Geesey, *Appl. Environ. Microbiol.* **66,** 3574 (2000).

[24] D. A. Caron, P. G. Davis, L. P. Madin, and J. M. Sieburth, *Science* **218,** 795 (1982).

[25] D. C. Smith, M. Simon, A. L. Alldredge, and F. Azam, *Nature (London)* **359,** 139 (1992).

[26] F. Azam, *Science* **280,** 694 (1998).

[27] T. Berman and L. Stone, *Microb. Ecol.* **28,** 251 (1994).

[28] H. G. Hoppe, *in* "Microbial Enzymes in Aquatic Environments" (R. J. Chrost, ed.), p. 60. Springer-Verlag, New York, 1991.

[29] G. W. Gooday, *Biodegradation* **1,** 177 (1990).

[30] M. Karner and G. J. Herndl, *Mar. Biol.* **113,** 341 (1992).

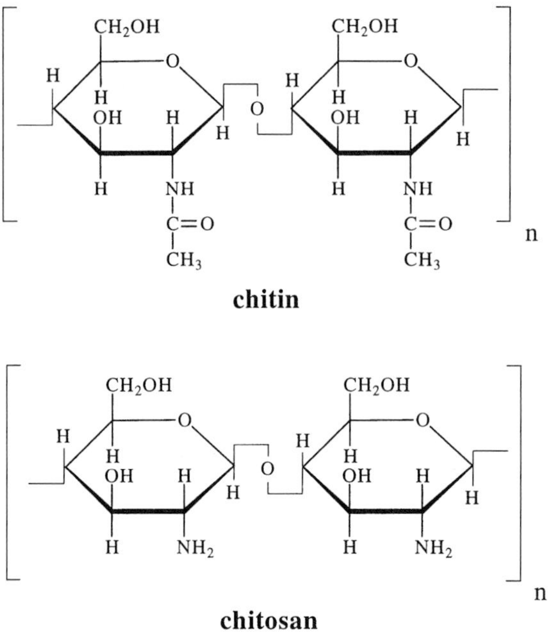

chitin

chitosan

FIG. 1. Haworth projections of structures of chitin and chitosan. Chitosan is soluble in aqueous solution. Thin chitin films are formed by spin casting chitosan followed by acetylation of the films (described in text).

associated with POM.[31] These bacteria synthesize and export chitinase enzymes that can cleave the β-1,4-linkage of the homopolymer chitin (Fig. 1).[32] High molecular weight chitin is degraded into chitin oligomers by extracellular endo-chitinase. Oligomers are acted on by chitobiase to produce dimers of chitin that can be incorporated into the periplasm of bacteria. In the periplasm these dimers are cleaved into monomers by N-acetylglucosaminidases. In addition, extracellular exo-chitinases cleave single sugars off the reducing ends of the polymer chains. Extra-cellular N-acetylglucosaminidase can also be secreted and cleave small oligomers into monomers outside the cell. A single chitinolytic bacterium may express all these enzymes in an effort to efficiently utilize high molecular weight solid chitin. Figure 2 shows the activity and possible location of different chitinase enzymes with respect to a chitinolytic bacterium. S91 is a genetic construct in which GFP reports promoter activity of a gene (chiA) encoding an extracellular chiti-nase. ELF-NAG is cleaved both by exochitinase and N-acetylglucosaminidase

[31] C. E. Zobell and S. C. Rittenburg, J. Bacteriol. **35,** 275 (1937).
[32] M. B. Brurberg, I. F. Nes, and V. G. H. Eijsink, in "Chitin Enzymology," Vol. 2 (R. A. A. Muzzarelli, ed.), p. 171. Atec Edizioni, Ancona, Italy, 1996.

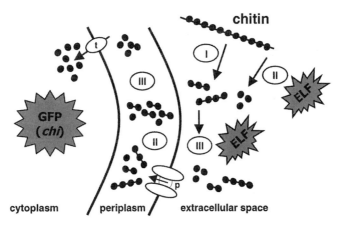

FIG. 2. Types of chitinolytic enzymes and possible locations with respect to a bacterium. I, Endochitinase; II, exochitinase; III, *N*-acetylglucosaminidase; p, porin; t, *N*-acetylglucosamine transporter; *chi*, gene for general chitinase enzyme. Sites of GFP and ELF-NAG fluorescence are indicated by GFP and ELF, respectively.

(Fig. 2). Although results suggest that *chiA* encodes an exochitinase,[21–23] this remains hypothetical because the enzyme has not been purified to homogeneity. Details of the genetic construct are given below and in the referenced literature.

Fabrication of Thin Films of Chitin

Published techniques were used to fabricate chitin films, with some modifications.[33–35] To our knowledge, chitin is not soluble in any volatile organic solvent. The protocol consists of casting films of chitosan onto a solid support and then converting the chitosan (Aldrich, Milwaukee, WI) to chitin by N-acetylation (see Fig. 1 for structures). Films are cast onto silicon coupons for the evaluation of bacterial gene expression, X-ray photoelectron spectroscopy (XPS) analysis, and profilometry; onto AT-cut quartz crystals for profilometry and mass determination by quartz crystal microbalance (QCM); and onto germanium (Ge) internal reflection elements (IREs) for infrared (IR) spectroscopy. Silicon coupons, 1×1 cm square (Harrick Scientific, Ossining, NY), are cleaned in "Piranha," which consists of a $70:30$ mixture of concentrated sulfuric acid and 30% hydrogen peroxide (Piranha solution reacts violently, even explosively, with organic materials).[36] The silicon coupons are then rinsed in distilled H_2O and dried under

[33] H. Bae and S. M. Hudson, *J. Appl. Poly. Sci. A Poly. Chem.* **35**, 3755 (1997).
[34] T. D. Rathke and S. M. Hudson, *J. Mater. Sci. Rev. Macromol. Chem. Phys.* **C34**, 375 (1994).
[35] Y. Qin, *J. Appl. Poly. Sci.* **49**, 727 (1993).
[36] D. A. Dobbs, R. G. Bergman, and K. H. Theopold, *Chem. Eng. News* **68**, 2 (1990).

a stream of pure dry nitrogen. Quartz crystals are cleaned with a cotton-tipped applicator successively in water, methanol, and chloroform and dried under a stream of dry nitrogen. Single-crystal, Ge IREs (EJ 2121; Harrick Scientific) are cleaned by ultrasonication in a base bath (saturated KOH in isopropyl alcohol) for 10 min followed by a series of rinses, all of which consist of ultrasonication in various liquids for 10 min. After the base bath are two rinses in "ultrapure" water followed by a gentle scrubbing with undiluted Micro cleaning solution, using cotton swabs. The cleaning solution is flushed off in a hard stream of ultrapure water. The IRE is then subjected to the following rinses: ultrapure water (twice), ethyl alcohol, and chloroform. IREs are air dried at 100°.

Chitosan solutions are prepared in aqueous 2.0% (v/v) acetic acid and then filtered through a 0.2-μm syringe filter to remove any insoluble debris. Except for cleaning, the spin-casting protocols for silicon coupons and quartz crystals are identical. Coupons are affixed to the middle of a centrifuge rotor with double-stick tape, and scrubbed with methanol followed by chloroform, using cotton-tipped applicators. Approximately 300 μl of the chitosan solution is placed on the surface of the coupon and spun with rapid acceleration at 4500 rpm for 5 min. While the surface is still spinning the chitosan thin film is washed successively with 1.0 ml of 5.0% (w/v) NaOH, 10.0 ml of distilled H_2O, and 1.0 ml of methanol. Films are then dried in a desiccating chamber overnight. For N-acetylation, coupons are placed in 10.0 ml of ice-cold, dry methanol and 2.5 ml of ice-cold acetic anhydride is added with stirring. The reaction is allowed to proceed for 18 hr at 4°. Coupons are then rinsed in methanol four times and placed on the spin caster, where they are rinsed with an additional 10.0 ml of methanol while spinning at 3000 rpm. Films are dried overnight in a desiccating chamber. The protocol for making thin chitin films on a Ge IRE is the same as for silicon coupons except that the IRE is fixed to the centrifuge rotor by a Teflon holder designed to accommodate the Ge IRE during casting. It consists essentially of a block of Teflon with an appropriately sized cavity.

The quality of the chitin thin films decreases with increasing chitosan casting solution concentrations. Chitin thin films prepared from solutions containing more than 4% (w/v) chitosan cannot be constructed because of the high viscosity of the chitosan solutions. Chitin thin films prepared from solutions containing more than 2% (w/v) chitosan trap air bubbles during casting, resulting in usable films but with visually observable film defects. On the basis of visual observation, 1% (w/v) chitosan-cast chitin thin films appear to be continuous and evenly distributed. The most critical step in the construction of these films is the cleaning of the silicon substratum. The slightest residue can result in major defects in the thin film. On N-acetylation of the chitosan films, the resulting chitin film always appears slightly darker in color.

TABLE I
ATOMIC CONCENTRATION OF CHITOSAN AND CHITIN THIN FILMS

	Atomic concentration (%)		
	Carbon	Nitrogen	Oxygen
Chitosan thin film	46.19 ± 1.12	5.56 ± 0.91	48.25 ± 1.01
Chitin thin film	62.19 ± 1.23	6.18 ± 0.74	31.63 ± 0.89
Chitin (theoretical)	57.14	7.14	35.7

Characterization of Thin Films

The fluorescence of the thin films is evaluated by epifluorescence microscopy. Differential interference contrast (DIC) microscopy and profilometry are used to evaluate the continuity of the chitin thin films. Surface chemistry is quantified spatially by XPS. The density of the chitin thin films is determined by thickness measurements obtained by profilometry and mass measurements determined by QCM. ATR-FTIR is used to characterize film chemistry and crystal structure.

Atomic Composition by X-Ray Photoelectron Spectroscopy

XPS is performed with a Physical Electronics (Eden Prairie, MN) model 5600 spectrometer. A 5-eV flood gun is used to offset charge accumulation on the samples. An aperture of 30 μm is used with a monochromatized aluminum K_α X-ray source at 350 W and a pass energy of 11.750 eV. Survey spectra and high-resolution spectra are gathered at a take-off angle of 45°. Abundance of carbon, nitrogen, and oxygen is calculated from high-resolution C1s, N1s, and O1s peak areas. The binding energy scale is referenced by setting the CHx peak maximum in the C1s spectrum to 285.0 eV.[37] The atomic composition of three 30-μm spots on each of three chitin thin films prepared from a 1.5% (w/v) chitosan solution ($n = 9$) is evaluated (Table I). The atomic composition is presented as percent carbon, nitrogen, and oxygen with the standard error.

The atomic concentrations of carbon, nitrogen, and oxygen of the chitin thin films are similar to the theoretical concentrations for pure chitin (Table I). Differences may be attributed to surface hydrocarbon contamination. The standard error in atomic concentration displayed between a total of nine different 30-μm spots on three separately prepared chitin thin films ranges from 0.74 to 1.23%. These standard errors are within the range of anticipated reproducibility for XPS.

[37] B. D. Ratner and D. G. Castner, in "Surface Analysis–Techniques and Applications" (J. C. Vickerman and N. M. Reed, eds.), p. 163. John Wiley & Sons, Chichester, UK, 1994.

Therefore, the XPS measurements indicate that the films have a high degree of spatial homogeneity.

Thickness by Profilometry

Film thickness is determined for chitin thin films cast onto quartz crystals and onto silicon substrata, using profilometry.[38] White light (200 nm) is simultaneously reflected from the test surface and a reference surface and is recombined to produce a fringence pattern. A vertical scanning piezoelectric transducer synchronized with a charge-coupled device (CCD) camera and frame grabber generates interferograms of the surface and a computer translates these interferograms into a quantitative three-dimensional (3-D) image of surface topography. To obtain a film thickness from the 3-D profile of the surface, the chitin thin films are scored with a 27-guage needle down to the substratum. Because the chitin thin films are transparent, it is necessary to sputter-coat the surfaces with approximately 150 Å of gold. The depth of the score is then measured with a Zygo (Middlefield, CT) NEW View 200 profilometer to obtain the film thickness. A 20× objective is used to obtain data with a vertical resolution of 0.1 nm and a lateral resolution of 0.64 μm. Data are collected at a scan rate of 2.4 μm sec^{-1} and analyzed with Zygo MetroPro software.

The thickness of films cast on silicon coupons from 0.5, 1.0, 1.5, and 2% (w/v) solutions of chitosan is determined (Fig. 3). Chitin film thickness ranges from approximately 3.6 nm for the 0.5% (w/v) chitosan casting solution to 230 nm for the 2% (w/v) solution. Within other constraints it is desirable to maximize film thickness in order to maximize the possible duration of experiments with degradative enzymes. The 2.0% (w/v) solution is difficult to cast consistently because of the high viscosity. Films cast from 1.5% (w/v) solutions are therefore selected for experiments. The average film thickness for chitin thin films cast from 1.5% (w/v) chitosan casting solutions onto the QCM electrodes is 120.0 ± 6.3 nm.

Mass by Quartz Crystal Microbalance

Film mass is measured by QCM.[39] Chitin thin films are prepared on AT-cut quartz crystals with a nominal frequency of 10 MHz (International Crystal Manufacturing, Oklahoma City, OK). The quartz crystals are 20 mm in diameter and 0.167 mm thick, with 100-Å-thick chromium (to enhance gold adhesion), and 1800-Å-thick gold bilayer electrodes. The electrodes on each side of the crystal are of different sizes and are plated prior to purchase. The ground electrode has

[38] F. C. Demarest, *Meas. Sci. Technol.* **9,** 1024 (1998).
[39] K. H. Stellenberger, M. Wolpers, T. Fili, C. Reinartz, T. Paul, and M. Stratmann, *Faraday Discuss.* **107,** 307 (1997).

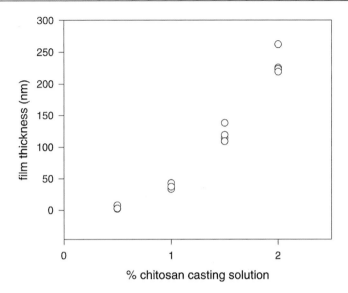

FIG. 3. Thickness of chitin films cast from solutions of various chitosan concentrations.

a diameter of 7 mm; the excited electrode has a diameter of 5 mm. The ground electrode is additionally coated with 1000 Å of silicon dioxide to match the surface chemistry of the solid silica substrata. The crystals are mounted in a spring mount HC-48/u base. A standard laboratory crystal oscillator, model 035360 (International Crystal Manufacturing), is used to excite the quartz crystal. The oscillator is equipped with a BNC output to a frequency counter. A Hewlett-Packard (Palo Alto, CA) 5314A frequency counter is used to measure the resonant frequency changes.

Mass is calculated from the change in frequency of the crystal resonance before and after thin film deposition. Mass measurements are combined with the thickness measurements obtained by profilometry to calculate density of the chitin thin films. The average mass of films is 24.6 ± 0.8 μg cm^{-2}, yielding an average density of 2.05 g cm^{-3}. Preliminary results suggest film thickness may increase as much as 30–40% for every 5° decrease in temperature. This may be caused by an increase in viscosity of the chitosan solutions at lower temperatures.

Intrinsic Background Fluorescence

Microbes attached to opaque materials are typically observed microscopically, using some form of epiillumination. Fluorescence associated with a material can interfere with resolution of micron-sized objects, especially when the underlying

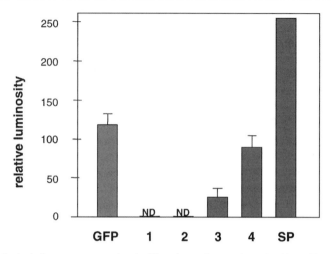

FIG. 4. Intrinsic fluorescence associated with various substrata determined by epiillumination of the sample with excitation at 481 nm and emission at 507 nm. Columns 1–4, chitin films cast from 1, 2, 3, and 4% (w/v) chitosan, respectively. SP, natural squid pen chitin; GFP, representative level of GFP fluorescence from upexpressed S91 cells in a biofilm on a chitin thin film [1.5% (w/v) casting solution] (more detailed information on biofilms is contained in Baty[21] and Baty et al.[22,23] ND, Not detected.

substratum exhibits surface features of a similar scale. Figure 4 shows the level of fluorescence emitted from various substrata, using epiillumination (details of microscopy are described below). Relative luminosity is a measure of total pixel brightness normalized to the maximum brightness possible. No detectable fluorescence is observed for the 1 and 2% (w/v) chitosan-cast chitin thin films (exposure of 6 sec and a gain of 4). Films cast from 3 and 4% (w/v) solutions exhibit some fluorescence. The fluorescence of these films is lower than that of natural squid pen chitin. Typical levels of fluorescence emission displayed by the GFP in biofilms of chitinolytic bacteria are shown for comparison. Details are given below.

Crystal Structure

Chitin is found in nature in three crystalline forms depending on whether the polymer chains are arranged in parallel (α-form), antiparallel (β-form), or as a mixture of parallel and antiparallel (γ-form).[40] The α-chitin form is the most abundant, followed by the β-chitin and γ-chitin forms. To determine the crystal structure thin film X-ray diffraction was performed, with inconclusive results. Infrared spectra of the films suggest the β-form (see the next section, on ATR-FTIR).

[40] R. A. A. Muzzarelli, *in* "Chitin" (R. A. A. Muzzarelli, ed.), p. 24. Permagon Press, Oxford, 1977.

Characterization by Attenuated Total Reflection-Fourier Transform
Infrared Spectroscopy

The infrared (IR) spectrophotometer is a Nicholet 740. Spectra are collected at 4 cm^{-1} resolution. Interferograms are apodized with a Happ–Genzel function. All spectra are collected in attenuated total reflection (ATR) mode. Briefly, the absorbance of IR within an evanescent field is measured. The field penetrates from the internal reflection element (IRE) into the adjacent medium. The penetration is on the order of 1 μm. IREs are germanium trapezoidal prisms with 45° angles, of dimension 0.2 × 5.0 × 2.0 cm (Harrick Scientific). A horizon cell mirror assembly (Harrick Scientific) is used to guide the IR beam into the prism. Exposure of chitin films to aqueous solutions is performed by inserting the prism into a flow chamber previously described.[20] The flow chamber is specially designed to allow microscopic observation of the surface.

Figure 5 shows an IR spectrum of a chitosan film on the IRE (Fig. 5a) and a spectrum of the corresponding chitin film prepared by acetylation (Fig. 5b). Band assignments are summarized in Table II.[41–46] Spectra are presented in units of wavenumbers (cm^{-1}). The composite band in the region from about 950 to 1200 in the spectra of both the chitin and chitosan films originates from pyranose ring vibrations that are, in general, complex.[41,42] This composite band is similar in structure for both the chitosan and chitin films. Because these vibrational modes involve couplings between many atoms in the pyranose ring, some alteration in band structure is expected on modification of a ring functional group through acetylation.

The most pronounced difference between the chitin and chitosan films is the appearance in the chitin spectrum of a composite band with components at 1660 and 1630 and bands at 1560, 1376, and 1308. These band positions are consistent with spectral changes expected on addition of a methyl group to the primary amine of chitosan via a secondary amide linkage (acetylation). The split of the amide I band into composite bands at 1630 and 1660 may be the result of hydrogen bonding of a proportion of the amide linkages, as is described frequently for proteins, with the band at 1630 resulting from a downshift in frequency due to hydrogen bonding of the carbonyl group.[45] With this assumption, using band areas of the component bands, the proportion of amide linkages involved in hydrogen bonding is 60%. The spectrum of cast films resembles β-chitin from *Oligobrachia ivanovi* tube

[41] J. J. Cael, K. H. Gradner, J. L. Koenig, and J. Blackwell, *J. Chem. Phys.* **62**, 1145 (1975).

[42] J. J. Cael, K. H. Gradner, J. L. Koenig, and J. Blackwell, *Biopolymers* **14**, 1885 (1975).

[43] C. J. Pouchert, "The Aldrich Library of Infrared Spectra," 2nd Ed., p. 1576. Aldrich Chemical Company, Milwaukee, Wisconsin, 1978.

[44] J. Blackwell, K. D. Parker, and K. M. Rudall, *J. Mol. Biol.* **28**, 383 (1967).

[45] M. Jackson and H. H. Mantsch, *Crit. Rev. Biochem. Mol. Biol.* **30**, 95 (1995).

[46] T. Miyazawa, T. Shimanouchi, and S.-I. Mizushima, *J. Chem. Phys.* **29**, 611 (1958).

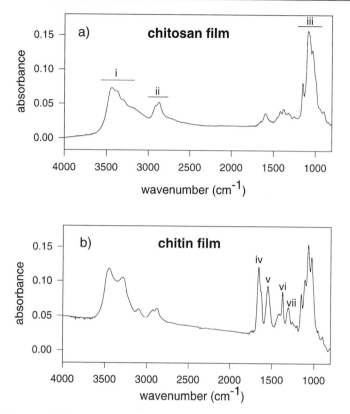

FIG. 5. Spectra of a thin film of chitosan (a) and chitin (b) acquired by ATR-FTIR. Spectral features labeled with roman numerals are referred to in Table II and in text. For structures see Fig. 1.

in some respects.[44] Bands having corresponding positions are evident at 3453, 3291, 1659, approximately 1635 (shoulder), 1550, 1376, 1202, 1154, 1071, 1030, and 953. The relative amplitudes of the main band and shoulder at, respectively, 1659 and 1635 are reversed. The position of the hydrogen-bonded amide I band at 1630 suggests that the pattern of hydrogen bonding resembles β-chitin more closely than α-chitin.[44] The protocol used to deacetylate the chitosan completely removes all acetyl groups according to the FTIR spectra, because there is no indication of amide I and II bands in the chitosan films. The band at 1600 in the chitosan spectrum may originate from NH_2 deformations of the primary amine group.[43]

There are some differences in the chitosan and chitin spectra in the region between 3000 and 3500. This region contains vibrational modes associated with N–H stretches of both primary amines and amides and also O–H stretches. Therefore, acetylation is expected to cause changes in band structure, and both chitosan and

TABLE II

BAND ASSIGNMENTS FOR SPECTRA PRESENTED IN FIG. 5

Figure label	Assignment	Ref.
i	NH stretch of amines and amides, OH stretch	41–43
ii	CH stretch	44
iii	Pyranose ring vibrations	41,42
iv	Amide I	45,46
v	Amide II	45,46
vi	CH bending mode of CH_3	46
vii	Amide III	46

chitin are expected to have overlapping bands in this region. It was not possible to make precise assignments for components of these composite bands.

ATR-FTIR enables direct detection of the acetylation process of chitosan films. Acetylation changes the band structure of the pyranose ring vibrations and the CH stretch region between 2800 and 3000 only slightly, whereas the amide I and II peaks are present only in the chitin films. Therefore, the ratio of the amide I or II bands to either the composite band associated with the pyranose ring, or that associated with the CH stretches, can be used to determine the extent of relative acetylation. This is shown in Fig. 6, where the region of pyranose ring vibrations was used for the normalization. Acetylation of the film is completed within about 24 hr of exposure to acetic anhydride.

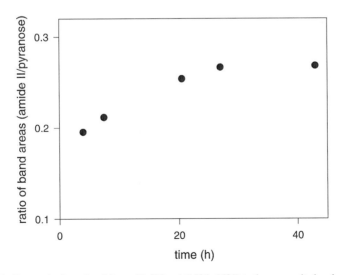

FIG. 6. Changes in the ratio of the amide II band (1504–1601) to the composite band originating from pyranose ring vibrations (981–1183) for times of acetylation in acetic anhydride.

Characterization of Chitinolytic Activity in Biofilms

The characterization of biofilms is described in three publications.[21-23]

Flow Cell

Biofilms are cultured on coupons placed in laminar flow cells (LFCs), based on a previous design.[47] LFCs are constructed of Teflon (McMaster Carr, Los Angeles, CA). The flow channel is 0.8 mm deep, 12 mm wide, and 48 mm in length. Two 1×1 cm squares are recessed, in series, into the floor of each LFC to accommodate silicon, chitin, or squid pen chitin coupons. A no. 2 coverslip (24×60 mm) is used as the viewing window and is sealed against the Teflon, using an oversized Viton gasket and an aluminum coverplate. Teflon influent and effluent lines are connected with Chemfluor PTFE gas fittings (Cole Parmer, Vernon Hills, IL). The LFCs are connected with silicon tubing (Cole Parmer) to both an air-sparged, continuously stirred tank reactor and a 20-liter carboy of artificial seawater. Influent lines are fed through a Buchler 12 roller Multistatic pump (Cole Parmer). The flow dynamics in the LFCs are characterized by injecting a slug dose of crystal violet dye. The flow of the crystal violet moved in a plug flow fashion, suggesting laminar flow of the aqueous phase across the coupons. The flow rate of artificial seawater through each LFC is 0.5 cm^3 min^{-1}, corresponding to a fluid flow velocity of 75 m day^{-1}. The shear stress at the surface under these conditions is equivalent to 0.004 pN μm^{-2} of force. The residence time of the bulk aqueous phase flowing through the LFCs is 55.3 sec.

Transcriptional Activity

GFP fluorescence is used to monitor the promoter activity of chiA. Pseudoalteromonas sp. strain S91 is derived from the wild-type strain S9, a chitinolytic marine bacterium isolated from the surface waters of Botany Bay, New South Wales, Australia in 1981.[48] Strain S91 contains plasmid pDSK519 with a complete chitinase gene (chiA) and a truncated chitin-binding gene (chiB).[49] Strain S91 is streptomycin resistant and the plasmid confers kanamycin resistance. The gfp gene is under the control of both the chiA and chiB promoters.[50,51] This gfp gene encodes the GFPmut2 variant.[52] GFPmut2 excites optimally at 481 nm with an emission maximum at 507 nm.

[47] D. G. Davies and G. G. Geesey, *Appl. Environ. Microbiol.* **61**, 860 (1995).

[48] B. Humphrey, S. Kjelleberg, and K. C. Marshall, *Appl. Environ. Microbiol.* **45**, 43 (1983).

[49] N. T. Keen, S. Tamaki, D. Kobayashi, and D. Trollinger, *Gene* **70**, 191 (1988).

[50] S. Stretton, S. Techkarnjanaruk, A. M. McLennan, and A. E. Goodman, *Appl. Environ. Microbiol.* **64**, 2554 (1998).

[51] S. Techkarnjanaruk and A. E. Goodman, *Microbiology* **145**, 925 (1999).

[52] B. P. Cormack, R. H. Valdivia, and S. Falkow, *Gene* **173**, 33 (1996).

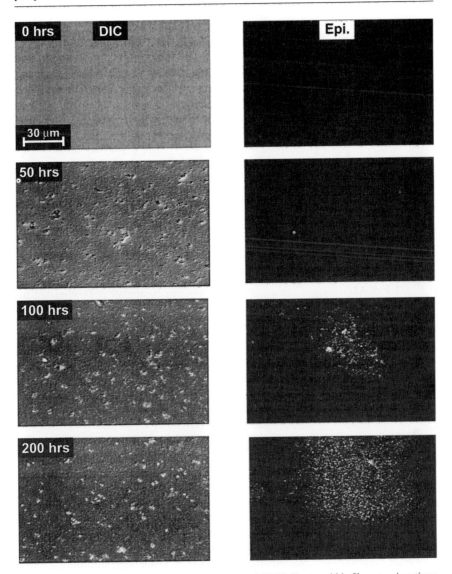

FIG. 7. DIC and epifluorescence (Epi.) micrographs of S91 biofilms on chitin films at various times after flow cell inoculation.

Micrographs of biofilms of S91 cultured on chitin thin films are shown in Fig. 7. DIC is used to visualize total cells, whereas epifluorescence is used to visualize the subset of the population expressing chitinolytic genes.

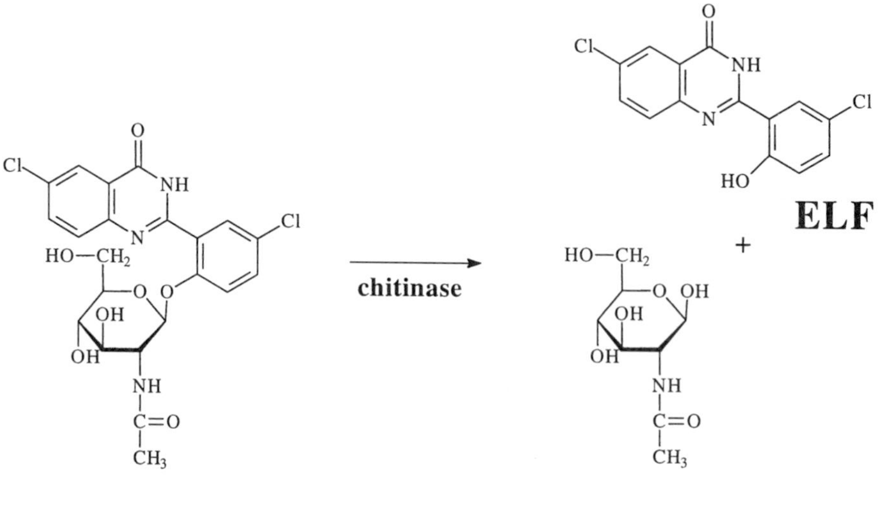

FIG. 8. Enzymatic hydrolysis of ELF-NAG that produces the fluorescent precipitating compound (ELF).

Catalytic Activity

The ELF-NAG fluorogenic probe is used to visualize regions of the coupons exhibiting chitinolytic activity. Enzymatic hydrolysis with pure *Streptomyces griseus* chitinase yields a bright yellow–green precipitate with the excitation (360 nm) and emission (540 nm) wavelengths of the fluorophore, 2-(2'-hydroxy-5'-chloro phenyl)-6-chloro-4(3*H*)-quinazolinone (ELF). The reaction is shown in Fig. 8.

ELF-NAG is dissolved in dimethyl sulfoxide at a concentration of 10 mM. A 50 μM solution of ELF-NAG is prepared by adding a 10-μl volume of the 10 mM stock solution to 2.0 ml of artificial seawater. The mixture is filtered through a 0.2-μm PTFE Millipore membrane to remove any residual crystals, injected into the LFC, incubated in the dark at 20° for 1 hr, and rinsed from the LCF. Areas where the probe has precipitated onto the surface are visualized by epifluorescence microscopy.

ELF-NAG and GFP excite in different ranges. This enables their discrimination in the same field of view. Figure 9 (see color insert) shows image overlays of three types of images. DIC images allow visualization of total cells; epifluorescent images acquired at different excitation wavelengths allow visualization of the level of promoter activity associated with individual cells, and the location of extracellular enzymatic activity, indicated by precipitation of the ELF fluorogenic

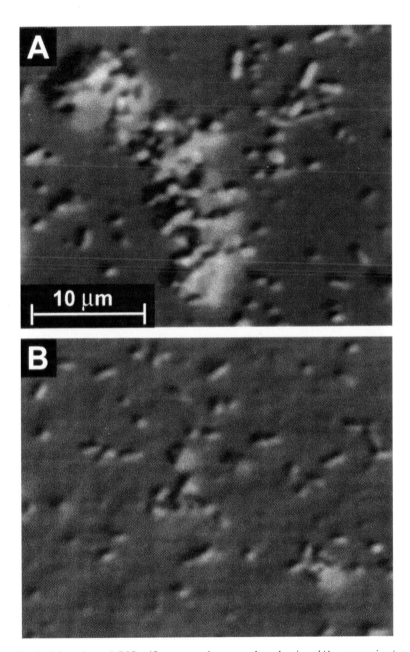

FIG. 9. Color-enhanced, DIC epifluorescence image overlays showing *chiA* upexpression (green pseudocolor) and ELF fluorescence (red pseudocolor) indicating chitinase activity. (A) Chitin film; (B) silicon substratum.

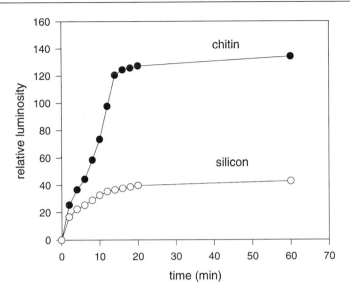

FIG. 10. Kinetics of appearance of fluorescence after exposure of S91 biofilms on chitin films (filled circles) and silicon coupons (empty circles) to the ELF fluorogenic probe. No fluorescence was detected from the abiotic control (data not shown). Fluorescence was detected by epifluorescence microscopy of biofilms cultured for 100 hr.[21–23]

probe onto the substratum. The images reveal that chitinase activity is closely linked to *chiA* upexpressed cells. However, not all cells that were *chiA* upexpressed produced active chitinase.

Epifluorescence microscopy and image analysis are used to monitor the relative rate of conversion of ELF-NAG to the ELF by measuring the relative luminosity of the enzyme active sites on chitin, silicon, and the abiotic silicon control (Fig. 10). The fluorescent product can be detected in as little as 2 min within a surface-associated population of S91 actively involved in chitin degradation. Even on a chitin-free silicon surface S91 exhibits some chitinase activity. The reactivity of ELF-NAG slows considerably by 20 min after inoculation of the enzyme substrate. It is possible that ELF is forming a physical barrier around the active site of the chitinase as it precipitates, thus limiting the mass transport of ELF-NAG to the enzyme.

Substrate Degradation Using Attenuated Total Reflection-Fourier Transform Infrared Spectroscopy

ATR-FTIR can be used to monitor the kinetics of degradation of a chitin film exposed to a solution of chitinase in a flow chamber. Difference spectra are computed as the ratio of all spectra acquired during the time course, to a

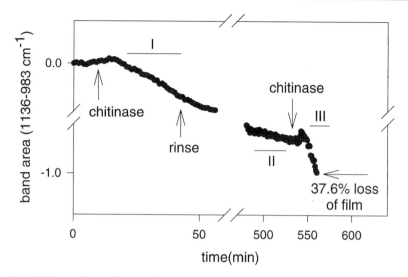

FIG. 11. Changes in negative area of the down-going band associated with the pyranose ring computed from ATR-FTIR difference spectra. Times of introduction of chitinase solution and rinse with PB are indicated. Roman numerals indicate intervals of computation of rates of degradation presented in Table III. Note that the time scale is condensed after the break. Loss of the film was computed on the basis of measurement of dried films as described.

spectrum taken immediately after the flow chamber is filled with buffer. As the film degrades, down-going bands appear in difference spectra as chitin is washed from the interface. The most prominent down-going bands are in the region of the pyranose ring vibrations.

For ATR-FTIR experiments a specially designed flow cell described previously is used.[20] The flow cell allows both ATR-FTIR measurements and microscopic observation of the substratum. The flow rate is 0.5 ml/min. Chitinase is from *S. griseus* (Fluka, Milwaukee, WI). The enzyme activity reported by the distributor is 1 μmol of *N*-acetylglucosamine per minute released per unit (U) at pH 4.4 at 37° with colloidal chitin as substrate (70 U/g). Chitinase solutions are made up in phosphate buffer (PB) consisting of sodium phosphate (pH 4.4) at 1 g/liter. All experiments are conducted at room temperature (approximately 20°).

Figure 11 shows the increase in negative area of down-going bands in the region from 983 to 1136 (cm^{-1}) as a chitin film is rinsed with PB and then exposed to chitinase (0.01 mg/ml). There is a significant delay between the time the enzyme is introduced into the flow chamber and the first indication that chitin is being lost from the substratum. The enzyme must diffuse to the substratum and bind to the chitin substrate before chitin degradation can begin. There may also be a significant delay between initiation of enzyme activity and significant chitin removal from the

TABLE III
RATES OF DEGRADATION OF CHITIN FILM BY CHITINASE

Time period (min)[a]	Rate of loss (μg/min)
15–40 (I)[a]	0.280
490–530 (II)	0.0230
548–560 (III)	0.3025

[a] See Fig. 11.

region of penetration of the evanescent field. The slight increase in the band area during this time period may result from a change in structure of the chitin film as the chitinase adsorbs and begins hydrolyzing the glycosidic linkages. The chitin film continues to degrade after the chitinase is rinsed from the flow chamber, with the rate of loss of the film only decreasing gradually. This may be due in part to chitin degradation associated with adsorbed chitinase that is only rinsed from the surface slowly as the film continues to decompose. The film is rinsed for approximately 8 hr (break in Fig. 11 ordinate and abscissa) and then exposed once again to a PB solution of chitinase (0.01 mg/ml). The film is then allowed to dry and a spectrum is acquired in air (without the flow chamber). According to the decrease in area of the composite band of the pyranose ring vibrations 37.6% of the chitin film is removed during the entire period of the experiment. Rates of degradation of the film during three time intervals were computed (indicated in Fig. 11). The slope of the linear regression for each time interval yields a rate in terms of change in area of the down-going band per minute. Area of the down-going band is converted to mass of the film by using the nominal density and thickness of the films prepared from a 1.5% (w/v) solution of chitin. This yields a rate of degradation in terms of mass of chitin lost from the film per minute. These data are presented in Table III for the three time intervals.

The first interval in Table III corresponds to the first exposure of the film to chitinase. The second interval is the rate of degradation after an 8 hr rinse period. This rate is an order of magnitude less than the rate of degradation during the first exposure to chitinase at 0.01 mg/ml. When the film is exposed a second time to chitinase at 0.01 mg/ml, the rate is increased to a level slightly higher than that of the first interval. The residual rate of degradation of the chitin film, measured for the rinse period between the first and second exposure to chitinase, makes up the balance of the difference between the rates for the first and second time intervals. This implies that there are two processes of degradation occurring during the second exposure to chitinase. Exposure to chitinase presumably disrupts the film by breaking glycosidic linkages and releasing monomeric N-acetylglucosamine. It may be that the film is destabilized by the first exposure to chitinase, and that loss of the film proceeds gradually even in the absence of the enzyme. For example, the packing of the chains may be disrupted by chitinase, allowing portions of the film to detach during exposure to flowing buffer even though the chitinase activity has become negligible.

The rates of degradation of the chitin film are comparable to the rate expected for degradation of colloidal chitin by chitinase from *S. griseus*. Assuming that activity is proportional to units of enzyme, the estimated activity of a 0.01-mg/ml solution of chitinase is 7×10^{-10} mol of *N*-acetylglucosamine per minute or 0.15 μg/min. The rates would not be expected to be exactly the same because experimental conditions differ. Most pronounced differences are the temperature (20 vs 37°), and the fact that enzyme is continually replenished in the present experiments by flowing the solution over the surface.

Synthesis of ELF-97-*N*-acetyl-β-D-glucosaminide Fluorogenic Probe

At the time the methodologies described here were being developed ELF-NAG was not available commercially. Therefore, a synthesis was designed and executed.[21] ELF-NAG is currently available (Molecular Probes, product E-22011). The major steps for probe synthesis are outlined in Fig. 12.

All organic chemicals are purchased from Aldrich (Milwaukee, WI), except for 5-chloro-2-aminobenzoic acid, which is purchased from Arcos Organics (Pittsburgh, PA). Nuclear magnetic resonance (NMR) spectra are obtained on a Bruker YLIV370.040.

Compound **I** (Fig. 12) (2-acetamido-2-deoxy-β-D-glucopyranose-1,3,4,6-tetraacetate) is prepared as follows. The O-acetylation and halogenation of *N*-acetyl-β-D-glucosamine are performed by a modification of published methods.[53] *N*-Acetyl-β-D-glucosamine (20 g) is dried and triturated and slowly dissolved in a mixture of acetic anhydride (100 ml) and concentrated sulfuric acid (13.0 ml) on ice. The mixture is held for 24 hr at room temperature and then poured into a cold solution of anhydrous sodium acetate (82 g) in water (520 ml) while stirring. The solution is then extracted with chloroform (3×100 ml). The chloroform extracts are combined and washed successively in ice-cold water (2×200 ml), ice-cold saturated sodium bicarbonate (3×200 ml), and ice-cold water (2×200 ml). The chloroform is removed with a rotary evaporator and dried over anhydrous sodium sulfate. The product is then further dried under reduced pressure over phosphorous pentoxide. The final product, 2-acetamido-2-deoxy-β-D-glucopyranose-1,3,4,6-tetraacetate (Fig. 12, compound **I**), is then dissolved and recrystallized from ethyl acetate.

The compound 2-acetamido-2-deoxy-β-D-glucopyranose-1,3,4,6-tetraacetate (10.0 g) is dissolved in a solution of acetic acid (120 ml) and acetic anhydride

[53] T. Inabe, T. Ohgushi, Y. Iga, and E. Hasegawa, *Chem. Pharm. Bull.* **32**, 1597 (1984).

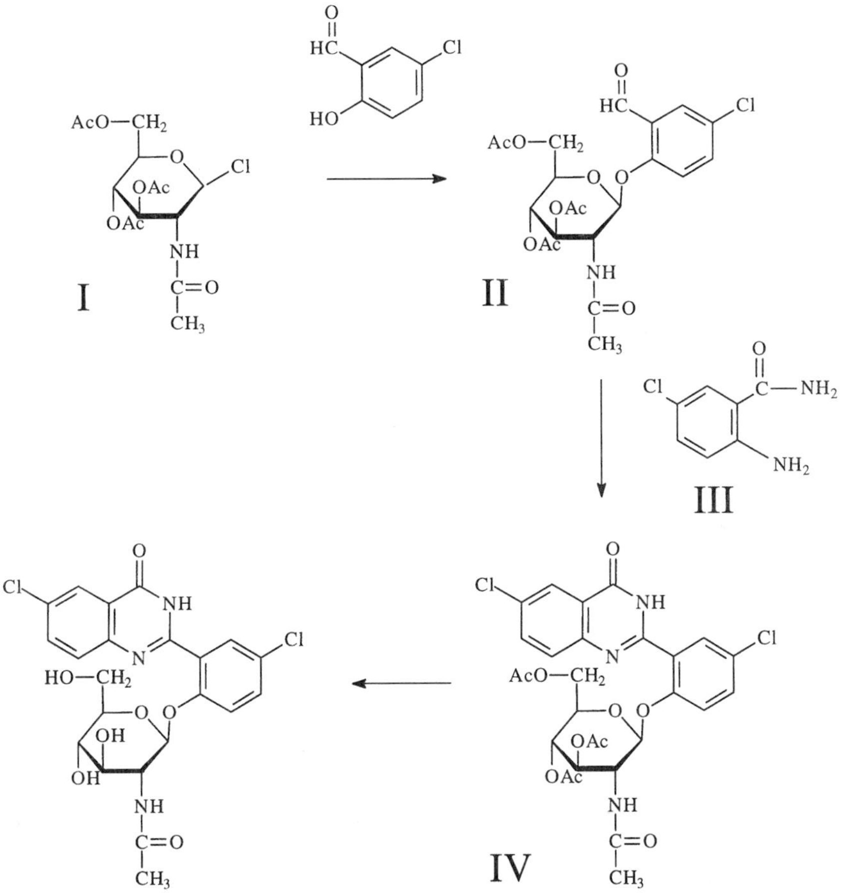

FIG. 12. Major steps in probe synthesis. Roman numerals refer to compounds mentioned in text. For simplicity hydrogen atoms are not shown. Ac, Acetyl group.

(1.0 ml) and cooled to 4°. The mixture is then saturated with dry hydrochloride gas, using a fine glass air stone while stirring on ice. The reaction mixture is then tightly sealed and held at room temperature for 48 hr while stirring. The reaction mixture is poured into chloroform (400 ml) and washed successively in ice-cold water (2 × 200 ml), ice-cold saturated sodium bicarbonate (3 × 200 ml), and ice-cold water (2 × 200 ml). The chloroform is removed with a rotary evaporator to a volume of 10–12 ml. The remaining solvent is removed under a gentle stream of dry nitrogen. The product is dried over anhydrous sodium sulfate and then further dried under reduced pressure over phosphorous pentoxide. The final product, 1-chloro-2-acetamido-2-deoxy-*b*-D-glucopyranose-3,4,6-tetraacetate, is redissolved and recrystallized from chloroform–diethyl ether.

Compound **II** (Fig. 12) [4-chloro-2-formylphenyl-(2-acetamido-2-deoxy-β-D-glucopyranose-3,4, 6-tetraacetate)] is prepared as follows. Compound **I** (3.7 g) and 4-chlorosalicylaldehyde (2.5 g) are dissolved in anhydrous dichloromethane (50 ml). To the solution is added at once Ag_2CO_3 (4.4 g) and 3-Å molecular sieves (5 g). 2,4,6-Collidine (2.1 ml) is added, dropwise, with stirring. The resulting mixture is stirred in the dark, at room temperature for 4 days under dry argon protection, and then diluted with chloroform (150 ml). The mixture is filtered through a pad of Celite diatomaceous earth and the residue is washed with chloroform (3 × 50 ml). The combined filtrates are washed successively with 1 M HCl solution (2 × 250 ml), water (1 × 250 ml), 0.1 M $Na_2S_2O_3$ solution (2 × 250 ml), saturated $NaHCO_3$ solution (2 × 250 ml), and water (2 × 250 ml). The organic layer is dried over anhydrous $MgSO_4$, and evaporated *in vacuo* to give an off-white solid. The crude material is purified on a silica gel column, using a gradient elution of chloroform and acetonitrile, to yield the final product, 4-chloro-2-formylphenyl-(2-acetamido-2-deoxy-β-D-glucopyranose-3,4,6-tetraacetate) (Fig. 12, compound **III**), a colorless solid (3.7 g; yield, 84%).

Compound **III** (Fig. 12) (4-chloroanthranilamide) is prepared by methods described previously.[54] 5-Chloro-2-aminobenzoic acid (anthranilic acid) (5.0 g) is dissolved in tetrahydrofuran (100 ml). The mixture is cooled to 0° and saturated with dry ammonia gas for 2 hr while stirring. The reaction mixture is then allowed to warm to room temperature and is stirred overnight. The tetrahydrofuran is removed by rotary evaporation. The product, 4-chloroanthranilamide, is recrystallized from toluene.

Compound **IV** (Fig. 12) (4-chloro-2-[2′-(6″-chloro-4(3*H*)-quinazolinonyl)]-phenyl-2-acetamido-2-deoxy-β-D-glucopyranose-3,4,6-tetraacetate) is prepared as follows. Compound **II** (741 mg), compound **III** (297 mg), and *p*-toluenesulfonic acid (20 mg) are dissolved in anhydrous methanol (30 ml) and stirred at room temperature for 2 hr. The reaction mixture is then refluxed for another 1 hr. The resulting solution is cooled to 0° and 2,3-dichloro-4,5-dicyano-1,4-benzoquinone (386 mg) is added in three portions over the course of 30 min. The mixture is stirred at 0° for 2 hr, and evaporated *in vacuo* to give a brown solid. The solid is washed with benzene (5 × 50 ml) to remove the hydroquinone formed during the reaction. The resulting residue is dissolved in methanol and precipitated by ether. The crude material is purified on silica gel, using a gradient elution of chloroform and methanol to yield the final product, 4-chloro-2-[2′-(6″-chloro-4(3*H*)-quinazolinonyl)]-phenyl-2-acetamido-2-deoxy-β-D-glucopyranose-3,4,6-tetraacetate, a colorless solid, (874 mg; yield, 76%).

[54] Z. Diwu, Y. Lu, R. H. Upson, M. Zhou, D. H. Klaubert, and R. P. Haugland, *Tetrahedron* **53**, 7159 (1997).

Compound **IV** is deprotected to yield the enzyme substrate ELF-97-*N*-acetyl-β-D-glucosaminide. Compound **IV** (800 mg) is dissolved in anhydrous methanol (10 ml) and cooled to 0°. Freshly prepared 0.1 *M* sodium methoxide in anhydrous methanol (3 × 0.1 ml) is slowly added to the solution in three portions over 2 hr under dry argon protection. The solution is stirred at 0° for 2 hr, warmed to room temperature, and stirred for 4 hr. The reaction mixture is then diluted with methanol (50 ml), and acidified (pH 6.0) with Amberlite IRC-50 ion-exchange resin (H$^+$ form). The mixture is filtered, and the resin is washed with methanol (2 × 25 ml). The combined filtrate is evaporated *in vacuo* to give an off-white solid. The crude material is dissolved in methanol, and precipitated by ether to afford the final product, ELF-97-*N*-acetyl-β-D-glucosaminide, an off-white solid (1.4 g; yield, 94%); ^{1}H-NMR (DMSO-d$_6$): 12.20 (1H, s); 8.1 (1H, s); 7.85, 7.28 (2H, dd); 7.70, 7.47 (2H, dd); 7.60 (1H, s); 5.10 (1H, m); 5.00 (1H, m); 4.92 (1H, m); 4.60 (1H, m); 4.30 (1H, m); and 3.10–3.90 ppm (9H, m).

Summary

Extracellular enzymes produced by bacterial biofilms tend to become an integral, permanent part of the biofilm/substratum system. Thus, characterizing extracellular enzyme activity is an essential component of understanding biofilm ecology. Methods have been presented for characterizing three aspects of extracellular enzyme activity in biofilms: promoter activity of the structural gene, local catalytic activity, and kinetics of collective substrate degradation. The abundance of intracellular transcript derived from a structural gene is only indirectly related to the magnitude of catalytic activity of the corresponding enzyme. This relationship may be particularly tenuous in the case of extracellular enzymes, which must be transported out of the cell in order to become active. Fluorogenic substrates that allow direct detection of an increasingly greater variety of enzyme activities are becoming available. There are technical problems, originating from surface roughness and intrinsic fluorescence, associated with microscopic examination of biofilms on natural materials. Thin films provide one option for acquiring data about biofilms colonizing relevant materials.

Acknowledgments

This work was sponsored by the National Science Foundation (OCE 9720151, Geesey and EEC 8907039, cooperative agreement), Office of Naval Research grant N00014-97-1-1062, and National Institute of Dental and Craniofacial Research grant N01-DE-62611. A portion of this work was performed in the Environmental Molecular Sciences Laboratory, a national scientific user facility sponsored by the Department of Energy's Office of Biological and Environmental Research located at Pacific Northwest National Laboratory.

[25] Isolation and Biochemical Characterization of Extracellular Polymeric Substances from *Pseudomonas aeruginosa*

By Jost Wingender, Martin Strathmann, Alexander Rode, Andrew Leis, and Hans-Curt Flemming

Introduction

Biofilms of *Pseudomonas aeruginosa* have been intensively studied during the last decade.[1] *Pseudomonas aeruginosa* offers several advantages as a model organism in biofilm research. This gram-negative bacterium is well characterized with respect to its molecular genetics, biochemistry, and physiology; it is also of hygiene relevance as an opportunistic pathogen, because biofilms harboring this bacterium in technical water systems and on medical devices may be a source of human infections.[1,2] In general, the extracellular polymeric substances (EPS) of *P. aeruginosa* predominantly consist of different polysaccharides and proteins.[3–5] Mucoid variants of *P. aeruginosa* are characterized by an overproduction of the viscous exopolysaccharide alginate, resulting in the production of large slimy colonies, when the bacteria are cultivated overnight on common agar-based media.[6,7] Alginates from *P. aeruginosa* are high molecular weight unbranched copolymers consisting of $(1 \rightarrow 4)$-linked uronic acid residues of β-D-mannuronate and α-L-guluronate.[8] These components are arranged in homopolymeric blocks of poly-β-D-mannuronate and heteropolymeric sequences with random distribution of mannuronate and guluronate residues.[9] *Pseudomonas aeruginosa* alginates are partially O-acetylated at the C-2 and/or C-3 positions of the mannuronate residues.[9] Mucoid strains of *P. aeruginosa* have been described to occur commonly in chronic infections of the lungs of cystic fibrosis patients,[10] where they grow in microcolonies on the host epithelium; aquatic biofilms in natural

[1] J. W. Costerton, P. S. Stewart, and E. P. Greenberg, *Science* **284**, 1318 (1999).
[2] K. Botzenhart and G. Döring, *in* "*Pseudomonas aeruginosa* as an Opportunistic Pathogen" (M. Campa, M. Bendinelli, and H. Friedman, eds.), p. 1. Plenum Press, New York, 1993.
[3] V. Sherbrock-Cox, N. J. Russell, and P. Gacesa, *Carbohydr. Res.* **135**, 147 (1984).
[4] N. W. Ross, R. Levitan, J. Labelle, and H. Schneider, *FEMS Microbiol. Lett.* **81**, 257 (1991).
[5] N. Marty, J.-L. Dournes, G. Chabanon, and H. Montrozier, *FEMS Microbiol. Lett.* **98**, 35 (1992).
[6] J. R. W. Govan, *in* "*Pseudomonas* Infection and Alginates" (P. Gacesa and N. J. Russell, eds.), p. 50. Chapman and Hall, London, 1990.
[7] S. Grobe, J. Wingender, and H. G. Trüper, *J. Appl. Bacteriol.* **79**, 94 (1995).
[8] L. R. Evans and A. Linker, *J. Bacteriol.* **116**, 915 (1973).
[9] G. Skjåk-Bræk, H. Grasdalen, and B. Larsen, *Carbohydr. Res.* **154**, 239 (1986).
[10] J. R. W. Govan and V. Deretic, *Microbiol. Rev.* **60**, 539 (1996).

ecosystems and in technical water systems represent habitats of mucoid bacteria.[7,11] The formation of biofilms is regarded as part of a natural "life cycle" of mucoid *P. aeruginosa*.[12] Alginates represent major components of the EPS of mucoid *P. aeruginosa* and have been implicated in the development as well as the maintenance of the mechanical stability of biofilms formed by *P. aeruginosa* on living and abiotic surfaces.[12]

Studies of the EPS from mucoid isolates frequently focus on the characterization of alginate. However, the role of other EPS components also must be considered for a comprehensive understanding of the physicochemical properties of biofilms, the interactions between EPS components, and the biological functions of EPS within the matrix of biofilms. In this chapter methods of isolation and biochemical characterization are described for both whole EPS and alginate from mucoid strains of *P. aeruginosa* grown as a confluent bacterial lawn on agar media as simple *in vitro* model systems for bacterial biofilms.

Isolation of Extracellular Polymeric Substances

Bacterial Strains

Most studies on agar-grown biofilms were performed using environmental mucoid strain *P. aeruginosa* SG81, which was isolated from the biofilm in a technical water system.[7] This strain is a stable alginate-producing strain that forms strongly mucoid colonies on standard agar media. In some experiments, the clinical mucoid strain *P. aeruginosa* FRD1 was compared with two mutant strains (FRD1152 and FRD1153) that were defective in O-acetylation of alginate[13]; these strains were used to investigate the role of acetyl groups in EPS properties and in biofilm formation.

Biofilm Growth on Agar Media

The growth medium for the cultivation of biofilms on agar surfaces is *Pseudomonas* isolation agar (PIA; Difco, Detroit, MI) containing 2% (v/v) glycerol as an additional carbon source that promotes alginate slime production. It is recommended that the glycerol concentration be reduced to 0.2% (v/v) when the biofilm material is to be analyzed by ^{13}C nuclear magnetic resonance (NMR) spectroscopy, because medium-derived traces of glycerol give signals that may interfere with the

[11] J. W. Costerton, M. R. W. Brown, J. Lam, K. Lam, and D. M. G. Cochrane, *in* "*Pseudomonas* Infection and Alginates" (P. Gacesa and N. J. Russell, eds.), p. 76. Chapman and Hall, London, 1990.

[12] D. G. Davies, *in* "Microbial Extracellular Polymeric Substances" (J. Wingender, T. R. Neu, and H.-C. Flemming, eds.), p. 93. Springer-Verlag, Berlin, 1999.

[13] M. J. Franklin and D. E. Ohman, *J. Bacteriol.* **175,** 5057 (1993).

analysis of biofilm components[14]; alginate production is only slightly reduced under these culture conditions. Each petri dish (diameter, 9 cm) contains 25 ml of agar medium. Careful predrying of the agar plates is necessary to avoid sticking of the cover to the bottom half of the petri dish during the incubation, caused by residual humidity. The cover should be loosely placed on the petri dish; we have observed that agar plates with firmly attached covers yield poor growth and weak slime formation, probably because of insufficient aeration.

The bacteria are pregrown overnight on PIA at 36°. Single colonies are suspended in 5 ml of sterile 0.14 M NaCl solution to a cell density of approximately 1×10^8 cells/ml. Aliquots (0.1 ml) are plated on PIA and incubated for 24 hr at 30 or 36°. Under these conditions, a confluent and heavily mucoid bacterial lawn is obtained.

Recovery of Extracellular Polymeric Substances

The confluent slimy growth on the surface of the agar medium is carefully removed with a sterile metal spatula and suspended in sterile 0.14 M NaCl (mass ratio of biofilm wet weight to NaCl solution of 1 : 16). This biofilm suspension is stirred vigorously on a magnetic stirrer for 30 min at room temperature. Complete dispersal of bacterial cells is confirmed microscopically. Cell numbers are determined in a Thoma counting chamber under a phase-contrast microscope. When biofilm material is recovered from calcium-containing agar media, the suspension medium additionally contains 1 mM EDTA to ensure complete dispersal of bacteria.

Extracellular material is removed from bacteria by centrifugation at 40,000g for 2 hr at 10°. High-speed centrifugation represents a physical method often used to separate soluble (slime) EPS from bacterial cells of pure cultures.[15] Using this method, ≥85% of uronic acids (alginate) as marker substances of EPS are removed from cells of mucoid environmental strain *P. aeruginosa* SG81 (Table I)[14] as well as of clinical mucoid strain FRD1 and its acetylation-defective mutants FRD1152 and FRD1153.[16] However, for removal of more firmly cell-associated EPS other separation methods must be considered. For example, EPS cross-linked by divalent cations in bacterial biofilms grown on calcium-containing agar media can be released with complexing agents (EDTA, EGTA) or cation-exchange resins such as Dowex 50 × 8 (sodium form), promoting the solubilization of the EPS matrix.[15] After the centrifugation step, residual bacteria are removed from the supernatants by filtration through cellulose acetate membranes (pore diameter, 0.2 μm). The

[14] C. Mayer, R. Moritz, C. Kirschner, W. Borchard, R. Maibaum, J. Wingender, and H.-C. Flemming, *Int. J. Biol. Macromol.* **26**, 3 (1999).

[15] P. H. Nielsen and A. Jahn, *in* "Microbial Extracellular Polymeric Substances" (J. Wingender, T. R. Neu, and H.-C. Flemming, eds.), p. 49. Springer-Verlag, Berlin, 1999.

[16] P. Tielen, J. Wingender, K.-E. Jaeger, and H.-C. Flemming, in preparation (2001).

clear filtrates represent sterile solutions of extracellular low molecular weight substances (e.g., pigments) and high molecular weight EPS. For the removal of low molecular weight material, the cell-free supernatants are dialyzed against 5 liters of deionized water for 1 hr at room temperature and subsequently against another change of 5 liters of deionized water overnight at 4°. The dialysis tubings (Visking dialysis tubing made of regenerated cellulose; Serva, Heidelberg, Germany) have a molecular weight cutoff of 12,000 to 14,000. Finally, EPS are concentrated by freeze-drying.

To evaluate the extent of cell damage or cell lysis resulting from a particular EPS isolation procedure, we recommend determining the activity of the intracellular marker enzyme glucose-6-phosphate dehydrogenase (G6PDH; EC 1.1.1.49) in the EPS preparations. We use a G6PDH assay that was originally described for a laboratory strain of *P. aeruginosa*.[17] The reaction conditions of maximal enzyme activity determined for this strain (pH 8.6, 37°, NADP as coenzyme) also proved to be optimal for mucoid strain *P. aeruginosa* SG81 used in our studies. G6PDH activity is assayed as follows: 0.05 ml of sample solution is added to 1.45 ml of substrate solution consisting of 120 mM Tris buffer, pH 8.6, 0.5 ml; 20 mM glucose 6-phosphate, 0.375 ml; 10 mM β-NADP, 0.25 ml; 250 mM MgCl$_2$, 0.09 ml; and deionized water, 0.235 ml. The mixture is incubated at 37° and the absorbance at 340 nm (A_{340}) is measured at regular intervals for up to 60 min. We use a commercial G6PDH from *Leuconostoc mesenteroides* (Sigma, St. Louis, MO) as a positive control that is run in parallel with the samples to be analyzed. In cell-free EPS solutions obtained by the centrifugation method described above, no increase in A_{340} was observed over 60 min, whereas in cell extracts obtained by ultrasonic disruption of the bacteria, significant G6PDH activity can be measured. Thus, the procedure of EPS isolation by centrifugation and subsequent membrane filtration gave no indication of the release of G6PDH into the extracellular material. In addition, only minor amounts of DNA ($\leq$0.2 μg/ml) were measured in cell-free EPS solutions, using the DNA-binding fluorochrome PicoGreen [PicoGreen double-stranded DNA (dsDNA) quantitation kit; Molecular Probes, Eugene, OR]. This confirmed that no substantial cell lysis occurred during the EPS isolation procedure described above.

Purification of Alginate

Various methods have been described in the literature for the isolation and purification of bacterial alginates. In general, crude exopolysaccharide is precipitated from culture supernatants by organic solvents and treated with nucleases and proteases to remove contaminating nucleic acids and proteins. Sometimes, alginate is purified by ion-exchange chromatography.[3] We use a selection of these

---

[17] F. M.-W. Ng and E. A. Dawes, *Biochem. J.* **132,** 129 (1973).

methods for the isolation of alginate from mucoid strains. Biofilm suspensions from agar-grown bacteria are prepared as described above. The confluent growth from 10 PIA plates is suspended in 100 ml of 0.14 M NaCl. After centrifugation of the biofilm suspension at 20,000g for 1 hr, followed by centrifugation of the supernatant at 40,000g for 2 hr, the crude exopolysaccharide is precipitated by addition of 3 volumes of ice-cold 95% (v/v) ethanol. After 30 min in an ice bath, the precipitate is collected on a glass-sintered filter, washed twice in 95% (v/v) ethanol and once in absolute ethanol, and dried over P_2O_5 in vacuum. The material is redissolved at a concentration of 2.5 mg/mL in 50 mM Tris-HCl buffer (pH 7.5) with 10 mM MgCl$_2$ and incubated in the presence of DNase I (10 μg/ml; Sigma) and RNase A (10 μg/ml; Sigma) at 36° for 4 hr. Alternatively, Benzonase (5 units/ml; Merck, Darmstadt, Germany) may be substituted for both nucleases, using a similar buffer with 2 mM MgCl$_2$; this endonuclease degrades both DNA and RNA. After addition of pronase E (12.5 μg/ml; Sigma) the mixture is incubated at 36° overnight, centrifuged at 40,000 g for 30 min, and dialyzed against two changes (5 liters each) of deionized water. The purified polysaccharide is precipitated with ethanol as described above, redissolved in deionized water, and recovered by freeze-drying.

Quantitation of Extracellular Polymeric Substance Components

Proteins

Essentially three different assays are applied for the determination of proteins in biological samples. A common method is the Lowry assay.[18] It is based on the reaction of peptide bonds with copper in alkaline solution (biuret reaction) followed by reduction of phosphomolybdic and phosphotungstic acids in the Folin–Ciocalteu phenol reagent. The bicinchoninic acid (BCA) assay[19] also involves the biuret reaction as the first step in protein quantitation, but BCA is used in the second step as detection reagent. The Bradford assay is based on the binding of Coomassie Brilliant Blue G-250 to proteins, resulting in a spectral shift of the absorbance maximum of the dye.[20] In all assays, color development is measured spectrophotometrically. The Lowry and Bradford assays are the most widely used methods for the quantitation of proteins in biofilms and EPS derived from biofilms. The BCA assay is also reactive with reducing sugars; thus, interference from carbohydrates (e.g., monosaccharides, reducing end groups of polysaccharides) as common constituents of EPS may occur when using the BCA assay. Many modifications of the Lowry and the Bradford methods have been described in the

[18] O. H. Lowry, N. J. Rosebrough, A. L. Farr, and R. J. Randall, *J. Biol. Chem.* **193**, 265 (1951).
[19] P. K. Smith, R. I. Krohn, G. T. Hermanson, A. K. Mallia, F. H. Gartner, M. D. Provenzano, E. K. Fujimoto, N. M. Goeke, B. J. Olson, and D. C. Klenk, *Anal. Biochem.* **150**, 76 (1985).
[20] M. M. Bradford, *Anal. Biochem.* **72**, 248 (1976).

TABLE I

COMPOSITION OF EXTRACELLULAR POLYMERIC SUBSTANCE FROM AGAR
MEDIUM-GROWN BIOFILMS OF *Pseudomonas aeruginosa* SG81

Component	Biofilm (μg/10^9 cells)	EPS[a] (μg/10^9 cells)	Proportion of EPS in biofilm (%)
Total carbohydrates	1005.8	766.6	76.2
Uronic acids (alginate)	473.8	402.8	85.0
Proteins	585.0	266.4	45.5

[a] The EPS components were quantitated in cell-free solutions obtained after removal of biofilm bacteria by centrifugation and membrane filtration of the supernatants.

literature; all three protein assays are commercially available in the form of kits or, alternatively, the single reagents can be obtained separately.

For protein quantitation in EPS from agar-grown *P. aeruginosa* biofilms, we routinely perform the Lowry assay with commercial reagents: modified Lowry reagent [alkaline cupric tartrate containing 78.8 m*M* sodium dodecyl sulfate (SDS); Sigma] and a Folin–Ciocalteu phenol reagent (Sigma). Working solutions of the reagents are prepared according to the manufacturer. Lowry reagent (0.5 ml) is added to 0.5 ml of sample solution. After 20 min at room temperature, 0.25 ml of Folin–Ciocalteu phenol working solution is added. After another 30 min at room temperature, the absorbance at 750 nm is measured against deionized water in a spectrophotometer. All samples are run in triplicate. Bovine serum albumin (BSA, fraction V; Sigma) is used as a protein standard. A linear standard curve (correlation coefficient ≥ 0.99) is obtained only at BSA concentrations ≤ 60 μg/ml. The deviation from linearity at higher protein concentrations was also observed with cell-free EPS solutions. Biofilm and EPS samples should be diluted to fall into the linear range of the calibration curve with BSA. An advantage of this procedure is the concomitant dilution of polysaccharides such as alginate to levels that do not interfere with the protein assay. Using the Lowry assay, substantial amounts of protein can be detected in the EPS of mucoid bacteria as shown for environmental strain *P. aeruginosa* SG81 in Table I.

The Bradford method (Bio-Rad protein assay; Bio-Rad, Hercules, CA) yielded considerably lower values of proteins in EPS from *P. aeruginosa* SG81 than the Lowry assay. Low protein concentrations determined by the Bradford assay in contrast to the Lowry method have also been described in EPS from activated sludge.[21] This difference may be due to the presence of small peptides, which are not measured by the Bradford assay; the Lowry assay allows the determination

[21] B. Frølund, R. Palmgren, K. Keiding, and P. H. Nielsen, *Wat. Res.* **30,** 1749 (1996).

of peptides/proteins with two or more peptide bonds, whereas the Bradford assay only measures proteins with a minimum of eight to nine peptide bonds.

Total Carbohydrates

Total carbohydrates in the EPS include both neutral and charged polysaccharides. They can be quantitated by using the phenol–sulfuric acid method according to Dubois et al.[22] After mixing 0.5 ml of sample solution with 0.5 ml of an aqueous 5% (w/v) phenol solution in a glass test tube, 2.5 ml of concentrated sulfuric acid is added rapidly. The tubes are vortexed and the mixtures are incubated for 10 min at room temperature followed by incubation for 15 min at 30° in a water bath. After cooling to room temperature for 5 min, absorbances of the solutions are measured at 490 nm (neutral polysaccharides) or 480 nm (acidic polysaccharides). All samples are run in triplicate. For the analysis of EPS from mucoid strains of *P. aeruginosa,* calibration curves are prepared with bacterial alginate (200 μg/ml stock solution) purified as described above.

Uronic Acids

A widely used method for the determination of uronic acids employs the *m*-hydroxydiphenyl reagent according to Blumenkrantz and Asboe-Hansen[23]; however, interference from neutral sugars may be substantial, resulting in a decrease in sensitivity of the assay and a deviation from linearity of standard curves at high ratios of neutral sugar to uronic acid residues in bacterial heteropolysaccharides such as xanthan. To avoid this interference from neutral carbohydrates, a modified sulfamate/*m*-hydroxydiphenyl assay according to Filisetti-Cozzi and Carpita[24] can be used for the quantitation of uronic acids (alginate) in EPS preparations from mucoid *P. aeruginosa.* In glass test tubes, 40 μl of 4 *M* sulfamic acid–potassium sulfamate reagent (aqueous solution of sulfamic acid with pH value adjusted to pH 1.6 with saturated KOH) is added to 0.4 ml of sample solution. After thorough mixing, 2.4 ml of sodium tetraborate reagent (75 m*M* sodium tetraborate in concentrated sulfuric acid) is added and the mixture is vortexed vigorously. The solutions are heated in a water bath at 100° for 20 min and then cooled in an ice bath for 5 min. Subsequently, 80 μl of 0.15% (w/v) *m*-hydroxydiphenyl in 0.5% (w/v) NaOH is added and the mixture is vortexed vigorously. After 10 min at room temperature, the absorbance of the pink-colored solution is read at 525 nm. All samples are run in triplicate. Calibration curves are prepared with bacterial alginate (200 μg/ml stock solution) purified as described above.

[22] M. Dubois, K. A. Gilles, J. K. Hamilton, P. A. Rebers, and F. Smith, *Anal. Chem.* **28,** 350 (1956).
[23] N. Blumenkrantz and G. Asboe-Hansen, *Anal. Chem.* **54,** 484 (1973).
[24] T. M. C. C. Filisetti-Cozzi and N. C. Carpita, *Anal. Biochem.* **197,** 157 (1991).

As shown in Table I uronic acids make up a large proportion of total carbohydrates in the EPS from the environmental mucoid strain *P. aeruginosa* SG81, but apparently additional polysaccharides are also present. This observation has also been reported for clinical mucoid strains, whose exopolysaccharides consisted of an alginate fraction and a fraction of non-uronic acid-containing polysaccharides.[3,5]

Acetyl Groups

The acetyl content of alginate is determined by a photometric assay according to Hestrin.[25] A working reagent is freshly prepared by mixing equal volumes of 2 M hydroxylamine and 3.5 M NaOH. Sample solution [0.5 ml of alginate (2 mg/ml) in 1 mM sodium acetate buffer, pH 4.5] is mixed with 1 ml of working reagent. After 1 min at room temperature, 0.5 ml of 4.2 M HCl is added. After mixing, 0.5 ml of a ferric chloride solution (0.37 M FeCl$_3$ in 0.1 M HCl) is added and the absorbance at 540 nm is read immediately. Calibration curves are prepared with acetylcholine chloride (5 mM stock solution) or glucose pentaacetate (1 mM stock solution). Using this assay, the acetyl content of alginate samples from 10 mucoid environmental isolates was shown to vary between 4.9 and 9.7% (w/w), corresponding to a molar ratio of acetate to uronate ranging from 0.17 to 0.33.[7] This method has been applied to identify acetylation-defective mutant strains of *P. aeruginosa* that produced alginates with a reduced acetyl content.[13]

Characterization of Extracellular Polymeric Substances

Gel Filtration

Gel filtration is a suitable method to gain information about the molecular weight distribution of biopolymers. Sephacryl S-500 HR (Amersham Pharmacia, Piscataway, NJ) reveals a fractionation range that allows the study of EPS and purified alginates from mucoid *P. aeruginosa*. For qualitative analysis, a column with a gel bed of 1 × 37 cm is used. The column is equilibrated at room temperature with 10 mM phosphate buffer (pH 7.5) containing 0.2 M NaCl. Samples (1 ml) are cell-free EPS solutions or solutions of purified alginate (1.5 mg/ml). Elution is performed at a flow rate of 40 ml/hr, using 10 mM phosphate buffer (pH 7.5) containing 0.2 M NaCl as eluent. Each fraction (0.6 ml) is assayed for total carbohydrates, uronic acids, or proteins as described above. The void volume of the column is determined with a high molecular weight dextran (mean molecular weight, 5 × 10^6– 40 × 10^6; Sigma).

Application of this method to the analysis of EPS from mucoid strains of *P. aeruginosa* shows qualitative differences in the elution behavior of the EPS

[25] S. J. Hestrin, *J. Biol. Chem.* **180**, 249 (1949).

components. Uronic acids (alginate) and proteins in the EPS from mucoid environ-mental and clinical strains reveal a heterogeneous elution profile; both components coeluted over a wide range of fractions; this may be an indication of interactions between the polymeric components. In contrast, EPS from acetylation-defective mutant strains could be fractionated into two distinct peaks, the first peak repre-senting high molecular weight alginate molecules and the second peak consisting of lower molecular weight proteins.[16]

Gel filtration on Sephacryl S-500 HR has also been used for the analysis of bacterial alginates treated with chlorine[26] and ultrasound[27]; the observed shifts of the elution profiles suggested a decrease in the molecular weight and/or change in the conformation of the polysaccharides. Sephacryl S-500 has also been applied in the study of EPS from other organisms such as *Pseudomonas fluorescens*.[28]

Thin-Layer Chromatography

Identification of alginate can be performed by high-performance thin-layer chromatography (HPTLC) of alginate hydrolysates.[29] In contrast to the more time-consuming method of descending paper chromatography that has been described for the analysis of acid and enzymatic hydrolysis products of bacterial alginate,[8,30] HPTLC has the advantage of being performed within 3 hr. For the acid hydrolysis of alginates according to Haug and Larsen,[31] 0.5 ml of 80% (v/v) sulfuric acid is added to 50 mg of the polysaccharide in an ice bath. The suspension is left for 18 hr at room temperature. After the addition of 6.5 ml of deionized water in an ice bath, the mixture is heated at 100° for 5 hr. The hydrolysates are neutralized by the addition of solid calcium carbonate and the precipitate is separated from the hydrolysate by filtration. Volumes (5 μl) of the hydrolysates are applied to HPTLC plates (10 × 10 cm) precoated with silica gel 60 (Merck). The plates are developed three times (1 hr for each development) in ethyl acetate–acetic acid–pyridine–water (5 : 1 : 5 : 3, v/v). After air drying the spots are visualized by spraying the plates with *p*-anisidine reagent[32] and subsequent heating at 130° for 10 min. Acid hydrolysates of seaweed alginate can be used as a reference for the uronic acids and their

[26] J. Wingender, S. Grobe, S. Fiedler, and H.-C. Flemming, *in* "Biofilms in Aquatic Systems" (C. W. Keevil, A. Godfree, D. Holt, and C. Dow, eds.), p. 93. Royal Society of Chemistry, Cambridge, 1999.

[27] K. A. Williams, H. A. Clark, and D. G. Allison, *J. Antimicrob. Chemother.* **36,** 463 (1995).

[28] B. Ruiz, A. Jaspe, C. SanJose, P. Gilbert, and D. G. Allison, *in* "Biofilms: The Good, the Bad and the Ugly" (J. Wimpenny, P. Gilbert, J. Walker, M. Brading, and R. Bayston, eds.), p. 269. BioLine, Cardiff, UK, 1999.

[29] J. Wingender and U. K. Winkler, *FEMS Microbiol. Lett.* **21,** 63 (1984).

[30] A. Linker and R. S. Jones, *J. Biol. Chem.* **241,** 3845 (1966).

[31] A. Haug and B. Larsen, *Acta Chem. Scand.* **16,** 1908 (1962).

[32] L. Hough, J. K. N. Jones, and W. H. Wadman, *J. Chem. Soc.* 1702 (1950).

corresponding lactones formed during hydrolysis. Mannuronate and guluronate are not available commercially; however, both uronic acids can be prepared by ion-exchange chromatography of alginate hydrolysates on Dowex 1 × 8 according to the method of Larsen and Haug.[33] In addition, commercial D-mannuronic acid lactone (Sigma) can be used as a reference both for the lactone and after addition of 0.1 M NaOH as a reference for mannuronate. In addition to the purpose of positive identification of bacterial alginates, the HPTLC technique has also been employed to quantitate the incorporation of radioactivity in mannuronate and guluronate from ^{14}C-radiolabeled precursors of alginate biosynthesis in *P. aeruginosa*.[34]

Infrared Spectroscopy

Infrared (IR) spectroscopy can be used to characterize whole EPS and to identify purified alginates. It is a convenient technique for the rapid and semiquantitative determination of alginate functional groups. Acetyl ester bonds of bacterial alginates give rise to IR absorption bands at 1250 and 1730 cm^{-1}.[3,8] O-Acetyl bands are not present in spectra of algal alginates or alkali-treated bacterial alginates. An absorption band at 893 cm^{-1} in bacterial alginate is likely to be characteristic of β linkages.[30] In contrast to the IR spectrum of purified alginate, the EPS of *P. aeruginosa* SG81 have a strong but broad IR absorption band at 1632 cm^{-1} that represents a composite of the 1603 cm^{-1} COO$^-$ asymmetric stretching band of alginate, and the amide I and amide II protein bands (approximately 1650 and 1550 cm^{-1}, respectively). This indicates a significant protein content in the EPS. Despite this spectral overlap, the amide II band is sometimes evident as a weak shoulder in the composite band. The ratio of D-mannuronate to L-guluronate (M : G) in dried alginate preparations provides an indication of the viscoelastic modulus of the alginate. High M : G alginates are relatively elastic, whereas low M : G alginates are more brittle.[35] Detailed assignments for the infrared absorption bands of algal sodium alginate were published by Sartori *et al.*[36] The attenuated total reflectance (ATR) mode is useful for analyzing hydrated alginate molecules, but standard transmission measurements are simpler and suffice for most qualitative and semiquantitative observations. Transmission measurements can be made with powdered (e.g., freeze-dried) alginate incorporated into pressed, alkali metal halide disks. IR-transparent disks are prepared by mixing approximately 100 mg of finely ground, IR-grade KBr or CaF$_2$ with 1–2 mg of accurately weighed, powdered alginate. The analyte-containing "window" is made by applying a pressure of 70–100 MPa to the matrix/analyte mixture *in vacuo,* using a purpose-designed

[33] B. Larsen and A. Haug, *Acta Chem. Scand.* **15,** 1397 (1961).

[34] J. Wingender, V. Sherbrock-Cox, P. Gacesa, and N. J. Russell, *Biochem. Soc. Trans.* **13,** 1148 (1985).

[35] N. J. Russell and P. Gacesa, *Mol. Aspects Med.* **10,** 1 (1988).

[36] C. Sartori, D. S. Finch, B. Ralph, and K. Gilding, *Polymer* **38,** 43 (1997).

die apparatus. Alternatively, spectra of similar quality can be obtained from aqueous suspensions dried under reduced pressure or gentle heating (40°) to produce thin films on nonhygroscopic, IR-transparent crystals (e.g., ZnSe). Infrared spectra of dried alginate films are collected by coadding 64 scans at a resolution of 4 cm^{-1}. For spectra of adequate intensity (>0.1 absorption units), using the ZnSe method, the alginate concentration should be approximately 3–4 mg/ml.

IR spectroscopy has also been used to determine the calcium content of algal alginate films.[36] IR absorption bands were correlated with cation concentrations that had been determined independently by atomic absorption spectroscopy. Calcium was shown to bind preferentially to G blocks, as inferred by the higher calcium content in low M : G alginates at equilibrium. Note, however, that the presence of impurities such as nucleic acids may suppress some alginate absorption bands or alter their intensities relative to one another, and that these impurities may not be obvious in IR spectra unless the contaminant-to-analyte ratio is extremely high. Second derivative spectra or Fourier self-deconvolutions are more useful to verify contamination, but other techniques such as NMR spectroscopy are more suitable for this purpose.

^{1}H Nuclear Magnetic Resonance Spectroscopy

^{1}H NMR spectroscopy can be used to determine the fractional composition of bacterial alginates.[3,7,9,37] The samples are pretreated by mild, acid hydrolysis and repeated washing in D_2O.[3,37] Alginate (10 mg) is dissolved in 3 ml of deionized water. The pH value is adjusted to 2.9 with 0.1 M HCl and the solution is heated at 100° for 30 min. After cooling to room temperature and neutralization of the mixture with 1 M NaOH, the sample is freeze-dried. Alginate (10 mg) and 3 mg of Na_2-EDTA are dissolved in 0.5 ml of D_2O (99.9%; Aldrich, Milwaukee, WI) and freeze-dried. Subsequently, 0.5 ml of D_2O is added to the sample and the mixture is freeze-dried again. This procedure is repeated three times. Finally, alginate is dissolved in 0.5 ml of D_2O. In our study of alginates from mucoid biofilm isolates of *P. aeruginosa,* recording of 300-MHz spectra is performed at 90° in a Bruker WM 300 spectrometer according to Grasdalen *et al.*[37] A common feature of alginates from mucoid environmental and clinical strains of *P. aeruginosa* is higher proportions of mannuronate than guluronate (e.g., ≥56% in alginates from mucoid environmental isolates)[7] and the presence of both blocks of polymannuronate and blocks of polymannuronate/guluronate, while blocks of polyguluronate are absent.[3,7,8,9,38]

[37] H. Grasdalen, B. Larsen, and O. Smidsrød, *Carbohydr. Res.* **68,** 23 (1979).
[38] B. K. Pugashetti, H. M. Metzger, Jr., L. Vadas, and D. S. Feingold, *J. Clin. Microbiol.* **16,** 686 (1982).

Lectin Staining

Biofilm Growth on Membrane Filters

Fluorescently labeled lectins can be used to visualize carbohydrate-containing EPS components in biofilms.[39] For lectin staining, biofilms are cultivated on membrane filters placed on PIA containing 2% (v/v) glycerol either with or without the addition of 0.1 M $CaCl_2$. Approximately 10^6 cells from an overnight culture (36°) on PIA in 6 ml of 0.14 M NaCl are vacuum filtered onto black polycarbonate membrane filters (filter diameter, 25 mm; pore diameter, 0.4 μm; Millipore, Bedford, MA). The filters are placed on the agar surface of PIA plates and incubated at 36° for 24 hr. In case of biofilms grown on PIA in the absence of calcium, fixation of the biofilm by treatment with 3% (v/v) formaldehyde for 30 min is necessary to stabilize the biofilm during the washing steps of the subsequent staining procedures.

Lectin-Binding Assay

Commercially available lectins labeled with fluorescein isothiocyanate (FITC) or tetramethyl-rhodamine isothiocyanate (TRITC) are used, such as a lectin from *Canavalia ensiformis* (concanavalin A, Con A) specific for D-(+)-glucose and D-(+)-mannose. The conjugated lectins are reconstituted with 10 mM phosphate buffer (pH 7.5) to stock solutions of 1 mg/ml, which are stored frozen in portions of 100 μl. For use, stock solutions are diluted with phosphate buffer to a lectin concentration of 10 μg/ml. Fifty-microliter samples of these staining solutions are applied directly to the top of the biofilms on the membrane filters. After an incubation period of 30 min in the dark at room temperature, excess staining solution is removed by four rinses with phosphate buffer. The stained biofilm preparations are examined either by epifluorescence microscopy or confocal laser scanning microscopy (CLSM) (FITC–lectins: excitation, 488 nm; emission, 500 nm; TRITC–lectins: excitation, 543 nm; emission, 580 nm).

Counterstaining of Bacteria

The determination of the position of bacteria in lectin-stained biofilms can be performed with various DNA-binding fluorochromes that specifically label the bacterial cells. The nucleic acid stains are chosen according to their emission maxima, which must differ from the fluorochrome-labeled lectins to enable differentiation. For example, samples with the green fluorescent FITC-labeled lectins can be counterstained with a red fluorescent propidium iodide (PI) staining solution. The biofilms are treated with 50 μl of a 30 μg/ml PI solution containing 5% (v/v)

---

[39] T. R. Neu and J. R. Lawrence, *in* "Microbial Extracellular Polymeric Substances" (J. Wingender, T. R. Neu, and H.-C. Flemming, eds.), p. 21. Springer-Verlag, Berlin, 1999.

TABLE II
INHIBITION OF BINDING OF LECTIN CONCANAVALIN A
TO BIOFILMS OF *Pseudomonas aeruginosa* SG81
BY VARIOUS SUGARS

Monosaccharide	Sugar concentration (mg/ml)				
	100	30	10	3	1
D-(+)-Glucose	0^a	0	1	2–3	3
D-(+)-Mannose	0	1	2–3	3	3
D-(+)-Galactose	2	3	3	3	3
N-Acetyl-D-glucosamine	2–3	3	3	3	3
N-Acetyl-D-galactosamine	3	3	3	3	3
D-(+)-Fucose	2–3	3	3	3	3
L-(−)-Fucose	3	3	3	3	3

[a] 0, Complete inhibition of lectin binding; 1, sporadic lectin binding; 2, lectin binding slightly lower than observed in the absence of inhibiting sugar; 3, lectin binding as obtained in the absence of inhibiting sugar.

ethanol and 0.01% (w/v) EDTA in phosphate-buffered saline after lectin binding. The biofilms are incubated for 30 min in the dark at room temperature and examined immediately by epifluorescence microscopy or CLSM techniques (FITC–lectins: excitation, 488 nm; emission, 500 nm; PI: excitation, 488 nm; emission, 635 nm). Samples stained with the red fluorescent TRITC-labeled lectins could preferably be counterstained with the green fluorescent proprietary stain SYTO 9 (Molecular Probes) as described by the manufacturer.

Competitive Inhibition Assay

The sugar-binding specificity of each lectin can be verified by a competitive inhibition assay using simple sugars. D-(+)-Glucose, D-(+)-mannose, D-(+)-galactose, D-(+)-fucose, L-(−)-fucose, N-acetyl-D-galactosamine, and N-acetyl-D-glucosamine (all from Sigma) are preincubated with the lectin-staining solutions at final concentrations of 1, 3, 10, 30, and 100 mg/ml for the sugars and 10 μg/ml for the lectins for 30 min in the dark at room temperature prior to the staining procedure, which is carried out as described above. Results of inhibition assays using Con A are shown in Table II. The inhibition of lectin binding by the various sugars is specified semiquantitatively by visual comparison of the binding pattern and fluorescence intensity of samples with and without added sugars. The inhibition effect can be described on a qualitative scale from 0 (no binding of lectin) to 3 (same binding of lectin as in the absence of sugar).

Section VII

Microbiological Aspects of Microbial Biofilm

[26] Approach to Analyze Interactions of Microorganisms, Hydrophobic Substrates, and Soil Colloids Leading to Formation of Composite Biofilms, and to Study Initial Events in Microbiogeological Processes

By Heinrich Lünsdorf, Carsten Strömpl, A. Mark Osborn, Antonio Bennasar, Edward R. B. Moore, Wolf-Rainer Abraham, and Kenneth N. Timmis

Introduction

Microorganisms are key agents in the cycling of elements and play important roles in a number of geological processes. Although the biochemistry of such processes has been studied intensively, the cellular activities and spatial interactions involved are poorly understood as a consequence of a paucity of adequate experimental systems and approaches. More recent and current developments, however, allow better experimental access to these interactions.

Many organic substrates that can be metabolized by microorganisms are hydrophobic in nature. Conventional wisdom has it that, because life processes predominantly take place in aqueous media, the microbial degradation of hydrophobic substrates is limited by their solubility in water. Microorganisms live predominantly in biofilms, either on an inorganic matrix such as soil particles, on organic aggregates composed of biological material with or without inorganic components, such as marine and lake "snow," or on the surfaces of higher organisms. In a heterogeneous medium such as soil, the essential lack of mobility of biofilm bacteria and poor mobility of hydrophobic substrates in pore water (restricted by low solubility and diffusion) are considered to constitute the principal rate-limiting factors in the turnover of such substrates. On the other hand, there can be significant mixing of soil materials through biological and abiological agents and influences, particularly hydrological activities, and microorganisms do detach from biofilms and move to other sites, either passively or actively by tactic responses, such that spatially separated hydrophobic substrates and the biological catalysts able to metabolize them may be brought together. Direct colonization of the substrate material, or of soil material to which the substrate has adsorbed, by relevant microorganisms can then occur. It is therefore possible that these direct microbe-substrate interactions are the most important for the metabolism of hydrophobic compounds, and that factors other than their water solubility and diffusion in pore water may determine degradation rates.

Microorganisms are able to metabolize some industrial xenobiotics, and are thus important agents of the removal of such pollutants from the environment. The degradation of toxic, hydrophobic xenobiotics, however, suffers both from poor solubility (low bioavailability) and toxicity, the latter of which may prevent or restrict growth directly on the substrate. Knowledge of microbial interactions with hydrophobic pollutants and soil components is scant but will be crucial to the development of effective bioremediation technologies. We have developed a close-to-nature, but experimentally amenable, approach to analyze interactions of autochthonous microorganisms with the components of polychlorinated biphenyl (PCB)-contaminated soils.[1] In this simple system, a composite biofilm, consisting of autochthonous microorganisms and soil components, and fed by PCB carbon, develops *de novo* on a support surface amenable to ultrastructural and microan-alytical procedures. The biofilms that develop consist of a dense lawn of clay aggregates, each one of which contains one or more bacteria, phyllosilicates, and grains of iron oxide material, all held together by bacterial extracellular polysac-charide (EPS). The clay leaflets are arranged in the form of a "houses of cards" and give the aggregates the appearance of "hutches" housing the bacteria. Because clay minerals have a high cationic exchange capacity and high absorbance poten-tial for polymers and bioorganics, they will act as high-capacity nutrient shuttles that transfer organic matter from its source to competent bacteria. The microoogan-isms that recruit PCB-loaded clay leaflets to create the biofilm of clay hutches essentially reflect those found in the PCB-polluted soil matrix.[2] We describe here the experimental set-up and procedures for analysis of the resulting biofilm.

To monitor the sequence of interactions ensuing after autochthonous microor-ganisms gain new opportunities to interact with soil particles and hydrophobic pollutants, existing interactions must first be disrupted by mechanical disintegra-tion of the soil matrix, by vigorous agitation of the soil sample with water. This liberates the soil microorganisms and colloids (inorganic and organic) into the bulk water phase and allows the formation of new associations. The suspension is then allowed to settle, and a low-energy substratum is floated on the water surface, creating a three-phase system, namely a fresh colonization surface on which the interactions will be studied, a soil/sediment as source of nutrients, and a water col-umn connecting the two. In a sense, the experimental set-up simulates the situation in a newly formed puddle.

Soil Preparation

Soil is composed mainly of macroaggregates (250 μm) and microaggregates (20–250 μm). The smallest particles are clusters of clay minerals stabilized by

[1] H. Lünsdorf, R. W. Erb, W.-R. Abraham, and K. N. Timmis, *Environ. Microbiol.* **2**, 161 (2000).
[2] B. Nogales, E. R. B. Moore, W. R. Abraham, and K. N. Timmis, *Environ. Microbiol.* **1**, 199 (1999).

organic and inorganic substances.[3,4] Clay particles are characterized by their high cation-exchange capacity, negative surface charge, and high content of soil organic matter. Soil aggregates form a matrix containing a capillary network filled with either air and/or water. The pore diameters of this network exclude or admit soil bacteria, which are found in pores ranging from 0.8 to 3.0 μm in diameter.[5-7] The experimental set-up described here involves a clayish glaciofluvial sandy soil[8] from a PCB-polluted site in Sachsen-Anhalt, Germany.

Pretreatment

Sieve 0.5–1.0 kg of soil material through a 4-mm-square mesh to remove small stones, coarse plant material, and larger arthropods and annelids, and then thoroughly mix with a spatula to obtain a batch of reasonably homogeneous material. Aliquot in 40- to 50-g quantities and store at 4° in plastic bags closed with cotton stoppers to allow sufficient aeration.

Experimental Set-Up

Mix in sterile 250-ml Erlenmeyer flasks 1 part soil (about 20 g) with 2 parts sterile, reverse osmosis-purified water (represents about four times the natural field capacity). Place the flasks on a rotary shaker and agitate at 200 rpm for 30 min at ambient temperature. Autoclave one flask three times, with intermittent incubation periods of 24 hr without shaking at ambient temperature, to serve as a control and reference for abiotic interactions of soil constituents with the substratum. Allow microcosms to equilibrate by gently shaking at 30 rpm overnight at ambient temperature. Initiate the experiment by gently layering a sterile Permanox slide,

[3] J. M. Tisdall and J. M. Oades, *J. Soil Sci.* **33,** 141 (1982).

[4] J. M. Tisdall, *Plant Soil* **159,** 115 (1994).

[5] E. A. Paul and F. E. Clark, *in* "Soil Microbiology and Biochemistry," p. 15. Academic Press, San Diego, California, 1996.

[6] A. G. Waters and J. M. Oades, *in* "Advances in Soil Organic Matter Research" (W. W. Wilson, ed.), p. 163. Royal Society of Chemistry, London, 1991.

[7] J. Hassink, L. A. Bouwman, K. B. Zwart, and L. Brussaard, *Soil Biol. Biochem.* **25,** 47 (1993).

[8] The site under study, situated north of Wittenberg, Germany, in the former German Democratic Republic, was a Soviet missile base and is contaminated with PCBs that leaked from damaged condensers. The site soil consists mainly of sand with some clay, and has a pH of 4–5. Soil PCB concentrations vary over a wide range, to a maximum of 28,000 mg of PCB kg^{-1} dry soil, but lie mostly in the range between 10 and 150 mg kg^{-1}. The PCB congener pattern and the relative amounts of the individual congeners—2.9% dichloro-, 44.7% trichloro-, 51.6% tetrachloro-, and 0.8% pentachlorobiphenyls—are uniform throughout the site. They are similar to those reported for the Russian PCB mixture trichlorophenyl, which is similar to Arochlor 1242,[28] except that they are depleted of the lowest chlorinated congeners. No other pollutants were detected in the site, and all concentrations of heavy metals were below critical thresholds (site C: As <5, Pb <10, Cd <0.5, Cr 10, Cu 6, Ni 4, Hg <0.1, and Zn 22 mg kg^{-1} dry soil).

face down, on the water surface, taking care not to resuspend the soil. Incubate the microcosms at ambient temperature for 14 days at 30 rpm.

Microscopic Analysis

The overall characteristics of the biofilm formed on the substratum, i.e., its topographical state, quality and approximate level of maturity, as well as the presence of non-biofilm-associated motile microbiota such as flagellates, ameba, and nematodes, are assessed by light microscopy (Fig. 1A and B), whereas its ultrastructure is examined by electron microscopic methods (Fig. 2).

Light Microscopy

Carefully remove the floating substratum with sterile tweezers and mount on the microscope stage with the biofilm side facing the lens. Add water from the microcosm to the biofilm surface to prevent it drying out during microscopic examination. Use low magnification ($\times 10$) for observation of protozoa. Take care not to disturb the biofilm with the lens. Immediately after examination by light microscopy, cut off an adequate part of the Permanox substratum, drain off residual surface water, and store in a screw-cap culture vessel at $4°$ for chemical and microbiological analyses. Place the other part of the biofilm in a fixation bath for electron microscopy.

Preparation for Electron Microscopy

Add fixative [1% (v/v) glutardialdehyde–10 mM HEPES, pH 7.5] to a petri dish to a level of 5 mm. Carefully place the substratum, biofilm side down, onto the fixative surface. Take care not to entrap air bubbles, because their surface tension can severely damage the biofilm surface. Allow fixation to proceed for 2 to 72 hr at $4°$.

Conventional Transmission Electron Microscopy

In addition, stain one part of the biofilm for acidic polysaccharides with 0.5% (w/v) ruthenium red in 0.5% (v/v) glutardialdehyde–10 mM HEPES, pH 7.5, at $4°$ overnight. Wash it twice in 100 mM cacodylate, pH 7.2, before placing in osmium tetroxide [1% (w/v) osmium tetroxide in 100 mM cacodylate, pH 7.2] at ambient temperature for 1 hr. Dehydrate the sample in an aqueous acetone series (10, 50, 70, 90, and 100%, and, again, 100%) on ice, by transferring the substratum with adhering solute from one vessel to the next, with 10 min in each vessel. At 50% acetone, the Permanox slide loses its floating property. Therefore, place the slide with the biofilm side up in the acetone.

For flat-bed embedding, immerse the biofilm in 1 part acetone: 1 part epoxy resin monomer [resin monomer composition 10.0 g of ERL-4206, 4.0 g of

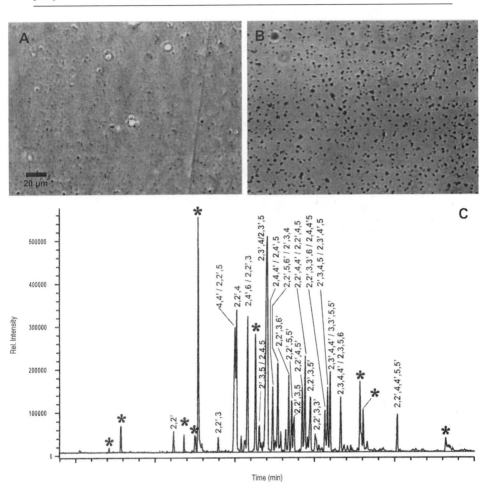

FIG. 1. A young composite biofilm of "clay hutches." (A) and (B) are low-magnification light microscope views of representative regions of a Permanox slide after incubation for 14 days on the surface of a water column of a water–sediment system. (A) Sterilized soil; (B) untreated soil. The dark spots are discrete aggregates of clay colloids and bacteria, so-called clay hutches, which are distributed fairly evenly on the substratum surface. The GC–MS spectrum of an extract obtained from the biofilm shown in (B) is shown in (C). Dichloro- to hexachlorobiphenyl congeners are identified by the relevant numbers, whereas unknown constituents of the substratum are indicated by asterisks.

DER-736, 26.0 g of NSA, 0.4 g of DMAE(S-1)][9] for 1 hr at room temperature, and then in pure resin monomer for 1 hr, followed by overnight incubation in pure monomer at room temperature. Thoroughly decant the epoxy resin and cover the biofilm with fresh resin monomer to a height of 1 mm. Polymerize the resin by

[9] A. R. Spurr, *J. Ultrastruct.* **26,** 31 (1969).

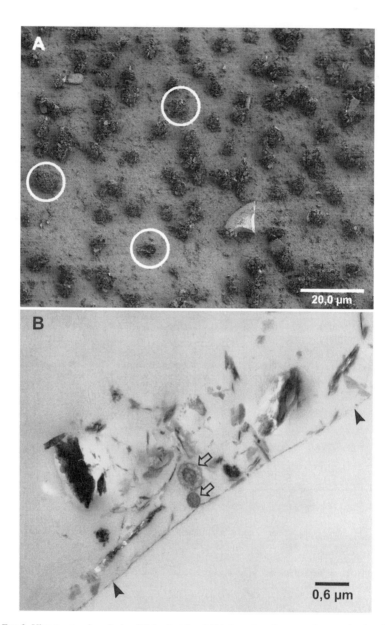

FIG. 2. Ultrastructural analysis of "clay hutches." (A) Scanning electron micrograph of a cluster of the clay hutches shown in Fig. 1B. Clay leaflets and other soil particles recruited to the surface of the aggregates can be clearly seen. The smallest aggregates are about 3–4 μm in diameter. Individual clay hutches selected for further study are circled. (B) Ultrathin section of a larger clay hutch cut perpendicularly to the substratum plane and treated as described in text, except for ruthenium staining. Filled arrowheads indicate the substratum surface. Individual bacteria are indicated by open arrows, and one can be seen to be in direct contact with the substratum surface. They are surrounded by a mass of loosely stacked clay leaflets and compact electron-dense granules.

incubating at 70° for 7 hr. Do not use excess resin monomer for this step, and maintain the container perfectly horizontal for polymerization, in order to avoid the resin running off the film before the onset of polymerization. After polymerization, carefully peel off the Permanox support, leaving the biofilm embedded in the epoxy resin layer. Inspect the embedded biofilm with a binocular for suitable areas and carefully cut them out with a mini-circular saw (Proxxon, Niersbach, Germany). Transfer suitable cut-outs to gelatin capsules or flat embedding molds and embed in epoxy resin as described above. Cut ultrathin sections 90 nm thick, using an ultramicrotome (Ultracut E; Leica, Bensheim, Germany) and a diamond knife (Diatome, Biel/Bienne, Switzerland). Pick up the sections on Formvar-coated 300 mesh hexagonal copper grids, stain with uranyl acetate [4% (w/v), 5 min] and lead citrate [0.3% (w/v), 5 min],[10] and examine with an energy-filtered transmission electron microscope, such as a Zeiss EM902 (LEO, formerly Zeiss, Oberkochen, Germany) at 80 kV under elastic bright-field settings (energy slit, 30 eV; objective aperture, 30 μm; condenser aperture, 400 μm) (Fig. 2B). Record images of magnifications between $\times 4400$ and $\times 50,000$ on Agfa Scientia film at an exposure dose of 45 electrons/μm^2. Develop films with full-strength Kodak (Rochester, NY) D19 for 4.5 min at 20°.

Scanning Electron Microscopy

Dehydrate a suitable part of the Permanox substratum in an acetone series, as described above. Transfer the substratum to the pressure chamber of a critical point drying unit (CPD030; Bal-Tec, Lichtenstein) filled with acetone at 10°. After three or four exchanges of acetone by liquid CO_2 raise the temperature to 40° and the pressure to 80 bar. Thereafter, slowly restore to atmospheric pressure over a period of 30 min. Mount dried biofilms on aluminum stubs and sputter-coat with gold (SCU040; Balzer Union, Lichtenstein) at 10 cm distance, in an argon atmosphere of a 0.06-mbar chamber pressure for 54 sec at a current of 45 mA. Examine the mounted biofilms with a scanning electron microscope, such as a Zeiss DSM 982 Gemini (LEO) equipped with a field emission gun, in a magnification range from $\times 100$ to $\times 10,000$ at 5 kV and 6-mm working distance (Fig. 2A). Record images either digitally or on type 120 roll film.

Element Mapping by Energy-Filtered Transmission Electron Microscopy

Key information in ultrastructural studies on microbe–mineral interactions is the distribution of elements in samples, determined through element analysis by energy-filtered transmission electron microscopy (EFTEM). Element-specific information is gained from inelastic scattering of primary beam electrons, accelerated

---

[10] E. S. Reynolds, *J. Cell Biol.* **17**, 208 (1963).

to an energy of 80 keV and a corresponding wavelength of 0.0042 nm, with inner shell electrons of atoms in the environmental sample.[11-17] To avoid multiple scattering of an already inelastically scattered primary electron during its passage through the sample, ultrathin sections (30 to 40 nm thick, corresponding to less than the order of the mean free path of plasmon losses, i.e., 80 nm for embedded biological matter and 40 nm for aluminum foil[18]) must be used. With the aid of an integrated electron spectrometer (in the case presented, a Castaing–Henri filter with a magnetic double prism and an electrostatic mirror), these inelastically scattered electrons can be used to produce both electron energy loss spectroscopy (EELS) spectra and high-resolution element distribution maps of the sample, for instance, of minerals in the immediate vicinity of autochthonous bacteria and, thus, within their ranges of physiological interactions.[1]

Moreover, if the energetic resolution of the electron gun and of the integrated spectrometer is sufficient, i.e., about 1.5 eV for the EFTEM Zeiss CEM 902, information about the elemental composition of a distinct structure in an environmental sample can be obtained from serial- or parallel-recorded EELS spectra within the energy range from 0 to 2,000 eV. Light elements, in particular C, N, P, O, and Ca, can be detected with high sensitivity. Even the chemical state of the elements in a given compound of interest can be studied by analysis of energy-loss near-edge fine structures (ELNES) within the EELS, by following the individual ionization edges of the elements within an energy range of about 50 eV[17] and comparing these edges with those of reference spectra of characterized compounds.[19,20] Information on

[11] L. Reimer, in "Transmission Electron Microscopy: Physics of Image Formation and Microanalysis." Springer-Verlag, Berlin, 1984.

[12] R. Bauer, Methods Microbiol. 20, 113 (1988).

[13] M. Thellier, C. Ripoll, C. Quintana, F. Sommer, P. Chevallier, and J. Dainty, Methods Enzymol. 227, 535 (1993).

[14] D. C. Joy, in "Introduction to Analytical Electron Microscopy" (J. J. Hren, J. I. Goldstein, and D. C. Joy, eds.), p. 223, Plenum Press, New York, 1979.

[15] C. Colliex, in "Advances in Optical and Electron Microscopy" (R. Barer and V. E. Cosslett, eds.), Vol. 9, p. 65. Academic Press, London, 1984.

[16] R. F. Egerton, in "Electron Energy-Loss Spectroscopy in the Electron Microscope," 2nd Ed., Plenum Press, New York, 1996.

[17] F. Hofer, in "Energy-Filtering Transmission Electron Microscopy" (L. Reimer, ed.), p. 225. Springer Series in Optical Sciences. Springer-Verlag, Berlin, 1995.

[18] L. Reimer, in "Energy-Filtering Transmission Electron Microscopy" (L. Reimer, ed.), p. 347. Springer Series in Optical Sciences. Springer-Verlag, Berlin, 1995.

[19] C. C. Ahn and O. L. Krivanek, in "EELS Atlas. A Reference Guide of Electron Energy Loss Spectra Covering all Stable Elements." ASU Center for Solid State Science (Temp, AZ) and Gatan (Warrendale, PA), 1983.

[20] L. Reimer, U. Zepke, J. Moesch, S. Schulze-Hillert, M. Ross-Messemer, W. Probst, and E. Weimer, in "EELSpectroscopy. A Reference Handbook of Standard Data for Identification and Interpretation of Electron Energy Loss Spectra and for Generation of Electron Spectroscopic Images." Institute of Physics, University of Münster, Germany, and Carl Zeiss, Electron Optics Division, Oberkochen, Germany, 1992.

the oxidative states and the coordination spheres of diverse elements, such as Al, Si, and Fe, and of oxides of S, P, and C, have thereby been obtained.[21-24]

Electron Energy Loss Spectroscopy and Electron Spectroscopic Imaging

Pick up 35-nm ultrathin sections of embedded samples on uncoated nickel grids (hexagonal, 460 mesh), blot, and air dry. Scan the sections at low magnification and low-beam intensity for the presence of bacterial–clay aggregates, i.e., "clay hutches," that have been cut perpendicular to the substratum surface. To analyze the elemental composition of a feature of interest by EELS, adjust the magnification to fit the structure into the opening of the spectrometer entrance aperture. At magnifications greater than $\times 50,000$, set the objective aperture to 12 mrad (equivalent to 60 μm); at lower magnifications set it to 6 mrad (equivalent to 30 μm). Set the energy selective slit in the energy dispersive plane to 5 eV and center it perfectly, with respect to the photomultiplier, for serial recordings in the spectrum mode. Before starting the serial EELS, preset and test acquisition parameters, i.e., the gain settings of the photomultiplier, the energy intervals between adjacent energy channels (1 to 2 eV), and the integration time per energy channel (1 to 3 sec). Record spectra (Fig. 3B; see color insert) for 5–20 min. Immediately thereafter, use the image mode to confirm that the position of the feature being analyzed has not changed relative to the spectrometer entrance aperture. Correct background according to the power law modes $I = AE^{-r}$,[16,25,26] because ionization edges are superimposed on the energy loss background intensities.

After analyzing a set of structures of interest by EELS, record distribution maps of characteristic sample elements by electron spectroscopic imaging (ESI), by means of a side-entry, cooled 14-bit change-coupled device (CCD) slow-scan camera system with a 1024–1024 chip as electron detector (Proscan CCD HSS 512/1024; Proscan Electronic Systems, Scheuring, Germany). Set the energy-selecting slit to an energy width of 15 eV, and optimize settings for ESI registration (i.e., beam current, 1 to 10 μA; integration time per acquisition step, 0.1 to 10 sec; condenser setting close to the cross-over with homogeneous illumination), for each structure of interest and element. In most cases, subtract two background images from the postionization edge (three-window method), in order to obtain pure element maps. Superimpose false color element distribution maps on the corresponding bright-field views, in order to obtain exact correlations of element distribution with ultrastructure features (Fig. 3A; see color insert).

[21] F. Hofer and P. Golob, *Ultramicroscopy* **21**, 379 (1987).

[22] K. M. Krishnan, *Ultramicroscopy* **32**, 309 (1990).

[23] R. Brydson, H. Sauer, W. Engel, and E. Zeitler, *Microsc. Microanal. Microinstruct.* **2**, 159 (1991).

[24] P. L. Hansen, D. McComb, and R. Brydson, *Micron Microsc. Acta* **23**, 169 (1992).

[25] R. F. Egerton, *Phil. Mag.* **31**, 199 (1975).

[26] C. Jeanguillaume, P. Trebbia, and C. Colliex, *Ultramicroscopy* **3**, 237 (1978).

Chemical Analysis

The soil used to illustrate the methods described here contains polychlorinated biphenyls (PCBs), which constitute the major source of carbon in the sample. The chemical analysis procedure described below is for this class of compound.

Gas Chromatography

PCBs are extracted by a modification of the standard procedure.[27] Add 10 μg of PCB 153, as internal standard, to 2 g of air-dried, homogenized soil. Suspend the sample in 5 ml of hexane, vortex for 1 min, and allow the hexane phase to separate. Repeat three times. Remove the solvent under reduced pressure and suspend the extract in 500 μl of octane. Analyze 1 μl of the extract by capillary gas chromatography, e.g., with a Hewlett-Packard (HP; Palo Alto, CA) 5890 Series II gas chromatograph equipped with an HP Ultra 2 capillary column (50 × 0.2 mm; film thickness, 0.11 mm) and flame ionization detector (FID), with hydrogen as the carrier gas (Fig. 1C). In this case, set the injection temperature to 250°, the detector temperature to 300°, and the oven program to 80° for 3 min and 90 to 288° at 6° min^{-1}, followed by an isothermal period of 20 min.

PCB congeners are identified by gas chromatography–mass spectrometry (GC–MS)[28] and by comparison with pure standards. Perform gas chromatography as described above, but with helium as the carrier gas, with the instrument connected to an HP 5989A quadrupole mass spectrometer. Set the electron impact ion source to 200°, the quadrupole temperature to 100°, and the electron energy to 70 eV. To isolate PCB from the "clay hutches" attached to the substratum, wash the substratum three times with 5 ml of hexane and analyze as described for soil.

Microbiological Analysis

Bacterial community structure and dynamics in biofilms can be studied by a variety of methods, including cultivation of individual organisms, analysis of isolated DNA/RNA, *in situ* analysis by fluorescently labeled hybridization probes (FISH), and microautoradiography. All these techniques are well described in the literature,[29–34] so the discussion here is limited to a brief, practical description of

[27] A. Büthe and E. Denker, *Chemosphere* **30**, 753 (1995).
[28] V. Ivanov and E. Sandell, *Environ. Sci. Technol.* **26**, 2012 (1992).
[29] R. Weller and D. M. Ward, *Appl. Environ. Microbiol.* **55**, 1818 (1989).
[30] D. M. Ward, *in* "Structure and Function of Biofilms" (W. G. Charaklis and P. A.Wilderer, eds.), p. 145. John Wiley & Sons, New York, 1989.
[31] N. B. Ramsing, M. Kühl, and B. B. Jörgensen, *Appl. Environ. Microbiol.* **59**, 3840 (1993).
[32] R. I. Amann, W. Ludwig, and K.-H. Schleifer, *Microbiol. Rev.* **59**, 143 (1995).
[33] S. Kalmbach, W. Manz, and U. Szewzyk, *Appl. Environ. Microbiol.* **63**, 4164 (1997).
[34] W. Manz, *Methods Enzymol.* **310**, 79 (1999).

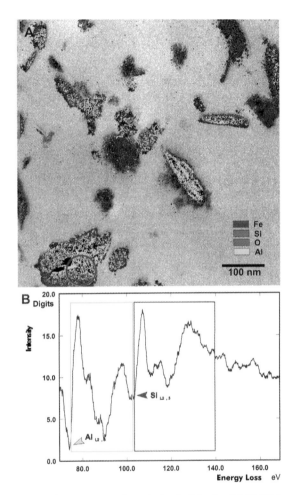

FIG. 3. Electron energy loss spectroscopic analysis of a "clay hutch." (A) High-resolution ESI element distribution map. False-color single-element maps of part of a clay hutch have been superimposed on the bright-field image. In addition to phyllosilicates (green, yellow, blue), granular iron oxides (red, blue) can be seen either isolated from or in contact with clay particles. (B) EELS spectrum of a clay leaflet. A 7300-nm^2 area of a clay leaflet [red circle, lower left-hand corner in (A)] was selected for EELS analysis. The intensity profiles of the ionization edges (arrowheads) and near-edge fine structures (ELNES) of the L2,3 energy levels of aluminum (boxed yellow) and silicon (boxed green) are shown.

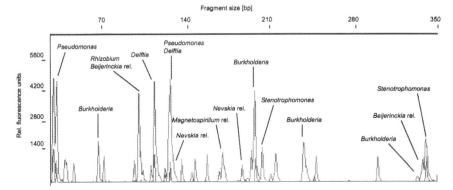

Fig. 4. T-RFLP fingerprints of bacteria extracted from clay hutches. DNA was isolated from clay hutch biofilms and a PCR product was obtained for T-RFLP analysis as described in text. The PCR-purified 16S rDNA was cleaved with *Alu*I, and the fragments were separated on a denaturing acrylamide gel. Blue peaks correspond to 6-FAM-labeled 5′-terminal fragments and green peaks to HEX-labeled 3′-terminal fragments. Red peaks correspond to the internal ROX-labeled GS500 DNA size standard. The minimal peak height was set at 100 fluorescence units. Phylogenetic assignments of peak pairs were made by comparison with a database of experimentally generated T-RFLPs, and by sequencing representative cloned 16S rDNA. The principal groups of organisms found to inhabit the clay hutches are, by and large, those previously shown to be abundant members of the PCB-contaminated soil community.

the use of terminal-restriction fragment length polymorphism (T-RFLP) analysis of amplified 16S rRNA genes,[35–42] in tandem with fingerprinting and sequence determination and analysis of cloned 16S rDNA, as applied to the study of biofilm communities and diversity and to monitor variation between samples and experiments.

DNA Extraction

Scrape the biofilm layer from the Permanox slide, using a sterile razor blade, and suspend in 50 μl of Tris–EDTA (TE) buffer, pH 8.0, in a sterile 0.5-ml Eppendorf tube. Extract DNA[43,44] immediately or store the sample at −20° until required. Transfer the sample to a lysing matrix of a FastDNA SPIN kit (for soil; QBIOgene, Carlsbad, CA) and extract DNA according to the manufacturer instructions. A Bead Beater (BioSpec, Bartlesville, OK) or a cuvette shaker may be used for mechanical disruption of bacteria, as an alternative to the QBIOgene FastPrep instrument recommended for use in conjunction with the kit. Disruption times of approximately 10–15 min are sufficient for most types of bacterial cells. Carry out sample homogenization, protein solubilization, and DNA extraction and purification onto a binding matrix, using the reagents of the kit and according to the instructions of the manufacturer. Add 20 μg of poly(I) · poly(C) (Pharmacia Biotech, Freiburg, Germany) prior to elution from the binding matrix to enhance DNA recovery. This extraction procedure yields enough DNA for several polymerase chain reactions (PCRs).

Community Terminal-Restriction Fragment Length Polymorphism Analysis

Amplify 16S rDNA from the extracted community DNA by PCR[45] with phosphoramidite dye 5′-end-labeled universal primers FAM63F (5′-CAG GCC TAA

[35] E. Avaniss-Aghajani, K. Jones, A. Holtzman, T. Aronson, N. Glover, M. Boian, S. Froman, and C. F. Brunk, *J. Clin. Microbiol.* **34**, 98 (1996).

[36] W.-T. Liu, T. L. Marsh, H. Cheng, and L. J. Forney, *Appl. Environ. Microbiol.* **63**, 4516 (1997).

[37] B. G. Clement, L. E. Kehl, K. L. DeBord, and C. L. Kitts, *J. Microbiol. Methods* **31**, 135 (1998).

[38] K. J. Chin, T. Lukow, and R. Conrad, *Appl. Environ. Microbiol.* **65**, 2341 (1999).

[39] M. M. Moeseneder, J. M. Arrieta, G. Muyzer, C. Winter, and G. L. Herndl, *Appl. Environ. Microbiol.* **65**, 3518 (1999).

[40] H.-P. Horz, J.-H. Rotthauwe, T. Lukow, and W. Liesack, *J. Microbiol. Methods* **39**, 197 (2000).

[41] H. Ludemann, I. Arth, and W. Liesack, *Appl. Environ. Microbiol.* **66**, 754 (2000).

[42] A. M. Osborn, E. R. B. Moore, and K. N. Timmis, *Environ. Microbiol.* **2**, 39 (2000).

[43] Because the samples available contain only small quantities of biomass, special precautions need to be taken to recover microorganisms and nucleic acids with optimal efficiency and reproducibility. In our hands, the application of the FastDNA SPIN kit for soil (QBIOgene, Carlsbad, CA) has provided the most effective means[44] of obtaining DNA from the approximately 100 μg of biomass present in the described biofilms. The DNA obtained is of such quality that it can be used, without further purification, for subsequent molecular applications.

[44] J. Borneman, P. W. Skroch, K. M. O'Sullivan, J. A. Palus, N. G. Rumjanek, J. L. Jansen, J. Nienhuis, and E. W. Triplett, *Appl. Environ. Microbiol.* **62**, 1935 (1996).

CAC ATG CAA GTC-3′) and HEX 1389R (5′-ACG GGC GGT GTG TAC AAG-3′) (20 pmol; Applied Biosystems, Foster City, CA), which hybridize at universally conserved target sites of 16S rRNA genes,[46] 2.5 U of AmpliTaq DNA polymerase (Applied Biosystems), and a 50 μM concentration of each dNTP (dATP, dCTP, dGTP, and dTTP) in 50 μl of 10 mM Tris-HCl (pH 8.3), 50 mM KCl, 1.5 mM MgCl$_2$, and 0.1 mg of gelatin. Start amplification reactions with a 2-min denaturation at 94°, followed by 30 cycles with the following profile: 1 min of denaturation at 94°; 1 min of primer annealing at 55°; and 2 min of primer extension at 72°; followed by a final 10-min extension step at 72°. Prepare all reagents as a master mix prior to addition to template DNA. Subject 3 μl of each PCR product to electrophoresis in 0.8% (w/v) agarose–TAE, and visualize by staining with ethidium bromide (0.5 μg/ml) for 15 min. Purify 45 μl of the PCR product on QIAQuick columns (Qiagen, Hilden, Germany), to remove unincorporated nucleotides and primers, and elute the DNA in a final volume of 50 μl. Digest 10 μl of purified 16S rDNA PCR amplicons with 2 μl of AluI or HhaI (New England BioLabs, Beverly, MA) in a total volume of 15 μl at 37° for 3 hr. Mix 2 μl of the restriction digest with 2 μl of deionized formamide, 0.5 μl of ROX-labeled GS500 internal size standard (Applied Biosystems), and 0.5 μl of loading buffer. Denature samples by heating at 95° for 5 min and transfer immediately to ice. Subject 1.5 μl samples of the denatured digests to electrophoresis in a 5% (w/v) polyacrylamide gel, containing 7 M urea, for 6 hr at 3000 V on an ABI Prism model 377 DNA sequencer (Applied Biosystems) with filter set A and a well-to-read length of 36 cm. Analyze T-RFLP profiles with GeneScan, version 2.1, software (Applied Biosystems). Estimate the sizes of terminal-restriction fragments (T-RFs), in base pair lengths, by comparison with the peak signals of the internal standard. See Fig. 4 (color insert) for an example.

Fingerprint and Sequence Analysis of 16S rRNA Gene Clone Libraries

In parallel with the community T-RFLP fingerprint analyses, generate clone libraries of 16S rDNA,[47] amplified from extracted DNA, using a modified pUC19 plasmid vector and ligation-independent cloning system[48] or a PCR-TA cloning system, such as that provided in the TOPO-TA kit (Invitrogen, Groningen, The Netherlands), for rapid and efficient cloning of complex mixtures of PCR-amplified DNA (e.g., community 16S rDNA). Harvest colonies of transformants and screen for the presence of full-length 16S rDNA inserts in preparation for typing the

[45] W. G. Weisberg, S. M. Barns, D. A. Pelletier, and D. J. Lane, *J. Bacteriol.* **173,** 697 (1991).

[46] J. R. Marchesi, T. Sato, A. J. Weightman, T. A. Martin, and J. C. Fry, *Appl. Environ. Microbiol.* **64,** 795 (1998).

[47] A. Felske, and R. Weller, *in* "Molecular Microbial Ecology Manual" (A. D. L. Akkermmans, J. D. van Elsas, and F. J. de Bruin, eds.), p. 3.3.3. Kluwer Academic Publishers, Dordrecht,The Netherlands, 2000.

[48] C. Aslanidis and P. J. de Jong, *Nucleic Acids Res.* **18,** 6069 (1990).

cloned 16S rDNA by fingerprinting and sequencing. Use those clones containing the correct-sized insert (approximately 1.6 kb) to prepare DNA extracts by a rapid colony-boiling method,[49] and PCR amplify the cloned insert, using vector-based primers and reagents and reaction conditions as described by Nogales *et al.*[2] Use the PCR-amplified 16S rDNA containing the expected full-length fragment for amplified ribosomal DNA restriction analysis (ARDRA)[50] for digestion with *Taq*I restriction enzyme (New England BioLabs). Determine the T-RFLP profiles of representative clones determined from *Taq*I restriction clusters, using *Alu*I or *Hha*I as described above, and compare the T-RF patterns of the individual clones with the profiles obtained from the biofilm communities. Directly sequence PCR-amplified 16S rDNAs from representative clones, using a DNA sequencer and the protocols of the manufacturer (Applied Biosystems) for "*Taq* cycle sequencing" with fluorescent dye-labeled dideoxynucleotides. The sequencing primers for 16S rRNA genes have been described previously.[51] Compare the sequence data, using the BLAST or FASTA search algorithms[52] for the GenBank and EMBL databases.[53,54] More detailed comparative sequence analyses may be carried out with conserved primary sequence regions and secondary structure characteristics as alignment reference[55] and sequence similarity calculations for unmasked, nearly complete, sequence pairs. In this manner, specific T-RFs within a complex community profile can be correlated with cloned sequence types, which in turn are identified by sequence determination.

Concluding Remarks

It is currently technically difficult to study microbial interactions in soil, to characterize spatial associations, and to determine structure–function relationships in soil biofilm communities. The experimental set-up described here[1] is technically simple, amenable to experimental manipulation, and allows interactions of autochthonous bacteria with soil colloids to be studied at the nanometer scale.

[49] E. Moore, A. Arnscheidt, A. Krüger, C. Strömpl, and M. Mau, *in* "Molecular Microbial Ecology Manual" (A. D. L. Akkermans, J. D. van Elsas, and F. J. de Bruin, eds.), p. 1.6.1. Kluwer Academic Publishers, Dordrecht, The Netherlands, 1999.

[50] A. A. Massol-Deya, D. A. Odelson, R. F. Hickey, and J. M. Tiedje, *in* "Molecular Microbial Ecology Manual" (A. D. L. Akkermans, J. D. van Elsas, and F. J. de Bruin, eds.), p. 3.3.2. Kluwer Academic Publishers, Dordrecht, The Netherlands, 1995.

[51] D. J. Lane, *in* "Nucleic Acid Techniques in Bacterial Systematics" (E. Stackebrandt and M. Goodfellow, eds.), p. 115. John Wiley & Sons, Chichester, UK, 1991.

[52] W. R. Pearson, *Methods Enzymol.* **183**, 63 (1990).

[53] D. A. Benson, I. Karsch-Mizrachi, D. J. Lipman, J. Ostell, B. A. Rapp, and D. L. Wheeler, *Nucleic Acids Res.* **28**, 15 (2000).

[54] W. Baker, A. van den Broek, E. Canon, P. Hingamp, P. Sterk, G. Stoesser, and M. A. Tuli, *Nucleic Acids Res.* **28**, 19 (2000).

[55] R. R. Gutell, B. Weiser, C. R. Woese, and H. F. Noller, *Prog. Nucleic Acid Res. Mol. Biol.* **32**, 155 (1985).

It seems to be close to nature in important respects, because the principal microorganisms that form the biofilm are similar to those abundant and active in soil samples taken from the same sites, namely bacteria belonging to taxa containing metabolically versatile strains known to degrade different xenobiotics.[2] Moreover, the set-up yields a community that seems to function in the same way as the soil community, because it must use PCBs as a main source of carbon and energy for growth, multiplication, biofilm formation, and production of the extracellular polysaccharide matrix. The set-up, and the analytical procedures for it that are described here, are generally applicable for the study of interactions of microbial communities with surfaces and hydrophobic substrates, and in particular the formation, composition, and functioning of composite biofilms, such as "clay hutches," consisting of microorganisms, their products, and particulate minerals.

The combination of morphological characterization, by light microscopy, ultrastructural analysis by scanning and transmission electron microscopy, determination of element distribution by ESI analysis of ultrathin sections, and characterization of community composition by 16S rRNA gene analysis (here, by T-RFLP, the generation of clone libraries, and the sequencing of selected clones) provides an overview of the structural features of the biofilm down to the nanometer scale and within the range of physiological interactions between individual bacteria and mineral components in the biofilm. The use of FISH permits *in situ* identification of individual bacteria revealed by microscopy. The introduction of radioactive substrates into the water column, and the combined use of *in situ* microautoradiography and FISH, enables structure–function relationships at the level of individual bacteria in the biofilm to be defined.

Because the set-up involves mechanical disruption of the soil structure, with liberation of the biotic and abiotic components into the water column, and the development of new associations from these components, kinetic studies may be carried out to determine the evolution of interactions and events in time and space during *de novo* formation of the biofilm. Although this will ordinarily involve the examination of multiple slides maintained for different periods of time on the surface of the water column, in the example experiments described here a single biofilm was analyzed. However, because the initiation of biofilm formation by individual bacteria is not tightly synchronized, biofilms in different stages of development are available for scrutiny on the same slide. This provides information about the approximate sequence of events in bacterial interactions with clay, other phyllosilicates, and iron–oxohydroxide aggregates, and the overall dimensions of clay mineral shells and their architecture. This in turn allows the estimation of diffusion parameters within the biofilms. One aspect to bear in mind when interpreting the sequence of events in the set-up described here is the low surface energy of the Permanox slide and the fact that different types of bacteria exhibit different degrees of surface hydrophobicity. This may impose a certain selectivity on the system with regard to the initial interactions that occur. Thus, some bacteria intrinsically capable of initiating biofilm formation may not do so in this system, although they may be able subsequently to integrate into the developing biofilm.

The described set-up has provided new insights into composite biofilm formation involving soil colloids, insights that suggest the active recruitment of clay leaflets loaded with hydrophobic soil organic material (in this case, mainly PCBs), which thereby function as nutrient shuttles, and other soil minerals, to form composite biofilm structures that initially represent minimal nutritional spheres and effective survival units. They may also represent key structures in microbiogeological processes.[1]

Acknowledgments

This work was supported by the BMBF (grant 0319433C). K.N.T. gratefully acknowledges generous support from the Fonds der Chemischen Industrie.

[27] Extracellular Polymers of Microbial Communities Colonizing Ancient Limestone Monuments

By Benjamín Otto Ortega-Morales, Alejandro López-Cortés, Guillermo Hernández-Duque, Philippe Crassous, and Jean Guezennec

Introduction

Biofilms are layered microbial communities growing on inert and living surfaces in a variety of terrestrial and aquatic environments, including submerged artificial substrata, invertebrate teguments, sediments, soils, and rock surfaces.[1–4] The exopolymers, mainly composed of polysaccharides, are an important component of such biofilm communities and their functional roles in terms of attachment, nutrient absorption, and protection against desiccation and antimicrobial

[1] J. W. Costerton, R. T. Irvin, and K. J. Cheng, *Annu. Rev. Microbiol.* **35,** 299 (1981).
[2] D. Prieur, F. Gaill, and S. Corre, *in* "Trends in Microbial Ecology: Proceedings of the Sixth Symposium on Microbial Ecology" (R. Guerrero and C. Pedrós-Alió, eds.), p. 207. Spanish Society for Microbiology, Barcelona, Spain, 1993.
[3] M. Kühl, R. Nøhr Gud, H. Plour, and N. Birger Ramsin, *J. Phycol.* **32,** 799 (1996).
[4] J. Guezennec, O. Ortega-Morales, G. Raguenes, and G. Geesey, *FEMS Microbiol. Ecol.* **26,** 89 (1998).

agents.[1,5-8] In addition, given the anionic substituents that these compounds may have, they can interact actively with dissolved ions in aqueous environments and with framework elements in mineral matrices.[9,10] The implications of such interactions between exopolysaccharides (EPS) and minerals may also include cation mobilization in soils, solubilization of toxic metals, and the dissolution of minerals.[6,11-13]

On other hand, studies carried out in Mayan archaeological sites (Yucatán, Mexico) have shown that thick microbial biofilms dominated by cyanobacterial populations contributed to the biodegradation of these buildings by supporting growth of organic acid-producing microorganisms and through active boring.[14,15] These studies have also demonstrated that these epilithic biofilms and their associated EPS were trapping dirt and other particulate materials, leading to chromatic alterations of these Mayan buildings. In addition, these microbial exopolymers may contribute directly to the deterioration processes by interacting through their polysaccharide fraction with metal cations,[16] which could lead to the complexation of soluble calcium, sequestering it from the limestone matrix. Therefore, the exopolysaccharides are thought to have a role in the biological weathering of stone monuments, although their mechanisms are not known as well as those related to the acidification by organic and inorganic acid-producing bacteria and fungi.[17,18]

This chapter describes procedures to extract and characterize the polysaccharidic fraction of naturally occurring microbial exopolymers associated with epilithic (growing on rock) biofilms. These procedures, which are currently used for the analysis of polysaccharides produced by pure bacterial cultures, can be successfully applied to analyze such exopolysaccharides.

[5] J. W. Costerton, G. G. Geesey, and K. J. Cheng, *Sci. Am.* **238,** 86 (1978).

[6] J. L. Geddie and I. W. Sutherland, *J. Appl. Bacteriol.* **74,** 467 (1993).

[7] R. P. Sinha, H. D. Kumar, A. Kumar, and D. P. Häder, *Acta Protozool.* **34,** 187 (1995).

[8] A. López-Cortés, *Precam. Res.* **96,** 25 (1999).

[9] T. J. Beveridge, *in* "Metal–Microbe Interactions" (R. K. Poole and G. M. Gadd, eds.), p. 65. IRL, Oxford, 1989.

[10] L. K. Jang and G. G. Geesey, *in* "Biohydrometallurgical Technologies" (A. E. Torma, L. Apel, and C. L. Brierley, eds.), p. 75. The Minerals, Metals and Materials Society, U.S.A., 1993.

[11] T. J. Beveridge, S. A. Makin, J. L. Kadurugamuwa, and Z. Li, *FEMS Microbiol. Rev.* **20,** 291 (1997).

[12] W. W. Barker and J. F. Banfield, *Geomicrobiol. J.* **15,** 223 (1997).

[13] S. A. Welch and P. Vandevivere, *Geomicrobiol. J.* **12,** 227 (1994).

[14] O. Ortega-Morales, G. Hernández-Duque, L. Borges-Gómez, and J. Guezennec, *Geomicrobiol. J.* **16,** 221 (1999).

[15] O. Ortega-Morales, J. Guezennec, G. Hernández-Duque, C. C. Gaylarde, and P. M. Gaylarde, *Curr. Microbiol.* **40,** 81 (2000).

[16] J. Corzo, M. León-Barrios, V. Hernando-Rico, and A. Gutiérrez-Navarro, *Appl. Environ. Microbiol.* **60,** 4531 (1994).

[17] G. Gómez-Alarcón, M. L. Muñoz, and M. Flores, *Int. Biodet. Biodeg.* **34,** 169 (1994).

[18] A. Gutiérrez, M. J. Martínez, G. Almendros, F. J. González-Vila, and A. T. Martínez, *Sci. Total Environ.* **167,** 315 (1995).

Materials and Methods

Stone samples from Mayan monuments (Yucatán, Mexico) are collected according to two criteria: exposure to direct sunlight (interior and exterior locations supporting biofilm growth), and the degree of degradation (sound nonbiofilmed stone and degraded surfaces). Samples are drawn aseptically with an ethanol-rinsed chisel, kept chilled on ice, and transported to the laboratory, where they are divided into portions for microbiological, microscopic, and biochemical analyses. The specimens for microbiological and microscopic studies are treated 24 hr after their collection, whereas the material for biochemical analysis was freeze-dried and stored under dark conditions for subsequent analyses.

Microscopic Analysis

Microscopic analyses are performed in order to identify the main cyanobacterial genera colonizing these surfaces and to determine the presence of exopolymers associated with natural biofilm samples. Microscopic analyses of scraped subsamples from each site, previously fixed in a buffered ($MgCO_3$) 3% (w/v) solution of formaldehyde in Z-8 medium,[19] are performed. Five micrographs are taken from each site at ×40 to determine relative proportions of the main cyanobacteria genera and at ×100 for identification (phase-contrast microscopy). Cyanobacteria are principally identified according to Geitler,[20] with additional reference to Komárek and Anagnostidis.[21]

Scanning electron microscopy (SEM) analyses are carried out on small fragments of biofilmed stone, which are fixed with 2.5% (v/v) glutaraldehyde for 1 hr, air dried overnight, and stored in a vacuum desiccator. This method has been successfully applied to analyze cyanobacteria-dominated marine epilithic biofilms without significant distortion of the microbial cells.[22] Specimens are then fixed to aluminium stubs, gold-sputtered, and examined with a Philips (Mt. Vernon, NY) XL 30 operating at 30 kV.

Exopolysaccharide Quantification

The exopolysaccharide (EPS) concentrations are measured according to the method proposed by Underwood et al.[23] Briefly, 2 ml of 2.5% (w/v) saline water (NaCl) is added to a known weight of dry ground stone (<30 mg), vortexed

[19] W. W. Carmichael, in "Fundamental Research in Homogeneous Catalysis" (V. Shilov, ed.), p. 1249. Gordon and Breach, New York, 1986.

[20] L. Geitler, in "Rabenhorst's Kryptogamenflora von Deutschland, Osterreich und der Schweiz. 14. Cyanophyceae" (R. Kolkwitz, ed.). Akademische Verlagsgesellschaft, Leipzig, Germany, 1932.

[21] J. Komárek and K. Anagnostidis, Arch. Hydrobiol. Suppl. **73**, 157 (1986).

[22] S. Nagarkar and G. A. Williams, Mar. Ecol. Prog. Ser. **154**, 281 (1996).

[23] G. J. C. Underwood, D. M. Paterson, and R. J. Parkes, Limnol. Ocean. **40**, 1243 (1995).

to suspend sediment, and incubated for 15 min at 20°. The supernatant containing the exopolysaccharide fraction is recovered after centrifugation (3620g) for 15 min and assayed, using glucose as standard.[24]

Monosaccharide Determination

The EPS fraction is obtained according to the method cited above, except that when the supernatant is recovered, 2 volumes of cold absolute ethanol is added and left overnight at 4° to precipitate the polymer. The solution is centrifuged, the supernatant is discarded, and the EPS pellet is dried, dialyzed against distilled water, and freeze-dried. The monosaccharide composition of extracted EPS is determined by forming trimethylsilyl derivatives.[25,26] The exopolysaccharide fraction is then analyzed after acidic methanolysis with 500 μl of methanol–HCl (3 N) and heated in screw-capped tubes at 100° for 4 hr. After cooling of methanolysates, the pH is adjusted to pH 7, the solution is re-N-acetylated, and the resulting methyl glycosides are converted into the corresponding trimethylsilyl derivatives by adding 50 μl of pyridine and 50 μl of silylation agent at ambient temperature for 2 hr. Subsequent gas chromatography (GC) analysis is performed on an HRGC 5160 instrument (Carlo Erba, Milan, Italy) equipped with a flame ionization detector and a fused silica CP-SIL-5CB capillary column (Chrompack) with a temperature gradient from 50 to 120°, at 20° min^{-1}, and from 120 to 250°, at 2° min^{-1}. *myo*-Inositol (50 μl) is used as internal standard.

Results

Microscopic Examination

The microscopy techniques (light microscopy and scanning electron microscopy) used in this study allow the identification and relative quantification of the main genera. *Xenococcus, Synechocystis,* and *Gloecapsa* were the dominant colonizers along with lower abundances of filamentous phototrophs and other bacteria (Fig. 1A and B). Large aggregates of cells belonging to the genus *Xenoccocus* were also seen within polymeric sheaths of variable thickness, using this microscopic technique (Fig. 1A and C).

A similar biofilm composition was found between the various types of surfaces, although their abundances were clearly different (Table I). Scanning electron microscopy images of air-dried samples showed with great detail the biofilm

[24] M. Dubois, K. A. Gilles, J. K. Hamilton, P. A. Reber, and F. Smith, *Anal. Chem.* **28,** 350 (1956).

[25] J. P. Kamerling, G. J. Gerwig, J. F. G. Vliegenthardt, and J. R. Clamp, *Biochem. J.* **151,** 491 (1975).

[26] J. Montreuil, S. Bouquelet, H. Debary, B. Fournet, G. Spik, and G. Strecker, *in* "Carbohydrate Analysis, a Practical Approach. Glycoproteins" (M. F. Chaplin and J. F. Kennedy, eds.), p. 143. IRL Press, Oxford, 1986.

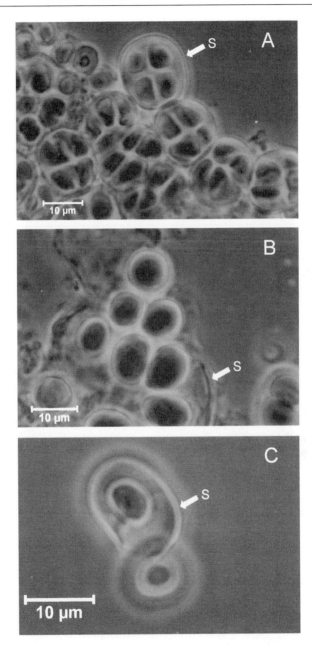

FIG. 1. Phase-contrast microscopy image shows the predominant epilithic unicellular cyanobacterium *Xenococcus* sp., colonizing the walls of these Mayan buildings (Yucatán, Mexico). (A) Groups of sister cells at different stages of division. (B) Liberation of nanocytes. (C) Nanocyte with sheath (S).

TABLE I

Microscopic and Biochemical Characterization of Various Microbial Biofilms and Associated Exopolysaccharides Covering Mayan Buildings in Yucatán, Mexico

	Stone interior ($n = 4$)	Stone exterior ($n = 3$)	Sound stone ($n = 2$)	Degraded stone ($n = 2$)
Macroscopic features	Thick green biofilm	Thick green biofilm	No visible biofilm cover	Thick dark biofilm and debris
Microbial biofilm	High numbers of *Xenococcus, Gloecapsa, Synechocystis*	Medium abundances of *Xenococcus, Gloecapsa*	Sparse coverage of *Xenococcus, Gloecapsa*	*Xenococcus,* filamentous phototrophs, other bacteria
EPS (μg g^{-1})	2103 (1646)^a	379 (154)	161 (112)	698 (186)

^{a} Values represent means ($\pm$SD).

distribution and the morphology of colonizers. These images showed that dense biofilms dominated by coccoid phototrophs were evenly distributed on interior surfaces and on exterior surfaces in shaded and moist locations, whereas the biofilm cover of sound stone surfaces, without any visible sign of biological colonization, was sparse and exhibited a patchy distribution (Fig. 2A). The scanning electron micrographs also revealed the presence of aggregated microcolonies of unicellular cyanobacteria as well as of filamentous cyanobacterial morphotypes embedded in an extensive exopolymeric matrix (Fig. 2B). Similar cyanobacterial populations belonging to the coccoid genera *Xenoccocus, Gloecapsa,* and *Synechocystis* were previously reported as main colonizers in the limestone buildings at this Mayan site.[15] It appears that chrococcalean cyanobacteria (i.e., *Gloecapsa* and *Gloethece*) are common inhabitants of calcareous substrata.[20] The use of both light and scanning electron microscopy provided complementary information and allowed in this study a preliminary description of the community structure and distribution of archaelogical epilithic biofilms, as previously reported for tropical marine cyanobacteria-dominated epilithic biofilms.[22]

Exopolysaccharide Measurement

The analysis of exopolysaccharides (EPS) extracted from these epilithic microbial communities showed that substantial and variable levels of this exopolymeric material were generally associated with these biofilms, with quantitative differences in EPS concentrations between interior and exterior surfaces, which were colonized by massive and scarce biofilms, respectively (Table I). A similar trend was observed in sound nonbiofilmed stone, where low amounts of EPS were detected, compared with the degraded surfaces. In general terms, the range of concentrations of EPS detected in these biofilms is smaller than those reported

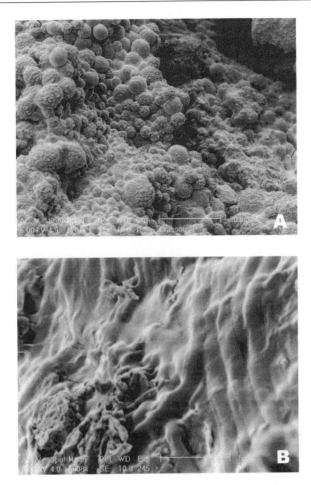

FIG. 2. (A) Scanning electron micrograph of air-dried stone specimens showing an epilithic biofilm dominated by coccoid phototrophs. (B) Exopolymeric material associated with the cyanobacteria-dominated biofilms colonizing the Mayan archeological site of Uxmal.

for sandy sediments colonized by epipelic diatoms, known to produce copious amounts of EPS.[23] However, these findings could indicate that cyanobacterial extracellular polymers consist primarily of more tightly bound polymers (capsular polysaccharides), seen as sheaths around cells in this study (Fig. 1A). This has been previously reported for certain cyanobacterial genera.[27] The occurrence of substantial amounts of exopolymeric material embedding microbial cells, as evidenced

[27] C. Bertocchi, L. Navarini, A. Cesàro, and M. Anastasio, *Carbohydr. Polym.* **12,** 127 (1990).

TABLE II
RELATIVE MONOSACCHARIDE COMPOSITION OF EXOPOLYSACCHARIDES RECOVERED
FROM EPILITHIC BIOFILMS COLONIZING MAYAN BUILDINGS IN YUCATÁN, MEXICO

	Type of surface			
Monosaccharide	Stone interior ($n = 4$)	Stone exterior ($n = 3$)	Sound stone ($n = 2$)	Degraded stone ($n = 2$)
Glucose	25.3 (16.0)[a]	47.8 (19.7)	67.8 (95.8)	18.7 (7.3)
Mannose	27.0 (20.0)	14.6 (2.3)	22.7 (7.3)	17.9 (4.7)
Galactose	22.2 (15.1)	14.4 (4.2)	9.5 (13.4)	12.1 (4.5)
Xylose	4.9 (2.8)	8.8 (3.8)	—[b]	11.8 (5.1)
Fucose	1.7 (1.3)	3.3 (3.0)	—	—
Arabinose	7.1 (3.8)	7.7 (2.6)	—	14.5 (3.2)
Rhamnose	5.6 (3.3)	3.4 (3.2)	—	7.4 (3.5)
Galacturonic acid	2.8 (3.3)	—	—	17.7 (4.4)
Glucuronic acid	3.4 (4.7)	—	—	—

[a] Values, expressed as mol%, represent means (±SD).
[b] —Not detected.

by microscopic and biochemical analyses, may represent ecological adaptations to the epilithic environments, including resistance to desiccation.[28,29]

Monosaccharide Composition of Polysaccharides

The microbial exopolysaccharides analyzed exhibited qualititative and quantitative differences in their monomeric composition, depending on the type of biofilm from which they were extracted, although the major sugars in all biofilms were the neutral sugars including glucose, mannose, and galactose, the remaining neutral and acid monosaccharides representing less than 25.5% (Table II). On the other hand, the EPS analyzed may have multiple sources of origin and it is not possible to link their monosaccharide composition to a specific microbial group or species. However, considering the predominance of cyanobacteria in these biofilms, it is likely that the monosaccharides detected reflect the exopolysaccharidic composition of cyanobacteria, which are known to produce highly variable polymers comprising neutral and acid sugars.[27,30] The presence of negatively charged functional groups (i.e., uronic acids) in the EPS analyzed suggests that these polymers may either enhance mineral dissolution by complexing the solubilized calcium, increasing the apparent solubility, or by forming metal–organic complexes at the

[28] G. Mazor, G. J. Kidron, A. Vonshak, and A. Abeliovich, *FEMS Microbiol. Ecol.* **21,** 121 (1996).
[29] B. De Winder, H. C. P. Matthijs, and K. J. Cheng, *Annu. Rev. Microbiol.* **35,** 458 (1981).
[30] A. Flaibani, Y. Olsen, and T. J. Painter, *Carbohydr. Res.* **190,** 235 (1989).

mineral surface, making reactive sites unavailable for further reaction.[13] Interestingly, in our study the monuments that showed degraded surfaces colonized by biofilms had the greatest amounts of galacturonic acid (17.7%). In addition, the presence of other anionic substituents, such as sulfate and pyruvate in microbial exopolysaccharides,[10] cannot be ruled out, as certain cyanobacterial genera produce sulfated exopolymers or sheaths.[31,32] Therefore, further studies are needed to determine the presence of these compounds in exopolysaccharides associated with these epilithic biofilms.

Concluding Remarks

The methods described in this chapter allowed a preliminary characterization of epilithic biofilms and their associated exopolysaccharides. The identification and the determination of the relative abundance of the main cyanobacterial genera were successfully carried out using phase-contrast microscopy. Scanning electron microscopy of air-dried samples provided clear images of the biofilm appearance and its distribution, without significant distortion of cells, typical of traditional dehydration procedures. The analysis of the exopolysaccharidic fraction allowed the determination of the monosaccharidic composition of these compounds, but further analyses to determine the presence of substituents (sulfate, pyruvate) and the protein and lipid content are needed to fully characterize these compounds. This information would be valuable for the understanding of potential role(s) of EPS in the deterioration processes of mineral matrices, such as stone surfaces in historic monuments.

Acknowledgments

This work was supported by the Consejo Nacional de Ciencia y Tecnología (Mexico), through a research grant, Biodeterioration of Mayan Monuments in Yucatán, Mexico (CONACYT 1997).

[31] Y. Bar-Or and M. Shilo, *Appl. Environ. Microbiol.* **53,** 2226 (1987).
[32] J. J. Ortega-Calvo, X. Ariño, M. Hernández-Marine, and C. Saiz-Jiménez, *Sci. Total Environ.* **167,** 329 (1995).

[28] Studying Phototrophic and Heterotrophic Microbial Communities on Stone Monuments

By Clara Urzì and Patrizia Albertano

Introduction

Paraphrasing a classic definition of soil, we can define stone monuments as the place in which art and (micro)biology encounter each other. The outdoor and indoor exposure of lithic surfaces leads, in fact, to the interaction of stone with its close environment and the consequent colonization by microorganisms, whose diversity and biodeteriorative activity are dependent on the carbon and energy sources as well as on the bioreceptivity of the substrata.[1,2]

Stone monuments dating back to prehistory and more recent history have been built up and decorated with a multitude of materials, the majority of which consisted of rocky and mineral components. Although organic compounds have been also used, particularly in the case of wall paintings and frescoes, the metabolic activity of microorganisms has resulted in the biomineralization of these types of substrata. Therefore, stone monuments are, in general terms, considered as built up by inorganic (both natural and artificial) materials.

In this type of manmade terrestrial environment, most of the organism settlement occurs at the surface and results in the formation of more or less thick biofilms that originate from the development of air-borne cells and spores. However, microbial colonization may produce different patterns and appear as a patchy distribution of cells that accumulate in fissures, cracks, or in subsurface and deep layers, depending on the porosity and state of conservation of the material as well as on the ecological requirement of individual species. Therefore, according to the distribution on or within the substratum, colonizing microorganisms can be distinguished in epilithobionts and endolithobionts, respectively.[3] Among the latter group, chasmoendolithic (living in preexisting cracks and fissures), cryptoendolithic (developing beneath the mineral surface layer), and euendolithic (actively boring the calcareous substratum) forms can be recognized.

In this chapter, we deal with the most recent methodologies that have been applied to the study of phototrophic and heterotrophic microorganisms by researchers working in the field of conservation of cultural heritage.

[1] O. Guillitte, *Sci. Total Environ.* **167**, 215 (1995).

[2] C. Urzì and M. Realini, *Int. Biodet. Biodeg.* **42**, 45 (1998).

[3] S. Golubic, I. E. Friedmann, and J. Schneider, *J. Sedim. Petrol.* **51**, 47 (1981).

How to Study Microbial Communities in "Monument" Habitat?

Most of the current methodological approaches to the study of microbial deterioration have been dealing with the quantification and identification of the different types of microorganisms within the stone community. A description of these methods is not included in this chapter.

Alternatively, two other aspects, both crucial in the assessment of the biodeteriorative action of microbes on monuments, are considered. The first involves the understanding of the biofilm architecture and the structural relationships existing between different functional groups of organisms. The second is based on the understanding of biofilm function and of the metabolic machinery that leads to the transformation of the lithic substrata and to a consequent change of the monument ecosystem. New methodologies are, therefore, focusing on these aspects in the attempt to provide the information needed for the establishment of control and prevention strategies against biodeterioration processes.

Components of Stone Microbial Community

It is well established that representatives of archaea, eubacteria, and eukarya can colonize stone materials and contribute to the deterioration of manufacts.[4–7] Among eubacteria and eukarya both photo- and heterotrophic organisms have been studied and techniques leading to their detection are widely employed. Chemoorganotrophic eubacteria of stone monuments are mainly gram-positive taxa, among which *Bacillus* (with low content of G+C) and actinobacteria representatives (*Micrococcaceae, Streptomycetaceae, Geodermatophilaceae, Micromonosporaceae*, etc., with high G+C content) are the most commonly detected. Archaea have also been found, and they are treated in [29] in this volume.[8]

Terrestrial phototrophs are able to colonize exposed rock surfaces outdoor and inside confined environments. Outdoor colonization of stone monuments is due to a variety of species belonging, mainly, to cyanobacteria and eukaryotic algae.[9] Cyanobacteria are the most important photosynthetic indoor organisms and, often together with a few green algae and diatoms, the only primary producers.[4] In

[4] P. Albertano and C. Urzì, *Microb. Ecol.* **38**, 24 (1999).

[5] S. Roelleke, A. Witte, G. Wanner, and W. Lubitz, *Int. Biodet. Biodegr.* **41**, 85 (1998).

[6] U. Wollenzien, G. S. de Hoog, W. E. Krumbein, and C. Urzì, *Sci. Total Environ.* **167**, 287 (1995).

[7] J.-J. Ortega Calvo, X. Ariño, M. Hernandez-Mariné, and C. Saiz-Jimenez, *Sci. Total Environ.* **167**, 329 (1995).

[8] G. Piñar, C. Gurtner, W. Lubitz, and S. Rölleke, *Methods Enzymol.* **336**, Chap. 29 (2001), this volume.

[9] J. J. Ortega-Calvo, M. Hernandez-Mariné, and C. Saiz-Jimenez, in "Recent Advances in Biodeterioration and Biodegradation" (K. L. Garg., N. Garg, and K. G. Mukerji, eds.), Vol. 1, p. 173. Naya Prokash, Calcutta, India, 1993.

hypogea, cyanobacteria form more or less thick biofilms close to the entrances and throughout the photic zone. The species composition is determined by light conditions and humidity. The calcifying *Scytonema julianum* and *Geitleria calcarea* and species of the genera *Chroococcus, Gloeocapsa, Aphanocapsa,* and *Nostoc* have been often reported for natural hypogea as well.[10]

Among fungi, mostly dematiaceous filamentous strains as well as meristematic, growing fungi seem to be dominant.

In general, oligotrophic behavior and high adaptability to different environmental conditions characterize these microbiota. It is noteworthy that most of the microorganisms developing on monuments appear unknown at the genus or species level, and some of them are presently studied in order to assess their phylogenetic and taxonomic positions.[11]

Methodology

Sampling Methods

In working with valuable objects of cultural and artistic relevance, nondestructive sampling methods are obviously preferred to destructive methods. The use of adhesive tape, as a noninvasive sampling technique, offers the possibility of sampling the organisms present on the rock faces, detecting them directly under the microscope, obtaining evidence of the existing connections with the substratum, and, eventually, cultivating the microorganisms for laboratory tests.[12] However, sometimes the removal of small portions of material (a few milligrams or grams of stone) is unavoidable in order to detect the extension, pattern, and depth of microbial colonization.

Materials and Reagents

4′,6′-Diamidino-2-phenylindole (DAPI)
Acridine orange (AO)
Adhesive scotch tape (fungi tape; DID, Milan, Italy)
Amman's lactophenol (20 g of pure phenol crystals, 20 g of lactic acid, 40 g of glycerol, 20 ml of distilled water)
Bioadhesive tape (Adhesive tabs; Electron Microscope Science, Washington, D. C.)
Coverslips

---

[10] L. Hoffmann, *Bot. Rev.* **55,** 77 (1989).

[11] C. Urzì, P. Salamone, P. Schumann, and E. Stackebrandt, *Int. J. System. Evol. Microbiol.* **50,** 529 (2000).

[12] C. Urzì, F. De Leo, and P. Salamone, *in* "Proceedings of 10th Euromarble Eurocare EU496 Workshop" (U. Lindborg, ed.), p. 36. NHB, Stockholm, Sweden, 2000.

Disposable filters (pore diameter, 0.22 μm)
Distilled water
Glass slides
Glass slides for fluorescence (SuperFrost; Forlab, Carlo Erba, Milan, Italy)
HCl
Metal slides

Nondestructive Sampling and Specimen Preparation for Microscopy. Adhesive tape is gently placed on the surface of the stone material and then transferred to a sterile glass slide. Small pieces of the tape are then cut and sorted for the following analyses.

LIGHT MICROSCOPY AND EPIFLUORESCENCE

1. Put one drop of sterile distilled water, Amman's lactophenol, or acridine orange on a slide.
2. Place the tape face down on the slide.
3. Cover with a coverslip in order to keep the tape as flat as possible.

CONVENTIONAL AND ENVIRONMENTAL SCANNING ELECTRON MICROSCOPY. No specific preparation procedure is used. The mounting of samples without previous chemical treatment has been shown to allow scanning electron microscopy (SEM) observation, although some shrinking of cells occurs. Environmental SEM (ESEM) (see below), however, allows direct observation of fresh, fully hydrated samples.

1. Mount, face up, a piece of the sampling tape directly on aluminum stubs, using biadhesive tape.
2. Keep in dry conditions.
3. Metalize by sputtering with gold or gold/palladium.
4. Observe at 15 kV.

CULTURE ANALYSIS. According to the result of the microscopic observation, cultivation and isolation of specific microorganisms can be required. In this case, a piece of the sampling tape is cut and placed directly into suitable cultural agarized or liquid media.[2,4,13]

Detection of Rock-Inhabiting Microorganisms on Stone Specimens. The alternative procedure to detect microorganisms from stone material is to sample the rock

[13] R. Rippka, J. Deruelles, J. B. Waterbury, M. Herdman, and R. Y. Stanier, *J. Gen. Microbiol.* **111,** 1 (1979).

(e.g., with a sterile scalpel or tweezers) and observe it directly by epifluorescence and/or confocal laser scanning microscopy (CLSM).

Analysis by Microscopy

Various microscopy techniques are currently used to answer various questions. Inverted, transmitted, and Nomarski difference interference contrast (DIC) light microscopes can provide information about microorganism morphology and structural relationships existing among different microbial species and between microbes and the colonized substratum. The combination of light microscopy (LM) with an epifluorescence system is generally suggested for easier discrimination of phototrophic from heterotrophic organisms and for identification through targeted fluorochromes of species present within biofilms.

Autofluorescent phototrophic microorganisms appear bright red, orange, or yellow, depending on the filter set and the prevailing photosynthetic pigments. By combining the appropriate fluorochromes, it is also possible to detect nonfluorescent organisms.

For the direct detection of total nonautofluorescent microorganisms, the most common fluorochromes used are acridine orange (AO) and 4',6'-diamidino-2-phenylindole (DAPI). The AO fluorochrome (excitation filter 450–490 nm) confers a uniform green stain to the bacteria, phototroph cytoplasm autofluoresces red, but the nuclei of eukaryotic cells appear green, while the cytoplasm of heterotrophs is orange. Stone particles emit a green/orange fluorescence. DAPI fluorochrome (excitation filter 340–380 nm) confers a blue coloration to DNA. Bacterial nucleoplasms are seen as blue uniform spots, and photosynthetic membranes appear in red and eukaryotic nuclei in blue.

Black-pigmented bacteria and fungi are not stained with fluorochromes, as they can be easily distinguished on bright backgrounds.

Procedure

EPIFLUORESCENCE

1. Scrape or powder a small amount of the rock surface.
2. Treat the material with a 4% (w/v) solution of HCl in order to minimize the amount of inorganic particles.
3. Eliminate excess HCl.
4. Suspend in distilled water.
5. Put one drop (20 μl) of suspension on a slide for fluorescence.
6. Add 20 μl of a sterilized solution of AO (0.1 mg/ml) or of DAPI (0.1 mg/ml). Sterilization of fluorochrome solution is carried out by filtration, using disposable filters.
7. Mount a coverslip.
8. Observe under a microscope (magnification ×40 and ×100).

CONFOCAL LASER SCANNING MICROSCOPE. When using CLSM, no special treatment of the rock sample is needed. However, nonautofluorescent cells must be labeled with fluorochromes. Examination of rock samples with an inverted CLSM microscope is preferred to the upright type because of the thickness of specimens. The application of confocal imaging does not need pretreatment or fixation of samples and provides in a nondestructive manner, effective information about the three-dimensional structure of thick intact biofilms, and evidence of microbial relationships and the distribution and colonization pattern of organisms on irregular substrata.

1. Place on the rock surface a drop of AO solution (0.2 mg/ml in water or 1:10–1:100 dilutions) for 1 min.
2. Rinse with a small amount of sterile distilled water.
3. Mount the rock sample face down on a metal slide with a hole over in which a coverslip is fixed with paraffin oil.
4. Observe under a microscope (magnification $\times 40 - \times 100$).

ELECTRON MICROSCOPY. Surface structure and relationships of organisms had usually been investigated by conventional scanning electron microscopy (SEM), until the development of the environmental scanning electron microscopy (ESEM). In addition, transmission electron microscopy (TEM), although unfortunately scarcely employed in this field of study, provides useful information about cell structural integrity and intercellular relationships. Ultrastructural features helpful for the identification of phototrophic microorganisms, i.e., thylakoid pattern and type of cell inclusions, have been revealed by the application of TEM.[14] In particular, the structure of the exopolymeric biofilm matrix is revealed in detail, showing differences in the arrangement of exopolysaccharide (EPS) fibrils, whereas cytochemical stains of samples offer the means to ascertain the presence of glucan reserves and complex capsular polysaccharides.[15,16]

Environmental Scanning Electron Microscopy Characteristics. The ESEM is a new type of microscope that allows the observation of fresh samples in the natural state, without modification or preparation. Samples are mounted directly on aluminum stubs, using biadhesive tape, and observed at 15 kV. However, the primary advantage of this method lies in its permitting variation in the sample environment through a range of pressures, temperatures, and gas compositions and in retaining at the same time the performance of a conventional SEM. Observations made

[14] P. Albertano, L. Kovacik, and M. Grilli Caiola, *Arch. Hydrobiol. Algological Studies* **75**, 71 (1994).
[15] P. Albertano, L. Luongo, and M. Grilli Caiola, *in* "Science, Technology and European Cultural Heritage" (N. Baer, C. Sabbioni, and A. I. Sors, eds.), P. 501, Butterworth-Heinemann, Oxford, 1991.
[16] P. Albertano, L. Luongo, and M. Grilli Caiola, *Nova Hedwigia* **53**, 369 (1991).

with the ESEM (Electroscan; Philips, Eindhoven, Holland) showed appreciable differences in the aspect of filamentous microorganisms; the surface of most of them appeared smooth and the cylindrical shape was never deformed. Calcite crystals, which are deposited on the polysaccharide sheath of calcifying filamentous cyanobacteria, also appeared different from those observed by conventional SEM.[3]

Sample Fixation for Scanning and Transmission Electron Microscopy. Prefixation of samples can be performed directly in buffered aldehyde fixatives, 2.0% (w/v) paraformaldehyde or 2.5% (w/v) glutaraldehyde, for 1–2 hr at room temperature or overnight at 4°. However, when biofilms are scraped from the rock surface, fixation can be conveniently performed by adding 2.5% (w/v) sodium-EDTA to the glutaraldehyde fixative in order to solubilize undesired carbonate particles. After three washings in buffer for 10 min, postfixation with buffered 1% (w/v) osmium tetraoxide follows for 1 hr. Dehydration is made in a graded ethanol series [70, 85, and 95% (v/v) and in anhydrous ethanol (100%) for 15 min each. Sample preparation can then proceed either for SEM or TEM according to conventional standardized protocols.

Recommended Cytochemical Methods

Staining of biofilms can facilitate the observation of particular cells or structures during microscopic observation, and in addition provide information about various types of molecules that react positively to specific chemical procedures. Some of these cytochemical stains can be applied directly to the stone material in order to determine the relationships between substrata and organisms; others can be applied to samples processed for TEM in order to localize molecules or structures at the subcellular level.

An important approach to the study of the microbial community colonizing rocks or stone surfaces consists of an assessment of the extension of the colonization process. Other than pigmented cells, which provoke an aesthetic deterioration, most of the organisms are not clearly detectable by eye. Thus, staining procedures help to detect organisms and/or their by-products. For example, the periodic acid–Schiff (PAS) technique is used to reveal acidic structural and reserve cell polymers.

Periodic Acid–Schiff's Reagent Staining for Light Microscopy

Using the following procedure, compounds such as extracellular polymeric substances (EPS), glycogen, starch, cellulose, chitin, mucin, protein–carbohydrates complexes, and glycolipids appear red. This technique was proposed for stone material by Warschied[17] and then successfully applied to estimate the extent of microbial colonization on/within stones.[2,18]

---

[17] T. Warscheid, Ph. D. Thesis, University of Oldenburg, Oldenburg, Germany, 1990.

Materials and Reagents

Distilled water
Ethanol
Glass vessels or glass petri dishes (diameter and height according to the rock
    sample size)
Periodic acid (Sigma-Aldrich, St. Louis, MO)
Schiff's reagent (Sigma-Aldrich)
Sodium metabisulfite (Sigma-Aldrich)

Procedure

1. Fix the rock sample with 70% (v/v) ethanol for 2 hr or overnight (according
to the size of the rock sample).
2. Transfer to 1% (w/v) periodic acid and shake for 5–8 min.
3. Transfer to 70% (v/v) ethanol and shake for 5 min.
4. Rinse in distilled water for 5 min.
5. Transfer to Schiff's reagent and keep shaking for 10 min.
6. Transfer to 0.6% (w/v) sodium metabisulfite and shake for 3 min.
7. Repeat step 6.
8. Wash twice in distilled water for 5 min each.
9. Transfer to 70% (v/v) ethanol and shake for 30 min.

Note: Stones whose composition is mainly carbonate can be partially destroyed
by the hydrochloric acid present in the Schiff's reagent. By keeping rock samples
for a longer time in ethanol, the red color of organic components is intensified. In
contrast, coloration of porous material is reduced.

Periodic Acid–Thiosemicarbazide–Silver Proteinate Reaction for
Ultrathin Section

The periodic acid–thiosemicarbazide–silver proteinate (PATAG) technique[19] is
based on the oxidation of polysaccharides and polysaccharide–protein complexes
by periodic acid and subsequent demonstration of dialdehyde oxidation products of cell-bound 1,2-glycols by thiosemicarbazide (TSC) and silver proteinate.
Excellent demonstration of glycogen and complex polysaccharides as electron–
dense particles and fibrils can be obtained by this method in glutaraldehyde- and
glutaraldehyde/OsO_4-fixed and Epon-embedded material.

[18] O. Salvadori, *in* "Proceedings of International Conference on Microbiology and Conservation"
    (O. Ciferri *et al.*, eds.), p. 73. CNR, Florence, Italy, 1999.
[19] J. P. Thièry, *J. Microsc.* **6,** 987 (1967).

Material and Reagents

Periodic acid (1%, w/v; Sigma-Aldrich): The solution can be stored up to
    1 month at 4°
Thiosemicarbazide (1%, w/v) in 10% (v/v) acetic acid, freshly prepared
Silver proteinate (1%, w/v; Merck, Darmstadt, Germany): It can be stored in
    the dark up to 1 month at 4°
Hydrogen peroxide (10%, v/v)
Glutaraldehyde (2%, w/v)-fixed, Epon-embedded samples
Gold grids (300 mesh) previously washed in acetone
Bidistilled water
Freshly prepared acetic acid series [10, 5, 2, and 1% (v/v)]
Glass vessels
Thin sections collected on gold grids (at least two for each sample)
Moisture and dark glass chambers.

Procedure

1. Float sections (1 grid as control) on 10% (v/v) hydrogen peroxide for 30 min
at room temperature.

2. Float the other sections on 1% (w/v) periodic acid for 30 min at room
temperature.

3. Wash all sections with three changes of bidistilled water, 10 min each.

4. Immerse all sections in 1% (w/v) thiosemicarbazide in 10% (v/v) acetic
acid for 1–24 hr in a moisture chamber.

5. Rinse three times, 1 min each, in 10% (v/v) acetic acid.

6. Rinse in an acetic acid series of 10, 5, 2 and 1% (v/v) for 20 min each.

7. Wash three times in bidistilled water for 10 min.

8. Immerse in 1% (w/v) silver proteinate in a dark chamber for 30 min.

9. Wash two times in bidistilled water for 10 min.

10. Drain off excess water with filter paper.

Note: Control sections should not show electron-dense stain of polysaccharide
compounds.

Studying Whole Single Cells

Fluorescence in Situ Hybridization on Glass Slide. Fluorescence *in situ* hybridization (FISH) is currently used to analyze complex communities as well
as for the "*in situ*" identification of microorganisms and for the phylogeny of
unculturable microorganisms.[20] To date, the application of this method to the

---

[20] R. Amann, W. Ludwig, and K.-H. Schleifer, *Microbiol. Rev.* **59,** 143 (1995).

"monument" communities has been restricted to a few case studies, reflecting the difficulties encountered in the use of molecular probes in this peculiar habitat and the poor knowledge of these microbial communities.

Hybridization protocols are similar to those described for other environments. Therefore, we do not discuss general problems, e.g., melting temperature (T_m), salt concentration, and hybridization stringency, that in any case must be carefully considered during the procedure. Appropriate values for each parameter should be, in fact, calculated for every single oligonucleotide used.

However, the following specific problems are often encountered in the analysis of stone communities:

Inorganic particles interference with the nucleic acid

Scarce taxonomic information about autochthonous species and consequent difficulty to build the right oligonucletide probe specific for the target microorganism

Interference with autofluorescent organisms in the same sample

Metabolic and morphological status of the bacteria and fungi (scarce or no permeability of cells and thus inaccessibility to the target rRNA, aggregate formations, etc.)

Interference of nucleic acids with inorganic particles and heavy metals is a well-known problem encountered in many natural terrestrial environments.[21] Protocols should consider the use of chelating agents in order to reduce the concentrations of disturbing elements.

The design of appropriate oligonucleotide sequences [mainly small subunit (SSU) rRNA] (length between 18 and 20 bp) is based on the knowledge of the strains previously isolated and subsequently taxonomically and phylogenetically characterized. For mixed populations in which autofluorescent organisms are also present (e.g., cyanobacteria and eukaryotic algae), the problem of undesired background fluorescence has still to be solved.

An additional problem is due to the tendency of microorganisms living on rocks to cluster in more or less thick aggregates that make observation by microscopy difficult. The classic procedure proposed by Amann[14] can then be followed, using either the appropriate adjustments set up for bacteria[22,23] or the protocol for fungi suggested by Sterflinger and Hain.[24]

Materials and Reagents

Antifading agent [Slow Fade antifade kit (Molecular Probes, Eugene, OR); *Citifluor AFI* (Citifluor, London, UK)]

[21] I. G. Wilson, *Appl. Environ. Microbiol.* **63,** 3741 (1997).

[22] V. La Cono, Master's thesis, p. 139. University of Messina, Messina, Italy, 2000.

Coverslips (preferably UV transparent and of high quality, such as Hybri-Slips, 22 × 60 mm; Sigma-Aldrich)

DAPI

Distilled water

Ethanol series of 50, 80, and 96% (v/v)

Ethylenediaminetetraacetic acid (EDTA) sodium salt, pH 7.2 (Sigma-Aldrich)

Fixation buffer: 4% (w/v) paraformaldehyde in phosphate-buffered saline (PBS)

Fluorescent probe(s)

Formamide (Sigma-Aldrich)

Gelatin solution: 0.1% (w/v) gelatin, 0.01% (w/v) $CrK(SO_4)_2 \cdot 12\ H_2O$, in ultrapure H_2O

Hybridization buffer: 0.9 M NaCl, 100 mM Tris-HCl (pH 7.2), 0.1% (w/v) sodium dodecyl sulfate (SDS), 35% formamide [concentration range between 10 and 60% (v/v)], in ultrapure H_2O

KOH ethanolic: 10% (w/v) KOH in ethanol

Lysozyme solution (1 mg/ml) in 100 mM Tris plus 5 mM EDTA, pH 7.5

Nonidet P-40 (Sigma-Aldrich)

PBS: 130 mM sodium chloride, 10 mM sodium phosphate buffer, pH 7.2

Slides: Teflon-coated slide, 12 well, 5 mm (Electron Microscope Science, Washington, D.C.)

SDS

Tris-HCl (pH 7.2)

Ultrapure water (Millipore, Bedford, MA)

Washing buffer: 180 mM NaCl, 20 mM Tris-HCl (pH 7.2), 0.1% (w/v) SDS, 5 mM EDTA (pH 7.2) in ultrapure H_2O

Washing solution: 100 mM Tris-HCl, 5 mM EDTA, 0.1% (w/v) Nonidet P-40, ultrapure H_2O, pH 7.2

β-1,4-Glucanase from *Bacillus subtilis* (Fluka Chemie, Buchs, Switzerland) (8.4 units in 400 μl of Tris buffer; 1 M, pH 6.0)

Aggregate Separation

A homogeneous microbial suspension is important for the optimal visualization of cell hybridization. Different methods can be applied to destroy cell aggregates present in the microbial films.

Procedure

1. Centrifuge the microbial suspension at 6000 rpm for 10 min.
2. Discard the supernatant.

[23] D. Hahn, R. Amann, and J. Zeyer, *Appl. Environ. Microbiol.* **59,** 1709 (1993).

[24] K. Sterflinger and M. Hain, *Studies Mycol.* **43,** 23 (1999).

3. Place 200 μl of the pellet in an Eppendorf tube and disrupt aggregates mechanically, using micropestles (Eppendorf, Hamburg, Germany).
4. Suspend in 200 μl of PBS.
5. Vortex with glass beads (diameter, 0.45–0.50 μm) for 30–60 sec.
6. Sonicate mildly for six cycles of 30 sec each.
7. Check the level of homogeneity of the microbial suspension.
8. Repeat, if necessary, step 5 or 6.

Bacterial Cell Preparation

1. Fix the cells at 4° for 3 hr by adding 3 volumes of fixation buffer in order to avoid enzymatic processes and allow a slight cell permeabilization.
2. Rinse in PBS.
3. Suspend in PBS–ethanol solution (1:1, v/v).
4. Store stock cell suspension at −20°.

Fungal Preparation

1. Collect the fungal mycelium on paper by filtration.
2. Wash with washing buffer.
3. Transfer 0.01 g of biomass into Eppendorf tubes.
4. Add 200 μl of β-glucanase.
5. Incubate at 55° for 6 hr.
6. Remove excess solution.
7. Rinse three times with PBS buffer.
8. Remove PBS.
9. Add 500 μl of fixation buffer.
10. Incubate overnight at 4°.
11. Eliminate excess solution.
12. Wash several times with PBS.
13. Store at −20° in PBS–ethanol (1:1, v:v).

Slide Preparation

1. Wash slides for 1 hr in ethanolic KOH.
2. Rinse in ultrapure water.
3. Air dry.
4. Immerse in gelatin solution at 70°.
5. Dry in a vertical position.

Fixation on Slides

The procedures suggested here are useful for most of the rock-inhabiting bacteria (mainly gram positive) and fungi (mostly melanized) and include cell wall permeabilization with lysozyme or glucanase, respectively. Contact time will depend on the thickness of the cell wall.

Procedure for Bacteria

1. Apply in the well of each slide 3 μl of a 10^9 -cell/ml suspension (O D of 1.0 at a wavelength of 550 nm).
2. Air dry in the dark.
3. Dehydrate in graded ethanol series [50, 80, and 96% (v/v)] for 3 min each.
4. Incubate at room temperature with lysozyme solution for 30–60 min.
5. Rinse with washing solution for 10 min.
6. Repeat steps 3 and 5.

Note: Step 4 is not necessary for gram-negative bacteria.

Procedure for Fungi

1. Place 5 μl of the fixed material in the well of a glass slide.
2. Air dry at room temperature.
3. Dehydrate in graded ethanol series [50, 80 and 96% (v/v), 3 min each step].
4. Eliminate the excess ethanol.

Oligonucleotide and Fluorochrome Probe Design

For the development of "*in situ*" identification methods for stone-inhabiting bacteria and fungi, the design of appropriate probes is required. However, it appears that most of the microbes living on rocks are at present unknown and thus the availability of probe sequences is still limited. For the study of a complex community, a top-to-bottom approach is suggested, in which probes are first used to distinguish within the domain level (archea, eubacteria, and eukarya), and then more specific probes should be used.[25] Oligonucleotides [Pharmacia (Uppsala, Sweden) or Life Technology (Rockville, MD)] are labeled in the 5' position with various molecules such as fluorescein isothiocyanate (FITC), tetramethylrhodamine isothiocyanate (TRITC), and Cy3 and Cy5 (indocarbocyanines). Table I present oligonucleotides already tested in the field of rock-inhabiting microorganisms. However, new probes should be available from more recent studies carried out in our laboratory as well as in others.

Hybridization

Hybridization should be performed in a moisture chamber to avoid concentration of the hybridization solution due to evaporation.

---

[25] R. Amann, W. Ludwig, and K.-H. Schleifer, *FEMS Microbiol. Lett.* **100**, 45 (1992).

TABLE I
OLIGONUCLEOTIDES TESTED FOR STONE-INHABITING MICROORGANISMS

Oligonucleotide	Type	Target position
EUB338 of 18 bp, 5'-GCTGCCTCCCGTAGGAGT-3'	Universal for bacteria[a]	16S rDNA, *Escherichia coli*
NOTEUB338 of 18 bp, 5'-ACTCCTACGGGAGGCAGC-3'	Universal for bacteria[a]	16S rDNA, *E. coli*
GPBHGC1901 of 18 bp, 5'-TATAGTTACCACCGCCGT-3'	Gram-positive bacteria with high G+C content[b]	23S rDNA, *E. coli*
Str950 of 18 bp, 5'-GCGTCGAATTAAGCCACA-3'	*Streptomyces* sp.[c]	16S rDNA, *E. coli*
AP665 of 21 bp, 5'-TTCGTTTAGTTATGAATC-3'	*Aureobasidium pullulans*[d]	18S rDNA, *Neurospora crassa*
S-K-Fungi-0521-a-A-17, TAAGGG(A/G)TTTA(A/G)ATTGT	Universal for fungi[e]	18S rDNA, *N. crassa*
S-G-Conio-1361-a-A-20, CAACCCACAAAAGTGAGTTG	*Coniosporium* sp.[e]	18S rDNA, *N. crassa*

[a] R. I. Amann, L. Krumholtz, and D. A. Stahl, *J. Bacteriol.* **172,** 762 (1990).

[b] C. Roller, M. Wagner, R. Amann, W. Ludwig, and K.-H. Schleifer, *Microbiology* **140,** 2849 (1994).

[c] E. Stackebrandt, D. Witt, C. Kemmerling, R. Kroppendstedt, and W. Liesack, *Appl. Environ. Microbiol.* **57,** 1468 (1991).

[d] S. Li, R. N. Spear, and J. H. Andrews, *Appl. Environ. Microbiol.* **63,** 3261 (1997).

[e] K. Sterflinger and M. Hain, *Studies Mycol.* **43,** 23 (1999).

Procedure for Bacteria

1. Add to each spot of cells 9 μl of hybridization buffer and 1 μl (50 ng) of fluorescent probe.

2. Incubate the slides at 37–46° for 2 hr.

3. Immerse the slides in a prewarmed (38–48°) washing buffer for 20 min.

4. Rinse in distilled water.

5. Air dry.

6. Add 10 μl of a DAPI solution (1 μg/ml)

7. Incubate at room temperature for 10 min.

8. Rinse in distilled water.

9. Air dry.

10. Mount in glycerol-based antifading agent.

11. Place a thin coverslip over material.

12. Squeeze gently excess antifade from the slide and remove with filter paper.

13. Observe with an epifluorescence microscope, magnification ×100, with various excitation filters (Table II) according to the fluorochrome used.

TABLE II
FLUORESCENT DYES EMPLOYED FOR FISH

Filter code (Leica)	Fluorochromes	Color of fluorescence	Excitation wavelengths (nm)
A	DAPI	Blue	(BP 340–380)
I$_3$	FITC and AO	Green	(BP 450–490)
Y$_3$	Cy3 and TRITC	Orange	(BP 535–550)
Y$_5$	Cy5	Far red	(BP 625–650)

As an alternative, slides can be kept at 4° in the dark before observation for up to 6 months.

Procedure for Fungi

1. Add 10 μl of hybridization solution and 1 μl of fluorescent probe (50 ng).
2. Incubate at 46° for 90 min.
3. Wash in 150 ml of prewarmed hybridization solution.
4. Rinse in ultrapure water.
5. Wash twice with prewarmed (48°) hybridization solution for 10 min.
6. Rinse in ultrapure water.
7. Dry at room temperature.
8. Mount with a coverslip with antifading agent.
9. Observe under a microscope (magnification ×40) with various excitation filters (Table II) according to the fluorochrome used.

Image Capture and Digital Image Processing

Light microscope (Leica DMR/HCS or Zeiss Axioplan) equipped with epifluorescence unit (100-W mercury lamp HBO 100) and various combinations of filters.

Image capture is carried out with a photocamera (on Fujicolor super HG 1600 ASAfilm, Kodak EES 1600 color reversal film) or with a black-and-white camera (Leica Quantimed, COHU high-performance CCD camera Q550IW), with the help of software (Leica QWin). After capture, black and white pictures can be transformed to color pictures with Adobe PhotoShop software, replacing the white with the corresponding original color of fluorochromes.

In Situ Assessment of Microbial Activity

Metabolic activity occurring within the microbial community can be measured *in situ* as well as *in vitro* by various methods. However, the need to apply nondestructive methods to the analysis of microbial communities on valuable archaeological and artistic surfaces led to the use of microsensors. Microsensors have been used to measure oxygen content, respiratory and photosynthetic activity, and pH analysis in the aquatic environment.[26,27] Amperometric oxygen microsensors have been assembled and applied for the measurement of photosynthesis and respiration at increasing irradiance on cyanobacterial biofilms in terrestrial environment.[28] In addition, potentiometric microsensors are presently being developed and applied for the measurement of H^+ concentration on cyanobacteria-dominated biofilms in order to quantify stone damage and gain new insights about microbial community function and biogeochemical fluxes. Preliminary results showed that alkalinization of the substrata during photosynthetic activity occurred to a sufficient extent to induce precipitation of carbonates, such as those present on calcareous surfaces of several monuments.[29]

Concluding Remarks

Several innovative methods have already been applied to the study of microbes in art and others are being developed for better and easier identification of species within the microbial communities that colonize stone monuments. Application of new techniques is strongly desirable because of the total lack of information about the mechanisms of biofilm development, and because of the scarcity of data about the ecophysiology of the organisms involved as well as about the biogeochemical fluxes of those ions and gases relevant to the process of stone deterioration.

Acknowledgments

This work was carried out with the financial support of the European Community Commission, contracts ENV4-CT98-0704, ENV4-CT98-0707, and EVK4-2000-00028, and of the National Research Council of Italy (CNR), contracts 99.03872.PF36 and 99.03688.PF36, and MURST.

The authors thank all the collaborators in Messina and Rome for their kind reading of the text.

[26] C. Lassen, R. N. Glud, N. B. Ramsing, and N. P. Revsbeck, *J. Phycol.* **34**, 89 (1998).
[27] N. P. Revsbeck, *in* "Microbial Mats" (L. J. Stal and P. Caumette, eds.), NATO ASI Series G-35, p. 135, Springer-Verlag, Berlin, 1994.
[28] D. Compagnone, V. Di Carlo, L. Bruno, P. Albertano, and G. Palleschi, *Anal. Lett.* **32**, 213 (1999).
[29] P. Albertano, L. Bruno, D. D'Ottavi, D. Moscone, and G. Palleschi, *J. Appl. Phycol.* **12**, 379 (2000).

[29] Identification of Archaea in Objects of Art by Denaturing Gradient Gel Electrophoresis Analysis and Shotgun Cloning

By GUADALUPE PIÑAR, CLAUDIA GURTNER, WERNER LUBITZ, and SABINE RÖLLEKE

Introduction

Microorganisms are found in most terrestrial and aquatic ecosystems as biofilms on surfaces. It is known that biofilms play an important role in almost all aspects of microbiology and may appear as either beneficial or potentially harmful populations of microorganisms.[1-3] However, only a few studies have been carried out describing the presence of archaea in biofims,[4] and most of them focus on biofilms formed by the methanogenic archaea group.[5,6] The detection of halophilic archaea on natural biofilms located on deteriorated wall paintings[7] prompted further studies to focus on this group of microorganisms, which were excluded in previous investigations using only eubacteria-specific primers. The native biofilms located on objects of art are difficult to access without the risk of damaging the original object, which makes direct analysis of the individual organisms difficult. Technological and methodological advances, especially those dealing with small amounts of sample, have facilitated the study of these complex microbial communities. In this field, new DNA-based techniques have been developed[8] that allow the identification of individual microbial species in sample material without the cultivation of the organisms, particularly for archaea, which are difficult organisms to cultivate.[9] Molecularly based studies have additional advantages in providing sequences that are directly comparable with existing sequence databases.[10] To

[1] K. T. Elvers, K. T. Leeming, C. P. Moore, and H. M. Lappin-Scott, *J. Appl. Microbiol.* **84,** 607 (1998).

[2] C. Dutkiewicz and H. Fallowfield, *J. Appl. Microbiol.* **85,** 597 (1998).

[3] J. W. Santo Domingo, C. J. Berry, M. Summer, and C. B. Fliermans, *Curr. Microbiol.* **37,** 387 (1998).

[4] P. L. Hartzell, J. Millstein, and C. LaPaglia, *Methods Enzymol.* **310,** 335 (1999).

[5] J. J. Godon, E. Zumstein, P. Dabert, F. Habouzit, and R. Moletta, *Appl. Environ. Microbiol.* **63,** 2802 (1997).

[6] C. S. Wu, J. S. Huang, J. L. Yan, and C. G. Jih, *Biotechnol. Bioeng.* **57,** 367 (1998).

[7] S. Rölleke, A. Witte, G. Wanner, and W. Lubitz, *In. Biodet. Biodeg.* **41,** 85 (1998).

[8] C. R. Woese, *Microbiol. Rev.* **51,** 221 (1987).

[9] W. B. Whitman, T. L. Bowen, and D. R. Boone, The methanogenic bacteria. *In* "The Prokaryotes" (A. Balows, H. G. Trüper, M. Dworkin, W. Harder, and K. H. Schleifer, eds.), p. 717. Springer-Verlag, New York, 1991.

[10] B. L. Maidak, J. R. Cole, C. T. Parker, Jr., G. M. Garrity, N. Larsen, B. Li, T. G. Lilburn, M. J. McCaughey, G. J. Olsen, R. Overbeek, S. Pramanik, T. M. Schmidt, J. M. Tiedje, and C. R. Woese, *Nucleic Acids Res.* **27,** 171 (1999).

identify microorganisms in sample material, extracted DNA can be used as a template to amplify ribosomal gene fragments by polymerase chain reaction (PCR), using primers for universal sequences; the fragments can then be cloned.[11] The result of such a strategy is a clone library, containing ribosomal sequences as inserts. By sequencing individual inserts and comparing the obtained sequence with sequences present in databases, it is possible to identify the phylogenetic position of the corresponding microorganisms without their cultivation. An alternative to this approach is the denaturing gradient gel electrophoresis (DGGE) of PCR-amplified gene fragments encoding rDNA. This method has been used for a variety of applications.[12] It allows the separation of DNA fragments of identical lengths but different sequences on the basis of their different melting behavior in a gel system containing a chemical gradient of denaturants. As result of such electrophoresis a band pattern is obtained that already visualizes the complexity of the microbial community and, in addition, individual members of the community can be identified. This approach has been applied successfully to investigate bacteria[13,14] and archaea[7] in biodeteriorated objects of art. This chapter describes the optimization of protocols used to work with archaeal sequences. For the identification of archaea, DGGE analysis of PCR-amplified rDNA with archaea-specific primers and shotgun clone libraries has been carried out in parallel to combine the advantages of both methods. The cloning of environmental PCR products was performed in order to avoid problems arising in parallel with excision of the bands from complex band patterns. In addition, more sequence information can be obtained, allowing a more reliable phylogenetic identification. Figure 1 shows the experimental procedure for analyzing the archaeal community present in natural samples. Through the combination of these methods we have identified members of the archaeal community in a biofilm located on a wall painting without prior cultivation.

Methodology

Description of Site, Sampling of Wall Paintings, and DNA Extraction

Description of Site. The samples used in this study were taken from mural paintings at the Necropolis of Carmona (Spain). The Necropolis was discovered and excavated at the end of the nineteenth century. Its various mausoleums, tombs, and sanctuaries were carved from bedrock and used during the first and second centuries A.D.

[11] D. M. Ward, R. Weller, and M. M. Bateson, *Nature (London)* **345,** 63 (1990).
[12] G. Muyzer, E. C. de Waal, and A. G. Uitterlinden, *Appl. Environ. Microbiol.* **59,** 695 (1993).
[13] S. Rölleke, G. Muyzer, C. Wawer, G. Wanner, and W. Lubitz, *Appl. Environ. Microbiol.* **62,** 2059 (1996).
[14] S. Rölleke, C. Gurtner, U. Drewello, R. Drewello, W. Lubitz, and R. Weissmann, *J. Microbiol. Methods* **36,** 107 (1999).

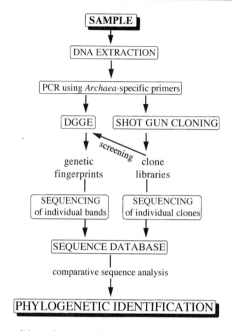

FIG. 1. Flow diagram of the various steps for the study of the structure of archaeal communities. Genetic fingerprinting by DGGE is the main focus of the strategy to study the presence of archaeal populations in a complex mixture. Furthermore, individual bands can be excised from the gels and/or independent clones from clone libraries can be sequenced to identify the community members. An additional advantage offered by this method is the use of DGGE for the screening of clone libraries.

Sampling. Three samples were taken from the Necropolis of Carmona, in different sites and at different heights. Sample S3 was taken from the wall of a small inner room, with high humidity and colonized by phototrophic microorganisms. This sample is a biofilm with a green color.

The samples were taken with sterilized scalpels and/or brushes by scraping off surface material to a depth of 1 to 3 mm, and kept in sterile tubes.

DNA Extraction. Small amounts of sample material (1–5 mg) are used for the extraction of DNA according to the method described by Schabereiter-Gurtner *et al.*[15] as follows.

1. The sample material is mixed with 100 μl of DNA extraction buffer I (150 mM Na$_2$EDTA, 225 mM NaCl; pH 8.5) and 45 μl of lysozyme (50 mg ml^{-1}), and incubated at 37° for 30 min.

[15] C. Schabereiter-Gurtner, G. Piñar, W. Lubitz, and S. Rölleke, *J. Microbiol. Methods,* in press.

2. Three microliters of 25% (w/v) sodium dodecyl sulfate (SDS) and 3 μl of proteinase K (20 mg ml^{-1}) are added and the samples are incubated at 37° for 60 min with agitation.

3. Fifty microliters of prewarmed (90°) DNA extraction buffer II [100 mM Na$_2$EDTA, 400 mM Tris-HCl, 400 mM disodium phosphate buffer (pH 8.0), 5.55 M NaCl, 4% (w/v) hexadecyltrimethyl ammonium bromide (CTAB); pH 8.0] and 9 μl of 25% (w/v) SDS are added and the samples are incubated at 65° for 60 min with agitation.

4. The samples are subjected to three freeze–thawing cycles (from −80 to 65°) in order to break the cells mechanically.

5. Further extraction is carried out with 200 μl of chloroform–isoamyl alcohol (24 : 1, v/v) to remove the CTAB–protein–polysaccharide complex. The aqueous phase is recovered by centrifugation (14,000 rpm for 5 min).

6. The DNA in the supernatant is cleaned with a QIAamp viral RNA minikit (Qiagen, Valencia, CA) as recommended by the manufacturer. The DNA is eluted three times through the silica-gel membranes with 60 μl of doubly distilled H$_2$O. The QIAamp viral RNA minikit is used (1) to avoid the risk of contamination that occurs during a common phenol–chloroform DNA extraction and further precipitation with ethanol, (2) to remove PCR inhibitors by using silica-gel membranes, and (3) to recover the small amounts of DNA available because carrier RNA is used.

From every sample the three DNA eluents are tested by PCR, using specific archaeal primers. A negative control is carried out during the procedure, in which water instead of sample material is used to exclude the possibility of cross-contamination.

Polymerase Chain Reaction Amplification of Archaeal 16S rDNA Fragments

Enzymatic amplification of the 16S rDNA[16] has been performed on DNA extracted directly from wall painting material. Two PCR protocols are necessary to obtain sufficient product for DGGE analysis and cloning experiments.

First Polymerase Chain Reaction Protocol. A primary amplification round using the archaea-specific primers ARC344[17] (forward) (5'-ACGGGGAGCAGCA-GGCGCGA-3') and ARC915[17] (reverse) (5'-GTGCTCCCCCGCCAATTCCT-3'), which amplify a PCR fragment of 590 bp, is performed in a Robocycler (Stratagene, La Jolla, CA). All reactions are carried out in 25-μl volumes containing 12.5 pmol of each primer, a 100 μM concentration of each deoxyribonucleoside

---

[16] L. Medlin, H. J. Ellwood, S. Stickel, and M. L. Sogin, *Gene* **71**, 491 (1988).
[17] L. Raskin, J. M. Stromley, B. E. Rittmann, and D. A. Stahl, *Appl. Environ. Microbiol.* **60**, 1232 (1994).

triphosphate, 2.5 μl of 10 × PCR buffer (100 mM Tris-HCl, 15 mM MgCl$_2$, 500 mM KCl; pH 8.3), 0.5 U of Taq DNA polymerase (Roche Diagnostics, Indianapolis, IN), filled up to 25 μl with sterile water. Volumes of 1.5–2.5 μl from the various elution steps of the DNA extraction are used as template DNA. The following thermocycling program is used: 5 min of denaturation at 95°, followed by 40 cycles consisting of 1 min of denaturation at 95°, 1 min of primer annealing at 60°, and 1 min of extension at 72°, with a final extension step of 72° for 5 min.

Second Polymerase Chain Reaction Protocol. For DGGE analysis, the second round of PCR is carried out as follows: 1.5 μl of the amplified product is transferred to a fresh reaction mixture as described above, but in addition 5% (v/v) dimethyl sulfoxide (DMSO) was used. The archaea-specific sequence ARC344 (forward) and the universal sequence 518[12] (reverse), which amplify a PCR product of 200 bp, are used. The reverse primer contains at its 5' end a 40-base GC-clamp (5'-CGCCCGCCGCGCGCGGCGGGCGGGGCGGGGGCACGGGGGG-3'), to stabilize the melting behavior of the DNA fragments.[12] The conditions of this second round of PCR were as follows: 5 min of denaturation at 95°, followed by 30 cycles of 1 min of denaturation at 95°, 1 min of primer annealing at 55°, and 1 min of primer extension at 72°, with a final extension step of 72°C for 5 min.

For constructing clone libraries, the second round of PCR is carried out as described for DGGE analysis, but no DMSO is used in the reaction and the universal consensus sequence 907 is used as reverse primer.[18] These primers amplify a PCR product of 547 bp.

A negative control is included in the whole procedure to prove that the system is not contaminated. A 10 μl volume of each PCR product is first analyzed by electrophoresis in 2% (w/v) agarose gels[19] before DGGE analysis.

Denaturing Gradient Gel Electrophoresis Analysis of rDNA Fragments

A 100 μl volume of each PCR product obtained by nested PCR is precipitated with 96% (v/v) ethanol and dissolved in 15 μl of TE buffer (10 mM Tris-HCl, 1 mM EDTA; pH 8.0). Gel electrophoresis is performed as previously described by Muyzer *et al.*[12] in 0.5× TAE (20 mM Tris, 10 mM sodium acetate, 0.5 mM Na$_2$-EDTA; pH 7.8). Gels are made with 8% (w/v) acrylamide stock solutions (acrylamide–N,N-methylene bisacrylamide, 37:1, v/v) containing 0 and 100% denaturant [100%: 7 M urea and 40% (v/v) formamide, deionized with AG501-X8 mixed bed resin (Bio-Rad, Hercules, Ca)].

The conditions are as follows: the linear chemical gradient ranges from 25 to 55% denaturant. Gels are run at 200 V with an optimum T_a of 60° in a

[18] G. Muyzer, A. Teske, C. O. Wirsen, and H. W. Jannasch, *Arch. Microbiol.* **164,** 165 (1995).

[19] J. Sambrook, E. F. Fritsch, and T. Maniatis, "Molecular Cloning: A Laboratory Manual," 2nd Ed. Cold Spring Harbor Laboratory Press, Cold Spring Harbor, New York, 1989.

minutes

10   60   120   180 C B A M

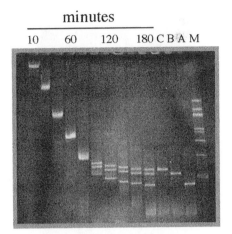

FIG. 2. Time travel experiment: Ethidium bromide-stained DGGE separation pattern of archaeal specific DNA fragments encoding 16S rRNA. A mixture of PCR products obtained from a *Halobacterium* sp., a *Naxos* sp., and *Natronobacterium gregorii* was applied every 20 min over a total of 3 hr and 10 min. The fragments stop running after 3 hr, when they have reached their melting point in the gradient of denaturants. Lane M, marker; lane A, DNA from *Halobacterium* sp.; lane B, DNA from *Naxos* sp.; lane C, DNA from *N. gregorii.*

D-Gene system (Bio-Rad). After completion of electrophoresis, gels are stained in ethidium bromide solution for 15 min and documented with a UVP (Upland, CA) documentation system. To allow comparative analyses of DGGE patterns obtained from different gels, a marker containing 16S rDNA PCR products derived from different bacterial species is used.[15]

Optimization for Denaturing Gradient Gel Electrophoresis Analysis. To establish the optimal DGGE running conditions for the archaeal DNA fragments, "time travel" experiments were carried out, in which a mixture of PCR products derived from genomic DNA of three different archaea species, i.e., *Halobacterium* sp., *Naxos* sp., and *Natronobacterium gregorii,* is applied to a denaturing gradient gel every 20 min over a total time of 3 hr and 10 min (Fig. 2). The optimal running time of the gel is determined as the time required before the three fragments migrate as independent single bands. As shown in Fig. 2, the fragments migrate as one band for the first 80 min. After this time, they start to separate from each other because of the denaturing conditions within the gel; the molecules partially melt and slow down. After 3 hr the maximum resolution of the three fragments is achieved. We have used these conditions for further experiments with archaeal sequences.

Denaturing Gradient Gel Electrophoresis Analysis of Polymerase Chain Reaction Products Derived from Wall Painting Samples. The three samples (S1–S3) taken from the Necropolis of Carmona were tested for the presence of

5   4   3   2   1   S3   M

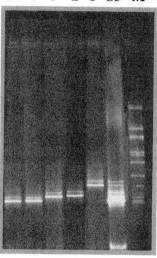

FIG. 3. Ethidium bromide-stained DGGE separation pattern of DNA fragments encoding 16S rRNA amplified from the S3 Carmona wall painting DNA by using the archaea-specific primer pair. M, Marker. Lane S3 shows the band pattern obtained with this original sample. Lanes 1 to 5 show the PCR products obtained from the clones containing the independent archaeal sequences of the original sample.

archaea-specific sequences with the archaea-specific primers ARC344 and ARC915. A 590-bp PCR product was obtained from one of the three samples from Carmona (the S3 sample). This demonstrates that archaea are present in this sample material. The amplified PCR product can be detected after the first round of PCR. However, a second round of PCR (as described above for DGGE) is carried out to obtain PCR products that can be used for DGGE analysis.

DGGE analysis is performed with the positive sample to obtain information about the diversity and complexity of the species composition. The DGGE band pattern obtained from Carmona sample S3 is shown in Fig 3. The band pattern contains a complex mixture consisting of five or six main individual bands. Each band represents an archaea taxon in the sample material.

Cloning of Environmental Polymerase Chain Reaction Products and Screening of Archaeal Clone Libraries

In parallel with DGGE analysis of 200-bp 16S rDNA archaeal fragments, cloning of environmental PCR products of 547 bp was carried out to avoid problems such as the excision of bands from complex band patterns and overlapping bands. These two effects produce mixed sequences difficult to be sequenced directly.[15] In addition, the phylogenetic identification is often limited by the short sequence obtained from DGGE analysis, because usually DNA fragments of short length are used. By sequencing clones more sequence information can be obtained, allowing a more reliable phylogenetic identification.

Cloning of Environmental Polymerase Chain Reaction Products. A 100 μl volume of each PCR product, obtained as described above for the construction of clone libraries, is pooled and purified by the QIAquick gel extraction kit protocol (Qiagen). The purified DNA is concentrated, dissolved in 15 μl of doubly distilled H_2O, and ligated into an ampicillin-resistant (Ap^r) pKS derivative (cut with the *Xcm*I enzyme). The ligation is transformed into $CaCl_2$-competent cells of *Escherichia coli* XL1-BLue (tetracycline resistant, Tc^r), which allows blue–white screening, on medium containing ampicillin (40 μg/ml), tetracycline (10 μg/ml), 5-bromo-4-chloro-3-indolyl-β-D-galactopyranoside (X-Gal, 0.1 mM), and isopropyl-β-D-thiogalactopyranoside (IPTG, 0.2 mM).

Screening of Archaeal 16S rDNA Clone Libraries. The clones were screened by PCR to confirm the presence of inserts bearing the archaeal DNA sequences, as follows: white colonies are picked, suspended in 40 μl of TE buffer (10 mM Tris-HCl, 1 mM EDTA; pH 8.0) and lysed by three freeze–thawing cycles. Samples (3 μl) are used as template DNA for PCR with the primers T3 (5′-AATTAACCCTCACTAA-AG-3′) and T7 (5′-TAATACGACTCACTATAGGG-3′). The conditions are as follows: 5 min of denaturation at 95°, followed by 35 cycles of 1 min of denaturation at 95°, 1 min of primer annealing at 46°, and 1 min of primer extension at 72°, with a final extension step of 5 min at 72°. A 10 μl volume of the PCR product is analyzed by electrophoresis in 2% (w/v) agarose gels. Positive clones produce an 800-bp PCR product, whereas the negative clones produce a 200-bp PCR product containing the two flanking regions of the plasmid.

In a second screening, the different positions of the clones in DGGE are discerned[15]. The 800-bp PCR product (1 μl) is used as a template for a nested PCR carried out as described for DGGE analysis. The obtained PCR product (15 μl) is subsequently analyzed by DGGE. The positions of the clones in the gel are compared with one another and with the band pattern obtained from the original sample. Figure 3 shows the positive clones, containing the archaeal rDNA inserts, obtained from sample S3 from Carmona, producing bands at different positions in the DGGE. Comparing the positions of the independent clones in the gel with the positions of bands in the original band pattern allows the identification and selection of the different clones for phylogenetic identification.

Sequencing of 16 rDNA Inserts and Identification of Archaea by Comparative Sequence Analysis

The rDNA inserts are sequenced as follows: PCR product (100 μl) obtained with primers T3 and T7 is purified with a QIAquick PCR purification kit (Qiagen) and sequenced with a Li-Cor (Lincoln, NE) DNA sequencer Long Read model 4200L.[20] Sequencing reactions are performed by cycle sequencing with the SequiTherm system (Epicentre, Madison, WI), with fluorescently labeled primers T3 and 5U (2 pmol each) and SequiTherm thermostable DNA polymerase.

The sequences obtained were compared with sequences of known organisms, using the Ribosomal Database Project (RDP)[10] and the EMBL nucleotide sequence database. The SIMILARITY-RANK tool of the RDP and the FASTA search option[21] for the EMBL database were used to search for close evolutionary relatives.

Sequence analyses of the different clones obtained from sample S3 from Carmona were carried out to obtain information about the identity of the corresponding organisms, as well as to verify that the amplified fragments were derived from archaea. The comparison of the sequences obtained from the Carmona sample with sequences in databases showed that the sequences grouped with those of the halophilic archaea group (89.5–95.8% similarity).

Concluding Remarks

Archaea are regarded as occurring in extreme habitats, such as hypersaline, anaerobic, alkaline, acidic, and hyperthermophilic environments. Further, they have not been previously regarded as important in global ecology. It is only more recently that this view has radically changed and the presence of archaea in a wider spectrum of habitats has been discovered.[22–33] Wall paintings can represent extreme habitats, favoring the growth of a variety of specialized microorganisms. In this "extreme" environment the formation of biofilms may enable archaea to survive exposure to suboptimal conditions. Such subpopulations may be part of

[20] L. R. Middendorf, J. C. Bruce, R. C. Bruce, R. D. Eckles, D. L. Grone, S. C. Roemer, G. D. Sloniker, D. L. Steffens, S. L. Sutter, J. A. Brumbaugh, and G. Patonay, *Electrophoresis* 13, 487 (1992).

[21] W. R. Pearson, *Methods Enzymol.* 183, 63 (1994).

[22] S. B. Bintrim, T. J. Donohue, J. Handelsman, G. P. Roberts, and R. M. Goodman, *Proc. Natl. Acad. Sci. U.S.A.* 94, 277 (1997).

[23] E. F. DeLong, *Proc. Natl. Acad. Sci. U.S.A.* 89, 5685 (1992).

[24] E. F. DeLong, K. Y. Wu, B. B. Prezelin, and R. V. M. Jovine, *Nature (London)* 371, 695 (1994).

[25] J. A. Fuhrman, K. McCallum, and A. A. Davis, *Nature (London)* 356, 148 (1992).

[26] G. Jurgens, K. Lindström, and A. Saano, *Appl. Environ. Microbiol.* 63, 803 (1997).

[27] R. Massana, A. E. Murray, C. M. Preston, and E. F. DeLong, *Appl. Environ. Microbiol.* 63, 50 (1997).

[28] M. A. Munson, D. B. Nedwell, and T. M. Embley, *Appl. Environ. Microbiol.* 63, 4729 (1997).

[29] J. G. Olsen, *Nature (London)* 371, 657 (1994).

[30] R-A. Sandaa, Ø. Enger, and V. Torsvik, *Appl. Environ. Microbiol.* 65, 3293 (1999).

[31] J. L. Stein and M. I. Simon, *Proc. Natl. Acad. Sci. U.S.A.* 93, 6228 (1996).

[32] M. J. E. C. van der Maarel, R. R. E. Artz, R. Haanstra, and L. J. Forney, *Appl. Environ. Microbiol.* 64, 2894 (1998).

[33] C. Vetriani, H. W. Jannasch, B. J. MacGregor, D. A. Stahl, and A-L. Reysenbach, *Appl. Environ. Microbiol.* 65, 4375 (1999).

a complex community, although the community can be dominated by completely different microorganisms. We demonstrate that these microorganisms occur in an ancient wall painting located in the Necropolis of Carmona (Andalucia, Spain) by using molecular means.

This chapter describes the development and application of a strategy combining DGGE analysis and shotgun cloning for the analysis of archaea in biodeteriorated materials of cultural heritage. There are few articles describing the use of DGGE to study archaea in different environments.[7,30,33–35] Some of these studies combine DGGE analysis with the excision of individual DGGE bands from the gel, in order to obtain sequence information about individual archaea members.[7,34,35] However, the information provided by phylogenetic analysis of sequences obtained directly from DGGE patterns is often not reliable because of the short sequence length. Furthermore, comigration of several different DNA sequences, which have the same melting behavior and therefore the same position in the gel, leads to overlapping DGGE bands that cannot be sequenced directly.[15] The strategy to overcome these problems is to construct DNA clone libraries and screen for different clones by DGGE. This combination offers the advantage of the original method of cloning DNA fragments with the advantage of DGGE, which allows both the screening for different clones and the analysis of the archaeal community together in one gel. The combination of DGGE analysis with the construction of clone libraries to study archaeal communities has already been described,[30,33] but instead of DGGE, restriction fragment length polymorphism (RFLP) was used for the screening of clones. RFLP analysis of 16S rDNA has been used for several years as a method for comparing rDNAs.[36,37] PCR products obtained with universal primers are digested with restriction enzymes. The typical analysis of restriction digests for isolates or clones is performed on relatively low-resolution agarose gels. This method has been used for the study of communities, where the potentially large number of fragments can be resolved by using polyacrylamide gels to produce a community-specific pattern.[38,39] However, RFLP analysis is of limited use for demonstrating the presence of specific phylogenetic groups or for estimating species abundance.[40] The combination of the methods described in the present study for the detection and identification of archaea offers advantages over other methods that have been

[34] A. E. Murray, C. M. Preston, R. Massana, L. T. Taylor, A. Blakis, K. Wu, and E. F. DeLong, *Appl. Environ. Microbiol.* **64,** 2585 (1998).

[35] L. Øvreas, L. Forney, F. L. Daae, and V. Torsvik, *Appl. Environ. Microbiol.* **63,** 3367 (1998).

[36] G. Laguerre, M. Allard, F. Revoy, and N. Amarger, *Appl. Environ. Microbiol.* **60,** 56 (1994).

[37] C. L. Moyer, F. C. Dobbs, and D. M. Karl, *Appl. Environ. Microbiol.* **60,** 871 (1994).

[38] A. J. Martinez-Murcia, S. G. Acinas, and F. Rodriguez-Valera, *FEMS Microbiol. Ecol.* **17,** 247 (1995).

[39] A. A. Massol-Deya, D. A. Odelson, R. F. Hickey, and J. M. Tiedje, *Mol. Microb. Ecol. Man.* **3.3.2.,** 1 (1995).

[40] L. Wen-Tso, T. L. Marsh, H. Cheng, and L. Forney, *Appl. Environ. Microbiol.* **63,** 4516 (1997).

previously described: first, the parallel application of DGGE analysis and clone libraries combines the potential of both methods and overcomes their limitations; second, the screening of the clones by DGGE analysis makes the procedure easier and faster than other molecular methods such as RFLP analysis.

This chapter will help to increase our knowledge about archaea, a group of organisms representing a third kingdom of the living world, which in the past has been overlooked in most environments because of the lack of suitable methods. The archaea detected in wall paintings may be important members of the community colonizing such objects of art, and could play a role in the damaging processes. The application of the strategy described in this chapter allows a more complete picture of archaea in natural environments and will help our understanding of the role they may play in biofilms.

Acknowledgments

This work is part of the European Union-funded project, Novel Molecular Tools for the Analysis of Unknown Microbial Communities of Mural Paintings and Their Implementation in Conservation/Restoration Practice (EU-ENV4-CT98-0705). This work was supported by the European Community Marie Curie Training Grant (contract BIO4-CT98-5057). We thank Prof. Cesareo Saiz-Jimenez for kindly supplying information and samples from the Necropolis of Carmona.

Section VIII

Probiotics

[30] Lactobacilli as Vehicles for Targeting Antigens to Mucosal Tissues by Surface Exposition of Foreign Antigens

By Peter H. Pouwels, Aldwin Vriesema, Beatriz Martinez, Frans J. Tielen, Jos F. M. L. Seegers, Rob J. Leer, Jan Jore, and Egbert Smit

Introduction

Lactobacillus strains have a number of properties that make them attractive candidates as delivery vehicles for the presentation to the mucosa of compounds of pharmaceutical interest, in particular vaccines and immunomodulators. Lactobacilli have been used for centuries in fermentation and preservation of food and feed, and are considered GRAS (generally regarded as safe) organisms. In addition, certain strains of *Lactobacillus* can colonize the gut and are believed to show health-promoting activities for humans and animals.[1-3] *Lactobacillus* species have the capacity to evoke a mucosal as well as a systemic immune response against epitopes associated with these organisms after oral or nasal administration.[4-6] These findings and earlier observations indicating that certain *Lactobacillus* species show a nonspecific immunoadjuvant effect by triggering of macrophages[7] have prompted research aimed at investigating the potential of these organisms to synthesize foreign antigens and present them to the immune system.[8,9]

In this chapter we present the construction and evaluation of a series of expression–secretion vectors with strong constitutive or regulatable promoters and

[1] R. Havenaar and J. H. J. Huis in't Veld, *in* "The Lactic Acid Bacteria in Health and Disease" (B. J. B. Wood, ed.), Vol. I, pp. 151–170. Elsevier Applied Science, London, 1993.

[2] S. Salminen, C. Bouley, M. C. Boutron-Ruault, J. H. Cummings, A. Franck, G. R. Gibson, E. Isolauri, M. C. Moreau, M. Roberfroid, and I. Rowland, *Br. J. Nutr.* **80**(Suppl. 1), S147 (1998).

[3] P. H. Pouwels, R. J. Leer, M. Shaw, M. J. Heijne den Bak-Glashouwer, F. D. Tielen, E. Smit, B. Martinez, J. Jore, and P. L. Conway, *Int. J. Food Microbiol.* **41**, 155 (1998).

[4] K. Gerritse, M. Posno, M. M. Schellekens, W. J. A. Boersma, and E. Claassen, *Res. Microbiol.* **141**, 955 (1990).

[5] K. Gerritse, M. Posno, M. M. Schellekens, W. J. A. Boersma, and E. Claassen, *Adv. Exp. Med. Biol.* **84**, 497 (1991).

[6] P. H. Pouwels, R. J. Leer, and W. J. A. Boersma, *J. Biotechnol.* **44**, 183 (1996).

[7] G. Perdigon, M. E. de-Macias, S. Alvarez, G. Oliver, and A. P. de-Ruiz-Holgado, *Immunology* **63**, 17 (1988).

[8] P. Slos, P. Dutot, J. Reymund, P. Kleinpeter, D. Prozzi, M. P. Kieny, J. Delcour, A. Mercenier, and P. Hols, *FEMS Microbiol. Lett.* **169**, 29 (1998).

[9] D. M. Shaw, R. J. Leer, C. Smittenaar, M.-J. Heijne den Bak-Glashouwer, F. J. Tielen, and P. H. Pouwels, *Immunology* **100**, 510 (2000).

efficient translation initiation regions designed for intra- and extracellular expression of homologous and heterologous proteins.

General Structure of Expression Vectors

The vectors can be amplified in *Escherichia coli* and in *Lactobacillus*. They comprise the *E. coli* vector pGEM, a broad host range replicon from *Lactobacillus pentosus,* and an antibiotic resistance marker, enabling their replication in a wide variety of lactic acid bacterial strains.[10] In addition, depending on whether the expressed protein should remain intracellular, be transported over the membrane and anchored to the cell wall, or be secreted into the culture medium, the vectors may harbor, in the order 5' to 3', the following elements: (1) a strong constitutive promoter or a strong regulatable promoter from *Lactobacillus,* (2) a translation initiation region of an efficiently expressed *Lactobacillus* gene, (3) a DNA sequence encoding the signal peptide of an efficiently secreted *Lactobacillus* protein, (4) the β-glucuronidase gene of *E. coli,* the product of which can be used as a reporter and/or to enhance the immunogenicity of the antigen that is expressed, (5) a DNA sequence encoding an anchor sequence from a *Lactobacillus* gene, allowing exposition of the protein at the bacterial surface, and (6) a transcription termination sequence of a *Lactobacillus* gene. Each of these elements constitutes a cassette flanked at both sides by unique restriction enzyme sites, facilitating easy manipulation and exchange of the cassettes from one vector to another. Although the vectors were initially designed for the expression of fusion proteins, they can be easily adapted, enabling the synthesis of nonfused proteins.[11,12] To expand further the host range of *Lactobacillus* strains in which the vector can replicate, additional vectors were constructed in which the replicon from *L. pentosus* was replaced by a replicon from a lactic acid bacterial species with an extended host range.[13] The nucleotide sequences of all parts of the expression vectors are known and are detailed in subsequent sections.

Strategy of Vector Construction

The strategy for the construction of the expression vectors was based on a novel method to construct the vectors in *E. coli* prior to their introduction in

[10] M. Posno, R. J. Leer, N. van Luijk, M. J. F. van Giezen, P. T. H. M. Heuvelmans, B. C. Lokman, and P. H. Pouwels, *Appl. Environ. Microbiol.* **57,** 1822 (1991).

[11] S. Chaillou, Y. C. Bor, C. A. Batt, P. W. Postma, and P. H. Pouwels, *Appl. Environ. Microbiol.* **64,** 4720 (1998).

[12] S. Chaillou, P. W. Postma, and P. H. Pouwels, *J. Bacteriol.* **180,** 4011 (1998).

[13] C. Platteeuw, G. Simons, and W. M. de Vos, *Appl. Environ. Microbiol.* **60,** 587 (1994).

Lactobacillus. This method circumvents the structural instability of *Lactobacillus* vector material in *E. coli* caused by cloning of strong *Lactobacillus* promoters and/or by cloning and expression in *E. coli* of *Lactobacillus* genes.[14] The construction route in *E. coli* involved the insertion of a nucleotide sequence containing a strong transcription terminator located downstream from the promoter, which is just upstream of the gene to be expressed. The newly introduced terminator will prevent transcription in *E. coli* of open reading frames (ORFs) that are to be inserted downstream from the promoter sequence. The newly introduced terminator sequence is flanked on both sides by a DNA sequence encoding a rare restriction enzyme site (*Not*I) that does not occur elsewhere in the plasmid, allowing easy removal of the terminator sequence from the vector prior to its introduction into *Lactobacillus.* After removal of the terminator sequence the cloned ORF will be transcribed from the promoter in *Lactobacillus.*

pLP400 and pLP500 Series of Vectors

The structure of the vectors is presented in Fig. 1A. A detailed description of their construction route and of the structure of the plasmids used for the construction of the expression vectors is depicted in Fig. 1B–D.

Promoters and Terminators

Expression of cloned genes in the prototype vectors is driven by the highly efficient, constitutive promoter of the *Lactobacillus casei* L-(+)-lactate dehydrogenase (L-*ldh*) gene (pLP500 series) or the regulatable promoter of the *Lactobacillus amylovorus* α-amylase (*amy*) gene (pLP400 series). The latter promoter, which was shown to be approximately 3-fold less efficient compared with the *ldh* promoter,[15] is active in medium containing cellobiose, galactose, or mannitol, but is repressed in a medium containing glucose.[16] At the end of the transcription unit the transcription terminator of the *Lactobacillus plantarum* conjugated bile acid hydrolase gene (T$_{cbh}$) is present. In between the promoter and the *L. plantarum cbh* terminator the vectors harbor the transcription terminator sequence of the *L. casei ldh* gene, to circumvent plasmid instability in *E. coli*. This terminator remains present only during manipulation of the plasmid in *E. coli*, but is removed before the plasmid is transferred to a *Lactobacillus* strain.

[14] P. H. Pouwels and R. J. Leer, *Antonie Van Leeuwenhoek* **64**, 85 (1993).
[15] H. J. Boot, C. P. A. M. Kolen, F. J. Andreadaki, R. J. Leer, and P. H. Pouwels, *J. Bacteriol.* **178**, 5388 (1996).
[16] B. C. Lokman, R. J. Leer, R. van Sorge, and P. H. Pouwels, *Mol. Gen. Genet.* **245**, 117 (1994).

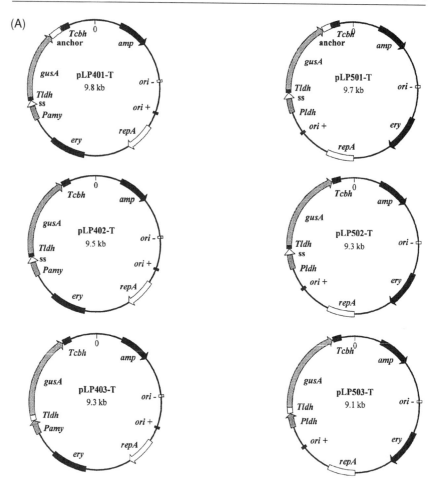

FIG. 1. (A) Structure of the expression vectors of the pLP400 and pLP500 series. (B) Construction route for the various vectors. Details of the construction are given in (C) and (D). As starting material for the construction of the expression–secretion vectors, pGEM-3 (Promega, Madison, WI) was used. A 260-bp *Hpa*I–*Hin*dIII fragment containing the last amino acid codon and transcription terminator sequence of the *cbh* gene of *L. plantarum* 80[26] was ligated between the *Hin*cII and *Hin*dIII sites of pGEM-3. Subsequently, the *Hin*dIII site was converted into an *Eag*I site, using a synthetic linker sequence. The remaining multiple cloning sequence was replaced by a chemically synthesized linker (GAATTCAGATCTACTAGTCTCGAGGCATGCGGATCCATTCAAGGTCACC-CTAGAGTCAAC) containing the following restriction enzyme sites: *Eco*RI, *Bgl*II, *Spe*I, *Xho*I, *Sph*I, *Bam*HI, and *Bst*EII. The *Bst*EII site is adjacent to the *Hin*cII site used for cloning of the *cbh* fragment. The resulting vector was designated pGTC3 (C). *Insertion of a second terminator sequence:* A synthetic linker with 5′ *Bgl*II and 3′ *Bst*EII termini encompassing the transcription terminator of the *L. casei ldh* gene flanked by two *Not*I sites and followed by a nucleotide sequence containing *Bam*HI, *Sst*I, and *Nco*I sites (cf. Fig. 4) was inserted between the *Bam*HI and *Bst*EII sites of pGTC3, yielding pTT3. Insertion of this linker destroyed the original *Bam*HI site but left the *Bst*EII site intact. *Insertion of E. coli gusA gene:* The *E. coli gusA* gene encoding β-glucuronidase was excised from pNOM2,[20] using *Nco*I and *Bst*EII, and cloned into a pUC19 derivative. The *gusA* gene was extended with two

(B)

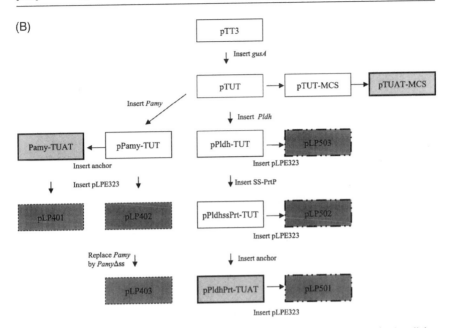

FIG. 1. (*continued*) codons by substituting nucleotides downstream from the last codon by a linker comprising an *Xho*I site, a stop codon, and a *Bst*EII site six nucleotides downstream from the stop codon (Fig. 2). An 1.8-kb *Nco*I–*Bst*EII fragment from this pUC-*gusA* plasmid with the modified *gusA* gene was cloned into the expression vectors, which were cut with *Nco*I and *Bst*EII between the *ldh* and *cbh* terminators. The resulting vector was denoted pTUT (C). *Insertion of anchor-encoding sequence:* An *Xho*I–*Bst*EII PCR fragment of the *prtP* gene from *L. casei* ATCC 393 encoding 117 C-terminal amino acids generated with primers based on the sequences of the *prtP* gene from *L. paracasei* was cloned between the *Xho*I and *Bst*EII sites of pTUT, yielding pTUAT (C). *Insertion of multicloning region:* A synthetic oligonucleotide, GGATCTTATCGATTTACGGATCCAACAATTGCAAGGTACCGAT-CAAGCTTACCCGGGAAGAAGACAGAATTCACAAGTCGACCGGGCCCTTGCCATGG, comprising restriction enzyme sites for the following enzymes, *Cla*I, *Bam*HI, *Mun*I, *Kpn*I, *Hind*III, *Sma*I, *Bbs*I, *Eco*RI, *Sal*I, *Apa*I, *Nco*I (MCS), was introduced between the *Bam*HI and *Nco*I sites of pTUT and pTUAT. The original *Bam*HI site was destroyed by this procedure, but a new *Bam*HI site was present within the newly inserted synthetic oligonucleotide, yielding pTUT-MCS and pTUAT-MCS [(D) and Fig. 4]. *Insertion of promoter-bearing DNA fragments:* Three promoter-bearing DNA fragments were inserted into pGTC3. The promoter, translation initiation region, and first eight codons of *L. casei ldh* (GenBank accession number AF351136), which are present on a 513-bp *Sau*3AI fragment, were cloned between the *Bgl*II and *Bam*HI sites of pGTC3 (yielding pGLC3), leaving the the *Bgl*II site 5' and the *Bam*HI site 3' from the promoter sequence intact (C). To obtain a P_{ldh}-directed expression–secretion system, a secretion signal sequence was cloned directly downstream of P_{ldhs}. The signal sequence of the *L. casei* ATCC 393 *prtP* gene was amplified via PCR with primers based on the *ptrP* genes of *L. lactis*[34] and *L. paracasei* subsp. *paracasei* NCDO 151.[25] A 239-bp fragment flanked by *Sph*I and *Bam*HI termini was obtained, which was shown to contain the Shine–Dalgarno sequence, translation start codon, and the first 70 codons of the *prtP* gene of *L. casei*. The fragment was cloned in pGEM, sequenced, excised with *Sph*I and *Bam*HI, and used to replace the 46-bp *Sph*I–*Bam*HI fragment containing the translation initiation region and first eight codons of the *ldh* gene of pGLC3. The resulting vector was named pGLC3-SS (C). To obtain a P_{amy}-directed expression–secretion system, a *Bgl*II–*Sph*I fragment (864 bp) containing the promoter, translation initiation region, and first 61 codons of the *L. amylovorus* α-amylase gene encoding the signal sequence was inserted between the *Bgl*II and *Sph*I sites of pGTC3. The vector constructed

(C)

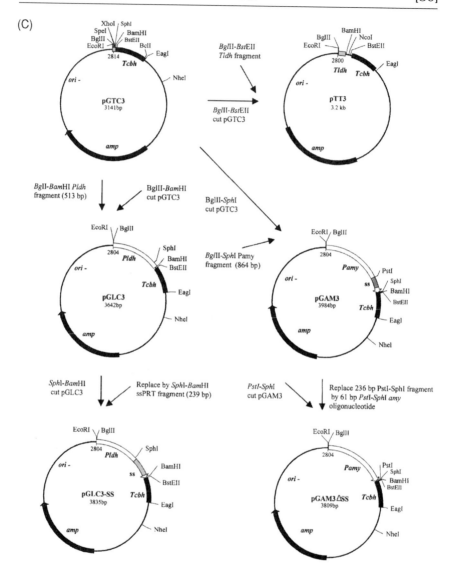

FIG. 1. (*continued*) in this way was called pGAM3 (C). To obtain a P_{amy}-directed expression system with its regulatable expression characteristics, but without secretion directing signals, the *Pst*I–*Sph*I sequence (236 bp) was replaced by a synthetic oligonucleotide with the same restriction enzyme termini. This oligonucleotide contained the entire sequence from *Pst*I to the translation start codon and the first seven amino acid codons of the *amy* gene. The shortened sequence comprises the AT-rich 5′ region but lacks the signal sequence peptidase site. The vector constructed in this way was called pGAM3ΔSS (C). Promoter-containing *Eco*RI–*Bam*HI fragments were isolated from pGLC3, pGLC3-SS, and pGAM3 and inserted between the *Eco*RI and *Bgl*II sites from pTUT and pTUAT. This resulted in pP$_{ldh}$TUT, pP$_{ldh}$-ss$_{prtP}$TUT, P$_{amy}$TUT, and pP$_{ldh}$-ss$_{prtP}$TUAT and P$_{amy}$TUAT. In these vectors ORF-encoding DNA fragments can be introduced in one of the restriction enzyme sites

(D)

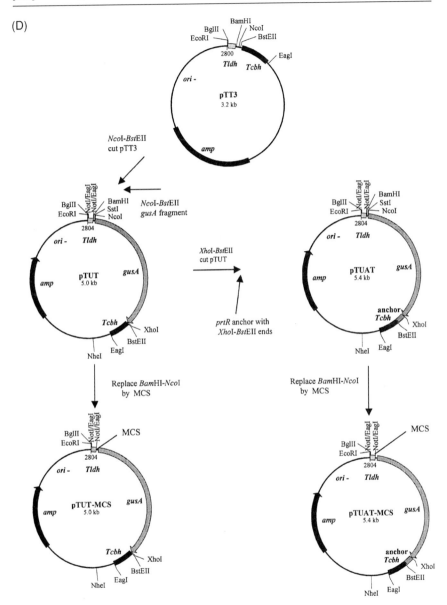

FIG. 1. (*continued*) between the two terminator sequences, allowing cloning in *E. coli* without expression. *Construction of hybrid Lactobacillus–E. coli vector:* The expression cassettes, obtained by digestion of the promoter-containing pTUT and pTUAT vectors with *Eco*RI, were ligated to pLPE323,[10] which was also cut with *Eco*RI, resulting in the following *E. coli–Lactobacillus* shuttle vectors, pLP401-T and pLP402-T, and pLP501-T, pLP502-T, and pLP503-T (A). Plasmid pLP403-T, which contains $P_{amy-\Delta SS}$ was obtained by substitution of $P_{amy-\Delta SS}$ for P_{amy} in pLP402-T. Removal of the T_{ldh} fragment from the various vectors resulted in pLP401, pLP402, etc.

Translation Initiation Region

Codons near the 5' end of *Lactobacillus* genes contain a decreased percentage of G and C residues compared with the remainder of these genes. This observation was explained by assuming that the presence of G and C residues in this region would favor the formation of stable RNA duplexes, which may interfere with translation initiation.[17] The expression and expression–secretion vectors comprise the entire nontranslated 5' region including the Shine–Dalgarno sequence,[18] translation initiation codon, and the A/T-rich N-terminal sequences of the lactate dehydrogenase-encoding gene *ldh* or the α-amylase encoding gene *amy*. These genes are efficiently expressed in *Lactobacillus*. The configuration around the translation initiation start site in the expression vectors was chosen so as to warrant translation of foreign proteins equally efficient as that of the authentic *Lactobacillus* L-lactate dehydrogenase and α-amylase, two proteins that are efficiently translated in *Lactobacillus*.

Site of Protein Expression

To enable secretion of the expressed proteins, the signal peptide-encoding sequences of the *L. amylovorus amy* gene or of the proteinase gene, *prtP,* from *L. casei* were used in expression–secretion vectors. For the synthesis of proteins that are to be secreted and surface bound vectors were made in which the anchor-encoding sequence of the *L. casei prtP* gene, the product of which links a protein to the peptidoglycan part of the cell wall,[19] was fused, in frame, downstream of the signal sequence.

Selected Cloning Sites

In prototype expression vectors multicloning site (MCS) regions were engineered into the vectors in three different reading frames (*Cla*I, *Bam*HI, *Mun*I, *Kpn*I, *Hin*dIII, *Sma*I, *Bbs*I, *Eco*RI, *Sal*I, *Apa*I, *Nco*I) between the two transcription terminators. Because the reading frame of the entire MCS has been modified in the vectors, any open reading frame (ORF) can be aligned in frame with that of the N-terminal sequence of the *amy* or *ldh* gene by selecting the appropriate vector. From the prototype vectors the MCS cassette can easily be transferred to the pLP400 and pLP500 vectors as all vectors contain unique *Bam*HI and *Nco*I sites. In the distal part of the MCS the present restriction enzyme sites can be easily manipulated to allow in-phase translation of inserted ORF sequences with the downstream reporter gene.

[17] P. H. Pouwels and J. Leunissen, *Nucleic Acids Res.* **22,** 929 (1994).
[18] J. Shine and L. Dalgarno, *Proc. Natl. Acad. Sci. U.S.A.* **71,** 1342 (1974).
[19] W. W. Navarre and O. Schneewind, *Microbiol. Mol. Biol. Rev.* **63,** 174 (1999).

Nucleotide sequence of the 3' end of the *gusA* gene

a.   GGC AAA TGAATCACA
        G   K   stop

                    XhoI   _Bst_EII   _Hind_III
b.   GGC AAA CTC GAG TAATGTAAGGTCACCTCGACAAGCTT
        G   K   L   E   stop   pUC cloning site

FIG. 2. Nucleotide sequence of the 3' end of the *gusA* gene before (a) and after (b) mutagenesis by PCR. An *Xho*I site was introduced immediately upstream of the translation stop codon to allow in-frame fusion of amino acid encoding sequences at the 3' end of the *gusA* gene.

Use of Escherichia coli β-Glucuronidase as Reporter

The vectors were designed so as to allow fusion of antigenic determinants of interest to a large protein moiety (i.e., *E. coli* β-glucuronidase), which serves two purposes. Because of its ability to hydrolyze the chromogenic substrate 5-bromo-4-chloro-3-indolyl-β-D-glucuronic acid (X-GlcU), β-glucuronidase can serve as a reporter protein for the synthesis of the fusion partner. Synthesis of antigenic determinants fused with β-glucuronidase can easily be visualized after growth of transformed bacteria on solid media by staining the colonies with X-GlcU. β-Glucuronidase can be quantitatively determined after growth of the bacteria in liquid media, either enzymatically or immunologically. Extension of β-glucuronidase with amino acid sequences at the N-terminal end did not significantly affect its enzymatic activity even if these sequences were 50–80 kDa in mass. An added benefit from the presence of the β-glucuronidase moiety is that fusion of an antigenic determinant to a large immunogenic protein (e.g., β-glucuronidase) enhances the immune response against the antigenic determinant, in particular if the determinant itself is small. The cassette structure allows deletion of the β-glucuronidase gene (*gusA*)[20] at will for expression of nonfused heterologous proteins, or fusion of ORFs to the 3' end of *gusA* (Fig. 2). When the reporter gene is omitted, in-phase translation with the *prtP* anchor sequence or the last amino acid codon of the *cbh* gene can be achieved. The presence of unique *Xho*I and *Bst*EII sites in the expression cassettes allows variations at will of the length of the anchor sequence to adjust the peptidoglycan spanning region to the thickness of the peptidoglycan layer of the host bacterium.

Synthesis of Nonfused Proteins

For the synthesis of nonfused proteins use can be made of the so-called two-cistron system, which was originally explored for *E. coli*[21] and later successfully

[20] I. N. Roberts, R. P. Oliver, P. J. Punt, and C. A. M. J. J. van den Hondel, *Curr. Genet.* **15,** 177 (1989).
[21] B. E. Schoner, *Methods Mol. Biol.* **62,** 89 (1997).

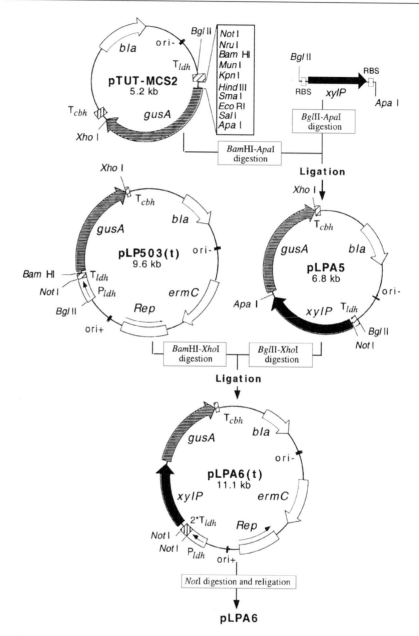

FIG. 3. Construction of the *Lactobacillus–E. coli* shuttle plasmid pLPA6 for expression of the isoprimeverose transport protein, XylP. The expression cassette of plasmid pLPA6 was derived from pLP503 by placing *xylP* upstream of *gusA*. Upstream of *xylP* a DNA fragment was inserted containing two *ldh* transcription terminators, flanked by *Not*I sites. The forward primer, (5'-TTCTAAGATCTA

applied in *Lactococcus*.[22] The rationale behind the system is the fusion of two genes in such a way that translation initiating at the first ATG is prematurely arrested at a site that overlaps the translation initiation region (TIR) of the gene of interest. It is assumed that ribosomes pausing at the stop signal of the first translated gene will not be released from mRNA but will reinitiate translation of the downstream gene. By selecting a TIR from an efficiently translated RNA as the first TIR, the RNA of the downstream gene is expected to be translated with equally high efficiency. Typically, a gene containing its genuine ribosome-binding site (RBS) is cloned at a distance of approximately 10 nucleotides downstream from a stop codon that is engineered into the reading frame of the *amy* or *ldh* gene. The approach has been successfully used for the expression in *L. plantarum* of the isoprimeverose transporter of *L. pentosus*[12] and the D-xylose-H^+ transporter from *Lactobacillus brevis*[11] (Fig. 3).

Detailed Description of Expression Vectors

Details of the construction of most of the vectors described above are given in the legend to Fig. 1. The nucleotide sequence of each vector can be assembled from the known nucleotide sequence of pLPE323,[10] the promoter region including the first 30 codons of the α-amylase gene of *L. amylovorus*,[23] or the promoter region including the first 8 specific codons of the *L. casei* L-*ldh* gene[24] (the sequence of a region of approximately 500 bp upstream of the L-*ldh* promoter was determined by Seegers and has been deposited in GenBank under accession number AF 351136), the transcription terminator sequence of the *L. casei* L-*ldh* gene,[24] the MCS (Fig. 4), the *E. coli gusA* gene,[20] the anchor sequence of the *prtP* gene from *L. casei* ATCC 393, which is the same as that from *L. paracasei*,[25] and the transcription terminator sequence of the *L. plantarum cbh* gene.[26]

[22] M. van de Guchte, J. Kok, and G. Venema, *Mol. Gen. Genet.* **227**, 65 (1991).

[23] A. Fitzsimons, P. Hols, J. Jore, R. J. Leer, M. O'Connell, and J. Delcour, *Appl. Environ. Microbiol.* **60**, 3529 (1994).

[24] S. F. Kim, S. J. Baek, M. Y. Pack, G. Perdigon, S. Alvarez, and A. Pesce de Ruiz Holgado, *Appl. Environ. Microbiol.* **57**, 2413 (1991).

[25] A. Holck and H. Naes, *J. Gen. Microbiol.* **138**, 1353 (1992).

[26] H. Christiaens, R. J. Leer, P. H. Pouwels, and W. Verstraete, *Appl. Environ. Microbiol.* **58**, 3792 (1992).

GGTACCA-**TTAA**TTGAATTCAGAAAGAAGGC-3′) used in the PCR amplification of *xylP*, generated *Bgl*II and *Kpn*I sites (underlined) and a stop codon (indicated in boldface) upstream of the original *xylP* ribosome-binding site (RBS). The reverse primer (5′-TTAGGGCCCTCCTTTCTTATCCCATC TTAC-3′) generated an *Apa*I site (underlined) downstream of the original *xylQ* RBS. *bla*, β-lactamase (ampicillin resistance) determinant; *ermC*, erythromycin resistance determinant. [Reproduced with permission from S. Chaillou, P. W. Postma, and P. H. Pouwels, *J. Bacteriol.* **180**, 4011 (1998).]

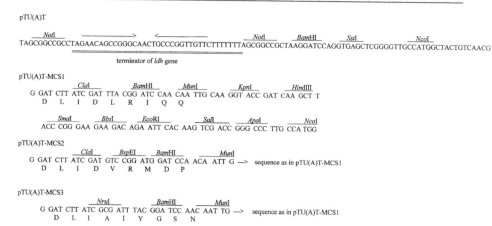

FIG. 4. Nucleotide sequence of the region between the translation initiation cassette and the *gusA* gene. The nucleotide sequence between the *Bam*HI and *Nco*I sites in pTUT-MCS and pTUAT-MCS is shown; this sequence is part of the synthetic oligonucleotide sequence encompassing the *L. casei ldh* transcription terminator signal and MCS. Two derivatives of pTU(A)T-MCS1 with different reading frames were generated by +1 and +2 frame shifts of the various restriction enzyme sites: (1) by replacing the *Cla*I–*Bam*HI sequence of pTU(A)T-MCS1 by a *Cla*I–*Bsp*EI–*Bam*HI sequence, the *Bam*HI site was shifted four nucleotides with respect to the first nucleotide (G), yielding pTU(A)T-MCS2; (2) pTUAT-MCS1 was digested with *Cla*I and the resulting DNA termini were filled in with dNTPs, using T4 DNA polymerase. After ligation the MCS contained a *Bam*HI site shifted two nucleotides with respect to G [pTU(A)T-MCS3]. DNA fragments can be cloned in the MCS in pTU(A)T-MCS1, -2, or -3 depending on the reading frame needed to align the reading frame of the cloned ORF and that of the *ldh* (or *amy*) sequence.

Vectors with Expanded Host Range

Vectors such as pLP323 containing an origin of replication similar to that of pC194 or pUB110[27] can replicate in a wide variety of gram-positive bacteria but not in thermophilic lactobacilli such as *Lactobacillus acidophilus* (our unpublished observations, 2000). To construct expression vectors for this class of bacteria, we have replaced the *L. pentosus* replicon and erythromycin resistance marker in the pLP400 and pLP500 plasmids by pNZ124,[13] which contains the origin of replication from the *Lactococcus* plasmid pSH71,[28] because our experiments have shown this plasmid could replicate in different strains of *L. acidophilus* and *Lactobacillus crispatus*. Because the replicon of pNZ124 is also functional in *E. coli,* the pGEM part of the vectors was removed as well. The new series of plasmids, containing the chloramphenicol resistance gene as selection marker,

[27] A. Gruss and D. Ehrlich, *Microbiol. Rev.* **53**, 231 (1989).
[28] W. M. de Vos, *FEMS Microbiol. Rev.* **46**, 281 (1987).

was designated pLP600 for vectors with an α-amylase promoter and pLP800 for vectors with an L-*ldh* promoter (Fig. 5).

Vectors with an Inducible *xyl* Promoter

The promoter of the *xylAB* operon of *L. pentosus* is tightly controlled. The *xyl* promoter, P_{xyl}, is strongly induced in the presence of xylose, and is strongly repressed by catabolite repression under inducing conditions.[16,29,30] Expression of a gene that was placed under the control of the *xyl* promoter on a multicopy plasmid was found to be 60-fold increased in the presence of xylose.[16] Four vectors containing the *xyl* promoter were constructed by replacing the P_{amy} of pLP601 and pLP602 by either an 130-bp fragment containing P_{xyl} from *L. pentosus* MD353 or by a 1.5-kb fragment containing P_{xyl} as well as the upstream located repressor gene *xylR* (*xylR*-P_{xyl}), yielding vectors pLP701, pLP7011, pLP702, and pLP7021 (Fig. 5). In bacteria containing pLP7011 or pLP7021, gene expression will be dependent on the presence of xylose in the medium, because of the presence on the plasmid of the repressor gene, *xylR*. In pLP701- or pLP702-carrying bacteria, which lack *xylR*, gene expression from the *xylA* promoter is constitutive. Expression from the *xylA* promoter in all vectors is subject to carbon catabolite repression. In the presence of carbon sources imposing carbon catabolite repression, expression will be strongly reduced. Table I summarizes the vectors described so far.

Manipulation of Vectors

All plasmid construction work can be carried out in *L. casei* ATCC 393 or *L. plantarum* 80 because the transformation efficiencies of these strains enable the cloning of ligation mixtures. However, because of the relative ease with which purified preparations of plasmid DNA can be obtained from *E. coli* compared with lactobacilli and enzymatically treated, we advise all construction work be performed in *E. coli*. In most cases plasmids pTUT containing P_{ldh} or pTUAT containing P_{amy} or $P_{ldh-ssPrt}$ (Fig. 1D) can be used as the starting material for the construction of the final expression vectors. When the reading frame of the gene to be expressed cannot be easily aligned with that of the vectors, an additional cloning step in pTUT-MCS1, -2, or -3, or in pTUAT-MCS1, -2, or -3, may be necessary (Fig. 4). After such a cloning step the cloned ORF may be exchanged, with or without flanking terminator sequences, for the corresponding region in

[29] B. C. Lokman, M. Heerikhuisen, R. J. Leer, A. van den Broek, Y. Borsboom, S. Chaillou, P. W. Postma, and P. H. Pouwels, *J. Bacteriol.* **179,** 5391 (1997).

[30] S. Chaillou, B. C. Lokman, R. J. Leer, P. W. Postma, and P. H. Pouwels, *J. Bacteriol.* **180,** 2312 (1998).

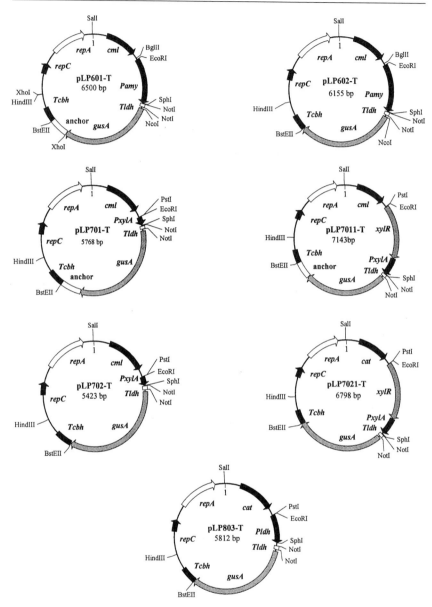

FIG. 5. Structure of the pLP600, pLP700, and pLP800 series of expression vectors. Xylose promoter fragments (with and without the repressor gene, *xylR*) were PCR amplified from the xylose operon of *L. pentosus* MD353 (GenBank accession number M57384). The fragments were cloned into the *Bgl*II and *Sph*I sites of pLP601 and pLP602, resulting in vectors pLP701, pLP7011, pLP702, and pLP7021.

TABLE I
EXPRESSION OF FOREIGN PROTEINS

Plasmid	Promoter	Replicon	Compartment	Selection marker	Phenotype on X-GlcU plates		
					Galactose	Glucose	Xylose/galactose
LP401	P_{amy}	pLP323	Cell wall	*ery*	Light blue	White	ND[a]
LP402	P_{amy}	pLP323	Medium	*ery*	Light blue	White	ND[a]
LP403	P_{amy}	pLP323	Intracellular	*ery*	Light blue	White	ND[a]
LP501	P_{ldh}	pLP323	Cell wall	*ery*	Dark blue	Dark blue	ND[a]
LP502	P_{ldh}	pLP323	Medium	*ery*	Dark blue	Dark blue	ND[a]
LP503	P_{ldh}	pLP323	Intracellular	*ery*	Dark blue	Dark blue	ND[a]
LP601	P_{amy}	pNZ124	Cell wall	*cml*	Light blue	White	ND[a]
LP602	P_{amy}	pNZ124	Medium	*cml*	Light blue	White	ND[a]
LP701	P_{xylA}	pNZ124	Cell wall	*cml*	Blue	Light blue	Blue
LP702	$XylR\text{-}P_{xylA}$	pNZ124	Cell wall	*cml*	White	White	Blue
LP703	P_{xylA}	pNZ124	Medium	*cml*	Blue	Light blue	Blue
LP704	$XylR\text{-}P_{xylA}$	pNZ124	Medium	*cml*	White	White	Blue
LP803	P_{ldh}	pNZ124	Intracellular	*cml*	Dark blue	Dark blue	ND[a]

[a] ND, Not determined.

one of the promoter-containing pTUT or pTUAT vectors. For the construction of expression vectors a number of *E. coli* strains can be used such as DH5α, JM101, and JA221. After verification of the structure of the recombinant plasmid by standard procedures,[31] the *ldh* terminator sequence is removed prior to transfer of the plasmid to *Lactobacillus*. For this purpose plasmid DNA is incubated overnight with excess *Not*I, followed by heat treatment (5 min at 65°) to inactivate the enzyme. Subsequently, the plasmid molecule is circularized with DNA ligase and used to electrotransform *Lactobacillus*.

Electroporation of Lactobacillus casei ATCC 393

Details of the procedure to genetically transform *L. casei* ATCC 393 are as follows.

1. Inoculate 100 ml of MRS (Difco) medium with 2 ml of an overnight culture of *L. casei* ATCC 393.
2. Incubate at 37° until the OD_{660} is 0.6.
3. Harvest the cells by centrifugation at 2500 rpm for 10 min.
4. Wash the pellet with excess 5 mM sodium phosphate buffer (pH 7.4) containing 1 mM MgCl$_2$ (EPWB) at 4°.

[31] J. Sambrook, E. F. Fritsch, and T. Maniatis, "Molecular Cloning: A Laboratory Manual," 2nd Ed. Cold Spring Harbor Laboratory Press, Cold Spring Harbor Laboratory, New York, 1989.

5. Wash twice with 1 ml of EPWB.

6. Wash once with 1 ml of electroporation buffer (EPWB with 0.3 M sucrose) at 4°.

7. Resuspend the bacteria in 1 ml of electroporation buffer.

8. Adjust electroporation settings to 100 W, 25 mF, and 1250 V.

9. Mix 50 μl of bacterial suspension with 5 μl of DNA ($\pm$1 μg) in a precooled 2-mm cuvette; apply pulse.

10. Add 450 μl of MRS medium immediately after the pulse, and incubate at 37° for 1 hr.

11. Plate various dilutions on MRS medium plates containing antibiotic (5– 10 μg of erythromycin or chloramphenicol) and incubate at 37° for 24 hr or more.

For further information on transformation procedures of other *Lactobacillus* species, including *L. acidophilus*, *L. pentosus*, and *L. plantarum*, the reader is referred to Refs. 10, 16, and 32.

A rapid procedure to identify which of the transformants contains a plasmid lacking the *ldh* terminator sequence involves a colony polymerase chain reaction (PCR) experiment. A colony is scraped from a plate, suspended in 20 μl of water, and heated in a microwave or kept for 5 min in boiling water. The lysate formed can be directly used for a PCR experiment with two primers derived from the regions that flank the terminator sequence at both sides. One of the primers usually is from the *ldh* or *amy* promoter region, whereas the second one is derived from the gene that was cloned. By choosing primers at an appropriate distance from the junctions with the terminator sequence, PCR products can be obtained that can be easily differentiated, as the product will be ∼110 nucleotides larger when the terminator is still present. It is our experience that between 10 and 50% of the transformants have the expected structure.

Further characterization of the plasmid vectors can be carried out on miniscreen DNA or after isolation and purification of larger quantities of DNA. The following procedure to isolate DNA from transformed lactobacilli, which can be applied at the miniscreen level and for the large-scale isolation of DNA, yields DNA that can be cut with most common restriction enzymes and can be used for sequencing reactions or PCR experiments.

Large-Scale Plasmid DNA Isolation from *Lactobacillus* Species

This protocol is a modified version of the protocol of the Jet Star purification kit from Genomed (Alameda, CA), which allows the use of this kit for the purification of plasmid DNA from lactic acid bacteria. The protocol can be adapted easily for other commercial kits by changing the volumes to the appropriate level. Inoculate

---

[32] D. C. Walker, K. Aoyama, and T. R. Klaenhammer, *FEMS Microbiol. Lett.* **138**, 233 (1996).

250 ml of MRS broth with the desired strain. Add appropriate antibiotics and incubate overnight. Best results are obtained, however, when cells are harvested during late exponential phase.

1. Collect cells by centrifugation.
2. Wash the cells once with 100 ml of 50 mM Tris-HCl, pH 8.0.
3. Resuspend the pellet in 20 ml of 50 mM Tris-HCl, pH 8.0, supplemented with lysozyme (5 mg/ml), mutanolysin (20 U/ml), and RNase (50 μg/ml).
4. Incubate on ice for 30 min.
5. Incubate for 20 min at 55°.
6. Add 20 ml of lysis solution, mix well (no vortexing), and incubate at room temperature for 5 min.
7. Add 20 ml of neutralization solution, mix well, and centrifuge for 10 min at >15,000g.
8. Equilibrate a column with equilibration solution.
9. Apply the supernatant to the column and allow the solution to pass through by gravitational flow.
10. Wash the column once with 30 ml of washing solution.
11. Elute the DNA with 10 ml of elution buffer.
12. Add 0.7 volume of isopropanol to the eluate and precipitate the DNA by centrifugation at 4° for 30 min at >15,000g.
13. Wash the pellet once with 70% (v/v) ethanol.
14. Air dry the pellet for 10 min and redissolve the DNA in a suitable volume.

Note: Using buffer E1 of the kit, supplemented with lysozyme and mutanolysin, results in poor lysis, most probably because of the presence of EDTA in this buffer. Lysozyme and mutanolysin should be added on the day of use because notably solubilized lysozyme loses its activity rapidly and does not withstand freezing well.

Expression of Foreign Proteins in *Lactobacillus*

The vectors described above permit the synthesis in lactobacilli of foreign proteins fused to β-glucuronidase or in a nonfused form. In this section we demonstrate the potential of the vectors, using β-glucuronidase and tetanus toxin fragment C (TTFC) as examples. The α-amylase promoter of *L. amylovorus* can be induced in the presence of carbon sources such as cellobiose, galactose, and mannitol, but is subject to catabolite repression by glucose. When *L. casei* transformants harboring plasmids of the pLP400 or pLP600 series were plated on indicator plates with X-GlcU, containing galactose as energy source, all colonies were light blue, but were white when glucose was used as energy source, indicating that expression of a foreign gene under the control of P$_{amy}$ in *L. casei* is under the same tight glucose repression as is the expression of the α-amylase gene in *L. amylovorus*[23]

(Table I). When transformants of the pLP500 or pLP800 series were plated on indicator plates all colonies were dark blue, irrespective of whether galactose or glucose was used as energy source. These results indicate that expression driven by P_{ldh} is more efficient than by P_{amy} (Table I). Essentially the same results were obtained when galactose was replaced by cellobiose or mannitol as energy source (not shown).

Lactobacillus transformants containing the pLP401,2,3-T or pLP501,2,3-T plasmids containing the T_{ldh} sequence displayed the white phenotype on selective medium. Reintroduction of pLP503 and pLP402 isolated from *L. casei* transformants into *E. coli* DH5α yielded transformants with blue (30%) or white (70%) phenotype on selective medium [L medium: 0.5% (w/v) glucose, 1.5% (w/v) agar, ampicillin (50 mg/ml), X-GlcU (40 mg/ml)]. The white phenotype was correlated with structurally unstable plasmids (size smaller than starting material). When *E. coli* transformants with a blue phenotype were suspended in physiological salt solution and streaked onto agar plates with selective medium, most colonies showed a white phenotype, indicating plasmid instability in this host.

Expression of the *gusA* gene in *L. casei* transformants carrying plasmid pLP7011 or pLP7021 with the *xylA* promoter of *L. pentosus*[16] was fully dependent on the presence of xylose in the medium, because colonies were white on galactose agar plates containing X-GlcU, whereas blue colonies were obtained when the plates also contained xylose. Blue colonies were found with plasmids that lacked the xylose repressor gene, *xylR* (pLP701 and pLP703) in galactose medium, confirming the constitutive expression from the *xylA* promoter. In glucose medium this expression was partly repressed, resulting in faint blue colonies (Table I).

The expression vectors have been successfully used to construct *Lactobacillus* strains from different species expressing a variety of antigens. Full-size proteins, such as the TTFC, several rotavirus surface proteins with sizes ranging from 37 to 87 kDa, urease A and B subunits from *Helicobacter pylori,* and some small peptide epitopes (e.g., from influenza virus and encephalitogenic myelin peptides) that were fused to β-glucuronidase have all been successfully expressed.[33] Production in *L. casei* of antigens targeted to the cytosol yielded levels of production ranging from 1 to 5%. Efficient production levels were also obtained in this and other *Lactobacillus* species when the products were directed to the bacterial surface. Up to 10,000 molecules can be found at the bacterial surface.

In some experiments instability of vector DNA was noticed when foreign antigens, only as tetanus toxin fragment C (TTFC)-encoding DNA fragments, were cloned in vectors with the strong constitutive *ldh* promoter together with the

---

[33] C. B. Maassen, J. D. Laman, M. J. den Bak-Glashouwer, F. J. Tielen, J. C. van Holten-Neelen, L. Hoogteijling, C. Antonissen, R. J. Leer, P. H. Pouwels, W. J. A. Boersma, and D. M. Shaw, *Vaccine* **17,** 2117 (1999).

prtP secretion signal sequence. This instability may be caused by interference with the protein transport system due to the combination of an efficient promoter and as efficient secretion signal.

Detection and Quantification of Antigen Expression on Lactobacilli, Using Flow Cytometric Techniques

Flow cytometry provides a straightforward and rapid method to detect and quantify antigens expressed on the cell surface of (gram-positive) bacteria. For the detection of the antigen, we used immunofluorescent staining with a specific antibody directed against the antigen of interest. To quantify the amount of fluorescing probe, which is directly correlated with the level of antigen expressed, the fluorescent signal was calibrated with a known standard. This enabled the calculation of the level of antigen expression by determining the mean fluorescence, using flow cytometry analysis, and comparing the values found with those of the calibration curve.

Fluorescence data from lactobacilli and microbeads were collected on a FACSCalibur (Becton Dickinson, San Jose, CA) equipped with a 15-mW, air-cooled, 488-nm argon ion laser. Forward and side light scatter detectors as well as the FL1 (FITC) detector used logarithmic amplification. To adjust cytometer settings lactobacilli were analyzed with two negative controls. One control consisted of *L. casei* expressing a nonrelevant cell wall-anchored antigen, using the primary as well as the secondary antibody. Nonspecific binding of the secondary antibody was checked by incubating *L. casei* expressing the antigen of interest with the secondary FITC-labeled antibody only. A live gate was set around all viable cells, 10,000 events were counted, and the mean fluorescense was determined with the Becton Dickinson research software program CellQuest.

Quantum 24P en Quantum 25P (FITC) beads (Flow Cytometry Standards, San Juan, Puerto Rico) were used as described by the manufacturer for the determination of the number of molecules of equivalent soluble fluorochrome (MESF). Fifty microliters of the beads was mixed with wild-type lactobacilli and analyzed with FACSCalibur equipment (Becton Dickinson). The peak channel for each of the beads was used for the calibration plot. Beads were collected, using the same cytometer settings as with lactobacilli; however, gating on forward and side scatter was set around the singlet population of the beads. Again, 10,000 events were collected and analyzed. Saturating levels of antibodies are necessary to obtain an accurate quantification. In addition, the fluorochrome to protein ratio (*F/P* ratio) must be determined (e.g., by using Quantum Simply Cellular Microbeads from Flow Cytometry Standards).

An example of the results of a typical experiment is shown in Fig. 6. Figure 6A shows the histogram plot and the statistics data, using the Quantum 24 P (FITC) kit.

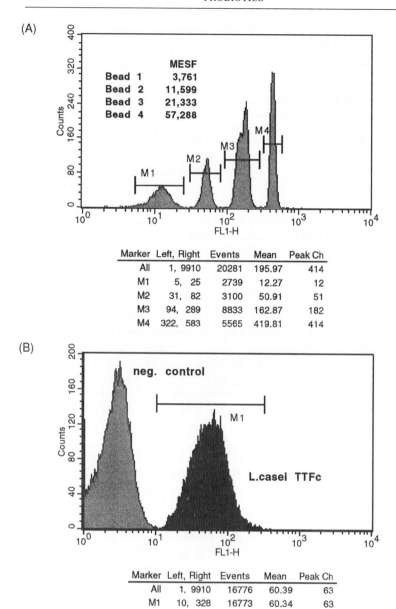

FIG. 6. Detection and quantification of antigen expression on lactobacilli, using flow cytometric techniques. Fluorescence data from lactobacilli and microbeads were collected on a FACSCalibur. (A) Histogram plot of the microbeads and the statistics data, using the Quantum 24 P (FITC) kit to quantify the number of molecules present. (B) Histogram of TTFC-expressing *L. casei* coated with rabbit anti-TTFC serum–FITC or as a negative control with normal rabbit serum, and stained with goat anti-rabbit.

Figure 6B shows a histogram of TTFC-expressing *L. casei* indirectly stained with goat anti-rabbit–FITC. The TTFC-expressing lactobacilli were coated with rabbit anti-TTFC serum or, as a negative control, with normal rabbit serum. Determination of antigen levels (in molecules per cell) can be calculated by dividing the MESF value (Fig. 6A), which corresponds to the mean fluorescence value of the positive sample in Fig. 6B, by the value of the *F/P* ratio.

[31] A Dot-Blot Assay for Adhesive Components Relative to Probiotics

By Maurilia Rojas and Patricia L. Conway

Introduction

Biofilms have been studied by researchers from a wide variety of disciplines. Over the last two decades the adhesive characteristics of bacteria and other cells as well as their surface components have assumed increasing significance in microbial ecology.[1] The adherent bacteria produce microcolonies, leading to the development of biofilms, which initially may be composed of only one bacterial type, but frequently develop to contain several species of bacteria living in a complex community.[2] The biofilm formed in the gastrointestinal tract is an example. In this living surface the best protection against attachment and invasion by pathogens could be the maintenance of the normal biota by adhering tenaciously to mucus overlaying the mucosa and resisting removal.[3] Numerous *in vitro* systems have been developed to examine and assess the adhesive properties of bacteria to gastrointestinal biofilm (to mucus[3–11] and to epithelial cells[12–18]). Adhesive traits of *Escherichia coli* strains have been extensively studied and have often been linked to virulence. The mechanism of adhesion of *Lactobacillus* strains is also being studied because of the beneficial role this group of microbes can exert on the

---

[1] I. Ofek and R. J. Doyle, "Bacterial Adhesion to Cells and Tissues." Chapman & Hall, New York, 1994.

[2] K. C. Marshal, "Microbial Adhesion and Aggregation." Springer-Verlag, Berlin, 1984.

[3] M. Rojas, "Studies on an Adhesion Promoting Protein from *Lactobacillus* and Its Role in Colonization of the Gastrointestinal Tract." Ph.D. Thesis. University of Göteborg, Göteborg, Sweden, 1996.

[4] T. Yamamoto and T. Yokota, *J. Clin. Microbiol.* **26**, 2018 (1988).

[5] P. L. Conway and S. Kjelleberg, *J. Gen. Microbiol.* **135**, 1175 (1989).

[6] G. W. Tannock, R. Fuller, and K. Pedersen, *Appl. Environ. Microbiol.* **56**, 1310 (1990).

[7] A. Henriksson and P. L. Conway, *Microbial Ecol. Health Dis.* **4**, 357 (1991).

host and because adhesion could favor colonization of the tract, thereby excluding pathogen. Laux *et al.*[12] have developed an *in vitro* method that has been widely used[18-22] and allows bacterial–mucosal interactions to be examined more closely. In that system, soluble mucous extracts prepared from mouse large and small intestine were immobilized on polystyrene, and the capacity of various radioactively labeled strains of *E. coli* to adhere to the immobilized components was assessed. In another method, the mucosal glycoproteins were radioactively labeled with ^{125}I prior to incubation with a standardized number of *E. coli* cells.[23,24]

Adhesion-promoting compounds on the surface of *Lactobacillus* cells that bind mouse stomach squamous epithelium have been studied by isolating them from either spent culture fluid after growth in a defined medium or from whole cells. Such compounds have been detected in fractionated extracts either by inhibiting adhesion of whole cells to tissue pieces in an *in vitro* adhesion assay[5] or by radioactively labeling the extracts, which can then be used in the *in vitro* assay instead of whole cells.[25] The former methods, which utilize mouse mucus preparations[12] or stomach squamous epithelia from mice,[5] require a considerable amount of material. Furthermore, the inhibition is an indirect method for demonstrating the presence of adhesion-promoting material. A disadvantage with the radiolabel method is that not all components will be labeled. Consequently, in studying the mechanism by which *Lactobacillus* and *E. coli* adhered to gastrointestinal mucus, a more convenient assay was required to confirm the presence of adhesion-promoting components on the cell surface of the bacteria, in extracts and in fractionated

[8] A. Henriksson, R. Szewzyk, and P. L. Conway, *Appl. Environ. Microbiol.* **57,** 499 (1991).

[9] S. Hicks, D. C. Candy, and A. D. Phillips, *Infect. Immun.* **64,** 4751 (1996).

[10] S. E. Craven and D. D. Williams, *Avian Dis.* **41,** 548 (1997).

[11] E. M. Tuomola, A. C. Ouwehand, and S. J. Salminen, *Lett. Appl. Microbiol.* **28,** 159 (1999).

[12] D. C. Laux, E. F. McSweegan, and P. S. Cohen, *J. Microbiol. Methods* **2,** 27 (1984).

[13] P. L. Conway, S. L. Gorbach, and B. R. Goldin, *J. Dairy Sci.* **70,** 1 (1986).

[14] T. Yamamoto and T. Yokota, *Infect. Immun.* **56,** 2753 (1988).

[15] A. Darfeuille-Michaud, D. Aubel, G. Chauviere, C. Ric, M. Bourges, A. Servin, and B. Joly, *Infect. Immun.* **58,** 893 (1990).

[16] M. H. Coconnier, T. R. Klaenhammer, S. Kerneis, M. F. Bernet, and A. L. Servin, *Appl. Environ. Microbiol.* **58,** 2034 (1992).

[17] D. A. Simpson, R. Ramphal, and S. Lory, *Infect. Immun.* **60,** 3771 (1992).

[18] P. Doig and T. J. Trust, *J. Microbiol. Methods* **18,** 167 (1993).

[19] D. C. Laux, E. F. McSweegan, T. J. Williams, E. A. Wadolkowski, and P. S. Cohen, *Infect. Immun.* **52,** 18 (1986).

[20] J. J. Nevola, D. C. Laux, and P. S. Cohen, *Infect. Immun.* **55,** 2884 (1987).

[21] S. U. Sajjan and J. F. Forstner, *Infect. Immun.* **58,** 860 (1990).

[22] P. L. Conway, A. Welin, and P. S. Cohen, *Infect. Immun.* **58,** 3178 (1990).

[23] M. Lindahl and I. Carlstedt, *J. Gen. Microbiol.* **136,** 1609 (1990).

[24] C. A. Wanke, S. Ronan, C. Goss, K. Chadee, and R. L. Guerrant, *Infect. Immun.* **58,** 794 (1990).

[25] A. Henriksson and P. L. Conway, *J. Gen. Microbiol.* **138,** 2657 (1992).

extracts. This assay was essential for the binding of components of the porcine mucus that are not retained on polystyrene when it is immobilized according to Laux et al.[12] The binding to polystyrene was more relevant for rodent mucus as used by Laux and co-workers than porcine mucus because rodent mucus is more hydrophobic.[26]

The aim of this work was to develop a qualitative in vitro assay for detection of binding of bacterial cell surface components to mucous extracts. A dot-blot assay was developed whereby extracts containing bacterial components and fractions were immobilized and then blotted with enzymatically labeled mucus. Results were compared with those obtained using the inhibition assay. In addition, whole cells of Lactobacillus and E. coli were tested in the dot-blot assay and results were compared with a modification of the method of Laux and co-workers.[22] The results obtained with the dot-blot assay provided further information about the binding of Lactobacillus and E. coli to gastrointestinal mucus, not only because adhesion-promoting compounds could be detected in fractionated extracts but also because porcine gastric mucin as well as small intestinal mucus could be used for blotting.

Methods for Adhesion Assays

Bacterial Strains

The strains used in this study and their origins are listed in Table I.

Primary Cultures

Lactobacilli and E. coli, stored at $-70°$ in 40% (w/v) glycerol, are grown overnight at $37°$ in a candle jar, using a 1% (v/v) inoculum in tubes containing 5 ml of MRS broth (Mann, Rogosa, and Sharpe; Oxoid, Basingstoke, UK) and tryptic soy broth (TSB; Difco, Detroit, MI), respectively.

Preparation of Mucus

Mucus is isolated from the small intestinal wall of 35-day-old pigs according to Conway et al.[22] Briefly, mucus is isolated from the small intestinal mucosa by gentle scraping into cool HEPES–Hanks' buffer (pH 7.4). Epithelial cells and large cellular components are removed by centrifugation once at $11,000g$ for 10 min and once at $26,000g$ for 15 min at $4°$. The protein concentration of the mucous preparation is determined by the Lowry method (kit from Sigma, St. Louis, MO). Mucus samples are kept in 1.5-ml culture tubes at $-70°$.

[26] M. Malmsten, E. Blomberg, P. Claesson, I. Carlstedt, and I. Ljusegren, J. Colloid Interface Sci. 151, 579 (1992).

TABLE I
BACTERIAL STRAINS USED IN STUDY

No.	Strain[a]	Origin[b]
1	*Lactobacillus* 1D	Pig
2	*Lactobacillus* 2D	Pig
3	*Lactobacillus* 3D	Pig
4	*Lactobacillus* 4J	Pig
5	*Lactobacillus* 5I	Pig
6	*Lactobacillus* 6I	Pig
7	*Lactobacillus fermentum* 104R	Pig
8	*L. fermentum* KLD	Human
9	*Lactobacillus murinus* C39	Pig
10	*Lactobacillus plantarum* 256	Ensilage
11	*Escherichia coli* K12 (K88ab)[c]	Laboratory strain
12	*E. coli* K12 pMK 005 (K88ac$^+$)[d]	Laboratory strain
13	*E. coli* K12 (K88ab$^-$)	Laboratory strain
14	*E. coli* K12 pMK 002 (K88ac$^-$)[e]	Laboratory strain
15	*E. coli* DH5 pMK 005 (K88ac$^+$)	Laboratory strain
16	*E. coli* DH5	Laboratory strain
17	*E. coli* Bd 4228/84I 0149 (K88ac)	Diseased pig
18	*E. coli* Bd 2221/75 08 (K88ac)	Diseased pig
19	*E. coli* Bd 1107/75 08 (K88ac)	Diseased pig
20	*E. coli* Bd 4545/84 0.149 (K88ac)	Diseased pig
21	*E. coli* Bd 1147/75 032 (K88ab)	Diseased pig
22	*E. coli* Bd 3027/75 0149 (K88ab)	Diseased pig

[a] Nos. 1–6, M. Rojas and P. L. Conway, *J. Appl. Bacteriol.* **81,** 474 (1996); no. 7, A. Henriksson, R. Szewzyk, and P. L. Conway, *Appl. Environ. Microbiol.* **57,** 499 (1991); no. 8, S. A. W. Gibson and P. L. Conway, *in* "Human Health: The Contribution of Microorganisms" (S. A. W. Gibson, ed.), p. 119. Springer-Verlag, London, 1994; no. 9, A. Henriksson and P. L. Conway, *Microbial Ecol. Health Dis.* **4,** 357 (1991); nos. 11 and 14 were generously supplied by K. Krogfeldt (Serum Institute, Copenhagen); nos. 12 and 13, L. Blomberg, Ph.D. Thesis, ISBN 91-86022-68-7, University of Göteborg, Göteborg, Sweden, 1992; nos. 15 and 16, M. Kehoe, R. Selwood, P. Shipley, and D. G., *Nature (London)* **291,** 122 (1981); nos. 10 and 17–23 were generously supplied by O. Söderlind (National Veterans Institute, Uppsala, Sweden).
[b] Except for laboratory strains and *L. plantarum*, all strains originated from the gastrointestinal tract.
[c] Contains entire wild-type plasmid in which K88 fimbrial gene is located.
[d] Contains only the K88 fimbrial gene.
[e] Contains the deficient K88 fimbrial gene.

Adhesion Assay

The routine adhesion assay is performed to compare its results with those of the dot-blot assay described here. The adhesion of *Lactobacillus* and *E. coli* cells to small intestinal mucus from 35-day-old pigs is studied by a modification of the method of Laux *et al.*[12] Briefly, *Lactobacillus* or *E. coli* cells are radioactively labeled by inoculating from primary cultures into MRS or TSB medium, respectively, to which is added 10 μCi of [*methyl*-1,2-^{3}H]thymidine (specific activity, 117 Ci mmol^{-1}; Amersham International, Little Chalfont, UK) per milliliter. Cells are collected when the cultures reach an optical density of about 0.5 at a wavelength of 600 nm during growth in a candle jar at 37°, and are washed twice prior to being suspended in HEPES–Hanks' buffer. The cell density is adjusted to an absorbance of 0.5 at 600 nm. Samples (200 μl) of small intestinal mucus (0.5 mg of protein per milliliter) and bovine serum albumin (BSA, 30 mg/ml) are immobilized in polystyrene tissue culture wells (Nunc, Roskilde, Denmark) by incubating the plates for 24 hr at 4°. The plates are washed twice with HEPES–Hanks' buffer to remove excess mucus or BSA, and then 200 μl of *Lactobacillus* suspension is added to the immobilized mucus or BSA. The tissue culture plates are incubated for 1 hr at 37°, and the wells are washed twice as described previously to remove unbound bacteria. Adhering bacteria are released by adding 0.5 ml of 5% (w/v) sodium dodecyl sulfate (SDS) to each well and incubating the plates overnight at 37°. SDS is collected from each well into scintillation vials and the level of radioactivity is determined by scintillation counting. In addition, 100-μl samples of the bacterial suspensions are also quantified for radioactivity and colony-forming units to allow for calculation of the number of bacterial cells adhering to the mucus-coated wells. All assays are performed (at least) in duplicate, and in two separate experiments.

Inhibition of Adhesion Assay

The adhesion of lactobacillus spent culture supernatant, cell surface extracts, and fractions from gel-filtration chromatography is studied by blocking the adhesion of lactobacilli cells to immobilized mucus in the adhesion assay as previously described.[27] Briefly, 200-μl samples of the supernatant or medium (treated exactly the same as supernatant) are added per well to tissue culture plates, already coated with immobilized mucus or BSA. The plates are then incubated at 37° for 1 hr prior to being washed with HEPES–Hanks' buffer. The adhesion assay is performed as described above, using these pretreated plates.

Adhesive Compound(s) from Lactobacillus fermentum 104R

Spent Culture Fluid. Primary culture is inoculated in a semidefined LDM medium[5] and grown at 37° for 48 hr. The culture is then centrifuged at 8000g for 15 min and the supernatant is retained. This supernatant is filtered through a nitrocellulose membrane (0.45 μm) and dialyzed against distilled water, using

---

[27] M. Rojas and P. L. Conway, *J. Appl. Bacteriol.* **81**, 474 (1996).

a 12-to 14-kDa cutoff membrane at 4° for 18 hr, with regular changes of water. This dialyzed supernatant is freeze-dried or frozen at −20° until tested for the presence of adhesion-promoting compound(s) by both adhesion inhibition and dot-blot assays.

Extraction of Cell Surface Compounds. LDM culture, after 14 hr of growth under the above-described conditions, is pelleted by centrifugation at 3000g for 10 min, washed with HEPES–Hanks' buffer, and suspended in 1 M lithium chloride or lysozyme solution [0.1 M Tris, 0.015 M NaCl, 0.05 M MgCl$_2$, lysozyme (40 μg ml^{-1})]. Cell suspensions are incubated with gentle stirring at 4° for 60 min. Bacteria are removed by centrifugation (8000g, 30 min and 5°). The supernatants are filtered through nitrocellulose membranes (0.45 μm), dialyzed against distilled water, freeze-dried, and kept at 4° until use.

Gel-Filtration Chromatography of LiCl Extracts from Lactobacillus fermentum 104R. The freeze-dried LiCl extract is dissolved in HEPES–Hanks' buffer and centrifuged at 13,000 rpm for 3 min to remove insoluble particles. A 4-ml volume of this solution (2.1 mg of protein) is applied to a Sephadex G-200 in an XK-26 column (Pharmacia-LKB, Uppsala, Sweden) for gel-filtration chromatography. HEPES–Hanks' buffer is used to equilibrate the column and elute the sample. The fractions in each 280-nm-absorbing peak are assessed for adhesion to small intestinal mucus of pig by the two different adhesion assays.

Dot-Blot Adhesion Assay

The dot-blot assay described in the present chapter is based on the formation of a complex between adhesion-promoting compound(s) from the cell surface of the bacteria and the enzymatically labeled receptor(s) in gastrointestinal mucus, followed by the visualization of bound components on a solid-phase matrix. The essence of this assay is the same as the common immunoassay, which involves the mixing of antigen with antibody followed by the discrimination of bound from free reactant.[28]

Enzymatic Labeling of Small Intestinal Crude Mucus and Partially Purified Mucin from Stomach of Pig. Small intestinal mucous preparations from 35-day-old pigs are supplemented with proteinase inhibitors phenylmethylsulfonyl fluoride (PMSF; 2 mM final concentration in dry ethanol), acetylleucylleucylarginine (leupeptin, 0.01 mM), and Na$_2$EDTA (1 mM). PMSF is sprinkled on the frozen sample to allow this inhibitor to percolate through the mucus during thawing, and the others are then added. After mixing, the protein concentration is adjusted to 4 mg ml^{-1} in 0.1 M sodium carbonate buffer (pH 9.5). The proteinase inhibitor-treated small intestinal mucus and the gastric mucin (Sigma; 4 mg/ml in carbonate buffer) are

[28] L. Hudson and F. C. Hay, *in* "Practical Immunology" (Elaine and Frances, eds.), p. 44. Blackwell Scientific, London, 1989.

labeled with horseradish peroxidase (HRP; Sigma). HRP (4 mg; approximately 1000 units/mg solid, dissolved in 2 ml of distilled water) is added to 400 μl of freshly prepared sodium periodate (0.1 M solution). The mixture is stirred gently for 20 min at room temperature and then dialyzed overnight at 4° against 0.001 M acetate buffer (pH 4.4). Sodium carbonate buffer (0.1 M, pH 9.5; 20 μl) is added in order to raise the pH to approximately 9–9.5 and immediately 1 ml is mixed with 1 ml of the proteinase inhibitor-treated small intestinal mucus or 1 ml of gastric mucin. The mixtures are held at room temperature for 2 hr with occasional stirring. Freshly prepared sodium borohydride solution (100 μl of a 4-mg/ml solution in distilled water) is added to reduce any free enzyme and the mixture is dialyzed against borate buffer (0.1 M, pH 7.4). Labeled small intestinal mucus and gastric mucin are mixed with equal volumes of 80% (v/v) glycerol and stored at −20°.

Cultivation of Bacteria. Cell suspensions are prepared by inoculating 1% of the lactobacillus primary cultures in MRS broth and 1% of *E. coli* primary cultures in TSB. Bacteria are grown for 8–10 hr, and then cultures are centrifuged for 5 min at 3000g at 5° and washed twice in HEPES–Hanks' buffer (0.01 M, pH 7.4). Optical density at 600 nm is adjusted to 0.8. Aliquots of serial dilutions are spread on MRS agar and incubated at 37° under anaerobic conditions for enumeration. These dilutions are also used to perform the dot-blot adhesion assay in order to determine the effect of the bacterial cell density in the assay.

Preparation of Solid-Phase Matrix. Polyvinylidene difluoride microporous membrane (Immobilon PVDF; Millipore, Bedford, MA), which is hydrophobic, mechanically strong and entirely accessible for protein adsorption, is used as the matrix to immobilize either bacterial cell suspensions or cell surface extracts. Membranes are prewetted in a small volume of 100% methanol for 2 sec, and then are placed in water for 5 min to remove the methanol. Subsequently, the membranes are equilibrated in HEPES–Hanks' buffer and put on a filter paper previously soaked in HEPES–Hanks' buffer in order to keep the membranes moist at all times.

Immobilization of Material. Cell suspensions, cell surface extracts, and fractions from gel-filtration chromatography (5, 10, or 20 μl drops) are spotted onto Immobilon-PVDF membranes. Membranes are incubated in 3% (w/v) BSA in HEPES–Hanks' buffer for 20 min at room temperature in order to block nonspecific sites. The membranes incubated in BSA are then washed twice in HEPES–Hanks' buffer. These membranes are incubated in 100 μl of HRP–mucus or HRP–mucin in 10 ml of HEPES–Hanks' buffer at room temperature for 2 hr. Membranes are washed three times in HEPES–Hanks' buffer (20 min each wash) and rinsed with 0.1 M sodium acetate (pH 5) prior to development with the substrate diaminobenzidine (2.5 mg of diaminobenzidine, 2.5 μl of hydrogen peroxide, and 10 ml of 0.1 M sodium acetate, pH 5). The reaction is stopped after 5 min by rinsing the membrane in 0.1 M sodium metabisulfite (an antioxidant).

Control Experiments

To ensure that the substrate is not cleaved by the immobilized bacteria, a control experiment is carried out. Whole cells of strains 1–7 (see Table I) are immobilized on the solid-phase matrix and treated with unlabeled mucin prior to blotting with HRP–mucin. When unlabeled mucin is added, no reaction is noted when the substrate is added. It can therefore be concluded that the cleavage of the substrate is attributable to the HRP on the mucin.

Volumes of 200 μl of HRP–mucus (0.3 mg of protein per milliliter) and HRP–mucin (0.2 mg of protein per milliliter) are immobilized in polystyrene tissue culture wells (Nunc) by incubation for 24 hr at 4°. The plates are washed twice with HEPES–Hanks' buffer to remove unbound material and 100 μl of substrate per well is added. After reaction of the HRP–mucus and HRP–mucin with the substrate, the solutions are transferred to a 96-well PolySorp plate (Nunc) and the optical density is read at 540 and 490 nm, respectively. The amount of mucus or mucin that has bound to the polystyrene is calculated, using standard curves. These standard curves are prepared by measuring the optical density in a microplate Bio-Kinetic reader (Bio-Tec Instruments) of samples (100 μl) of HRP–mucus (3.37 μg of protein per microliter) and HRP–mucin (2.0 μg of protein per microliter) as well as serial dilutions of each (1–1000) reacted with 100 μl of substrate.

When the dot-blot assay is performed with bacterial cells, HEPES–Hanks' buffer and BSA are found to retain the cells on the membrane better than phosphate-buffered saline (PBS) added with Tween 20 (0.05%, v/v).

Results

It has been reported that the adhesion of *L. fermentum* 104R to small intestinal mucus from piglet is mediated by proteinaceous compound(s).[27] To study the adhesin(s) and receptor(s) that were involved in this binding, a direct assay was needed for detecting the presence of these compounds during the steps of the purification and characterization.

Initially, whole cells of *L. fermentum* 104R were used to perform the dot-blot adhesion assay. As the density of the cells increased, so too did the intensity of the color of the dot. For the density range 1.5×10^3 to 3.0×10^6 CFU, a clear positive reaction was obtained. At higher cell numbers, the resultant dot was somewhat smeared (results not shown).

Cell wall extracts and spent culture fluid of *L. fermentum* 104R were assessed with the dot-blot adhesion assay in order to determine whether the method could be used to detect adhesion-promoting compound(s) (Table II). Results indicate that the adhesion-promoting compound(s) are in the lithium chloride and lysozyme extracts and in the spent culture fluids. The detection of adhesion-promoting compound(s) with affinity for HRP–mucin was dependent on both the complexity of the protein mixture and on the total amount of protein in the assay. In addition, the concentration of the adhesion-promoting compound(s) relative to the total protein

TABLE II

DOT-BLOT ADHESION ASSAY OF *Lactobacillus fermentum* 104R CELL
EXTRACTS AND DIALYZED AND CONCENTRATED SPENT CULTURE FLUID
DOTTED ON MEMBRANES AND THEN BLOTTED WITH HORSERADISH
PEROXIDASE-LABELED, PARTIALLY PURIFIED PORCINE GASTRIC MUCIN[a]

Extract dotted on membrane	Protein concentration (μg) dotted on membrane	Reaction with HRP–substrate
Spent culture fluid	4.15	+
LiCl extract	0.45	+
	0.90	++
Lysozyme extract	1.65	−
	0.08	++

[a] Results are expressed as ++, +, and −, where ++ is a strong positive
reaction, + is a positive reaction, and − represents no detectable reaction.

concentration influenced the assay to some extent as shown in Table II, because
when LiCl extract was diluted by 50%, the concentration of the reaction product was less as judged by a less intense color reaction. Lysozyme extract was a
more complex mixture of proteins than LiCl extract, as was visualized by SDS–
polyacrylamide gel electrophoresis (PAGE) (results not shown). When 1.65 μg
(10 μl) of total protein was immobilized on the membrane, the adhesion-promoting
compound(s) were not detected; however, when the sample was diluted twice a
dot was visible.

The adhesion pattern of the fractions from the spent culture fluid of *L. fermentum* 104R, eluted by Sephadex G-200 gel-filtration chromatography, is also shown
in Table III. Compounds with affinity for mucus were found in fractions 38, 40,
45, 50, 72, 74, 76, 78, 80, 85, 90, and 95 as detected by the dot-blot adhesion
assay (Table III). When 5-, 10-, and 20-μl volumes of fraction 80 were dotted
on the membrane and tested for adhesion to HRP–mucus, visible differences in
the intensity of the color were not observed. BSA effectively blocked background
binding on the membrane in the dot-blot assay.

Spent culture fluids and fractions from *L. fermentum* 104R were also used in
the inhibition assays. The adhesion of strain 104R cells to immobilized mucus
pretreated with its spent culture supernatant fluid decreased to 62%, relative to the
untreated control mucus (Table III). When the immobilized mucus was pretreated
with the various fractions of the spent culture fluid from strain 104R, varying
degrees of adhesion were noted. When the dot-blot assay results were compared
with the adhesion inhibition assay results in Table III, it was interesting to note
that when adhesion was less than 80% of the control, the dot blot gave a positive reaction. Approximately 85% adhesion gave a weak dot-blot reaction and
at more than 90% adhesion, inhibition was not detectable at all by the dot-blot
assay.

TABLE III

COMPARISON OF DOT-BLOT ASSAY WITH ADHESION INHIBITION ASSAY
FOR STUDYING AFFINITY FOR PIG SMALL INTESTINAL MUCUS OF
Lactobacillus fermentum 104R SPENT CULTURE FLUID FRACTIONS[a]

Fraction	Adhesion inhibition assay (%)[b]	Dot-blot assay[c]
Spent culture fluid	62 ± 0.78	+
1	100 ± 0.09	−
26	95 ± 0.49	−
30	85 ± 1.25	±
34	83 ± 1.72	±
38	70 ± 0.58	+
40	70 ± 0.77	+
45	75 ± 1.67	+
50	79 ± 1.72	+
55	114 ± 2.67	−
60	111 ± 2.25	−
65	92 ± 0.89	−
70	95 ± 2.17	−
72	62 ± 0.35	+
74	71 ± 1.25	+
76	77 ± 0.88	+
78	72 ± 1.17	+
80	68 ± 3.13	+
85	78 ± 0.14	+
90	77 ± 1.09	+
95	73 ± 0.19	+
101	84 ± 0.37	±

[a] The spent culture fluid was fractionated by gel filtration, using
Sephadex G-200.

[b] Adhesion of *L. fermentum* 104R whole cells to immobilized mucus
treated prior to the assay with either buffer (control), spent culture
fluid, or fractions eluted from gel-filtration chromatography is presented. Results are expressed relative to the buffer-treated control
as a percentage of adhesion ±SD.

[c] (−) No, (±) weak, and (+) strong dark dot indicating the amount
of HRP-labeled mucus binding to the individual fractions that were
dotted on membranes.

The results of the adhesion of HRP–small intestinal mucus and HRP–mucin to
a range of *Lactobacillus* strains presented in Table I, using the dot-blot assay and
the adhesion to immobilized mucus assay, are shown in Table IV. Both mucin and
mucus bound to all those *Lactobacillus* strains originally isolated from intestinal
tracts and that were investigated. According to the control experiment the binding
could be blocked by nonlabeled mucus or mucin, but not by BSA. When the

TABLE IV

PORCINE GASTROINTESTINAL MUCUS AND MUCIN BINDING PATTERN OF
Lactobacillus STRAINS OF GASTROINTESTINAL ORIGIN, USING TWO METHODS[a]

Lactobacillus strain	Dot-blot assay[b]			Adhesion assay using immobilized[c]	
	Control[d]	Mucus	Mucin	Mucus	BSA
1D	−	+	+	7.1 ± 5.2	1.1 ± 0.5
2D	−	+	+	9.5 ± 3.4	1.4 ± 0.4
3D	−	+	+	6.3 ± 1.5	4.5 ± 1.8
4J	−	+	+	7.3 ± 2.0	3.8 ± 2.7
5I	−	+	+	6.0 ± 1.2	6.5 ± 0.4
6I	−	+	+	5.0 ± 0.7	3.9 ± 0.2
104R	−	+	+	31.0 ± 2.0	2.8 ± 1.3
KLD	NT	+	+	1.2 ± 5.5	0.5 ± 0.7
C39	NT	+	+	24.0 ± 1.7	24.0 ± 1.3
256	NT	+	+	1.6 ± 1.6	1.1 ± 1.4

[a] The dot-blot assay, in which the lactobacilli were immobilized on a membrane, and a modification of the method of Laux *et al.,* in which the porcine mucus was immobilized on a polystyrene surface.

[b] The bacterial suspensions were initially dotted on the membrane and overlaid with BSA to prevent background binding prior to being blotted with HRP–mucus or HRP–mucin.

[c] Adhesion of *Lactobacillus* whole cells to immobilized mucus or BSA (control). Results are expressed relative to the number of cells added and are presented as a percentage ±SD.

[d] Control refers to experiments in which nonlabeled mucin in addition to BSA was used to block the binding prior to addition of HRP-labeled mucus or mucin in the dot-blot assay. NT, Not tested.

modification of the method of Laux *et al.*[22] was used to study adhesion of the same strains to porcine mucus that was immobilized in polystyrene wells, only *L. fermentum* 104R bound to mucus to a greater extent than to BSA. Mucin was not used when performing the routine method because, according to the control experiment, more than 3.5 μg of HRP–mucus adhered to polystyrene when an excess (60 μg) was added, whereas only 0.15 μg of mucin adhered when also added in excess (40 μg).

By means of the dot-blot adhesion assay, *E. coli* cell suspensions immobilized on the solid-phase matrix were blotted with HRP–mucus and HRP–gastric mucin (Table III). All strains bearing functional K88 fimbriae bound mucus, as well as mucin, when compared with *E. coli* K12 lacking the K88 gene or with a nonfunctional K88 gene or when compared with the DH5 strain, which lacks the K88 gene. In addition, wild-type isolates of *E. coli* that express K88 fimbriae also bound mucus and mucin in the dot-blot assay. Adhesion of *E. coli* to immobilized mucus,

TABLE V

ADHESION OF HORSERADISH PEROXIDASE–MUCUS AND HORSERADISH PEROXIDASE–MUCIN
TO *Escherichia coli* STRAINS, DETERMINED BY DOT-BLOT ASSAY AND ADHESION
OF *Escherichia coli* STRAINS TO IMMOBILIZED MUCUS AND BOVINE SERUM ALBUMIN

E. coli strain	Control *E. coli* strain[c]	Dot-blot assay[a]		Adhesion assay[b]	
		Mucus	Mucin	Mucus	BSA
K12K88ab	K12	+	+	35 ± 0.5	5.2 ± 0.8
K12pMK	K12	+	+	33 ± 2.1	1.6 ± 1.0
005 K88ac	pMK 002				
DH5 pMK005 K88ac	DH5	+	+	44 ± 0.5	0.0 ± 0.4
Bd 4228/84I 0149 K88ac	K12	+	+	56 ± 2.0	2.9 ± 0.3
Bd 2221/75 08 K88ac	K12	+	+	46 ± 0.6	3.3 ± 0.2
Bd 1107/75 08 K88ac	K12	+	+	33 ± 0.9	6.5 ± 0.5
Bd 4545/84 0.149 K88ac	K12	+	+	32 ± 1.1	2.2 ± 0.2
Bd 1147/75 032 K88ab	K12	+	+	29 ± 1.0	2.2 ± 0.1
Bd 3027/75 0149 K88ab	K12	+	+	41 ± 1.2	3.2 ± 0.3
K12	DH5	−	−	2 ± 1.9	1.4 ± 0.8

[a] The bacterial suspensions were initially dotted on the membrane and overlaid with BSA to prevent background binding prior to being blotted with HRP–mucus or HRP–mucin.

[b] Adhesion of radioactively labeled *E. coli* whole cells to immobilized mucus or BSA (control). Results are expressed relative to the cells added and are presented as a percentage ±SD.

[c] Results are expressed as a positive reaction (+) or no response (−) when compared with the response by the control strain. For the Bd series wild-type strains, no syngeneic strains were available as a control and therefore (+) represents a strong reaction compared with *E. coli* K12 control strains.

using a modified method of Laux *et al.*[22] (Table V), is consistent with that noted when using the dot-blot assay. The assay was not carried out for immobilized mucin, because less mucin was retained on the polystyrene as reported in the control experiments.

Discussion

Both *Lactobacillus* and *E. coli* have the capacity to adhere to immobilized porcine small intestinal mucus.[12,22,27,29] The dot-blot assay presented here gave direct evidence when it was used to evaluate whole bacterial cell suspensions of *Lactobacillus* as well as the well-defined adhesion of *E. coli* K88 fimbriated cells.

From the results of experiments in which serially diluted *L. fermentum* 104R cells were dotted onto the membranes and blotted with HRP–mucus or HRP–mucin, it was noted that the strength of the color in the dot-blot assay increased as

[29] H. W. Smith, *J. Pathol. Bacteriol.* **89,** 95 (1965).

the cell density increased. Consequently, the dot-blot assay has the potential to be used for quantifying adhesion by scanning the membranes with a densitometer.

Whole cell extracts and spent culture fluids were immobilized on membranes and the capacity to bind to HRP–mucus was assessed. It was possible to visualize the interaction of the adhesion-promoting compound(s) with the mucus. Binding, however, was not detected when the adhesion-promoting compound(s) were part of a complex protein mixture (Table II). When the complex protein mixture was diluted, the adhering compound(s) were visualized (Table II). When dots (5, 10, and 20 μl) of one adhering fraction from spent culture fluid of *L. fermentum* 104R were immobilized on the membrane and tested for adhesion of HRP–mucus, differences in the strength of the color were not visualized. For more concentrated samples, however, the color on the dot was related to the dilution (Table II). This was consistent with the results obtained when bacterial cell suspensions were diluted and supports the suggestion that the method has the potential to be used for quantifying adhesion if the intensity of the dot were measured.

HRP–small intestinal mucus and HRP–gastric mucin bound to all *Lactobacillus* strains isolated from the intestinal tract (Table IV) when assayed by the dot-blot method. These *Lactobacillus* strains bound poorly to immobilized mucus and the binding was as poor as the control (immobilized BSA), except for *L. fermentum* 104R, which adhered better to mucus than to BSA. In the dot-blot assay the immobilized bacterial cells were treated with BSA prior to blotting with HRP–mucus or HRP–mucin to reduce background binding and therefore a positive reaction in the dot-blot assay indicated binding other than to BSA. While the results from the dot-blot assay suggested a common mechanism of adhesion in these *Lactobacillus* strains with the same origin and suggest that mucin acts as a receptor for *Lactobacillus* adhesion-promoting compound(s), the low percentage of adhesion to immobilized mucus is difficult to understand. At this stage, it may be speculated that all lactobacilli tested have surface components with affinity for mucin and, in addition, that strain 104R is able to bind other components in the crude mucus. It is possible that for some strains, the adhering mucous components may be exposed only in the dot-blot assay. Alternatively, it may be speculated that not all porcine mucous components bound to the polystyrene or that when the mucus was immobilized on the polystyrene it may have been in a configuration that did not permit some *Lactobacillus* cells to bind to it. The Laux *et al.*[12] assay was developed for rodent mucus and because rodent and porcine mucin differ in hydrophobicity,[26] it is probable that this component from porcine mucus binds more poorly to the polystyrene. On the other hand, in the dot-blot assay the HRP-labeled mucous or mucin compounds are free to bind the adhesin(s) on the cell surface of the immobilized bacteria.

Results presented in Table V illustrate that the dot-blot assay can be used to study the binding of *E. coli* K88 fimbriae to piglet ileal mucus because results were consistent with those obtained by the routinely used method of Laux *et al.*[19] In addition, using the dot-blot assay it was possible to show that *E. coli* bearing K88 fimbriae also bind to gastric mucin. This could not be studied by immobilizing the mucin on polystyrene because mucin bound more poorly to the polystyrene. The dot-blot assay could therefore be valuable for studying the receptor(s) for K88 fimbriae. To date, several compounds have been reported to bind K88 fimbriae.[30-33] From the results presented in Table V, it appears that mucin contains a receptor site for K88 fimbriae, which mediates colonization of the ileal mucosa.[30,33,34] Interestingly, Wanke *et al.*[24] reported that *E. coli* binds to mucin but suggested that no fimbriae were involved in colonization. It has been also suggested that *E. coli* type 1 fimbriae bound to oligomannosyl residues of the mucin link glycopeptide.[21,35]

It was concluded that the dot-blot assay was valuable for detecting adhesion-promoting compounds in fractionated and crude soluble extracts thereof and was an improvement over existent methodology because 20-fold less material was required when compared with an adhesion inhibition assay. If required, it has the potential to be used quantitatively. Furthermore, it has been shown to be valuable for studies of receptors because it has also been tested with a mucous constituent, mucin. The conventional methodology of Laux *et al.*[12] for studying adhesion to immobilized mucus could not be used for mucin because the porcine gastric mucin did not immobilize well. Additional advantages with the dot-blot assay are that the assay is a direct assay rather than the indirect adhesion inhibition assay and radioactive labeling is not required. It was shown to correlate well with the conventional method of Laux *et al.*[12] for *E. coli* K88 fimbriae-mediated adhesion to porcine ileal mucus, but not for a range of *Lactobacillus* strains.

Acknowledgments

This work was supported by Stiftelsen Lantbruksforskningen and by the National Council for Science and Technology, Mexico.

[30] J. W. Metcalfe, K. A. Krogfelt, H. C. Krivan, and P. S. Cohen, *Infect. Immun.* **59**, 91 (1991).
[31] R. A. Gibbons, G. W. Jones, and R. Sellwood, *J. Gen. Microbiol.* **86**, 228 (1975).
[32] P. T. J. Willemsen and F. K. de Graaf, *Microb. Pathog.* **12**, 367 (1992).
[33] L. Blomberg, "Factors Regulating Porcine Colonization by *Escherichia coli* K88." Ph.D. Thesis. University of Göteborg, Göteborg, Sweden, 1992.
[34] G. W. Jones and J. M. Rutter, *Infect. Immun.* **6**, 918 (1972).
[35] S. U. Sajjan and J. F. Forstner, *Infect. Immun.* **58**, 868 (1990).

[32] Understanding Urogenital Biofilms and Potential Impact of Probiotics

By GREGOR REID, CHRISTINE HEINEMANN, JEFFREY HOWARD,
GILLIAN GARDINER, and BING S. GAN

Introduction

The urogenital tract (vagina, cervix, perineum, and urethra) in females is host to approximately 50 species of bacteria and yeasts. While studies have documented the aerobic and anaerobic composition of the microbiota, relatively little is known about the biofilm dynamics. Organisms such as uropathogenic *Escherichia coli,* yeast, and *Gardnerella* can be introduced via sexual intercourse, while other routes of seeding include the skin and intestine. The microbiota and the overall microenvironment can change on an hourly basis[1] and they are influenced by micturition and urinary components, as well as by hormones such as estrogen, exposure to spermicides such as nonoxynol-9, antibiotics, semen, presence of a diaphragm or intrauterine device, menstruation and other factors.[2] Indeed, the fluctuating nature of the urogenital microbiota is such that only a minority of women maintain a so-called "normal" *Lactobacillus*-predominant microbiota for extended time frames.[3]

An estimated 1 billion urogenital infections occur in women each year, worldwide. Many of these lead to complications such as preterm labor, increased susceptibility to sexually transmitted diseases, and infertility. Antimicrobial agents are used to treat infections, but increasing numbers of women have shown an interest in, and desire to try, alternative therapies such as probiotics (nonpathogenic, host-derived microorganisms that beneficially affect the host by improving microbial balance of the target niche[4]). The concept is that as lactobacilli are the predominant organisms in a healthy urogenital tract, they could be applied exogenously to restore and maintain a normal microbiota. The present chapter provides suggestions on methodologies that may be used to examine the impact of probiotic organisms on the microbiota.

Basic bacteriology is useful to identify the organisms present, but as many are strict anaerobes and adherent to epithelial cells, special handling (such as avoidance of exposure to oxygen, sonication to remove bacteria from cells) should be undertaken to ensure recovery. In addition, as many of the microorganisms in

---

[1] J. M. Seddon, A. W. Bruce, P. Chadwisk, and D. Carter, *Br. J. Urol.* **48,** 211 (1976).

[2] G. Reid, "Old Herborn University Seminar Monograph" (P. J. Heidt, P. B. Carter, V. D. Rusch, and D. van der Waaij, eds.), Vol. 12, p. 1, 1999.

[3] J. R. Schwebke, C. M. Richey, and H. L. Weiss, *J. Infect. Dis.* **180,** 1632 (1999).

[4] S. A. W. Gibson (ed.), "Human Health: The Contribution of Microorganisms" Springer-Verlag, London, 1994.

the microbiota cannot be cultivated by standard procedures, molecular techniques that detect nucleic acids rather than viable cells offer many advantages over classic culture techniques.

The following methods focus on ways to study urogenital biofilms attached to cells recovered from patients as well as those being examined *in vitro*.

Methods Used to Study Urogenital Biofilms

Gram Stain

Epithelial cells can be recovered from the cervix, deep vagina, vaginal introitus, or vulva by using a cotton swab (and from urethra via first stream urine, which is centrifuged at 500g). Vaginal epithelial cells are the most commonly evaluated for the status of the urogenital flora. These cells are recovered by streaking the swab (of vaginal discharge or taken by rotating the swab around the vaginal wall) onto a glass slide, air drying, and staining with classic Gram stain with crystal violet, iodine, and safranin as the counterstain. The smears are examined under ×1000 oil immersion and scored for gram-positive *Lactobacillus* rod-shaped morphotypes and small curved gram-variable and gram-negative *Gardnerella* and *Bacteroides* morphotypes and gram-positive cocci on cells and in their vicinity (Nugent score[5]). Five or more lactobacilli morphotypes and fewer than five other morphotypes per oil immersion field constitute a "normal" vaginal microbiota (score of 0–3) as illustrated in Fig. 1. When the smear has fewer than five lactobacilli morphotypes per oil immersion field and five or more *Gardnerella* morphotypes together with five or more other morphotypes (gram-positive cocci, curved gram-variable rods, or fusiforms), the patient is regarded as having "bacterial vaginosis" (BV) (score 7–10) (Fig. 2). Equal numbers of both morphotypes represent an "intermediate" status of the microbiota (score 4–6). Yeast can also be detected, and if they are present in high numbers, and the patient has white vaginal discharge, then a yeast infection would be diagnosed.

BV is perhaps one of the most poorly diagnosed diseases found in women. The traditional criteria include the presence of three of the four following signs: thin, homogeneous vaginal discharge that adheres to the vaginal walls, an elevated pH > 4.5, a positive KOH "whiff" test, and the presence of clue cells (packed with gram-negative rods) on microscopic examination. In practice, few physicians carry out these tests. Furthermore, clue cells are difficult to find and pH can be elevated for reasons other than BV. The Nugent system provides several advantages and is sensitive and specific. An example of the usefulness of this test can be found in examining pregnant women. In our experience, up to 70% of women presenting with signs and symptoms of preterm labor have an abnormal vaginal microbiota as scored by the Nugent system. This then provides a practical and valuable tool in assessing patients.

[5] R. P. Nugent, M. A. Krohn, and S. L. Hillier, *J. Clin. Microbiol.* **29**, 297 (1991).

Vaginal epithelial cell

Lactobacillus morphotypes

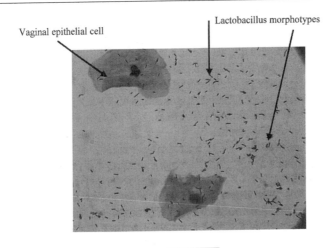

Bar Approximately 20 microns

FIG. 1. Photomicrograph of gram-stained vaginal smear from woman with normal vaginal microbiota. Image clearly shows gram-positive lactobacilli (arrows) dominating the field of view, both adherent to the cells and in the surrounding milieu.

Bar Approximately 20 microns

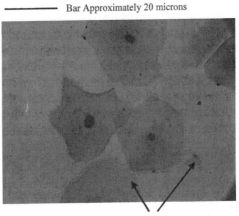

BV organisms in microcolonies on smear and
adherent to vaginal epithelial cells

FIG. 2. Photomicrograph of gram-stained vaginal smear from woman with bacterial vaginosis microbiota. Image clearly shows clue cells and surrounding areas covered with small gram-negative rods (arrows).

The examination of uroepithelial cells, harvested from bladder urine, can also provide a useful method to assess the presence of pathogenic biofilms. Normally, the bladder is regarded as being sterile. However, it is likely that bacteria enter from the urethra on a regular basis. When the numbers of bacteria reach over 100,000/ml urine, the patient is deemed to have asymptomatic bacteriuria. This is a common finding in elderly, spinal cord-injured, and neurogenic bladder patients. For the most part, this colonization without symptoms is left untreated. However, urine does not always demonstrate the presence of infecting organisms. In addition, if the patient is treated with antibiotics, the urine is invariably clear of bacteria, but the bladder itself may still be colonized. An examination of the transitional bladder cells (collected by centrifuging (500g) the urine; usually 50 cells per sample are examined) by Gram stain can provide proof of colonization and the existence of dense gram-negative rods and/or gram-positive cocci. Studies of spinal cord-injured patients have shown that such pathogenic bacterial biofilms do indeed exist, when urine culture is negative.[6]

The Nugent test and the urine biofilm test are both useful but require practice by the person reading the smears. In the former, some lactobacilli can be almost coccoid in shape, while the Gram stain itself may stain rods as red and purple, thus making scoring difficult. In the latter, biofilms by their nature are multilayered, and this is not always easy to see by one-dimensional microscopy. The use of deconvolution microscopy, described below, improves the detection.

To identify specific species, antibody can be raised against urogenital organisms and conjugated to fluorescein label, and then viewed by fluorescence microscopy.[7,8] Currently, there are no commercially available labels for the species that colonize the urogenital tract, and so this test would be time consuming to set up, and may be better as a research tool.

It should be pointed out that the recovery methods may themselves disrupt microbial biofilms on the cells. The very movement of a swab across an epithelial surface may dislodge organisms as well as cells, while washing in buffer can do likewise. An ideal solution would be to obtain a small biopsy, but ethically this cannot be justified in most cases. Thus, one must be aware of the limitations of the material being analyzed.

Additional Microscopy and Other Analytical Tools

Epithelial cells can be recovered from swabs or urine, fixed in 5% (v/v) glutaraldehyde and cacodylate buffer (0.1 M, pH 7.2) with 0.15% (w/v) ruthenium red for 2 hr at room temperature, and then embedded, sectioned, and examined under

[6] G. Reid, R. Charbonneau-Smith, D. Lam, M. Lacerte, Y. S. Kang, and K. C. Hayes, *Paraplegia* **30**, 711 (1992).

[7] R. L. Cook, G. Reid, D. G. Pond, C. A. Schmitt, and J. D. Sobel, *J. Infect. Dis.* **160**, 490 (1989).

[8] N. Yuki, K. Watanabe, A. Mike, Y. Tanaka, M. Ohwaki, and M. Morotomi, *Int. J. Food Microbiol.* **48**, 51 (1999).

a transmission electron microscope. This can show details of microbes attached to the cell surface and to each other and any glycocalyx material surrounding the microcolonies. The plunging of the specimens into liquid propane at $-195°$, prior to any fixation, is referred to as *freeze substitution*[9] and is a useful method to preserve the glycocalyx if the biofilm structure is of interest. Cell wall and membrane structure indicates whether the organisms are gram positive or negative. However, there are a number of disadvantages with this method. This process is extremely time consuming, and it is often difficult to isolate a section that contains organisms (as not all cells have bacteria and sometimes only a few have extensive biofilms). Unless labeled probes are used (such as antibody conjugated to fluorescein or immunogold secondary antibody[10]), it is not possible to confirm the species or the viability of the organisms present.

More sophisticated microscopy tools have evolved. Two of these, namely confocal laser scanning microscopy (CLSM)[11] and deconvolution microscopy, have proved to be particularly useful for imaging urogenital biofilms. Both systems allow researchers to generate three-dimensional images of biofilms that in turn provide important structural information. In our studies we have used deconvolution microscopy (Delta Vision; Applied Precision, Issaquah WA) because it offers researchers some advantages over the confocal method that are better suited for imaging thin biofilms. Among its most salient features is the use of conventional lamp illumination, such as a mercury arc lamp. Unlike confocal microscopy, which uses a pinhole system to physically reject blurred fluorescence from other optical planes, the deconvolution system uses a computer algorithm to deal with "out-of-focus" fluorescence. This more efficient use of the fluorescence provides researchers a gentle imaging environment that is crucial for live cell studies. When combined with the system's fast image acquisition times it can dramatically reduce sample photobleaching and phototoxicity. However, perhaps the greatest feature of the deconvolution system is the image restoration software. The Delta Vision software (IRIX-based softWoR$_x$, version 2.5), like other deconvolution systems, is capable of dramatically reducing out-of-focus fluorescence. Although a considerable amount of computational time is required the resulting spatial resolution is impressive. A good example of this is shown in Fig. 3 (see color insert). Serial optical z sections of a *Staphylococcus aureus* biofilm were acquired, computationally deconvolved,[12] and then collectively volume rendered. Both top and side (180°) views of one of the bacterial microcolonies are shown. Fluorescent dyes FM 4–64 and DAPI (Molecular Probes, Eugene, OR) were used to visualize both the plasma membrane surfaces and DNA of the bacteria, respectively.

[9] G. Reid, R. L. Cook, R. J. Harris, J. D. Rousseau, and H. Lawford, *Curr. Microbiol.* **17**, 151 (1988).

[10] L. Hawthorn and G. Reid, *J. Biomed. Mater. Res.* **24**, 1325 (1990).

[11] J. R. Lawrence, D. R. Korber, B. D. Hoyle, J. W. Costerton, and D. E. Caldwell, *J. Bacteriol.* **173**, 6558 (1991).

[12] D. A. Agardi, Y. Hiraoka, P. Shaw, and J. W. Sedat, *Methods Cell. Biol.* **30**, 353 (1989).

In future we plan to examine how biofilms, containing multiple species of bacteria (urogenital microbiota), colonize surfaces and whether probiotic strains of lactobacilli can disrupt established biofilms of pathogenic bacteria. The use of live cell cultures combined with time-lapse imaging will permit direct monitoring of these bacterial interactions. In addition we plan to employ green fluorescent protein (GFP) and its spectral variants in studies of biofilms containing multiple species of bacteria. The fluorescent protein technology will be a particularly useful molecular tool for carrying out intracellular and extracellular protein trafficking studies (e.g., *Lactobacillus*-produced adhesins).

Molecular Techniques

Other methods allow strain-specific identification of organisms within the urogenital biofilm. Nucleic acid-based techniques detect microbial macromolecules rather than viable cells, and thus cells that may be difficult or impossible to recover by culture can still be identified. Such molecular techniques offer the advantages of rapidity, reliability, and sensitivity over classic strain identification methods.[13] A brief review of these methods is presented to draw the reader's attention to important techniques that can aid in identifying bacteria within urogenital biofilms.

Pulsed-field gel electrophoresis (PFGE) involves the use of rare cutting restriction enzymes to digest the microbial genome into relatively few (5–50) large DNA segments. In this way, a DNA fingerprint, which is highly characteristic of the particular organism, is obtained and referred to as a restriction fragment length polymorphism (RFLP). This pattern is representative of the entire genome. So far, there have been no studies of vaginal epithelial cells in which PFGE was applied. In theory, it could differentiate applied probiotic lactobacilli from indigenous strains.

Ribotyping involves restriction of the total genome with an endonuclease, separation of the DNA fragments by agarose gel electrophoresis, and subsequent hybridization with a probe for either 16S, 23S, or 5S rRNA genes. The basis of the technique is that bacteria generally contain multiple copies of rRNA genes throughout their genome and thus a number of fragments of different sizes will hybridize to the probe, giving a characteristic fingerprint. Ribotyping has been used successfully to differentiate *Lactobacillus* strains,[14] whereby genomic DNA was isolated from a number of culture collection strains and clinical isolates, restricted with relevant enzymes, and the restriction fragments transferred to nitrocellulose membranes and hybridized with 16S and 23S rRNA ^{32}P-labeled probes to yield distinct ribotype patterns. This method has also been used to show that a probiotic strain can colonize the vagina[15] as well as for tracking colonization of the intestine by lactobacilli and bifidobacteria.[16]

---

[13] R. I. Mackie, A. Sghir, and H. R. Gaskins, *Am. J. Clin. Nutr.* **69**, 1035S (1999).

[14] W. Zhong, K. Millsap, H. Bialkowska-Hobrzanska, and G. Reid, *Appl. Environ. Microbiol.* **64**, 2418 (1998).

[15] G. Reid, K. Millsap, and A. W. Bruce, *Lancet* **344**, 1229 (1994).

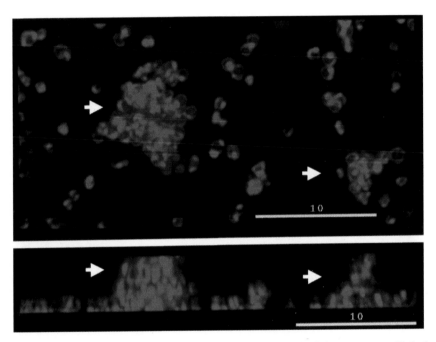

FIG. 3. Deconvolution image of *Staphylococcus aureus* biofilm. *Staphylococcus aureus* (Oxford strain) was cultured in six-well culture dishes containing 1-mm glass coverslips previously coated with bovine skin collagen type III (Sigma). After 3 days of culture in BH medium the coverslips were gently washed (PBS) and stained (20 min at 22°) with a PBS solution containing 5 M FM 4-64 [*N*-(3-triethylammoniumpropyl)-4-(6-[4-(diethylamino)phenyl]hexatrienyl) pyridinium dibromide] and 1 M DAPI (Molecular Probes). FM 4-64 (Molecular Probes) is a water-soluble vital dye that fluoresces only on insertion into the outer leaflet of the plasma membrane, which makes it particularly useful for delineating the morphology of the bacteria (top view) and the biofilm microcolonies (arrows). The bottom figure is a side view that illustrates the depth of the biofilms (arrows). Images were acquired with a Nikon Eclipse TE200 inverted fluorescence microscope and a ×60 Nikon Pan Apo objective (1.4NA).

DNA probes are labeled synthetic oligonucleotides of defined sequence that can specifically hybridize to a target complementary sequence. To obtain species-specific probes, oligonucleotides are commonly directed toward regions of 16S and 23S rRNA. Once a suitable probe has been designed and labeled, DNA or RNA is extracted from the microorganism or sample of interest. Developments in simple and rapid nucleic acid extraction methods for use directly on biological specimens mean that it is not necessary to rely on culture of the organisms. The nucleic acid probe is then hybridized to the extracted nucleic acids, which are immobilized on nylon or nitrocellulose filters, or hybridization can be performed directly on living cells. The label, which can be radioactive, enzymatic, or fluorescent, is then used to detect specific targets. Using such an approach a genus-specific DNA probe, based on 16S rRNA sequences, has been used to identify the prevalence of lactobacilli within a gastrointestinal population in pigs.[17] By using fluorescently labeled rRNA probes, it is possible to detect microorganisms *in situ* (fluorescent *in situ* hybridization, or FISH) with minimal disruption to microbial biofilms. Oligonucleotide probes can also be used as primers in a polymerase chain reaction as outlined below.

Polymerase chain reaction (PCR) is one of the most useful molecular tools for rapid and specific detection of a wide range of bacterial species. It can also be used to subtype bacteria within a genus or species by employing 16S rRNA, and intergenic spacers, among others, as primer targets. Direct extraction of PCR-quality bacterial DNA from vaginal samples is possible and avoids the need for time- and labor-intensive culture procedures. Probes directed toward specific microbial sequences can be used as PCR primers to detect and enumerate probiotic strains within complex microbial populations such as those found in the vaginal tract. We have developed a primer pair for the specific detection of *Lactobacillus fermentum* RC-14 via PCR in individuals administered this strain. These primer pairs are directed toward a specific nucleotide sequence of the 16S–23S rRNA spacer region, which was identified in the RC-14 strain by PCR amplification and subsequent sequencing of this region. The creation of a "checkerboard" system comprising a single support membrane with different DNA probes and applying denatured DNA from a multispecies sample such as vaginal cells, will, in future, allow (after hybridization) the simultaneous identification of multiple bacterial species in the urogenital tract.[18]

Alternatively, randomly amplified polymorphic DNA (RAPD) PCR employs short arbitrarily chosen primers that bind to genomic DNA at randomly occurring complementary target sequences under low-stringency annealing conditions. It is essentially a survey of the genome for sites with which a primer shares full or partial homology. Resultant discrete DNA fingerprints allow differential identification

[16] A. L. McCartney, W. Wenzhi, and G. W. Tannock, *Appl. Environ. Microbiol.* **62,** 4608 (1996).

[17] A. Sghir, D. Antonopoulos, and R. I. Mackie, *System. Appl. Microbiol.* **21,** 291 (1998).

[18] S. S. Socransky, C. Smith, L. Martin, B. J. Paster, F. E. Dewhirst, and A. E. Levin, *Biotechniques* **17,** 788 (1994).

of strains in a multispeciated specimen.[19] Pure cultures of the microorganism in question are required for RAPD analysis. Genomic DNA can be isolated by one of a number of available procedures and the extracted DNA used in subsequent PCR amplifications with the chosen random primer(s). A method that has previously been employed to successfully identify probiotic *Lactobacillus* strains in a complex microenvironment uses 1× *Taq* polymerase buffer, 5 mM MgCl$_2$, a 200 μM concentration of each dNTP, a 1 μM concentration of a single primer of arbitrary sequence, 0.625 U of *Taq* DNA polymerase, and 1 μM template DNA in a 25-μl PCR.[19] DNA is amplified for 35 cycles, using the following temperature profile: denaturing at 93° for 1 min, annealing at 36° for 1 min, followed by polymerization at 72° for 1 min. The amplified PCR products are then analyzed by agarose gel electrophoresis. Each strain generates a unique banding pattern that can be subsequently used to track an administered probiotic strain in the urogenital tract. However, problems can arise if the probe is too sensitive, as it may lose a degree of specificity.[20]

Cell Signaling within Urogenital Biofilms

With respect to quorum sensing and microbial communication between organisms in the urogenital microbiota, no information has yet been reported. Cell signaling has already been reported for organisms such as *Pseudomonas aeruginosa,* which can be found in the urogenital tract (particularly in the bladder of spinal cord-injured patients and those who are catheterized). The most interesting endeavor will be to see how signals are affected when pathogens and nonpathogens are present within the same urogenital biofilm. It is hoped that this information, and that acquired through microscopy studies, will help us better understand how the normal microbiota is maintained, and what factors change the balance to an infected state.

In summary, direct microscopy, in combination with molecular probes, can identify organisms within a biofilm adherent to urogenital tissues. These techniques can provide vital information about the dynamics of a medically important microbial niche. They will also provide tools to follow the progress of selected probiotic organisms as they colonize the vagina and potentially restore and maintain a normal microbiota.[21]

Acknowledgment

The support of NSERC is appreciated.

[19] G. Gardiner, R. P. Ross, J. K. Collins, G. Fitzgerald, and C. Stanton, *Appl. Environ. Microbiol.* **64,** 2192 (1998).
[20] S. C. Ricke and S. D. Pillai, *Crit. Rev. Microbiol.* **25,** 19 (1999).
[21] G. Reid, A. W. Bruce, N. Fraser, C. Heinemann, J. Owen, and B. Henning, *FEMS Immunol. Med. Microbiol.* **30,** 49 (2001).

[33] Surface Characterization and Adhesive Properties of Bifidobacteria

By Rodrigo Bibiloni, Pablo F. Pérez, Graciela L. Garrote,
E. Anibal Disalvo, and Graciela L. De Antoni

Introduction

The importance of intestinal microbiota in the health status of both children and adults is generally recognized. The colonization of different portions of the intestinal tract by beneficial microorganisms constitutes the first defensive barrier against the invasion of pathogenic microorganisms or toxic substances.[1] The ability to adhere to the intestinal epithelia and to compete with other microorganisms seems to be crucial for a probiotic strain to colonize the gastrointestinal tract.[2,3] Bacteria of the genus *Bifidobacterium* are normal inhabitants of the gut of humans and animals, where they play an important role in the prevention of gastrointestinal disorders.[4,5] Bifidobacteria have been successfully employed to prevent antibiotic-related diarrhea and acute infant diarrhea. Protection of intestinal epithelia could be attributed to the adherent properties of some strains and to the liberation of antimicrobial substances such as lactic and acetic acids.[6–8]

One approach to obtaining information regarding adherent behavior in the intestinal ecosystem is to study the surface properties of the selected strains. The adhesion of microorganisms to natural environments can be inferred by studying the surface properties of bacteria, employing a combination of methodologies.[6,9,10] This procedure includes several determinations such as partitioning of bacteria in hydrocarbon–aqueous interfaces, ζ potential, binding to solid surfaces, capacity to aggregate particles, and adhesion to monolayers of cell cultures. We applied these methodologies to a collection of bifidobacteria isolated at CIDCA (Centro de Investigación y Desarrollo en Criotecnología de Alimentos, La Plata, Argentina)

[1] G. L. Simon and S. L. Gorbach, *in* "Infections of the Gastrointestinal Tract" (M. J. Blaser, P. D. Smith, J. I. Rardin, H. B. Greenberg, and R. L. Guerrant, eds.), p. 53. Raven, New York, 1995.

[2] M. F. Bernet, D. Brassart, J. R. Neeser, and A. L. Servin, *Gut* **35**, 483 (1994).

[3] S. Elo, M. Saxelin, and S. Salminen, *Lett. Appl. Microbiol.* **13**, 154 (1991).

[4] A. S. Naidu, W. R. Bidlack, and R. A. Clemens, *Crit. Rev. Food Sci. Nutr.* **38**, 13 (1999).

[5] J. A. Kurmann and J. L. Rasic, *in* "Therapeutic Properties of Fermented Milks" (R. K. Robinson, ed.), p. 117. Elsevier Applied Science, London, 1983.

[6] G. Kociubinski, P. Pérez, and G. L. De Antoni, *J. Food Protect.* **59**, 739 (1996).

[7] P. F. Pérez, J. Minnaard, E. A. Disalvo, and G. L. De Antoni, *Appl. Environ. Microbiol.* **64**, 21 (1998).

[8] M. F. Bernet, D. Brassart, J. R. Neeser, and A. L. Servin, *Appl. Environ. Microbiol.* **59**, 4121 (1993).

[9] I. Ofek and R. J. Doyle, *in* "Bacterial Adhesion to Cells and Tissues" (I. Ofek and R. J., Doyle, eds.), p. 16. Chapman & Hall, New York, 1994.

[10] R. J. Doyle and M. Rosenberg, *Methods Enzymol.* **253**, 542 (1995).

and to reference strains. This procedure allowed the selection of adherent strains on the basis of their capacity to bind to different surfaces and to correlate these properties with the nature of the bacterial surface.

When designing a strategy to study the bacterial surface, the methods should be carefully standardized. It is important to report the suspending buffers because ionic strength and pH may have an effect on the results obtained. Temperature of assays as well as the conditions in which the bacteria are grown are also important. When comparisons between laboratories are to be made, it is recommended that reference strains with different adherent properties be included.

Cells and Culture Media

Bacterial Strains

Strains *Bifidobacterium bifidum* NCC 189 (formerly CIDCA 536), NCC 200 (formerly CIDCA 538), NCC 235 (formerly CIDCA 533), CIDCA 537, CIDCA 539, CIDCA 5310, CIDCA 5311, CIDCA 5313, *Bifidobacterium breve* CIDCA 532, CIDCA 5312, CIDCA 5314, and *Bifidobacterium adolescentis* CIDCA 5315, CIDCA 5316, and CIDCA 5317 were isolated from infant feces. Strain CIDCA 531 was isolated from a commercial fermented milk product. Strains are identified by gram staining, fructose-6-phosphate phosphoketolase activity, sugar fermentation profile, and whole-cell protein analysis using reference strains belonging to the American Type Culture Collection (ATCC, Manassas, VA). Four strains were obtained from Morinaga Milk Industry (Higashihara, Zama City, Japan).[7,11] Stock bacterial cultures are maintained at $-80°$ in 0.3 M sucrose. Experiments are performed after reactivation of microorganisms by two consecutive subcultures in liquid medium. Bacteria are cultured in MRS (Difco Laboratories, Detroit, MI) or TPY medium.[12] Cultures are incubated at $37°$ under anaerobic conditions. All cultures are harvested in stationary phase: 24 hr for MRS cultures and 48 hr for TPY cultures. Bacteria are harvested by centrifugation at 10,000 g for 10 min at $4°$, washed twice, and suspended in phosphate-buffered saline (PBS), pH 7, to a suitable density according to the assay. Surface properties may vary after successive subculturing or if cells are stored at low temperatures (4 to $-20°$). Consequently, it is preferable to perform the experiments with freshly reactivated bacteria. Medium composition as well as origin of medium components are also variables that can affect bacterial surface properties. Reproducibility regarding surface properties of bifidobacteria can be achieved only if the composition of the culture medium and conditions of incubation and storage are carefully standardized.

[11] A. Gómez Zavaglia, G. Kociubinski, P. Pérez, and G. De Antoni, *J. Food Protect.* **61,** 865 (1998).

[12] V. Scardovi, *in* "Bergey's Manual of Systematic Bacteriology" (P. H. A. Sneath, N. S. Mair, M. E. Sharpe, and J. G. Holt, eds.), Vol. 2, p. 1418. Williams & Wilkins, Baltimore, Maryland, 1986.

Enterocyte-Like Cells

Caco-2 cells (ATCC HTB-37) are purchased from the ATCC. Cells are maintained in liquid nitrogen according to the manufacturer instructions. Cells are routinely grown in Dulbecco's Modified Eagle's Medium (DMEM, with glucose at 4.5 mg/ml; ICN Biomedicals, Aurora, OH), supplemented with nonessential amino acids, penicillin (12 IU/ml), streptomycin (0.012 mg/ml), gentamicin (0.047 mg/ml), amphotericin B (0.0025 mg/ml), and 20% (v/v) heat-inactivated (30 min, 56°) fetal calf serum (Gen, Buenos Aires, Argentina). Antibiotics are obtained from GIBCO-BRL (Life Technologies, New York, NY). Cells are cultured at 37° in a 5% CO_2–95% air atmosphere. The culture medium is changed every 2 days. Cells are used at postconfluence, after 21 days of culturing.

Physicochemical Properties of Bacterial Surface

The hydrophobic–hydrophilic character of the bacterial surface depends on the relative distribution of the net charge, polar and nonpolar groups.

Hydrophobicity Assays

Surface hydrophobicity originates from the distribution of nonpolar residues on the bacterial surface. These nonpolar residues enhance the bacterial affinity for a nonpolar solvent: the higher the exposure of the residues, the higher the affinity for the solvent. Several methods have been described to measure microbial adhesion to nonpolar substrata and cell surface hydrophobicity.[9] Even though no single method will satisfy all requirements, some of them could be employed to attempt a screening over a large number of strains. Partitioning by *p*-xylene or *n*-hexadecane seems to be a good alternative to achieve this end. When bacterial suspensions are mixed with a volume of xylene or hexadecane, the microorganisms will adhere to the surfaces of liquid droplets, forming a "creamy" upper layer consisting of cell-coated hydrocarbon droplets. Partitioning by the solvent is quantified by measuring the drop in turbidity in the aqueous phase after the mixing procedure.[10] The relative amount of bacteria in both solvents is an index of their hydrophobicity. The hydrophobicity index could be expressed as

$$H\% = [(A_0 - A)/A_0] \times 100$$

where A_0 and A are the absorbance at 600 nm before and after mixing with solvent. Highly hydrophobic strains can be distinguished from those having low hydrophobicity through the proposed assay as depicted in Table I.

On the basis that adhesion to hydrocarbon is time dependent, it is recommended that the kinetic approach, in which adhesion to different hydrocarbon volumes is followed as a function of time, be employed.[10] In this method, hydrophobicity is expressed through the removal coefficient. This typical kinetic MATH (microbial

TABLE I
Surface Characterization and Adhesive Properties of Some Strains of Bifidobacteria

Strain of *Bifidobacterium*	$H\%$	ζ potential (mV)	AT (min)	AI	HA	% Adhesion to polystyrene	% Adhesion to Caco-2
B. breve CIDCA 532	4.8 ± 3.0	$-(33.0 \pm 1.9)$	>20	0.40 ± 0.09	$-$	9.41 ± 0.90	1.94 ± 1.30
B. bifidum NCC 189 (formerly CIDCA 536)	96.0 ± 3.9	$-(38.2 \pm 2.0)$	8	0.64 ± 0.18	$+$	33.01 ± 0.29	14.65 ± 0.07
B. bifidum CIDCA 537	95.0 ± 4.0	$-(34.9 \pm 1.8)$	8	0.58 ± 0.14	$+$	40.54 ± 5.32	15.61 ± 0.84
B. infantis NCC 200 (formerly CIDCA 538)	3.3 ± 2.3	$-(24.6 \pm 0.5)$	>20	0.32 ± 0.09	$-$	4.55 ± 1.21	5.53 ± 0.47
B. bifidum CIDCA 5310	95.0 ± 0.9	$-(35.4 \pm 1.8)$	12	0.78 ± 0.21	$+$	32.74 ± 1.39	15.61 ± 0.41
B. bifidum CIDCA 5313	91.0 ± 4.2	$-(30.6 \pm 0.6)$	3	0.48 ± 0.03	$+$	58.23 ± 0.71	ND

Abbreviations: $H\%$, hydrophobicity index; AT, autoaggregation time; AI, autoaggregation index; HA, hemagglutination; ND, not determined.

adhesion to hydrocarbon) assay seems to be more suitable for differentiating bifidobacterial strains of similar high hydrophobicity. Single-point determination of hydrophobicity correlates well with the kinetic assay. However, strain CIDCA 531, which shows high hydrophobicity by single-point determination, has a low removal coefficient compared with other highly hydrophobic strains.

ζ Potential and Electrophoretic Mobility

ζ Potential is a measure of the electrical potential resulting from the net charges distributed on the bacterial surface. It is defined as the drop in the electric potential between the bacterial surface and the shear plane between the bacteria and the liquid phase in which the bacteria migrate under the effect of an electrical constant field. ζ potential can be calculated by determination of the electrophoretic mobility (μ) of bacteria when a constant electric field is applied. Viscosity, pH, and ionic strength of the media are variables that contribute to the total mobility and should be standardized. The electrophoretic mobility is measured in a bacterial suspension placed in a capillary H-cell with two Pt or Ag/AgCl electrodes connected to an external direct current source.

Procedure

1. Calibration: Fill the H-cell with a suspension of phosphatidylserine liposomes in 0.01 M NaCl, pH 7.4 (buffer, 1 mM HEPES). Fix the optical set-up considering that the ζ potential of phosphatidylserine liposomes is −120 mV. After this calibration the optical set-up must be kept unchanged. Wash thoroughly with water.

2. Replace the suspension with a suspension of bacteria in 1 mM KC1.

3. Connect the electrodes to a direct current source and fix the potential at 40 V.

4. Select a single bacterium, moving uniformly and rectilinearly along a graduated reticule in the optical plane.

5. Count the number of squares displaced per unit time.

6. Repeat the measurement at least 10 times by changing alternatively the polarity of the electrodes.

7. Repeat the procedure with different bacteria in the same sample until reproducible results are obtained.

8. Calculate the bacterial mobility as the ratio between the bacteria velocity (number of squares per unit time) and the imposed electric field. Calculate the zeta potential (ζ), using the Helmholtz–Smoluchowski equation $\zeta = \eta\mu/\varepsilon\varepsilon_0$ where ε and ε_0 are the dielectric constants of the aqueous solution and the free space, respectively, η is the viscosity of the medium, and μ is the mobility determined as described above. The temperature was maintained at 25° in all cases. Values for the constants are taken from handbooks.

FIG. 1. Macroscopic view of typical bifidobacterial cultures. *Left:* Culture of the agglutinating strain *B. bifidum* CIDCA 5310 grown in TPY. *Middle:* Same culture after mixing, *Right:* Culture of the nonagglutinating strain *B. breve* CIDCA 532 grown in TPY.

Usually, the variation coefficients of data obtained with bacterial suspensions are between 2 and 5%. It is recommended that all solutions be prepared with Milli-Q water. ζ potential data for some isolated strains are reported in Table I.

Bacteria–Bacteria and Bacteria–Erythrocyte Aggregation

Autoaggregation

Some strains of the genus *Bifidobacterium* have a tendency to grow adherent to glass tube, forming aggregates or clumps in liquid media. The clumps are commonly formed by large clusters of cells as determined by optic and electronic microscopy. If a bacterial culture is allowed to stand undisturbed after mixing for a period of time, aggregating bacteria can be distinguished from nonaggregating bacteria because the former sediment clarifies the medium whereas the latter persist dispersed in the bulk (Fig. 1). The clumps are relatively stable after vigorous mixing or washing with buffer, but they can be disrupted by washing with detergents. Differences between autoaggregating strains can be assessed semiquantitatively by the agglutination index (or autoaggregation index) (AI)[7] and the autoaggregation time (AT).

Autoaggregation Index

AI is defined as the ratio between the bacterial density (dry weight in milligrams per milliliter) and the turbidity at 600 nm (A_{600}) of a homogenized culture.

Procedure

1. Grow bacteria as indicated previously.
2. Homogenize bacterial culture and read immediately the turbidity at 600 nm (A_{600}). Dilute the sample, if necessary, with culture medium.
3. Centrifuge a fixed volume, accurately measured, of the same culture at 10,000 g for 10 min at 4° and wash twice with PBS. The third wash should be done with distilled water. Dry the pellet at 100° until constant weight. Calculate the dry weight in milligrams per milliliter.
4. Calculate the AI as follows:

$$AI = \text{dry weight}/A_{600}$$

AI data for some isolated strains are reported in Table I.

Autoaggregation Time

Changes in the turbidity of a bacterial suspension that sediments spontaneously can be monitored as a function of time. Typical curves reflecting this behavior are shown in Fig. 2A. From these data relative changes in sedimentation (A_{rel}) can be calculated as follows:

$$A_{rel} = [(A - A_i)/(A_{sat} - A_i)]$$

where A is the turbidity of the sample at a given time, A_i is the initial turbidity of the sample, and A_{sat} is the saturating turbidity obtained at long times. A_{rel} values plotted as a function of time for various bacterial dilutions give curves similar to those shown in Fig. 2B. Note that over a particular dilution all curves overlap. At this point, it is possible to obtain graphically the autoaggregation time (AT). The AT is defined as the time at which the relative absorbance (A_{rel}) reaches 50% of the saturation value, regardless of the dilution.

Procedure

1. Grow and harvest bacteria as indicated previously. Adjust the optical density to 1.0–2.0 at 600 nm.
2. Dispense the suspensions in cuvettes and evaluate the time course of turbidity for 20 min at timed intervals until no change is observed.
3. Repeat the procedure with serial dilutions of bacterial suspensions.
4. Calculate A_{rel} at each time, using the formula given above, and plot A_{rel} as a function of time for each dilution.
5. Extrapolate the straight line defined by A_{sat} and determine $A_{sat}/2$. Using those curves in which sedimentation is independent of the dilution, find the corresponding AT.

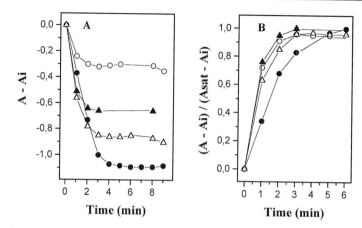

FIG. 2. (A) Sedimentation profile monitored at A_{600} for strain *B. bifidum* CIDCA 5313 grown in TPY. The assay was performed at four serial dilutions of a single stock suspension. (B) Analysis of sedimentation curves: A_{rel} versus time. Key: (●) stock suspension, (△) 1 : 2 dilution; (▲) 1 : 4 dilution; (○) 1 : 8 dilution.

The rate of sedimentation depends on the initial density of bacteria, suspending medium, temperature, and the bacterial surface composition. Nonautoaggregating strains show the same phenotype regardless of the growth medium. In contrast, autoaggregating strains clearly show this phenotype when they are grown in TPY but not in MRS. For this reason, TPY is the recommended medium for these types of assays. The AT can be used to determine the degree of autoaggregating ability among autoaggregating strains. AT data for some isolated strains are reported in Table I.

Hemagglutination

Erythrocytes are a convenient source of cells to test bacterial surface properties, because they have a myriad of external domains for bacterial adhesins. Most bacterial adhesins are detectable as hemagglutinins when a positive hemagglutination (HA) is obtained with intact bacterial cells. In many cases the exposed sites for bacterial adhesins are antigens that are also expressed in epithelial intestinal cells. For this reason the HA is an adequate experimental system to predict specific interactions between bacteria and enterocytes.

Procedure

1. Grow and harvest bacteria as indicated previously.
2. Follow the methods described by Pérez et al.[7]

It is important to use U-shaped microtiter plates. In the absence of hemagglutination the erythrocytes are able to sediment to the bottom of the well. When HA occurs, the erythrocytes and bacteria form a diffuse net of clumped cells covering the entire well. Always check HA by optical microscopy. Bifidobacterial strains are dissimilar regarding hemagglutinating activity: some of them agglutinate red blood cells even at low densities whereas others do not hemagglutinate at all. Some results are reported in Table I.

Adhesion to Solid Surfaces

Adhesion to Polystyrene

Bacterial adhesion to a surface is considered the first step in colonization. Strains forming clumps may also grow by adhering to the walls of a container (Fig. 1). Therefore, quantification of the ability to adhere to glass or polystyrene can be employed to characterize bifidobacterial strains. Quantification of bacteria adhering to solid surfaces may be performed by various methods [e.g., optical microscopy, staining, radiolabeling, enzyme-linked immunosorbent assay (ELISA), bioluminescence, enzymatic activity.[13]

Procedure

1. Grow and harvest bacteria as indicated previously. Adjust the optical density to 1.0–2.0 at 600 nm.
2. Add 500 μl of bacterial suspension to a 24-well tissue culture plate (Corning, Corning, NY) and incubate for 1 hr at 37°.
3. Wash three times with PBS to eliminate nonadherent bacteria.
4. Quantify adherent bacteria by crystal violet staining[14] or by fructose-6-phosphate phosphoketolase (F6PPK) assay.[15]

Fructose-6-phosphate Phosphoketolase Activity

To have a specific method for quantification of bifidobacteria, we developed an assay based on the activity of F6PPK, a key enzyme in the fermentation pathway of Bifidobacterium.[12] F6PPK cleaves fructose 6-phosphate into acetyl phosphate and erythrose 4-phosphate. The acyl part of the acid anhydride is converted to hydroxamic acid by hydroxylamine. The hydroxamic acid forms a brightly purplish complex with trivalent iron that can be quantified spectrophotometrically at 500 nm. Color development is improved by using Triton X-100 in the reaction mixture, allowing the quantification of low bacterial densities. The F6PPK assay detects as little as 0.05 mg of bifidobacteria and shows a linear relationship in

[13] G. D. Christensen, L. Baldassarri, and W. A. Simpson, Methods Enzymol. 253, 477 (1995).
[14] S. McEldowney and M. Fletcher, Appl. Environ. Microbiol. 52, 460 (1986).
[15] R. Bibiloni, P. F. Pérez, and G. L. De Antoni, J. Food Protect. 63, 322 (2000).

terms of absorbance per milligram up to 1 mg for most of the strains tested. Higher bacterial densities should be diluted for better quantification. Because the specific enzymatic activity may vary among different strains, a standard curve should be generated for each strain.[15]

Procedure

1. Perform the adhesion assay as described in the previous section.
2. Add to each well 90 μl of a reaction mixture containing fructose 6-phosphate (80 mg/ml), NaF (6 mg/ml), sodium iodoacetate (10 mg/ml), and 0.25% (v/v) Triton X-100 in 0.05 M phosphate buffer supplemented with 0.05% (w/v) cysteine, pH 6.5. Incubate for 40 min at 37°.
3. Add 70 μl of 13.9% (w/v) hydroxylamine-HCl, freshly neutralized with NaOH (pH 6.5), and incubate for 10 min at room temperature.
4. Add 40 μl of 15% (w/v) trichloroacetic acid, 40 μl of 4 M HCl, and 40 μl of $FeCl_3 \cdot 6 H_2O$ (50 mg/ml) in 0.1 M HCl. Centrifuge for 3 min at 14,000g.
5. Read the absorbance of the supernatant at 500 nm in a microplate reader. This value corresponds to adherent bacteria.
6. Quantify total added bacteria to each well as follows: centrifuge 500 μl of the bacterial suspension for 2 min at 14,000 g and discard the supernatant. Proceed according to steps 2 to 5. Read the absorbance of the supernatant at 500 nm in a microplate reader. This value corresponds to the total added bacteria.
7. Calculate the percentage of adherent bacteria:

$$\text{Adh}\% = (A_{500} \text{ of adherent bacteria}/A_{500} \text{ of total added bacteria}) \times 100$$

To improve sensitivity, the adhesion assay can be performed with six-well plates. Volumes should be adjusted according to the size of the plate to maintain the correct concentration of the reagents. Assays performed with the same bacterial suspension showed a variation between 1 and 10%. Each assay under standardized conditions includes at least two reference strains, one adherent and another nonadherent, in each plate. This method allows a comparison between strains with different adhesion properties. This simple and reproducible assay can also be employed in the quantification of adherent *Bifidobacterium* to eukaryotic cells even in the presence of other bacteria. The results obtained by quantification of F6PPK correlate well with those obtained by the nonspecific method of crystal violet in terms of adhesion in comparison with reference strains. Adhesion studies can also be performed with bacteria suspended in their own spent culture supernatant. However, it should be remembered that metabolic end-products may affect adhesion. Data about adhesion to polystyrene of some bifidobacterial strains are reported in Table I.

Interaction with Enterocyte-Like Cells

Adhesion to Caco-2 Cells

Intestinal microbiota play a main role in gut homeostasis. Interaction between bacteria and intestinal cells is important, not only because attached microorganisms are more resistant to cleansing mechanisms but also because they may trigger cellular responses.[16,17] Caco-2 cell cultures have been widely accepted as an *in vitro* model to study interactions between bacteria and enterocytes.[2,18] This cell line, derived from a human colon adenocarcinoma, shows morphological and functional differentiation similar to that of enterocytes. Caco-2 monolayers in late postconfluence (21 days in culture) display apical microvilli, functional tight junctions, and brush border enzymatic activity resembling that of intestinal epithelial cells.[19–22] Other intestinal cell lines (e.g., HT-29, T84) can be used to study bacteria–cell interactions.[23–25] On this basis experimental protocols should be carefully adapted.

Procedure

1. Seed Caco-2 cells (passage 23 or above, about 2×10^4 cells/cm^2) in either 6- or 24-well tissue culture plates. If microscopic examination is to be performed, cells are grown on glass coverslips.

2. Coverslips are prepared as follows: boil coverslips with 5% (v/v) nonionic detergent for 5 min. Rinse exhaustively with tap water. Wash five times with 96° ethanol. Rinse five times with Milli-Q water. Sterilize at 121° for 15 min and dry at 60°.

3. Incubate seeded plates as described previously.

[16] S. Blum, Y. Delneste, S. Alvarez, D. Haller, P. F. Pérez, C. Bode, W. P. Hammes, A. M. A. Pfeifer, and E. J. Schiffrin, *Int. Dairy J.* **8**, 63 (1999).

[17] E. J. Schiffrin, F. Rochat, H. Link-Amster, J. M. Aeschlimann, and A. Donnet-Hughes, *J. Dairy Sci.* **78**, 491 (1995).

[18] M. F. Bernet-Camard, V. Liévin, D. Brassart, J. R. Neeser, A. L. Servin, and S. Hudault, *Appl. Environ. Microbiol.* **63**, 2747 (1997).

[19] M. Pinto, S. Robine-Leon, M. D. Appay, M. Kedinger, N. Traidou, E. Dussaulx, B. Lacroix, P. Simon-Assmann, K. Haffen, J. Fogh, and A. Zweibaum, *Biol. Cell.* **47**, 323 (1983).

[20] P. H. Vachon and J. F. Beaulieu, *Gastroenterology* **103**, 414 (1992).

[21] I. J. Hidalgo, T. J. Raub, and R. T. Borchardt, *Gastroenterology* **96**, 736 (1989).

[22] H. Matsumoto, R. H. Erickson, J. R. Gum, M. Yoshioka, E. Gum, and Y. S. Kim, *Gastroenterology* **98**, 1199 (1990).

[23] Y. Delneste, A. Donnet-Hughes, and E. J. Schiffrin, *Nutr. Rev.* **56**, S93 (1998).

[24] M. F. Bernet, D. Brassart, J. R. Neeser, and A. L. Servin, *Appl. Environ. Microbiol.* **59**, 4121 (1993).

[25] M. H. Coconnier, T. R. Klaenhammer, S. Kerneis, M. F. Bernet, and A. L. Servin, *Appl. Environ. Microbiol.* **58**, 2034 (1992).

(A)   (B)

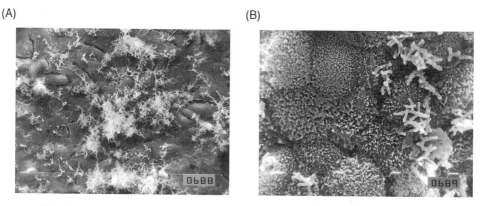

FIG. 3. Scanning electron micrographs of adhesion assays performed with strain *B. bifidum* CIDCA 5310 on Caco-2 cells. (A) Adhesion pattern of clumps of bifidobacterial cells on monolayers. Original magnification: ×7000. (B) Close observation (original magnification, ×10,000), showing undamaged enterocyte-like cell microvilli.

4. Wash the monolayers the times with GKN (NaCl, 8 g/liter; KCl, 0.4 g/liter; glucose, 2 g/liter; $NaH_2PO_4 \cdot H_2O$, 0.69 g/liter; Na_2HPO_4, 1.57 g/liter; pH 7.2–7.4) or tissue culture medium without fetal bovine serum at room temperature.

5. Add 0.5 ml (24 wells) or 1 ml (6 wells) of bacterial suspension (approximately 0.5–1.0 absorbance unit at 600 nm) to monolayers containing 1 ml of GKN. Homogenize.

6. Incubate at 37° for 1 hr without shaking.

7. Wash three times with GKN to remove nonadherent bacteria.

8. Assess adhesion by optical microscopy, scanning microscopy (see Fig. 3), radiolabeling bacteria, or F6PPK assay.

Some authors reported adhesion assays using monolayers between passages 40 and 90.[2,8] However, we found similar results working with cell passages as low as 23.[7] DMEM could be replaced by RPMI 1640 (ICN Biomedicals) or MEM (ICN Biomedicals) with no significant differences in the results. Care should be taken in regard to the supplementation of cell culture media, because high concentrations of nonessential amino acids in RPMI 1640 medium leads to rounding and detachment of Caco-2 cells. To protect monolayers, all washings should be done with solutions at room temperature or at 37°.

The minimum bacterial density that can be used to perform adhesion assays depends on the detection method used to assess adhesion. On the other hand, suspensions over 2.0 absorbance units may produce damage to the underlying monolayer. Adhesion assays of bifidobacteria can be developed on the basis of different techniques.

Optical Microscopy

1. Staining can be performed after fixation of monolayer with methanol for 5 min or without fixation.

2. Incubate monolayers at room temperature with 0.5 ml of Giemsa solution (1 : 10 dilution of Giemsa stock solution in phosphate buffer) for 5 min (dilution 1 : 10 from stock solution in phosphate buffer, pH 7).

3. Wash with KH_2PO_4 (10 mM, pH 7).

4. Leave samples to dry at room temperature.

Scanning Electron Microscopy. This procedure is based on Bibiloni *et al.*[26]

1. Fix cells by adding 1 ml of cold 2.5% (v/v) glutaraldehyde in PBS. Samples can be fixed for 1 hr or may be stored at 4° overnight.

2. Dehydrate samples by immersing in graded ethanol series [10 min each in 10 to 100% (v/v) ethanol]. Samples can be stored at 4° in 70% (v/v) ethanol.

3. After a treatment with isoamyl acetate for 10 min, critical-point dry samples in liquid CO_2.

4. Mount samples on scanning electron microscopy (SEM) stubs and coat with gold 24 for 10 min by vacuum evaporation.

Radiolabeling. Proceed according to Granato *et al.*[27]

Fructose-6-phosphate Phosphoketolase Assay. This procedure is based on Bibiloni *et al.*[15]

1. Perform the adhesion assay as described above, and after the last washing step proceed to steps 2–6.

2. Lyse the monolayers by incubating with distilled water for 1 hr at 37°.

3. Collect material by centrifugation at 14,000g for 10 min.

4. Suspend the pellets in PBS containing 0.25% (v/v) Triton X-100.

5. Proceed as stated in steps 3 to 7 in the F6PPK assay.

6. Adhesion is calculated as follows:

$$\text{Adh}\% = (A_{500} \text{ of adherent bacteria}/A_{500} \text{ of total added bacteria}) \times 100$$

The F6PPK assay appears to be the most suitable method, considering its specificity. It gives more reproducible results than viable counts and optical microscopy. It does not require special equipment, as does radiolabeling. Moreover, it can be applied to assess adhesion of bifidobacteria in combination with other

[26] R. Bibiloni, P. L. Sarmiento, and G. L. De Antoni, *Scanning* **20**, 243 (1998).
[27] D. Granato, F. Perotti, I. Masserey, M. Rouvet, M. Golliard, A. Servin, and D. Brassart, *Appl. Environ. Microbiol.* **65**, 1071 (1999).

microorganisms on the basis that F6PPK activity is considered a distinctive feature of bifidobacteria. Bioluminescence is not recommended because eukaryotic ATP interferes with the results.

Typical adherent bifidobacterial strains show values of about 10%. Interexperiment variability may be high, so controls of adherent and nonadherent strains in each assay should be included. Adhesion studies performed with the F6PPK reaction show a good correlation with microscopic observation and radiolabeling studies in terms of adhesion of the same strains to Caco-2.[7,16] Data concerning adhesion to Caco-2 of some bifidobacterial strains are reported in Table I.

Interaction of Bifidobacteria with Intestinal Pathogens on Caco-2

The Caco-2 system has been largely employed to study the effect of *Bifidobacterium* against intestinal pathogens including displacement, competition, or exclusion, both in adhesion or invasion of pathogenic strains.

Exclusion of Associated Pathogens

1. Incubate 0.5 ml of a bifidobacterial suspension on Caco-2 monlayers for 1 hr at 37°.
2. Add 0.5 ml of the suspension of the pathogen (10^5 or 10^6 CFU/ml) and incubate for 1 hr at 37°.
3. Wash three times with GKN to remove nonadherent bacteria.
4. Add 1 ml of sterile distilled water and incubate for 1 hr at 37° to lyse monolayers.
5. Plate appropriate dilutions on nutrient agar [meat extract (3 g/liter) and meat peptone (5 g/liter)].
6. Calculate the percentage of associated pathogens in each well:

$$\%Ass = (CFU\ associated\ pathogens/CFU\ total\ added\ pathogens) \times 100$$

Pathogen Invasion Assay. Intestinal cell invasion can be measured by quantitative determination of bacteria located within the Caco-2 cells, using gentamicin. Because the antibiotic does not diffuse into eukaryotic cells, bacteria adhering to the intestinal border are killed, while those that invade Caco-2 are not killed.

1. Incubate 0.5 ml of a bifidobacterial suspension on Caco-2 monlayers for 1 hr at 37°.
2. Add 0.5 ml of the suspension of the pathogen (10^5 or 10^6 CFU/ml) and incubate for 1 hr at 37°.
3. Wash three times with GKN to remove nonadherent bacteria.
4. Add 1 ml of gentamicin (200 μg/ml in PBS) and incubate for 1 hr at 37°.
5. Wash the monolayers twice with GKN and add 1 ml of sterile distilled water. Incubate for 1 hr at 37°.

6. Plate appropriate dilutions on nutrient agar [meat extract (3 g/liter and meat peptone (5 g/liter)].

7. Percent invasion in each well is calculated as follows:

$$\% \text{ invasion} = (\text{CFU after gentamicin treatment}/\text{CFU total added}) \times 100$$

Note that gentamicin concentration should be adjusted according to the susceptibility of individual pathogenic strains.

Preliminary studies show that *Bifidobacterium bifidum* CIDCA 537 and CIDCA 5310 are unable to prevent adhesion of *Salmonella arizonae* to Caco-2. However, CIDCA 5310 seems to be effective in reducing the cytopathic effect produced by this pathogen.[28,29]

Approach to Chemical Characterization of Bacterial Surface

The chemical composition of the bacterial surface could be inferred by subjecting bacteria to several treatments prior to any of the aforementioned assays. Treatments with chaotropic agents allow one to determine whether noncovalently bound molecules are involved in the surface properties. Proteinaceous factors could be evidenced after a proteolytic treatment. Metaperiodate may be useful to assess whether oxidizable carbohydrates are responsible for the so-called properties.

Treatment with Chaotropic Agents

1. Grow and harvest bacteria as indicated previously. Adjust to appropriate optical density.

2. Centrifuge 1 ml of the bacterial suspension at 14,000g for 2 min.

3. Discard the supernatant and suspend the pellet in 1 ml of 3.0 M guanidinium chloride or 5.0 M LiCl.

4. Incubate for 1 hr at 37° in a shaker.

5. Wash twice with PBS and suspend the pellet in 1 ml of the same buffer.

6. Determine as previously described: hydrophobicity index, ζ potential, aggregation time, hemagglutination index, adhesion to polystyrene, adhesion to Caco-2, etc.

Treatment with Metaperiodate

1. Grow and harvest bacteria as indicated previously. Adjust to appropriate optical density.

2. Centrifuge 1 ml of the bacterial suspension at 14,000g for 2 min.

[28] R. Bibiloni, P. F. Pérez, and G. L. De Antoni, *Anaerobe* **5,** 519 (1999).

[29] R. Bibiloni, P. F. Pérez, and G. L. De Antoni, *in* "Food Microbiology and Food Safety into the Next Millenium" (A. C. J. Tuijtelaars, R. A. Sanson, F. M. Rombouts, and S. Notermans, eds.), p. 827. Ponsen & Looyens, Wageningen, The Netherlands, 1999.

3. Discard the supernatant and suspend the pellet in 1 ml of sodium metaperiodate (50 mM) in PBS.

4. Incubate for 30 min at room temperature.

5. Wash twice with PBS and suspend the pellet in 1 ml of the same buffer.

6. Determine as previously described: hydrophobicity index, ζ potential, aggregation time, hemagglutination index, adhesion to polystyrene, adhesion to Caco-2, etc.

Treatment with Proteolytic Enzymes

1. Grow and harvest bacteria as indicated previously. Adjust to appropriate optical density.

2. Prepare enzyme solutions to a final concentration of 2.5 mg/ml in the following buffers: trypsin and chymotrypsin in Tris-HCl (50 mM), NaCl (100 mM), pH 8 and pepsin in glycine-HCl (50 mM), NaCl (100 mM), pH 2.2.

3. Centrifuge 1 ml of the bacterial suspension at 14,000g for 2 min.

4. Discard the supernatant and suspend the pellet in 1 ml of the enzyme solutions.

5. Incubate for 1 hr at 37°.

6. Inactivate trypsin and chymotrypsin with 100 μl of fetal calf serum.

7. Wash twice with PBS and resuspend the pellet in 1 ml of the same buffer.

8. Determine as previously described: hydrophobicity index, ζ potential, aggregation time, hemagglutination index, adhesion to polystyrene, adhesion to Caco-2, etc.

Application of the above-described treatment to some bifidobacterial strains indicates that surface components, easily eliminated with chaotropic agents, are involved in adhesion to Caco-2 and to a certain degree in hemagglutination. These surface determinants could be glycoproteins or carbohydrate chains because metaperiodate treatment affects adhesive properties. Some bifidobacterial strains are more sensitive to pepsin than to the other proteolytic enzymes in regard to adhesion to Caco-2 and hemagglutination. Moreover, the effect to trypsin and chymotrypsin is different in each strain, highlighting the strain-dependent trait of the surface properties.[30]

Comments and conclusions

Adhesion to Caco-2 seems to be the most suitable *in vitro* model to infer the behavior of bifidobacteria in the intestinal environment. However, other models can provide valuable information in this regard. One of these methods is the determination of hemagglutinating activity on the basis that erythrocytes possess on their surface antigens similar to those found in enterocytes.

[30] R. Bibiloni, P. F. Pérez, and G. L. De Antoni, *Anaerobe* 5, 483 (1999).

A correlation between physicochemical properties (hydrophobicity and ζ potential) and the described models showed that all highly hydrophobic strains ($H\%$ = 90–98) produced positive HA with human erythrocytes of groups AB, O^+, A^+, and B^+. Nonhydrophobic strains ($H\%$ <40) were negative for HA. Among 20 tested strains, only 2 exceptions to this behavior were found (*Bifidobacterium pseudolongum* CIDCA 531 and *B. bifidum* CIDCA 5311). All hemagglutinating strains were also autoaggregating and adherent to Caco-2. On the other hand, it was not possible to correlate ζ potential with the ability to interact with erythrocytes and Caco-2.

Studies performed on four *B. bifidum* strains [NCC 189 (formerly CIDCA 536), CIDCA 537, CIDCA 5310, and CIDCA 5313] showed that autoaggregation as well as adhesion to Caco-2 cells were abolished by protease treatment (trypsin, chymotrypsin, and pepsin), whereas HA involved only pepsin-sensitive factors. Although autoaggregation is eliminated only by proteases, HA and adhesion to Caco-2 are also abolished by chaotropic agents and metaperiodate.

Special attention should be given to the selection of the culture medium because bacterial composition is strongly influenced by medium composition. In our case, we observed that autoaggregation was higher in TPY than in MRS, suggesting that other interactions may also vary through the modification of surface properties.

Acknowledgments

The authors are grateful to Lic. J. Minnaard, Dr. G. Kociubinski, Dr. A. Gómez Zavaglia, L. Brandi, and M. Tadei, who have worked in our laboratory on *Bifidobacterium*. This work was financially supported by Consejo Nacional de Investigaciones Científicas y Técnicas (CONICET), Comisión de Investigaciones Científicas de la Provincia de Buenos Aires (CIC-PBA), and Agencia Nacional de Promoción Científica y Tecnológica (ANPCyT).

Author Index

A

Abeliovich, A., 338
Abraham, S. N., 183, 196, 197(14)
Abraham, W.-R., 317, 318, 324(1), 329(1), 330(2), 331(1)
Acheson, C., 156
Acinas, S. G., 365
Ackermann, H. W., 172
Adair, C. G., 181
Adams, J. L., 165, 167(12), 174(12)
Adams, M. H., 172, 233
Adler, J., 100
Aeschlimann, J. M., 421
Afione, S., 74
Agardi, D. A., 407
Ahmer, B. M. M., 53, 54
Ahn, C. C., 324
Alavi, M., 29
Albertano, P., 340, 341, 343(4), 345, 349(14), 355
Alberti, L., 29
Albizu, I., 178, 186(21), 196, 197(13)
Aldea, M., 54, 57(24)
Allard, M., 365
Alldredge, A. L., 281
Allison, C., 29, 30
Allison, D. G., 310
Almendros, G., 332
Altebaeumer, M., 76, 165, 167(14a)
Altman, A. E., 49
Alvarez, S., 369, 379, 421, 424(16)
Amann, R. I., 133, 134, 135(32), 136(32), 204, 272, 326
Amarger, N., 365
Amman, R., 348, 352, 353
Ammendola, A., 130, 131(13), 139(13), 140(13), 143(13)
Ammendolia, M. G., 178, 193(22)
Amorena, B., 178, 186(21), 196, 197(13)
Anagnostidis, K., 333
Anastasio, M., 337, 338(27)

B

Andersen, J. B., 61, 108, 112, 115(29), 130, 131, 149, 169, 182, 280
Andrade, J. D., 279
Andreadaki, F. J., 371
Andreasen, K. H., 266, 273(22), 274(22), 276(22)
Andrews, J. H., 353
Anthoni, U., 30
Antonissen, C., 386
Antonopoulos, D., 409
Aoyama, K., 384
Appay, M. D., 422
Arbeit, R. D., 203
Archer, G. L., 195, 197(1), 214, 215, 217, 230, 231, 232(43; 49), 255
Arciola, C. R., 178, 193(22)
Ariño, X., 339
Arnscheidt, A., 329
Aron, G. M., 163, 165, 173
Aronson, T., 326
Arrieta, J. M., 326
Arth, I., 326
Artz, R. R. E., 364
Asboe-Hansen, G., 308
Ashdown, N., 230
Aslanidis, C., 328
Aubel, D., 389(15), 390
Augustin, J., 230
Austin, J. W., 48
Ausubel, F. M., 56, 57(29)
Avaniss-Aghajani, E., 326
Ayala, F. J., 53
Azam, F., 281

B

Baars-Lorist, C., 265
Babitzke, P., 54, 57(24)
Baca-DeLancey, R. R., 102, 104, 107(11)
Baddour, L. M., 178, 180, 183, 184, 186, 188(43), 191, 192(43), 193(43), 196, 197, 197(14), 199, 207, 216, 218(5), 240

Subject Index

A

Acylated homoserine lactones
assays
  bioassays, 42, 116–117
  gas chromatography–mass spectrometry
    advantages, 117–118
    applications, 127–128
    derivatization, 119
    3-oxo *N*-acylhomoserine lactone assay in
      Pseudomonas aeruginosa biofilm,
      121–122
    peak identification, 120–121
    positive electron impact mass
      spectrometry, 128
    running conditions, 119
    sensitivity, 121
    standards, 118
  Pseudomonas aeruginosa assays
    applications, 47
    biofilm culture apparatus, 45
    radiolabel assays, 43–44, 46–47
    sample preparation and inoculation, 46
furanone antagonists
  extraction from surface, 125–126
  gas chromatography–mass spectrometry,
    126–127
  host defense in *Delisea pulchra,* 110, 116,
    122–124
  localization with fluorescence microscopy,
    124–125
  mechanism of action, 123–124
  surface concentrations, 127
green fluorescent protein reporter
  acylated homoserine lactone antagonist
    evaluation, 116
  advantages, 112
  design of fusions, 113–114
  flow system for biofilm studies, 115
  fluorescence measurement
    calibration, 114–115
    confocal microscopy, 115
    epifluorescence microscopy, 114

  modification of native reporter, 112–113
  quorum sensing in gram-negative bacteria, 41,
    109–110
  reporter constructs for live bacterial monitors,
    overview, 111–112
  structural classes, 117–118
  synthesis, 41, 109
  transcriptional regulatory protein interactions,
    41–42, 109
Adhesion assays, bacteria
  adherent mutant isolation from nonadherent
    strains, 156–157
  bead assays, 156
  Congo red staining, 153
  crystal violet staining, 153–154
  fluorescence microscopy in quantitative
    analysis, 154, 156
  probiotics, *see Bifidobacterium; Lactobacillus*
  screening for altered adhesion mutants,
    153–154
  selection of system, 152–153
agfD, see Salmonella typhimurium multicellular
  behavior
algD, see Pseudomonas aeruginosa biofilm
Alginate, *see Pseudomonas aeruginosa* biofilm
atlE, see Staphylococcus epidermidis biofilm
ATP, biofilm content assay, 265
ATR-FTIR, *see* Attenuated total
  reflection-Fourier transform infrared
  spectroscopy
Attenuated total reflection-Fourier transform
  infrared spectroscopy
  comparison of chitin and chitosan films,
    289–292
  ELF-97-*N*-acetyl-β-D-glucosaminide
    cleavage assay, 295–298

B

Bifidobacterium
  adhesion
    determinants, 411, 427
    enterocyte-like cell adhesion

457